Integrated Nanomaterials and their Applications

Deepa Suhag • Atul Thakur • Preeti Thakur
Editors

Integrated Nanomaterials and their Applications

Springer

Editors
Deepa Suhag
Amity Centre for Nanotechnology
Amity University Haryana
Gurgaon, Haryana, India

Atul Thakur
Amity Centre for Nanotechnology
Amity University Haryana
Gurgaon, Haryana, India

Preeti Thakur
Amity School of Applied Sciences
Amity University
Gurugram, Haryana, India

ISBN 978-981-99-6104-7 ISBN 978-981-99-6105-4 (eBook)
https://doi.org/10.1007/978-981-99-6105-4

This Springer imprint is published by the registered company Springer Nature Singapore Pte Ltd.
The registered company address is: 152 Beach Road, #21-01/04 Gateway East, Singapore 189721, Singapore

Paper in this product is recyclable.

Preface

In today's day and age, the world exists between the captivating realm of integrated nano-biomaterials. In this book, we delve into the exciting intersection of nanotechnology and biology, wherein the boundaries between the synthetic and the living converge.

Integrated nano-biomaterials have emerged as a groundbreaking field, offering innovative solutions to address the complex challenges in biomedical research and healthcare. By harnessing the unique properties of nanoscale materials and their interactions with biological systems, scientists and engineers have unlocked a multitude of applications that hold immense potential to revolutionize our understanding and treatment of diseases, enhance diagnostic tools, and create novel therapeutic interventions.

In this book, we embark on a comprehensive journey through the world of integrated nano-biomaterials, guided by leading experts and pioneers in the field. Together, we will explore the fundamental principles that underpin the design, synthesis, characterization, and application of nanomaterials in the context of biological systems. We will delve into the intricate mechanisms by which nano-biomaterials interact with cells, tissues, and organs, and how they can be tailored to exhibit desired functionalities for specific biomedical applications.

From nanoscale drug delivery systems and bioimaging agents to tissue engineering scaffolds and biosensors, the potential applications of integrated nano-biomaterials are vast and diverse. Through this book, we aim to equip researchers, students, and professionals with a comprehensive understanding of the field, providing them with the necessary tools and knowledge to push the boundaries of innovation.

We invite you to embark on this enlightening journey, where you will witness the convergence of disciplines, the power of interdisciplinary collaborations, and the sheer beauty of nature-inspired design principles. Prepare to be inspired, amazed, and empowered by the incredible potential of integrated nano-biomaterials to shape the future of medicine, biology, and human well-being.

Last but not least, the editors wish to thank all the authors for their time and efforts, along with Springer Nature Publishing, who have collectively made this book see the light of day. Without each one of you, this book wouldn't have been possible.

Gurgaon, Haryana, India Deepa Suhag
Gurgaon, Haryana, India Atul Thakur
Gurugram, Haryana, India Preeti Thakur

Contents

1 Introduction to Nanotechnology 1
Deepa Suhag, Preeti Thakur, and Atul Thakur

2 Everyday Nanotechnology 19
Raksha Rathore, Deepa Suhag, Fayu Wan, Atul Thakur, and Preeti Thakur

3 Classification of Nanomaterials (Carbon, Metals, Polymers, Bio-ceramics) 37
Fayu Wan, Atul Thakur, and Preeti Thakur

4 Structural Properties of Nanoparticles 57
Atul Thakur, Preeti Thakur, and Sahbi Baccar

5 Carbon Nanomaterials in the Field of Theranostics 81
Fayu Wan, Atul Thakur, and Preeti Thakur

6 MXene Nanomaterial for Medical Application 107
Anand Salvi, Neetu Dhanda, Saarthak Kharbanda, Abhilash Pathania, Preeti Thakur, and Atul Thakur

7 Metal Nanoparticles in the Field of Medicine and Pharmacology 127
Muhammad Ehsan, Deepa Suhag, Raksha Rathore, Atul Thakur, and Preeti Thakur

8 Polymers and Polymeric Composites in Nano/Bio-Medicine 153
Aarti Singh and Deepa Suhag

9 Bio-Ceramics and Bio-Glasses for Orthopedics and Tissue Engineering 177
A. C. Sun, Atul Thakur, and Preeti Thakur

10 Biocompatibility, Bio-Clearance, and Toxicology 201
Hui-Min David Wang, Preeti Thakur, and Atul Thakur

11 Functionalization of Biomaterials 223
Raksha Rathore, Deepa Suhag, Fayu Wan, Ritesh Verma, Atul Thakur, and Preeti Thakur

12 Bioprinting for Therapeutics 245
Ritesh Verma, Neetu Dhanda, Raksha Rathore, Deepa Suhag, Fayu Wan, Atul Thakur, and Preeti Thakur

13 Nanomaterials for Precision Medicine 269
Abhigyan Satyam and Deepa Suhag

14 Nanotechnology for AI in Healthcare 291
Swati Kaushik and Deepa Suhag

15 Nanomaterials in Bioimaging and Diagnostics 311
Adeeba Shakeel and Sonali Rawat

16 Bioinspired Nanomaterials 329
Fayu Wan, Saarthak Kharbanda, Preeti Thakur, and Atul Thakur

17 Nanomaterials for Environmental Applications 349
Yassine Slimani and Essia Hannachi

18 Nanotechnological Applications in Food and Agriculture 393
A. V. Trukhanov, Preeti Thakur, and Atul Thakur

19 Nano/Mesoporous Materials as State-of-the-Art Military Applications 419
Larrisa V. Panina, Atul Thakur, and Preeti Thakur

20 Toxicology and Toxicity Studies of Nano-Biomaterials 433
Raksha Rathore, Deepa Suhag, Fayu Wan, Atul Thakur, and Preeti Thakur

21 Nano-Remediation Perspectives 457
Fayu Wan, Atul Thakur, and Preeti Thakur

22 Impact of Nano-Biomaterials on the World 477
Irina Edelman, Sergey Ovchinnikov, Atul Thakur, and Preeti Thakur

Correction to: Nanomaterials for Environmental Applications C1
Yassine Slimani and Essia Hannachi

Introduction to Nanotechnology

1

Deepa Suhag, Preeti Thakur, and Atul Thakur

Abstract

The potential for a varied range of applications makes a basic understanding of nanotechnology important to us. The technological and industrial sectors including electrical, agricultural, medical, and electronic technologies, homeland security and food safety, transportation, and many more have benefited greatly from the development and revolutionization brought about by nanotcchnology. For the creation of novel materials with distinctive properties after the nanoscale specifies their architecture, nanotechnology now makes use of advancements in chemistry, mechanical, materials science, and biotechnology making it a multidisciplinary field. This book examines the recurrent applications of nanotechnology in recent decades. As a ground-breaking science, nanotechnology is defined by the impact it has on technological advancements. Nanotechnology is anticipated to offer

D. Suhag
Centre for Nanotechnology, Amity University Haryana, Gurugram, India

NanoLatticeX, Amity University Haryana, Gurugram, Haryana, India

P. Thakur (✉)
Centre for Nanotechnology, Amity University Haryana, Gurugram, India

NanoLatticeX, Amity University Haryana, Gurugram, Haryana, India

Nanotechnology Wing, Innovative Science Research Society, Shimla, India
e-mail: pthakur@ggn.amity.edu

A. Thakur
Centre for Nanotechnology, Amity University Haryana, Gurugram, India

NanoLatticeX, Amity University Haryana, Gurugram, Haryana, India

Nanotechnology Wing, Innovative Science Research Society, Shimla, India

School of Electronics and Information Engineering, Nanjing University of Information Science & Technology, Nanjing, China

D. Suhag et al. (eds.), *Integrated Nanomaterials and their Applications*,
https://doi.org/10.1007/978-981-99-6105-4_1

solutions to many present issues. Thus, nanotechnology has a wide range of uses, from altering the physics of currently used devices to developing entirely new tactics based on molecular self-assembly, from developing novel materials with nanoscale dimensions to investigating whether we can directly control matter at the atomic level.

Keywords

Nanotechnology · Materials science · Nanoscale dimensions · Atomic level

1.1 Introduction

The most popular word for the study of materials at nanoscale is "nanotechnology." When the bulk is divided into smaller size, the material's characteristics alter [1]. In the areas of electrical, optical, or mechanical properties, among others, nanoscale materials have special qualities. Materials that have been reduced to the nanoscale can suddenly exhibit properties that are very different from what is seen on a macroscale. For instance, inert substances such as platinum can serve as catalysts, opaque compounds such as copper can become transparent, stable substances such as aluminum can catch fire, solids such as gold can change into liquids at room temperature, and insulators such as silicon can transform into conductors [2]. Rather than the composition of the particle, the surface area of the nanomaterials determines how they behave. Nanomaterial refers to the incredibly small materials used to create the particles' distinctive physical and chemical features. This material has a single dimension that is less than or equal to 100 nanometers. The nanoparticles can be found in nature. Any mineral, including silver, or carbon can be used to make it. They must fall within the range of 1 nm and around 100 nm in size. These materials cannot be seen with the human eye; thus, laboratory microscopes are employed to observe them. As sizes are lowered, the materials' quantum characteristics are impacted. These materials are free because of their small size; as the size is decreased, the materials' quantum characteristics are impacted. Due to their small size, these materials are immune to internal flaws found in structures and are shielded from mechanical failure. The system becomes more reactive when the area of the surface-to-volume ratio rises. On dispersion, these substances can rearrange the crystal structure. Self-assembly is demonstrated by crystals generated for the semiconductor industry and by the chemical synthesis of big molecules. In spite of the fact that this "positional assembly" allows for more control during building, it is currently exceedingly labor-intensive and unsuitable for industrial applications. The field of technology is arguably making the greatest efforts to apply the various physical features found at the nanoscale to produce compact, potent creations with a variety of possible environmental effects.

The benefits of nanomaterials could potentially alter the future of practically all industrial sectors. The field of nanotechnology has opened new horizons for the development of innovative materials with unique and unprecedented properties. In particular, the integration of different types of nanomaterials has led to the creation of a new class of materials known as integrated nanomaterials. These materials have the potential to revolutionize a wide range of industries, including electronics, medicine, energy, and environmental science. The impact of integrated nanomaterials on the present and the future is expected to be significant. In the present, these materials are already being used in various applications, such as drug delivery systems, biosensors, and energy-storage devices. Nanotechnologies have revolutionized advances in medicine, specifically in diagnostic methods, imaging, and drug delivery. To provide cancer patients with a personalized treatment plan, IBM scientists are also focusing on leveraging nanotechnology to examine DNA in just a few minutes, instead of weeks. Other researchers in medical technology are looking at using nanotechnology to deliver medications like chemotherapy or vaccines to target particular cell types in the body. There are several distinct nanotechnology efforts in the experimental stages that seek to hyper-target medication to cancer cells, and experimental nano-sponges are being developed to absorb poisons in the body. It is also being investigated as an early diagnostic tool to find infectious diseases and tumors much earlier than our existing technology can. One of the guiding principles of nanotechnology is a tiny device that is injected into the body as a sensor or medical delivery device. Furthermore, the development of integrated nanomaterials is expected to lead to the creation of new and more efficient technologies that will address some of the world's most pressing challenges, such as climate change and sustainable energy. The discovery and use of carbon nanomaterials has allowed the introduction of many new areas of technology in nanomedicine, biosensors, and bioelectronics. Various investigations have focused on the incorporation of distinct types of nanomaterials, for example, oxide metals, carbon, magnetic and industrial/agro-industrial wastes [3–5].

This book will provide an overview of the current state of research in integrated nanomaterials, including their synthesis, properties, and applications. It will also discuss the challenges and opportunities associated with the development of these materials and their potential impact on various industries. Overall, the book will demonstrate the exciting potential of integrated nanomaterials and their role in shaping the future of technology and innovation (Fig. 1.1).

1.2 Nanotechnology in Daily Life

As of now it is challenging to determine the precise timeline of human use of nanoscale items. The use of nanomaterials, however, has a long history, and for a very long time, humanity inadvertently exploited them for a variety of purposes. Around 4500 years ago, humans employed asbestos nanofibers to strengthen pottery mixtures. Around 4000 years ago, the ancient Egyptians used Pbs nanoparticles in a hair coloring solution. Another noteworthy historical example is the Lycurgus Cup. It is a dichroic cup made in the fourth century AD by the Romans. In direct light, it has a jade-like appearance, yet in transmitted light, it has a translucent ruby hue. It displays varied colorations depending on the incident light. These color variations

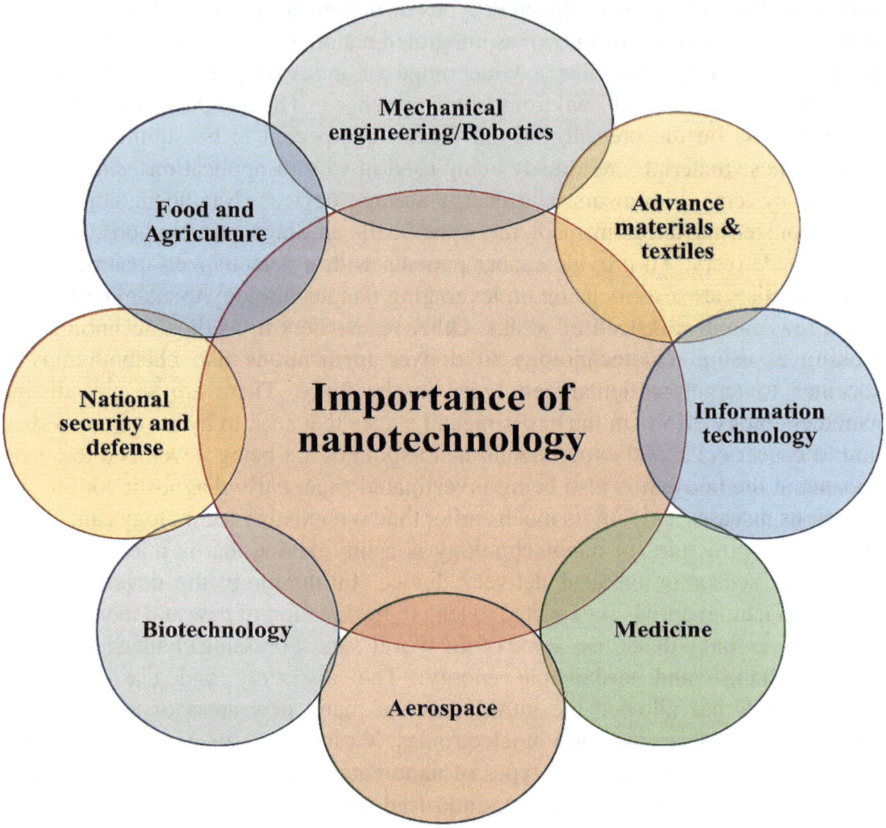

Fig. 1.1 Importance of nanotechnology in every aspect of life

can be seen because Ag and Au nanoparticles are present. The word "nanometer" was first used by Richard Adolf Zsigmondy in 1914. The term "nanotechnology" was originally suggested in 1959 by American physicist and Nobel Prize winner Richard Feynman during a speech at the annual meeting of the American Physical Society. The methods, recipes, and later science of creating and testing tools, works of art, and weaponry are the primary criteria for characterizing and classifying human historical ages, from the Stone Age to the current Silicon Age. Nanomaterials have been used steadily throughout human history, and this field is rapidly expanding.

Now let us discuss nanoceramics. Nanomaterials have been used as ceramic dyes since prehistoric times. The most widely used artificial substance today is nanoceramic cement, which the Romans frequently used to construct aqueducts, bathhouses, and buildings. Recent archeological findings also show that the Macedonians used cement centuries before the Romans; therefore, they were not the first. Even the study of nanoparticles is not entirely new. Only 15 years ago, it seems, did the phrase start to appear in the names of scholarly articles. However, under the word "ultrafine," earlier literature included the topic of today's nanotechnology. The aforementioned instances show that nanomaterials have a long history. However, they also have a lot of future in them. Currently, nanomaterials play commercial roles in scratch-free paints, surface coatings, electronics, cosmetics, environmental remediation, sports equipment, sensors, and energy-storage devices [6, 7]. Nanotechnology is a great example of an emerging technology because it offers engineered nanomaterials with the great potential to produce products with significantly improved performances. Upholstered pieces of furniture can be made drip-proof and stainproof using nanotechnology, just like waterproof and stainproof clothes. Also, manufacturers can minimize flammability by up to 35% by coating the foam used in upholstered furniture with carbon nanofibers. This is made possible by nanotechnology. Nanotechnology advancements have produced novel surfaces, materials, and shapes with biomedical applications [8]. Nanotechnology is being used to improve traditional biomedical applications in a number of fields, including biosensors, tissue engineering, controlled release systems, intelligent systems, and the use of nanocomposites in implant design [9]. Nanomaterials have become a fascinating class of materials with a vast array of real-world uses. Five silicon atoms or ten hydrogen atoms lined up is equal to the length of a nanometer, and with at least one dimension falling between 1 and 100 nm can be categorized as nanomaterial. Nanoparticles' functional characteristics differ from those of conventional materials, associated technologies have developed quickly and have a wide range of uses. Nanomaterials offer numerous additional specific qualities that traditional materials do not have in addition to the characteristics of traditional materials. Different types of nanomaterials have been vying for entry into competitive sports in recent years, which has considerably aided in raising the level of play and developing scientific and humanistic sports. Nanomaterials are currently widely employed in sports engineering disciplines, including stadiums, sporting goods, and athletic apparel [10]. In today's scientific and commercial research facilities, there is a lot of genuine global interest in all the different types of nanomaterials. Macroscopic and nanoscale materials science's major objective is to give engineers working in all sectors new and improved building blocks. However, compared to the better developed science and engineering of macroscopic materials, nanomaterials research has unique characteristics, the most notable of which is its revolutionary nature. For example, the logical design of nanoparticles allows for exceptionally high surface area. It is possible to create nanomaterials with exceptional magnetic, electrical, optical, mechanical, and catalytic capabilities that differ significantly from those of their bulk counterparts [11–13]. By carefully regulating the size, shape, synthesis conditions, and appropriate functionalization, the properties of nanomaterials can be tailored to meet specific needs [14]. This book describes advances in nanomaterials, specifically carbon nanomaterials, MXene-nanomaterials, Bioceramics, Bioinspired nanomaterials, nanomaterials applications in the environment, food and agricultural fields, toxicology, and biocompatibility of nanobiomaterials [15–17]. Given the possibility that future research will produce a large number of additional nanomaterials with unique features and the continual exponential development in the number of uses, it seems certain that nanomaterials

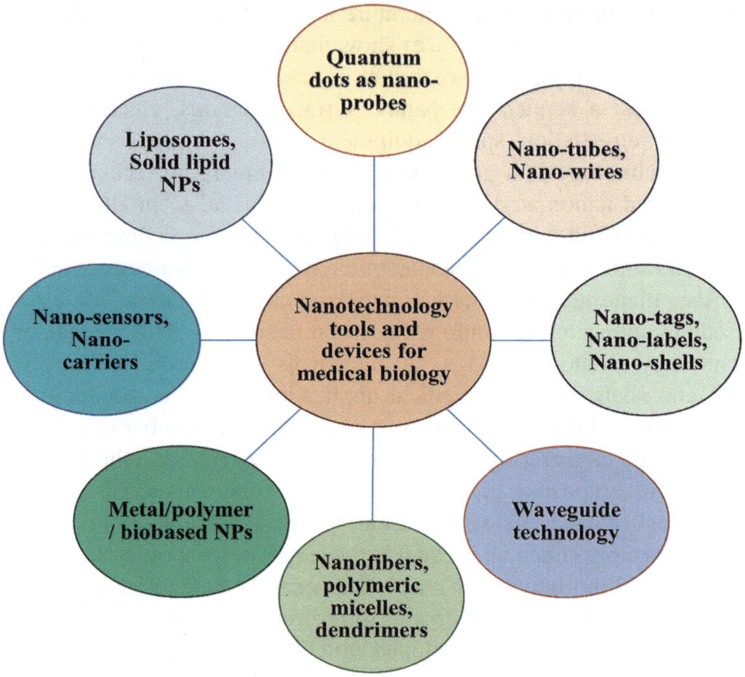

Fig. 1.2 Nanotechnology tools and devices used in medicine

will, in some way, affect almost every area of our lives. Nanomaterials play a crucial part in many of the high-tech items we use every day. Without these amazing materials, it would be impossible to make stamp-sized sound recording devices, contemporary passenger and combat jets, spacecraft and space stations, extremely large buildings, and lengthy bridges. We today find it impossible to envisage our lives without these devices, as one could not have predicted 50 years ago due to the rapid development that made it possible for them to be realized. Nanotechnology can spur economic growth and boost capacity and quality in industrial sectors [18]. The development of nanotechnology includes manipulating conventional materials to change their microscopic makeup. In silicon chips, this technology is frequently utilized. Nanocomposites are made by combining specific conventional materials with nanoparticles. For instance, titanium and zinc oxide nanoparticles are used in nanotech sunscreens. Additionally, cleaning and medical items are made with silver nanoparticles. The fact that carbon nanotubes are 115 times stronger than steel is another astonishing revelation. From this point forward, these chemicals are added to glass, steel, and polymers to strengthen them [19] (Fig. 1.2).

It is not too bold to say that those who control nanomaterials and nanotechnology also control technology as a whole. This is because of the impact that the development of nanomaterials and nanotechnology has had on our global infrastructure. Think about the nanoscale. It is one billionth of a meter in length. A human hair strand has a diameter of about 50,000 nanometers. Think about how precise a nanometer is throughout machine construction and use. Nanotechnology is atomic and molecular engineering. What about smaller-sized computers, processors, motors, or robots? While it may sound like science fiction, this is actually physics. Numerous businesses all over the world are creating goods based on nanotechnology, including microprocessors, batteries, textiles, medications, and more. The era of nanotechnology began when researchers all across the world could employ instruments like the atomic force microscope and scanning tunneling microscope. The discovery of the substance known as graphene by physicists Andre Geim and Kostya Novoselov, which won them the 2010 Nobel Prize in Physics, is well known to everyone. In this nanomaterial, the carbon atoms are organized in a single layer and in a honeycomb pattern.

Electrochemical energy storage and conversion systems have found its use because of this material's strength, flexibility, transparency, and electrical conductivity. Nanomaterials are already used in a wide range of engineering fields, so they are not a holy grail of nanoscience. Nanomaterials are widely used in modern computers, stain-resistant clothing, and suntan lotions, as well as in mechanical and bioengineering fields. Other applications, such as nanomaterial-filled composites and drugs delivered via nano-capsules, are developing at a fast pace and will be widely used in the near future [20]. For scientific communications, numerous applications in a variety of domains have been proposed. Numerous chapters in the book go into great detail about the applications of nanomaterials. Nanotechnology can also be used to optimize adhesives. Interestingly, most glues lose their stickiness at high temperatures, but a powerful "nano-glue" not only withstands high temperatures—it gets stronger as the surrounding temperature increases.

Nanoparticles are used in countless aspects of daily life. Adding to this usage are some of the following: Nanoscale additives are used to create polymer composite materials like bumpers, bats, and rackets. Nanoscale thin-films for glasses are water-repellent and antibacterial as well as for LCD displays. Nanoparticles are used in the cosmetics sector in shampoos, creams, lotions, sunscreens, and specialized makeup kits. Nanocomposites in food keep things fresh for longer by preventing bacterial growth. Rechargeable batteries, effective sensors and electronics, solar panels, warning systems, thermoelectric materials, and many more things are all products of nanotechnology. Specifically, nanoparticle catalysis is used in vehicle catalytic converters and petroleum refineries (Fig. 1.3).

1.3 Nanotechnology in Medicine

The creation of nanoscale materials for nanomedicine has, up until a few years ago, been relatively straightforward; however, at the moment, research is concentrating on the development of more complicated materials that optimize its clinical usage. An illustration would be the development of materials that release active components in response to the application of a stimulus, allowing for precise

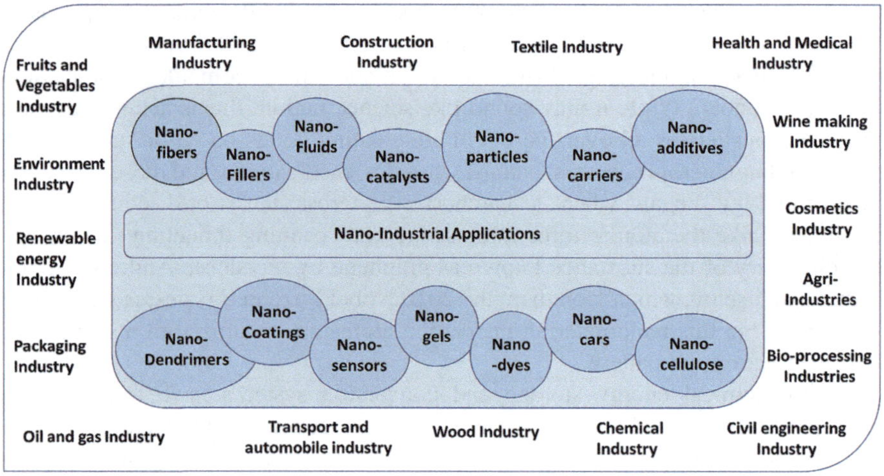

Fig. 1.3 Nano-industrial applications [10]

administration of the target compound to the required organ at the appropriate time. These substances may be activated by internal cues, such as pH changes that alter their porosity or the presence of particular enzymes in a particular tissue that cause them to degrade. Applying an "external" stimulation, such as an alternating magnetic field or a certain wavelength of light, could also cause them to become active. The functionalization of nanoparticles with various biomolecules is another example of material development [21]. The difficulty in this situation is the wide range of materials and biomolecules that can be combined, each of which has a unique ideal joining technique. Finally, research is being done on hybrid materials that let many functionalities to be performed simultaneously. Despite the complexity this diverse field of research entails, it is still possible to outline certain broad trends in the evolution of nanomedicine. On the one hand, it is crucial to keep researching the usage of novel materials, synthetic or semi-synthetic, that can offer fresh or improved features that enable optimizing their use in various applications [22, 23]. In order to manufacture the materials, whether they are new or already in existence, it is necessary to investigate alternative manufacturing methods. This includes creating processes that are simple to scale up and economically viable as well as using synthesis techniques that produce the least amount of waste. Additionally, we must create nanomaterials with the proper structural organization. Nanomaterials need to be stable at pH $= 7$ and in a physiological environment, as well as not causing toxicity or an immunological reaction in the body, in order to be used in vivo [24]. To improve their biocompatibility and stability under physiological settings, nanoparticles are typically coated with a substance. In turn, this coating makes it easier to combine various biological components (such as antibodies, medications, etc.) depending on the demands of each application [25]. As one would expect, the limitations are less severe in the case of in vitro applications, allowing the use of different materials or particle sizes. To successfully plan for a therapeutic usage, it is crucial to understand how nanomaterials and biological systems interact [26]. Understanding the ability of nanomaterials to cross biological barriers is crucial because doing so may make it possible to cure diseases like ophthalmic or neurodegenerative disorders where it is difficult for drugs to reach the desired site for treatment [27]. We also need to comprehend the risks that nanoparticles represent to our health. The wide range of results in toxicological studies that result from the lack of standardization of the procedures for the research of nanomaterial toxicity during its preclinical evaluation is one of the problems we currently face.

All of the upcoming improvements must always be done so while maintaining the safety of using these materials during their synthesis, use, and eventual recycling or disposal. The negative effects of nanomaterials on the environment must be kept to a minimum. The development of fully biodegradable materials for biomedical purposes that the body could excrete or metabolize is thus one of the problems of the future. The study of nanomedicine has benefited from better understanding of nanoparticle characteristics. The use of gold nanoparticles to combat lymphoma, a cancer kind that targets lipid cells, is one hopeful advancement in nanomedicine [28]. A gold nanoparticle with the appearance of a cholesterol cell has been created by researchers. The cancer is prevented from "feeding" on genuine cholesterol cells when this nanoparticle binds to a lymphoma cell, starving it to death. Finally, in order to facilitate the manufacture of novel nanomaterials and make their clinical use easier to attain, it is crucial to convey the knowledge developed in their development to the industry. As a result of these special qualities, nanoscale materials have considerably benefited the medical industry. Currently, a wide range of medical applications use nanoscale materials; nevertheless, research must continue to find new uses for materials that already exist, such as the use of super adhesives for the treatment of vascular malformations or the advancement of methods for diagnosing tropical diseases. In order to deliver medications specifically to particular cells, researchers are exploring methods to employ nanoparticles. Given that chemotherapy and radiation therapies can harm both healthy and sick tissue, this is particularly encouraging for the treatment of cancer. Dendrimers, nanomaterials having several branches, might boost drug delivery's effectiveness and speed. To stop the growth of cerebral palsy in rabbits, researchers have used dendrimers that transport drugs. The list goes on. Fullerenes' anti-inflammatory properties can be enhanced to minimize or even halt allergic responses. Nanomaterials may reduce bleeding and speed up clotting. The ability of nanoparticles to identify and cling to certain proteins or diseased cells enhances diagnostic imaging and testing. Nanomaterials in therapeutics, medicine, and drug delivery are discussed in detail in upcoming chapters of the book (Fig. 1.4).

Although the subject of dangers posed by nanoparticles cannot be overstated, there are a growing number of commercial goods in the market that contain nanoparticles, although only a handful of substances are now utilized in significant quantities. As a result, they secretly enter our daily lives, mostly in the form of commercial personal care items. These nanoparticles are excellent prospects for the cosmetics and personal care industries because of their modified properties. Color,

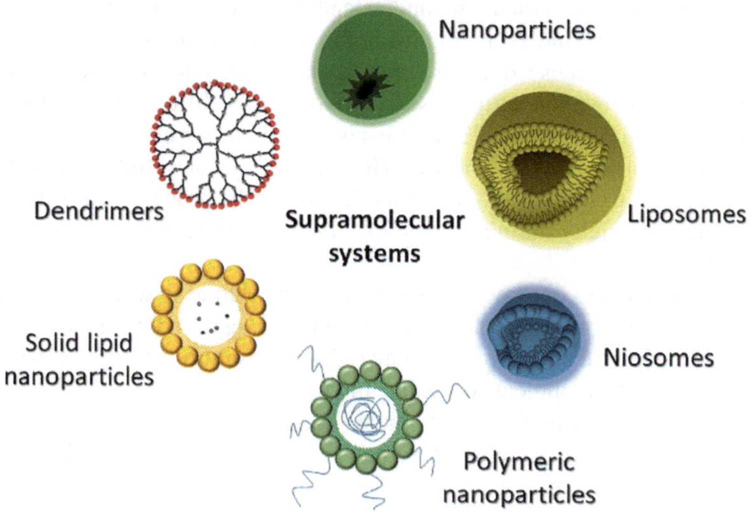

Fig. 1.4 Drug delivery with supramolecular systems [26]

transparency, solubility, and chemical reactivity are a few of these characteristics. They might facilitate the absorption of cosmetics into the skin by promoting diffusion from the cosmetic carriers into the epidermal layer. However, stable and insoluble nanoparticles including polymers, gold, silver, and titanium dioxide may directly enter the body and pose health risks. Toxicology and toxicity studies of nanomaterials are discussed in length in this book (Fig. 1.5).

Talking about regenerative medicine, the goal of regenerative medicine is to restore the functionality of tissue or cells that have been damaged by aging, disease, or injury. According to recent research, nanotechnology can provide sophisticated biomaterials with specific morphologies that can generate an extracellular environment at the nanoscale that can speed up stem cell differentiation in tissue engineering while also increasing stem cell adhesion and proliferation. Regenerative medicine and stem cells are relatively young fields in biomedicine, yet they have great clinical significance. Regenerative medicine is not a miracle cure, while regenerative medicine holds great promise, it is important to remember that it is not a panacea. Many of the approaches are still in the experimental stages, and there are no guarantees that they will be effective for all patients or all conditions. Regenerative medicine is a rapidly evolving field. We may anticipate the emergence of new techniques and technology as our knowledge of regenerative medicine deepens, which could fundamentally alter how we treat illness and injury. Through stem cell transplantation, differentiation, and tissue regeneration, they can be used to treat disease. Annually, an increasing number of orthopedic surgeries utilizing orthopedic implants are carried out because of an increase in trauma and an aging population. Traditional orthopedic implants do have a number of issues, including infection, poor host tissue conformance, and implant loosening that leads to treatment failure. By simulating

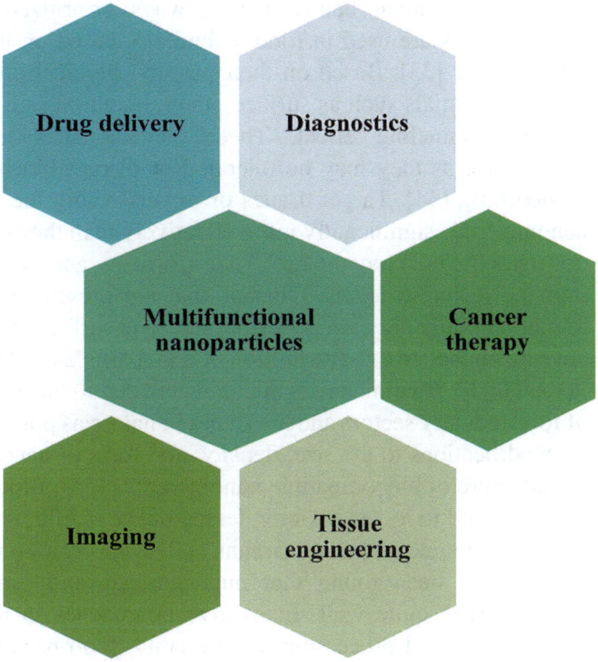

Fig. 1.5 Applications of multifunctional nanoparticles

the extracellular matrix settings in culture, the advent of nanoscale technology has made it possible to understand stem cell therapy clearly in vivo. Nanotechnology also appears to hold great promise for opening up new avenues for stem cell research [29]. For the preservation, multiplication, and differentiation of stem cells needed to progress tissue engineering, nanotechnology has introduced dynamic three-dimensional nano-environments or nano-scaffolds with controlled nano-morphologies [30]. Biomaterials and the scaffold that supports them can be created using nanohydroxyapatite, titanium, calcium phosphate, graphene oxide, carbon nanotubes, or a combination of them, for tissue engineering in orthopedic surgery. In view of their combination of nanoscale properties, great biocompatibility, and low toxicity, these composite scaffolds offer a very broad range of potential applications in orthopedic surgery and regenerative medicine. These include increasing internal fixation, skeletal defect repair, and bone tissue regeneration, as well as regulating the biological activity of stem cells (such as attachment, proliferation, and differentiation). Despite several impediments to stem cell nanotechnology research in orthopedic surgery and regenerative medicine, its potential applications have raised serious concerns. With the advancement of nanotechnology and stem cell research, the study of stem cells based on nanotechnology will become more detailed, presenting novel strategies for tissue engineering and stem cell therapy in orthopedic surgery and regenerative medicine [31, 32]. Nanomaterials in regenerative medicine and bone tissue engineering are discussed in further chapters of this book in detail.

As said, human beings have always prioritized innovation. Biocompatible nanomaterials are used in food technology based on their properties and the goals they achieve [33]. Based on their physicochemical properties, nanomaterials have unique potentials such as surface effect, quantum size effect, small size effect, and quantum tunneling effect. These characteristics influence their behavior in biosystems, as they may be tolerated or disrupt biochemical and/or physiological homeostasis [34]. Target tissues or organs within the organism can be reached by nanomaterials significantly more effectively than they can by their counterparts. It is understood that biocompatible nanoparticles act as foreign stimuli that increase immune response levels. Glucans that are present in nature have anticancer and immunomodulatory effects. Shape, size, and surface physicochemical characteristics have an impact on the distribution of nanoparticles and their pharmacokinetics in the human body. Nanomaterials are being created to have amazing properties in a range of food industry sectors and are found in nature as potential biocompatible materials.

Modifications to the size, density, and form of nanoparticles are required for the manufacture of biocompatible nanomaterials [35]. Biocompatible nanomaterials are being created to lessen toxicity, lessen negative effects on the digestive system, and enhance immune response. Organs and tissues can be targeted using nanomaterials [36]. It has been found that nutraceuticals and functional foods interact with nanomaterials quite well. Innovative approaches in the food business, including food safety, food processing, food quality, food packaging, and food labeling, are made possible by the applications of these nanomaterials. Nanotechnologies produce food with higher quality and longer shelf lives than traditional methods of processing food. They also avoid infection. The goal of food processing is to produce high-quality food while also ensuring its safety through improved food sensing and nanostructured ingredients [37]. Because silver, zinc, and calcium nanoparticles have antimicrobial activity and biocompatibility, they are used as edible films [38]. Nanomaterials are commonly combined with polymers in food packaging and processing. These nanomaterials improve the flexibility and durability of food contents. Nanomaterials, like food, enter the body through human consumption. The study of nanomaterial toxicity mechanisms aids in the development of biocompatible nanomaterials. Nanotechnology is being used to create innovative products with novel properties. Cellulosic nanomaterials (CNMs) are the name given to organic nanoparticles. Green CNMs are produced by bacteria, algae, wood pulp, tunicates, and other renewable resources. Nanomaterials such as metal nanoparticles, carbon nanotubes, carbon nanomaterials, nanocomposites, and nanowires have enhanced biosensing and sensing design systems for use in food analysis [39]. These nano-biosystems are also helpful in the creation of fresh food detecting methods. Nanomaterials serve a range of purposes in the detection of foodborne pathogens to assure the safety of food. Nano-sensors are used in food packaging to help track the physical, chemical, and biological changes made to food during processing [40]. Toxins, pollutants, and food pathogens can all be found using specially created nano-sensors and nanodevices used in smart packaging. During storage and transportation, the quality of the packaged meals is monitored and reported using the intelligent packaging systems with sensors and indicators [41, 42]. The nanotechnological applications in food and agriculture will be discussed in length in the coming chapters of this book.

1.4 Green Nanotechnology

The biosynthesis of nanomaterials from natural bioactive agents, such as plant materials, microbes, and a variety of biowastes, such as agricultural leftovers, eggshells, vegetable waste, fruit peels, and others, is a key component of green nanotechnology's quest for sustainability. It is an economical, straightforward, risk-free, safe, nontoxic, and environmentally responsible strategy [43]. Green nanotechnology is a key component of clean technologies intended to clean up the environment and turn extra-bioactive items into green nanomaterials that are more profitable and environmentally benign [44, 45] (Fig. 1.6).

Thinking about the future, it is predicted that soon the entire computer will be built on a single microchip, thanks to nanotechnology. The RAM and nanotransistors would be strong and sophisticated in protecting against data loss incidents. The nanostructures of future OLED displays will increase their energy, weight, density, and lifespan efficiency. Other electronic products, such as smart

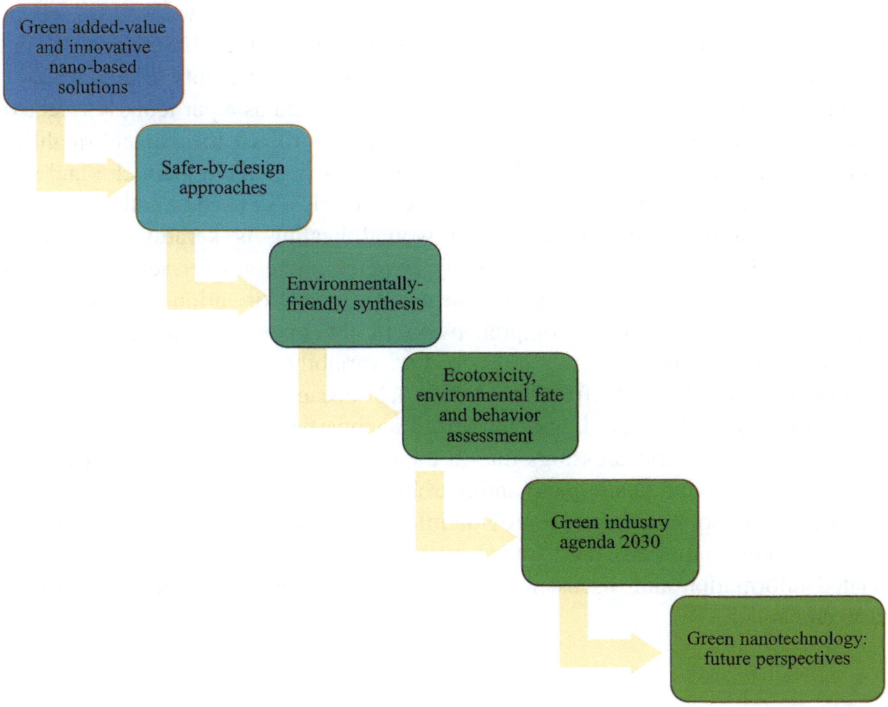

Fig. 1.6 Outlook of green nanotechnology

cards, RFID, smart packaging, flash memory, and antimicrobial coatings, will be updated with higher standards. Furthermore, such widespread use has minimal health hazards. In the modern world, the environment has grown to be a serious concern. Any additional deterioration could have negative effects on our future. Nanotechnology will contribute to a more ecologically friendly future through mechanical filtering and the neutralization of toxic chemicals [46]. According to futurists, the years between 2020 and 2050 will be a "period of relative scarcity." Nanotechnology is beneficial to humanity. It will contribute to the fabric of life, ushering in lives that are cleaner, safer, more advanced, expanding, and comfortable.

Because of their small size, NPs find their way easily to enter the human body and cross the various biological barriers and may reach the most sensitive organs. Toxicity assessment of these engineered products is required to understand possible adverse effects and their fate inside the human body [47].

1.5 Conclusion

An increasing number of innovative therapies have been created in the field of wound healing. Nanostructured systems have been employed to speed up wound healing at various stages. The drug may be created at the nanoscale level to act as its own "carrier" or nanomaterials may be used to deliver the medication. Different natural and synthetic substances and their mixtures have been developed for wound healing. Due to their important biological roles, natural polymers have been used in a wide range of wound healing techniques. As a result of their ability to give tissues a moist and protective environment, collagen is recognized as a particularly effective healing scaffold with outstanding and comparatively robust mechanical qualities. Another illustration is fibrin, which collects platelets to create a main clot and stop bleeding. Since it is mechanically strong, silk fibroin is also used. Another common example of a nanomaterial utilized in wound healing is keratin. The use of nanomaterials has the potential to greatly advance processes for self-healing that can resemble regeneration. To tailor these nanomaterials for various wound healing applications, still, a deeper comprehension of the underlying mechanisms and cellular reactions is necessary due to the variability of the injured tissues. Nanoparticles can be used effectively in this field because of their excellent surface-to-volume ratio. Silver, gold, and zinc metal nanoparticles are excellent candidates for inclusion in wound dressings due to their remarkable features, which include stimulation of wound healing and antimicrobial activity.

At last, this book attempts to provide information at a single platform about the basic concepts, advances, and trends relating to nanomaterials via covering the related information and discussing almost every field in which nanomaterials are now employed.

References

1. Thakur P, Thakur A (2022) Introduction to nanotechnology. In: Thakur A, Thakur P, Khurana SP (eds) Synthesis and applications of nanoparticles. Springer, Singapore. https://doi.org/10.1007/978-981-16-6819-7
2. Krukemeyer MG, Krenn HF, Wagner W, Resch R (2015) History and possible uses of nanomedicine based on nanoparticles and nanotechnological Progress. J Nanomed Nanotechnol 6:1–7
3. Taneja S, Thakur P, Ravelo B, Thakur A (2022) Nanocrystalline samarium doped nickel-zinc-bismuth ferrites: Investigation of structural, electrical and dielectric properties. Mater Res Bull 154:111937
4. Verma R, Thakur P, Chauhan A, Jasrotia R, Thakur A (2023) A review on MXene and its' composites for electromagnetic interference (EMI) shielding applications. Carbon 208:170–190
5. Dhanda N, Thakur P, Sun A-C, Thakur A (2023) Structural, optical and magnetic properties along with antifungal activity of Ag-doped Ni-Co nanoferrites synthesized by eco-friendly route. J Magn Magn Mater 572:170598
6. Thakur A, Punia P, Dhar R, Aggarwal RK, Thakur P (2022) Separation of cadmium and chromium heavy metals from industrial wastewater by using Ni-Zn nanoferrites. Adv Nano Res 12:457–465
7. Ivanets A, Prozorovich V, Kouznetsova T (2022) Advancement in water purification by nanomaterials. In: Thakur A, Thakur P, Khurana SP (eds) Synthesis and applications of nanoparticles. Springer, Singapore
8. Klebowski B, Depciuch J, Parlinska-Wojtan M, Baran J (2018) Applications of noble metal-based nanoparticles in medicine. Int J Mol Sci 19(12):4031
9. Suhag D, Bhatia R, Das S, Shakeel A, Ghosh A, Singh A, Sinha OP, Chakrabarti S, Mukherjee M (2015) Physically cross-linked pH-responsive hydrogels with tunable formulations for controlled drug delivery. RSC Adv 5(66):53963–53972
10. Li P (2022) Nanomaterials in sports training and its biological safety. Sci Program 2022:1–9
11. Thakur P et al (2018) Influence of cobalt substitution on structural, optical, electrical and magnetic properties of nanosized lithium ferrite. J Mater Sci Mater Electron 29:16507–16515
12. Chandel S et al (2018) Effect of non-magnetic Al^{3+} doping on structural, optical, electrical, dielectric and magnetic properties of $BiFeO_3$ nano-ceramics. Ceram Int 44:4711–4718
13. Chahar D, Kumar D, Thakur P, Thakur A (2023) Visible light induced photocatalytic degradation of methylene blue dye by using Mg doped Co-Zn nanoferrites. Mater Res Bull 162:112205
14. Suhag D, Chauhan M, Shakeel A, Das S (2020) Emerging trends in nanotheranostics. NanoBioMedicine:335–368
15. Chalia S, Bharti MK, Thakur P, Thakur A, Sridhara SN (2021) An overview of ceramic materials and their composites in porous media burner applications. Ceram Int 47:10426–10441
16. Sharma P et al (2017) Nanomaterial fungicides: in vitro and in vivo antimycotic activity of cobalt and nickel nanoferrites on phytopathogenic fungi. Global Chall 1(9):1700041
17. Thakur P, Thakur A (2022) Nanomaterials, their types and properties. In: Thakur A, Thakur P, Khurana SP (eds) Synthesis and applications of nanoparticles. Springer, Singapore. https://doi.org/10.1007/978-981-16-6819-7_2
18. Malik S, Muhammad K, Waheed Y (2023) Nanotechnology: a revolution in modern industry. Molecules 2023(28):661
19. Hasirci V, Vrana E, Zorlutuna P, Ndreu A, Yilgor P, Basmanav FB, Aydin E (2006) Nanobiomaterials: a review of the existing science and technology, and new approaches. J Biomater Sci Polym Ed 17(11):1241–1268
20. Sabir F et al (2022) Functionalized nanoparticles in drug delivery: strategies to enhance direct nose-to-brain drug delivery via integrated nerve pathways. In: Thakur A, Thakur P, Khurana SP (eds) Synthesis and applications of nanoparticles springer. Singapore

21. Shahdeo D, Roberts A, Kesarwani V, Horvat M, Chouhan RS, Gandhi S (2022) Polymeric biocompatible iron oxide nanoparticles labeled with peptides for imaging in ovarian cancer. Biosci Rep 42(2):BSR20212622
22. Kaur S, Sharma K, Singh R, Kumar N (2022) Advancement in crops and agriculture by nanomaterials. In: Thakur A, Thakur P, Khurana SP (eds) Synthesis and applications of nanoparticles. Springer, Singapore
23. Bharti MK, Gupta S, Chalia S, Garg I, Thakur P, Thakur A (2020) Potential of magnetic nanoferrites in removal of heavy metals from contaminated water: mini review. J Supercond Nov Magn 33:3651–3665
24. Yang Z, Xie J, Zhu J, Kang C, Chiang C, Wang X, Kong L, Mi H, Chen F, Chen Z, Zhang A, Yu B, Lee RJ, Teng L, Lee LJ (2016) Functional exosome-mimic for delivery of siRNA to cancer: in vitro and in vivo evaluation. J Control Release 243:160–171
25. Zhou C, Meng-Qi C, Jiaming H, Zhaogang Y, Zhaofeng C (2016) Preparation of functionalized TiO2 nanotube arrays and their applications. Sci Adv Mater 8(6):1231–1241
26. Grande F, Ioele G, Occhiuzzi MA, De Luca M, Mazzotta E, Ragno G, Garofalo A, Muzzalupo R (2019) Reverse transcriptase inhibitors nanosystems designed for drug stability and controlled delivery. Pharmaceutics 11(5):197
27. Suhag D, Shahdeo D and Gandhi S (2019) Emerging trends and advances in targeted nanotheranostics. SMC Bull 10(1):17–22
28. Shahdeo D, Kesarwani V, Suhag D, Ahmed J, Alshehri SM, Gandhi S (2022) Corrigendum to" self-assembled chitosan polymer intercalating peptide functionalized gold nanoparticles as nanoprobe for efficient imaging of urokinase plasminogen activator receptor in cancer diagnostics"[Carbohydrate Polymers, 266,(2021) 118138]. Carbohydr Polym 284:119239
29. Pan S, Yu H, Yang X, Yang XR, Wang Y, Liu Q, Jin L, Yang Y (2017) Application of nanomaterials in stem cell regenerative medicine of orthopedic surgery. J Nanomater 2017:1–12
30. Kaushik S, Gandhi S, Chauhan M, Ma S, Das S, Ghosh D, Chandrasekharan A, Alam B, Parmar AS, Sharma A, Santhoshkumar T, Suhag D (2020) Water-templated, polysaccharide-rich bioartificial 3D microarchitectures as extra-cellular matrix bioautomatons. ACS Appl Mater Interfaces 12(18):20912–20921
31. Kala D, Gupta S, Kaushal A (2022) Nanotechnology in healthcare. In: Thakur A, Thakur P, Khurana SP (eds) Synthesis and applications of nanoparticles. Springer, Singapore
32. Hafner A, Lovrić J, Lakoš G, Pepić I (2014) Nanotherapeutics in the EU: an overview on current state and future directions. Int J Nanomedicine:1005
33. Ranjha MMAN, Shafique B, Khalid W, Nadeem HR, Mueen-Ud-Din G, Khalid MZ (2022) Applications of biotechnology in food and agriculture: a mini-review. Proc Natl Acad Sci India Sect B Biol Sci 92(1):11–15
34. Ranjha MMAN, Shafique B, Batool M, Kowalczewski PŁ, Shehzad Q, Usman M, Manzoor MF, Zahra SA, Yaqub S, Aadil RM (2021) Nutritional and health potential of probiotics: a review. Appl Sci 11(23):11204
35. Khanra A, Sangam S, Shakeel A, Suhag D, Mistry S, Rai MP, Chakrabarti S, Mukherjee M (2018) Sustainable growth and lipid production from Chlorella pyrenoidosa using N-doped carbon nanosheets: unravelling the role of graphitic nitrogen. ACS Sustain Chem Eng 6(1): 774–780
36. Adabi M, Naghibzadeh M, Adabi M, Zarrinfard MA, Esnaashari SS, Seifalian AM, Faridi-Majidi R, Aiyelabegan HT, Ghanbari H (2017) Biocompatibility and nanostructured materials: applications in nanomedicine. Artif Cells Nanomed Biotechnol 45(4):833–842
37. Amenta V, Aschberger K, Arena M, Bouwmeester H, Botelho Moniz F, Brandhoff P, Gottardo S, Marvin HJ, Mech A, Quiros Pesudo L, Rauscher H, Schoonjans R, Vettori MV, Weigel S, Peters RJ (2015) Regulatory aspects of nanotechnology in the agri/feed/food sector in EU and non-EU countries. Regul Toxicol Pharmacol 73(1):463–476
38. Nadeem M, Ghaffar A, Hashim M, Murtaza MA, Ranjha MMAN, Wang C, Riaz MN (2021) Sonication and microwave processing of phalsa drink: a synergistic approach. Int J Fruit Sci 21(1):993–1007

39. Bains A, Narang D, Chawla P, Dhull SB (2020) Nanoparticles and antibiotic drug composite. CRC Press eBooks, pp 165–186
40. Huang Y, Mei L, Chen X, Wang Q (2018) Recent developments in food packaging based on nanomaterials. Nanomaterials (Basel) 8(10):830
41. Silvestre C, Duraccio D, Cimmino S (2011) Food packaging based on polymer nanomaterials. Prog Polym Sci 36(12):1766–1782
42. Nile SH, Baskar V, Selvaraj D et al (2020) Nanotechnologies in food science: applications, recent trends, and future perspectives. Nano-Micro Lett 12:45
43. Srivastava N, Mishra G, Saxena SK (2022) Environmental nanomedicine. In: Thakur A, Thakur P, Khurana SP (eds) Synthesis and applications of nanoparticles. Springer, Singapore
44. Thakur A, Chahar D, Thakur P (2022) Synthesis of nanomaterials by biological route. In: Thakur A, Thakur P, Khurana SP (eds) Synthesis and applications of nanoparticles. Springer, Singapore
45. Alqarni LS, Alghamdi MA, Alshahrani AA, Nassar A (2022) Green nanotechnology: recent research on bioresource-based nanoparticle synthesis and applications. J Chem 2022:1–31
46. Gupta R, Xie H (2018) Nanoparticles in daily life: applications, toxicity and regulations. J Environ Pathol Toxicol Oncol 37(3):209–230
47. Thakur, P., Verma, V., Thakur, A. Toxicity of nanomaterials: an overview. Synthesis and applications of nanoparticles, (2022) pp. 535-544. In: Thakur, A., Thakur, P., Khurana, S.P. (eds) Synthesis and applications of nanoparticles. Springer, Singapore https://doi.org/10.1007/978-981-16-6819-7_25

Everyday Nanotechnology

2

Raksha Rathore, Deepa Suhag, Fayu Wan, Atul Thakur, and Preeti Thakur

Abstract

One of the most exciting twenty-first-century technologies is nanotechnology. Nanotechnology focuses on the creation and application of materials of nanoscale dimensions in several facets of life. Due to their small size, nanoparticles have a high surface-to-volume ratio and a variety of unique characteristics. The development of novel techniques for creating new products, formulating new chemicals and materials, and replacing the current generation of equipment with improved performance equipment are some of the main applications of the young and developing science known as nanotechnology. These techniques also provide environmental remediation options. The potential impact on society, business, and the economy will become increasingly clear as nanotechnology transitions from theoretical to practical, encouraging essential responses to existing issues. Ultimately, the chapter underscores the important role that

R. Rathore
Amity Centre for Nanotechnology, Amity University Haryana, Gurugram, India

D. Suhag (✉) · P. Thakur
Amity Centre for Nanotechnology, Amity University Haryana, Gurugram, India

nanoLatticeX, Amity University Haryana, Gurugram, Haryana, India
e-mail: dsuhag@ggn.amity.edu

F. Wan
School of Electronics and Information Engineering, Nanjing University of Information Science & Technology, Nanjing, China

A. Thakur
Amity Centre for Nanotechnology, Amity University Haryana, Gurugram, India

nanoLatticeX, Amity University Haryana, Gurugram, Haryana, India

School of Electronics and Information Engineering, Nanjing University of Information Science & Technology, Nanjing, China

D. Suhag et al. (eds.), *Integrated Nanomaterials and their Applications*,
https://doi.org/10.1007/978-981-99-6105-4_2

nanotechnology is likely to play in shaping our future, and the need for individuals to remain informed and engaged as these technologies continue to evolve. In addition to catalysis and environmental remediation, the chapter also examines the uses of nanotechnology in the fields of animal, agricultural, nutritional, medical, and pharmaceutical sciences. The chapter gives summaries of the most recent nanotechnology applications that have an impact on several facets of human life.

Keywords

Nanotechnology · Agriculture · Environmental remediation · Pharmaceuticals

2.1 Introduction

Nanotechnology involves the manipulation, designing and developing materials/ matter at the atomic or nano-meter or molecular level [1]. A material can be regarded as nanomaterial when it has at least one dimension or internal structure in nano-meter size. Nanomaterials have a very high surface-to-volume ratio due to which they possess some unique physiochemical properties such as solubility, color, diffusivity, etc. Their optical [2], magnetic [3], and thermodynamic properties are also very different as compared to their macroscopic counterparts [4]. People of the modern era are not the pioneers in using nanotechnology because nano-dimensional structures and their utility within organisms have always been in existence from the very beginning [5]. Nucleic acids, i.e. DNA and RNA, proteins such as enzymes are some nano-dimensional structures that are essential for sustaining life and are found even in primitive unicellular organisms like blue green algae (BGA), archaebacteria, etc. Also, nanotechnology is not confined to just one domain of science such as biology, chemistry, physics, engineering, or toxicology, instead, it can be considered as an interdisciplinary domain of science [6]. Nanoparticles were known to even ancient people of the fourth century. Roman people of the fourth century used nanoparticles made up of metals to adorn cups and glasses so that they could look fancier. The marvelous cups which are named as Lycurgus were decorated with gold and silver nanoparticles, looked green in white light but when enlightened from within they looked red in appearance [7]. With the help of nanotechnology, nanoparticles of desirable and characteristic features [8] can be synthesized using conventional synthesis techniques such as hydrothermal [9], co-precipitation [10], citrate precursor [11], and green synthesis.

2.2 Basic Understanding of Nanotechnology Involves the Following Themes

(i) The scale which measures nano-dimensional structures;
(ii) The dynamics and the types of bonds/interactions existing between nano-dimensional objects;

(iii) Significance of surface-to-volume ratio in nanoparticles;
(iv) Using properties of nano-dimensional objects for developing nanotechnology [12].

2.3 Applications of Nanotechnology in Day-to-Day Life (Fig. 2.1)

2.3.1 Medicine/Nanomedicine

Nanomedicine refers to the application of nanotechnology for medical needs and is defined as the use of nanoparticles for illness diagnosis, monitoring, control, prevention, and therapy. For clinical purposes, nanomaterials can be used in three different areas of nanomedicine: diagnosis (nano-diagnosis), controlled drug delivery (nano-therapy), and regenerative medicine. Another field that unifies diagnosis and treatment is called theranostics, and it is a promising methodology that holds

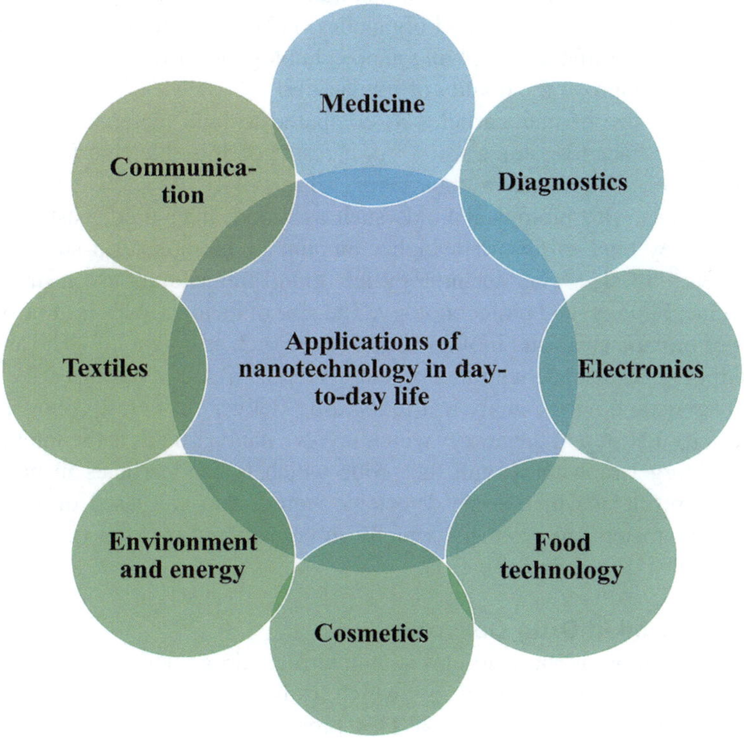

Fig. 2.1 Applications of nanotechnology in day-to-day life

both the finding/imaging specialist and the drug in a comparable framework. Nanomedicine is holding out hope for positive advances in clinical practice. Nanomaterials can be created to have specific biological targets in mind by changing their size, shape, compound arrangement, and surface. A careful molecular plan must be used to provide an effective biological result. For two main reasons, it is therefore envisaged that there would be significant information on how nanomaterials interact with biological frameworks. The first is associated with the physio-pathological notion of infections. The organic processes that cause diseases take place at the nanoscale and can be influenced by several factors, such as altered properties, improperly folded proteins, contamination by bacteria or infections. The prudent design of nanomaterials to concentrate on the specific site of activity desired in the body would be made possible by an improved understanding of the sub-atomic cycles. The interaction between the surface of nanomaterials and the climate in natural liquids is another issue. In this distinctive scenario, the accurate depiction of the biomolecular crown is of utmost importance for comprehending how the nano-particle interacts with cells and influences biological processes. This point of interaction comprises dynamic systems, such as the exchange between the surfaces of nanomaterials and the surfaces of naturally occurring components (proteins, layers, phospholipids, vesicles, and organelles) [13]. Knowing that all tissues have a limited functional lifetime, applying nanotechnology to medicine has the potential to significantly improve the quality of human life.

The smaller size of nanoparticles as compared to bulk materials leads to their unique high surface-to-volume ratio. As a result of this, their surface energy is enhanced which in turn leads to their increased reactivity. As soon as nanoparticles come in contact with biological fluids such as blood, lung fluid, gastro-intestinal fluid, etc., they tend to absorb a higher amount of biomolecules such as lipids, proteins, etc. [14, 15]. Nanotechnology has found tremendous use in the field of biomedicine, biology, and biotechnology. The size of nanoparticles is almost similar to those of biomolecules and biological structures such as organelles. The amalgamation of biological science with that of nanomaterials has revolutionized the designing of diagnostic devices, analytical tools, drug delivery systems, contrast agents, etc. The very first step in the assays which involve nanoparticles is the interaction or binding of target molecules with the probe which is nanoparticles in nano-based assays as a result of which some detectable signals that are based on optical and magnetic parameters are produced which can be captured by MRI (magnetic resonance imaging) (Fig. 2.2, Table 2.1).

2.3.1.1 Controlled Drug Delivery

Another application of nanoparticles is in controlled drug release. The drugs can be manipulated at nanoscale dimensions which leads to the alteration in fundamental properties and bioactivity of drugs. This process allows good control over the different aspects of the drugs which are mentioned as under:

- Can change the solubility and blood pool retention time;
- Can cause controlled drug release for shorter and longer time duration;

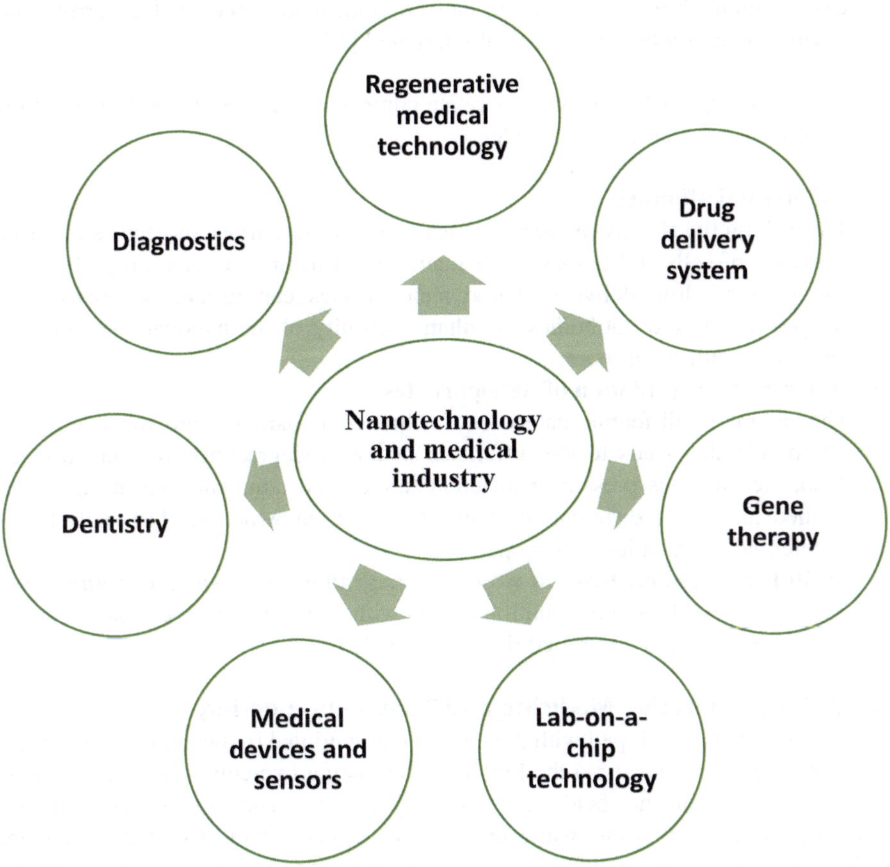

Fig. 2.2 Applications of nanotechnology in medical industry [9]

Table 2.1 Various types of nanodevices which are used in clinical applications are tabulated [16]

Modality	Potential applications
Cantilevers	• High-throughput screening • Disease protein biomarker detection
Carbon nanotubes	• DNA mutation detection • Disease protein biomarker detection
Dendrimers	• Image contrast agents
Nanocrystals	• Improved formulation for poorly soluble drugs
Nanoparticles	• Targeted drug delivery, permeation enhancers • MRI and ultrasound image contrast agents
Nano-shells	• Tumor-specific imaging
Nanowires	• High-throughput screening • DNA mutation detection (SNPs)
Quantum dots	• Optical detection of genes and proteins in animal models and cell assays • Tumor and lymph node visualization

- Brings about drug release which can be controlled based on the surrounding environment or which is specifically targeted [17].

The basic approaches which can cause controlled drug release with the help of nanoparticles are mentioned as under:

- **Differential affinities**
  The differential affinity of nanoparticles with various surface molecules of different types of cells and tissues is the main reason behind targeted drug efficiency and targeted drug delivery. These nanoparticles can further be coated with antigen-recognized antibodies to enhance affinity of the nanoparticles with the target molecule even more.
- **Differential degradation of nanoparticles**
  The nature of different materials of which nanoparticles are made up of is different which leads to the difference in the biodegradation of nanoparticles. Also, the enzymes present in different tissues and adjuvants which have been applied also have different mechanisms of action which further leads to the differential degradation of nanoparticles.
- **Multi-layer nanoparticles in which the degradation curve can be adjusted**
  Nanoparticles which have multiple layers with adjustable degradation curve will be quite efficient in controlled drug release [18].

2.3.1.2 Regenerative Medicine and Tissue Engineering

The biomaterials developed with the help of conventional tissue engineering are not as promising as expected but the limitations are being overcome by the coalition of nanotechnology in the field of tissue engineering. The size of conventional biomaterials is in microns, while the surface features of most of the tissues and cells are nano-meter sized. In view of this fact, nanostructured biomaterials are considered more promising in the field of tissue engineering and regenerative medicine because using nano-meter scale surface features enhances tissue regeneration greatly. Some common examples in tissue engineering and regenerative medicines include use of nanostructured titanium implants which enhance calcium deposition leading to increased bone cell responses. But enhanced bone tissue regeneration is not the only example of nanomaterial-based regenerative medicine. Nanosized scaffolds for skin tissue regeneration are also being used greatly. One of the examples of skin tissue regeneration scaffolding material includes nanostructured chitosan (CS) grafted with poly (ε-caprolactone) (PCL) [19].

2.3.2 Diagnostics

Nanomolecular diagnostics, which provides new alternatives for clinical diagnostic procedures, is a phrase used to describe nanobiotechnology and its applications in the life sciences, particularly in molecular diagnostics. Molecular diagnostics is a crucial aspect of the development of personalized medicine, which incorporates point-of-care performance of diagnostic techniques [20]. Nanobiotechnology combined with molecular diagnostics increases the clinical diagnosis substantially. "Nanoscale probes" are best suited for detailed examination of receptors, pores, and other nanoscale-sized components of living cells in this type of diagnosis. For combination diagnostics and therapies, nanodevice can be implanted as a preventive and prophylactic approach in early illness diagnosis. Small sample sizes and diagnostic procedures that employ nanoscale particles as tags or labels are two benefits of utilizing nanotechnology in molecular diagnostics.

The diagnostic techniques based on nanomaterials are helpful in the detection of infectious diseases such as tuberculosis by making use of a diagnostic kit named as BioMEMS and also for cardiovascular diseases, genetic diseases such as cancer, monitoring of glucose with the help of nano-sensors, detection of single nucleotide polymorphism such as diseased state during sickle cell anemia. One of the nanomanipulation techniques known as optical tweezers is being utilized to detect tissue growth, abscess, cysts, calculi, etc. [16].

2.3.3 Electronics

Nanoelectronics is an emerging field that makes use of nanodevices and nanomaterials so that certain characteristics of conventional electronics can be improved including chips with enhanced memory [21], miniaturized antennas [22], portability (lab-on-chip), and lesser electricity consumption. Nanoelectronics is widely being used as nano-biosensors so that they can detect disease-causing agents with greater efficiency. Electronic devices are an integral part of biosensors which assist in the detection of biomolecules by making use of ligands and analytes. The surface of various metals like gold (Au) is the basic substrate on various electronic devices. Gold (Au) is considered as the most prominent material for this purpose because it is quite sensitive, stable, and easy to use for surface modification. Apart from gold, silver (Ag) and silicon are also commonly used for modifying surfaces of nanoelectronics. Silicon (Si) is commonly used in semiconductor-based electronics because it is highly specific and extremely stable. Some other metals used for coating of nanoelectronics surfaces are nickel (Ni) and platinum (Pt) [23]. Nanoelectronics have better screen displays and the memory chips in such devices are highly dense which helps in reducing size. The transistors which are used in integrated circuits nowadays have reduced size because we employ nanotechnology to develop them [24]. In contemporary nanotechnology-based computers, the transistors are developed in such a manner that their size is reduced to fit into the silicon chips, i.e. they are compact [1]. This is in accordance with Moore's law which states that on a single chip the number of transistors placed are increasing enormously. It not only adds to the enhanced performance and energy efficiency but also contributes to the size reduction of the computers [25].

2.3.4 Food Technology

The twenty-first century's new frontiers are nanoscience and nanotechnology. In comparison to their use in pharmaceuticals and drug delivery, their applications to the agriculture and food industries are relatively new. Some of the growing areas of nanotechnology for food and agriculture include smart nutrient delivery, bioseparation of proteins, a quick sampling of biological and chemical pollutants, and nanoencapsulation of nutraceuticals. The potential of nanotechnology for food applications will be realized, thanks to developments in technologies like DNA microarrays, microelectromechanical systems, and microfluidics (Fig. 2.3).

2.3.4.1 Preservation with the Help of Nano-Sensors

The shelf life of food products can be enhanced greatly by encapsulating them with the help of nanoencapsulation. Nanoencapsulation aids in slowing the degradation process until the food has reached the target site. The food materials which are coated by nanomaterials remain unexposed to moisture and various gases. The

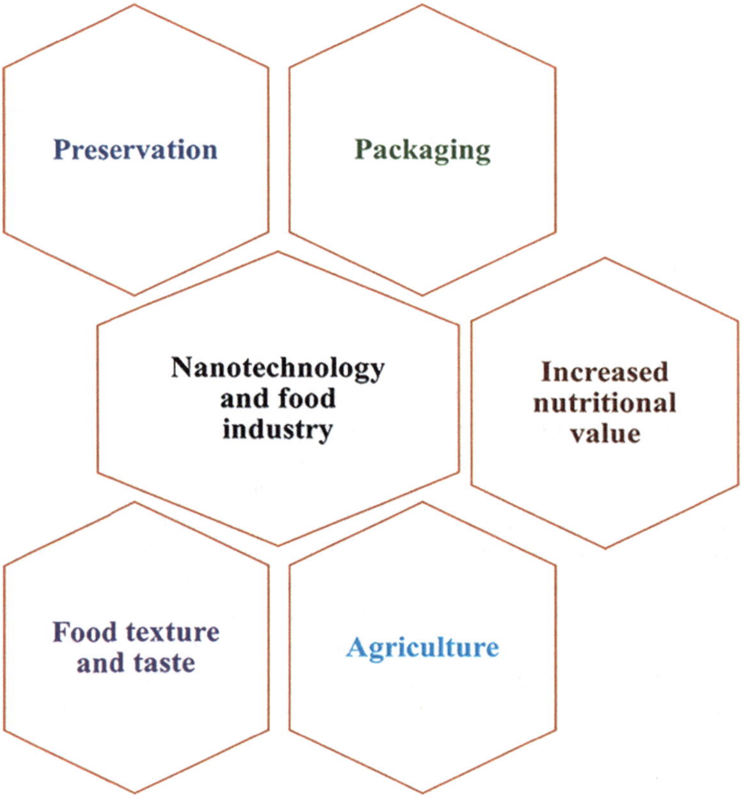

Fig. 2.3 Applications of nanotechnology in the food industry

chemical degradation is slowed down with the help of encapsulation of functional components which surround the droplets which in turn is accomplished by manipulating the interfacial layer surrounding them. For example, nano-encapsulated curcumin exhibits reduced antioxidant activity [4].

2.3.4.2 Packaging

Nanofillers which may contain nano-whiskers, nanotubes, or nanofibers can be used in food packaging industry because it can not only help in reducing pollution but also protects the food in a much better way. Conventional packaging materials such as plastics not only cause environmental pollution but also their manufacturing process requires a lot of fossil fuels for developing them. Nanotubes made up of α-lactalbumin which is a milk protein is being considered as the most promising substance for food packaging because it has many desirable characteristics such as great viscosity as well stiffness. Their 8 nm sized diameter cavities enable efficient binding of food constituents such as enzymes or vitamins. In addition to this, they are also capable of encapsulation which counters unwanted flavors and odor of the compounds [26]. Nanocomposites are quite efficient in providing aegis against microbial infections because these types of packing materials make interaction with the food components to retard the growth of microbes [4].

2.3.4.3 Food Quality Enhancement in Terms of Nutritional Value

Various food products in the market are already available which make use of nanoparticles to improve the nutritional value in the food. Some of these food products include enhanced iron content in nutritional drinks, vitamin rich micelles, oils that have phytochemicals and minerals, breakfast cereals with zinc oxide [27].

2.3.4.4 Food Texture, Taste, and Appearance

One of the methods used for enhancing flavor retention and release is nanoencapsulation. Silicon dioxide (SiO_2) and titanium dioxides are mostly used as flow agents or coloring agents in the food items. SiO_2 even imparts fragrance to food products [2]. Without altering the flavor or quality of a food, more nutrients can be added using nanoencapsulation methods. Producers could be able to incorporate nutrients that do not occur naturally or that, up to now, could not be incorporated for chemistry-related reasons. Such encapsulation techniques have the added benefits of extending shelf life, shielding encapsulated ingredients from the environment and unwanted interactions during food preparation and delivering precise contents to a particular target place within the body.

2.3.5 Agriculture

Agriculture is another sector in which nanotechnology is successfully being incorporated. The main area within the agriculture industry which makes use of nanoparticles is the production of fertilizers by making use of nanotechnology. Nano-fertilizers are the main focus because crop production mainly depends upon fertilizers [28]. Nano-fertilizers are almost three times more efficient in nutrient uptake. In addition to this, they also enhance stress tolerance in the plants in a much better way than conventional fertilizers. The nano-encapsulated fertilizers are able to release growth factors and nutrients such as nitrogen (N), potassium (K), and phosphorus (P) in a much more controlled and targeted way. Nano-sensors are widely used for the detection of plant pathogens. Nanotechnology has also made it very easy to detect pesticides [29] (Fig. 2.4).

2.3.5.1 Nano-fertilizers

The enormous quantities of fertilizers that were given considerably increased production, but they also harmed the beneficial soil bacteria. Plants cannot access fertilizers because of runoff and pollution. Fertilizers using nanomaterial coatings can remedy this problem. The drawbacks of conventional chemical fertilizers can be solved with a nanohybrid construction like nano-fertilizers (NFs). The NFs transfer nutrients to the plants in an intelligent way, demonstrating their superiority over bulky chemical fertilizers in terms of crop output and environmental sustainability. Depending on the methods of application and the characteristics of the particles, plants can absorb NFs through their roots or foliage. NFs help plants better withstand biotic and abiotic stressors. It lowers the cost of production and lessens the impact on the environment.

2.3.5.2 Supply of Micronutrients

Micronutrients including zinc, iron, manganese, boron, copper, molybdenum, etc. are known to be crucial for growth and development. Due to the enormous increase in crop yields and innovative agricultural practices brought about by the green

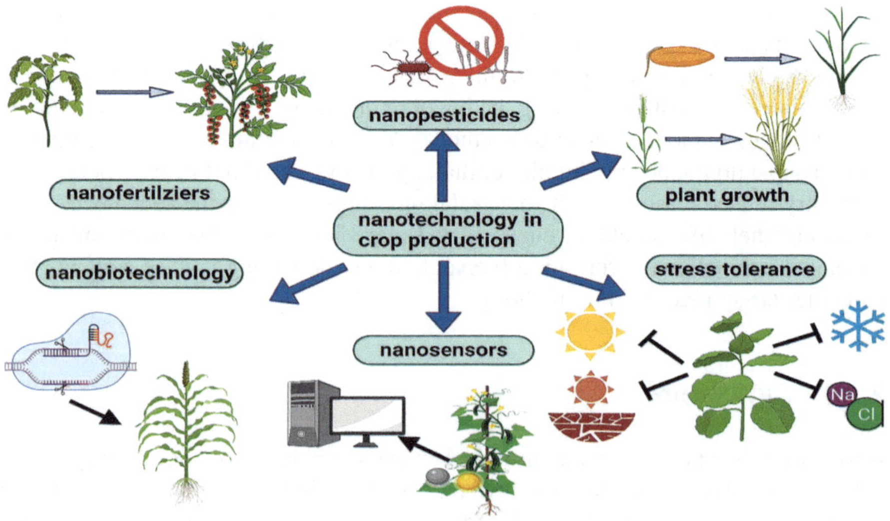

Fig. 2.4 Applications of nanotechnology in agriculture [29]

revolution, soil micronutrients including zinc, iron, and molybdenum have been progressively falling. Absorption can be improved by treating the leaves with micronutrients. To strengthen and improve the quality of soil, nanoformulations of micronutrients can be sprayed on plants or added to the soil for root absorption [28].

2.3.5.3 Management of Insect Pests

The use of synthetic agrochemicals has altered agriculture, but it has also given rise to a new problem known as insect pest resistance. Nanoparticles have immense potential for monitoring and controlling insect pests in modern agriculture. Garlic oil's insecticide effectiveness against Tribolium castaneum (the red flour beetle) has been enhanced by PEG-coated nanoparticles.

2.3.5.4 Nano Fungicides

Crop fungal infections significantly reduce crop output. Although there are many fungicides that are sold commercially, utilizing them also harms plants. The use of nanotechnology can greatly aid in solving this problem. Nanoparticle-based antifungal medications have been tried against dangerous fungus. Silver nanoparticles outperformed zinc oxide and titanium dioxide nanoparticles in terms of their ability to inhibit the growth of fungi at lower concentrations [30].

2.3.5.5 Nano-herbicides

Nano-based particles are utilized in the production of herbicides to create nano-herbicides, which are then applied successfully using a delivery method based on nanoparticles. Nano-herbicides have several benefits, including a decrease in the amount of synthetic pesticides used, an efficient delivery mechanism, increased personal safety, and quicker reaction times. Nano-herbicides are a cutting-edge technology that solves all the problems with conventional herbicides. The development of weed resistance, which necessitates strategic planning for its ultimate eradication, is one of the main issues that conventional herbicides are currently confronting. Weeds pose the biggest threat to agriculture, as they lower yields by scavenging nutrients that would otherwise be accessible to crop plants. Nano-herbicides can have a significant impact on the ecologically safe weed control of crops [31]. Such issues may be resolved with the use of nano-herbicides with qualities like a high penetration capability and an efficient delivery system.

2.3.5.6 Biosensors

Biosensors are tools that locate environmental elements, assess crop growth efficiency, pinpoint the type and location of issues, and use computers, sensors, remote sensing gear, global positioning systems, and precision farming. Worldwide issues that humanity is experiencing include climate change, a growing population that has an impact on food production and safety, and biosensors have long been recognized as one of the most effective tools for finding solutions. By providing ongoing monitoring or early identification of disease outbreaks that can be stopped, biosensors can support sustainable agriculture. Additionally, it is crucial for keeping an eye on food danger factors such as pesticides, veterinary drugs, heavy metals, viruses, toxins, and unauthorized additives.

2.3.6 Cosmetics

Nowadays nano-emulsions are extensively being used in making cosmetic products because certain properties of nanoparticles make them the most promising candidate for this purpose. Nanoparticles not only provide better UV protection, color, finish, and long-lasting effects but also get incorporated deeper into the skin to give better results. Owing to their smaller size, nano-emulsions have greater stability, enhanced surface area, and uniformity in the distribution of cosmetics on the skin. Their thermodynamics and kinetic stability make the cosmetic formulation more suitable because it restricts undesirable phenomena such as Ostwald ripening, sedimentation, and flocculation. Many international cosmetic brands such as L'Oréal, Garnier, Bioderma, Avenue, Laroche-posay have claimed to use micellar facial cleansers which are much more efficient in cleaning as compared to conventional cleansers (Fig. 2.5).

Anti-aging cream, facials, skin moisturizers, etc. are certain other cosmetic products in which nano-emulsions are being used. Apart from this other commodity products in which nano-emulsions have been incorporated are deodorants,

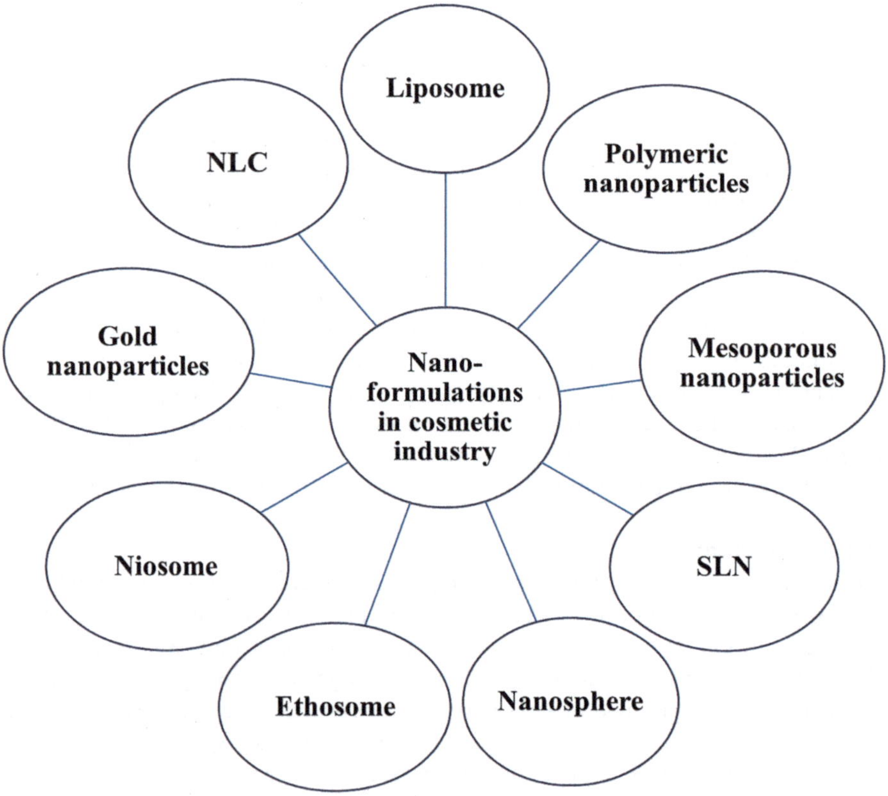

Fig. 2.5 Different nanotechnological formulations used in cosmetic industry

shampoos, toothpastes, hair dyes, etc. [19]. Besides nano-emulsions some other nanosized materials such as fullerenes, liposomes, nanotubes, quantum dots, etc. are also being used in cosmetic products [32].

2.3.7 Environment and Energy

Further, nanotechnology is also being used for the betterment of the environment. Nanotechnology is being used for controlling pollution. Several industries are switching their conventional manufacturing methods into nanotechnology-based methods so as to curb generation of new type of pollution. Nanotechnology is also being used to reduce the cost of making products with the help of alternative energy sources. One example is the utilization of silver as a catalyst in the production of propylene oxide. Propylene oxide is extensively used in the manufacturing of plastics, detergents, brake fluid, paints, etc.

Nowadays windmill blades which generate electricity are being incorporated with Epoxy containing carbon nanotubes (CNTs), such CNTs have less weight and they are more strong than the conventional material which is used for making windmill blades. Being stronger and lighter in weight makes them superior to other materials because it helps them to generate more electricity. Nanomaterials are also being used in solar cells because their usage not only enhances efficiency but also reduces the cost of electricity production. For this purpose, silicon nanowires are incorporated inside that polymer which is required for designing solar cells [16]. Nanotechnology is also being employed in water conservation and purification. For inference, several nano-filters have been designed recently which can filter out pesticides out of the drinking water with a greater efficiency. These nano-filters are almost 20 times more efficient than the conventional filters. Some nano-dimension structures have also been developed which are capable of removing arsenic from potable water in a cost-effective manner [33].

2.3.8 Communication

In simple words, nano-communication means the exchange of information by making use of nanodevices. Nanomachines interconnected via nanonetworks are used for this purpose and the communication may occur via wired or wireless mode [34]. Nanochips were produced for the first time in 2000 by semiconductor industries which were subsequently incorporated into computers and smartphones successfully. Using nanochips leads to compaction and therefore reduction in the size of the computers and smartphones because now more transistors can be accommodated on a single chip. It also adds to the efficiency of the system because with the help of nanochips, a microprocessor can have at least 10 times more transistors [35]. It is pertinent to mention here that the use of nanotechnology in proper communication is still limited because such devices still lack certain processing capabilities and the ability to manage usage of power [34].

2.3.9 Textiles

The textile industry is another area that heavily utilizes nanotechnology. Recent years have seen the emergence of "smart textiles," which are made by combining smart nanoparticles with more conventional materials. A smart textile can respond to changes in the environment by altering one or more of its features in order to perform a purpose. Three generations have passed since the invention of smart fabrics. The first generation of smart textiles, also referred to as "passive" textiles, are able to detect changes in their environment but are unable to change their own qualities. For instance, fabrics coated with various metal oxide nanoparticles can provide apparel that is IR/UV resistant and cotton impregnated with silver nanoparticles has antimicrobial properties [36].

Nanoparticles not only add to the durability of the fabric but also enhance their functionality. Nanomaterials impart some other desirable features to the fabrics including resistance to wrinkles, stain repellent, optical display, UV light blocking [37], antimicrobial resistance, odor resistance, etc. [16]. In future, nano-moieties might be used to develop fabrics which will be able to respond to the external environment by making use of external stimuli such as color, electricity, and other physiological signals. Such intelligent and smart clothing have been displayed in prominent fashion weeks such as London fashion week, New York fashion week, Paris fashion week, etc. [38]. Nowadays, functional clothing is being designed which is integrated with communication devices, electronics, and nanomaterials. One of the best examples of nanomaterial-based dresses is Bubelle, this dress can change colors as per the mood swings of the person wearing it.

Second-generation, or "active," smart fabrics are constructed from substances that first recognize environmental changes or stimuli before responding appropriately. Examples include thermochromic materials that change color in reaction to temperature changes and shape-memory fabrics that can respond to mechanical deformations. Third-generation active textiles, sometimes known as "super-smart" textiles, incorporate soft and smart electronics, such as sensors, nano-generators, and systems for storing energy. On-body electronics, for instance, can give sensing for various diseases, risks, or pollutants. Additionally, attractive optical parts on smart textiles can be made possible by nano-generators and energy storage devices. It is now possible to take advantage of the unique characteristics of nanomaterials to give fabrics and clothing a high extra value functionality while keeping other desirable qualities like comfort, flexibility, lightness, and esthetic appearance. This is possible because the ability to incorporate microelectronic devices with textiles has advanced to the point where it is possible to do so. Nanomaterials have been incorporated into textiles using a variety of methods. The process of producing the fibers from which the fabrics are made employs a "bottom-up" strategy. The "top-down" strategy, in contrast, is used during the finishing stages, such as with printing, spray coating, or impregnation techniques. It has been demonstrated that the relatively new technique of electrospinning, which creates fibers and fabrics from processed raw materials, is perfect for creating nanofibers [39–42] (Fig. 2.6).

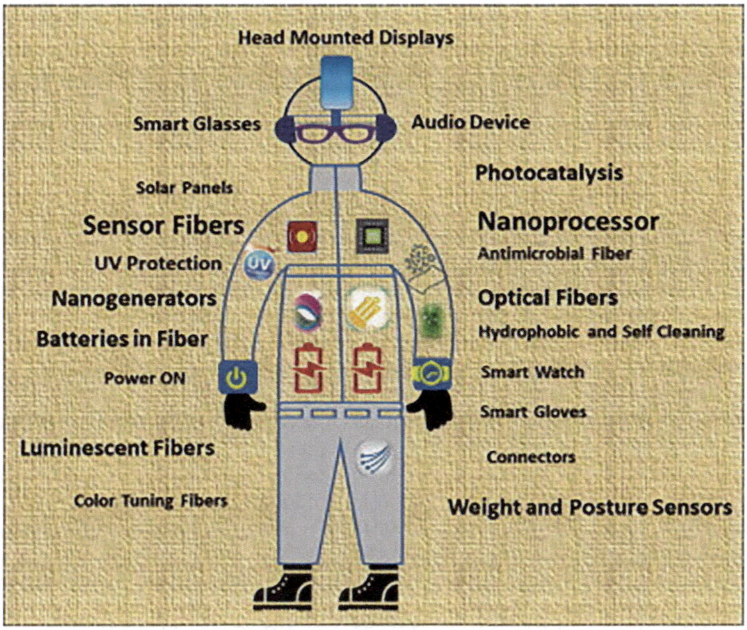

Fig. 2.6 Smart clothing made up of nanomaterials [43]

2.4 Conclusion

In conclusion, nanotechnology has the potential to revolutionize many aspects of our everyday lives. From improving the efficiency of energy production and storage to enhancing medical diagnosis and treatment, the applications of nanotechnology are vast and diverse. One of the most exciting areas of nanotechnology is its potential to address some of the world's most pressing challenges, such as climate change and disease. Nanotechnology can help reduce our reliance on fossil fuels by developing more efficient solar cells and energy storage devices. It can also aid in the development of new medicines and therapies, such as targeted drug delivery and nanosensors for early disease detection. As with any new technology, there are also potential risks and challenges associated with nanotechnology. Research in this field must be conducted with caution to ensure that the benefits outweigh the risks. There must also be a focus on developing safe and sustainable practices for the manufacture and disposal of nanomaterials. Overall, nanotechnology has the potential to transform our world in many positive ways, but it is important that we approach this technology with a responsible and ethical mindset. By investing in research and development, promoting transparency and accountability, and prioritizing safety and sustainability, we can ensure that nanotechnology is used for the betterment of all. Overall, this chapter "Everyday Nanotechnology" provides an overview of the various ways in which nanotechnology is currently being used in our daily lives, from consumer products to medical applications. It highlights the potential benefits and risks associated with these applications and emphasizes the need for continued research and development to ensure the safe and responsible use of nanotechnology.

References

1. Thakur A, Thakur P, Khurana SMP (2022) Synthesis and applications of nanoparticles, vol 1-544. Springer, Singapore. https://doi.org/10.1007/978-981-16-6819-7
2. Chahar D, Kumar D, Thakur P, Thakur A (2023) Visible light induced photocatalytic degradation of methylene blue dye by using Mg doped Co-Zn nanoferrites. Mater Res Bull 162:112205
3. Naagar M, Chalia S, Kharbanda S, Thakur P, Thakur A (2023) Correlating tailored saturation magnetization of normal-inverse spinels with noncollinear magnetic ordering. J Mater Sci Mater Electron 34:473
4. Singh TK, Shukla S, Kumar P, Wahla V, Bajpai VK (2017) Application of nanotechnology in food science: perception and overview. Front Microbiol 8
5. Thakur P, Thakur A (2022) Introduction to nanotechnology in synthesis and applications of nanoparticles. Springer, Singapore, pp 1–17. https://doi.org/10.1007/978-981-16-6819-7_1
6. Albrecht MA, Evans CW, Raston CL (2006) Green chemistry and the health implications of nanoparticles. Green Chem 8(5):417
7. Logothetidis S (2012) Nanotechnology: principles and applications. In Nanoscience and Technology (pp. 1–22). Springer Nature
8. Kamble NS, Bhosale NKV, Mohite NR, Navale NP (2022) Methods of preparation of nanoparticles. Int J Adv Res Sci Commun Technol:640–646
9. Verma R, Thakur P, Sun ACA, Thakur A (2023) Investigation of structural, microstructural and electrical characteristics of hydrothermally synthesized $Li_{0.5-0.5x}Co_xFe_{2.5-0.5x}O_4$, $(0.0 \leq x \leq 0.4)$ ferrite nanoparticles. Phys B Condens Matter 661:414926
10. Thakur P, Thakur A, Singh M (2008) Low temperature synthesis of $Mn_{0.4}Zn_{0.6}In_{0.5}Fe_{1.5}O_4$ nanoferrite for high-frequency applications. J Phys Chem Solids 69(1):187–192
11. Kumari S, Dhanda N, Thakur A, Singh S, Thakur P (2023) Investigation of calcium substitution on magnetic and dielectric properties of Mg–Zn nano ferrites. Mater Chem Phys 297:127394
12. Schönborn KJ, Höst GE, Palmerius KL (2015) Measuring understanding of nanoscience and nanotechnology: development and validation of the nano-knowledge instrument (NanoKI). Chem Educ Res Pract 16(2):346–354
13. Shinde MA, Patwekar M, Patwekar FI, Bajaber MA, Medikeri A, Mohammad FS, Mukim M, Soni SK, Mallick J, Jawaid T (2022) Nanomaterials: a potential Hope for life sciences from bench to bedside. J Nanomater 2022:1–13
14. Soares S, Sousa J, Pais A, Vitorino C (2018) Nanomedicine: principles, properties, and regulatory issues. Front Chem 20(6):360
15. Malik S, Muhammad K, Waheed Y (2023) Nanotechnology: a revolution in modern industry. Molecules 28:661
16. Alharbi KK, Al-sheikh YA (2014) Role and implications of nanodiagnostics in the changing trends of clinical diagnosis. Saudi J Biol Sci 21(2):109–117
17. Saini R, Saini S, Sharma S (2010) Nanotechnology: the future medicine. J Cutan Aesthet Surg 3(1):32–33
18. Cai XJ, Xu YY (2011) Nanomaterials in controlled drug release. Cytotechnology 63(4): 319–323. https://doi.org/10.1007/s10616-011-9366-5. Epub 2011 Jul 1
19. Khang D, Carpenter J, Chun YW, Pareta R, Webster TJ (2008) Nanotechnology for regenerative medicine. Biomed Microdevices 12(4):575–587
20. Kala D, Sharma TK, Gupta S et al (2020) AuNPs/CNF-modified DNA biosensor for early and quick detection of O. tsutsugamushi in patients suffering from scrub typhus. 3 Biotech 10:446

21. Chandel S, Thakur P, Tomar M, Gupta V, Thakur A (2017) Investigation of structural, optical, dielectric and magnetic studies of Mn substituted BiFeO3 multiferroics. Ceram Int 43(16): 13750–13758
22. Saini A, Thakur A, Thakur P (2016) Effective permeability and miniaturization estimation of ferrite-loaded microstrip patch antenna. J Electron Mater 45:4162–4170
23. Gopinath SC, Anbu P, Arshad MK, Lakshmipriya T, Voon CH, Hashim U, Chinni SV (2017) Biotechnological processes in microbial amylase production. Biomed Res Int 2017(2017): 1272193
24. Chaurasia N (2017) Nanotechnology and nanomaterials in everyday life. Int J Sci Res (IJSR) 6(4):1560–1562
25. Modarressi M, Sarbazi-Azad H (2018) Topology specialization for networks-on-chip in the dark silicon era. Advances in computers. Elsevier BV
26. Sozer N, Kokini JL (2009) Nanotechnology and its applications in the food sector. Trends Biotechnol 27(2):82–89
27. Aziz ZDA, Nasir HM, Ahmad A, Setapar SHM, Peng WL, Chuo SC, Khatoon A, Umar K, Yaqoob AA, Ibrahim MNM (2019) Role of nanotechnology for design and development of cosmeceutical: application in makeup and skin care. Front Chem 7
28. Peteu SF, Oancea F, Sicuia OA, Constantinescu F, Dinu S (2010) Responsive polymers for crop protection. Polymers 2(3):229–251
29. Liu C, Zhou H, Zhou J (2021) The applications of nanotechnology in crop production. Molecules 26(23):7070
30. Shyla KK, Natarajan N, Nakkeeran S (2014) Antifungal activity of zinc oxide, silver and titanium dioxide nanoparticles against Macrophomina phaseolina. J Mycol Plant Pathol 44(3):268–273
31. Manjunatha SB, Biradar D, Aladakatti Y (2016) Nanotechnology and its applications in agriculture: a review. J Farm Sci 29:1–13
32. Raj S, Jose S, Sumod US, Sabitha M (2012) Nanotechnology in cosmetics: opportunities and challenges. J Pharm Bioallied Sci 4(3):186
33. Nagar A, Pradeep T (2020) Clean water through nanotechnology: needs, gaps, and Fulfillment. ACS Nano 14(6):6420–6435
34. Galal A, Hesselbach X (2018) Nano-networks communication architecture: Modeling and functions. Nano Commun Networks 17:45–62
35. Dan Hutcheson G (2004) The first nanochips. Sci Am 290(4):76–83
36. Kumar S, Bhanjana G, Sharma A, Sarita, Sidhu MC, Dilbaghi N (2015) Herbicide loaded carboxymethyl cellulose nanocapsules as potential careers in agri nanotechnology. Sci Adv Mater 2015(7):1143–1148
37. Zhou Y, Tang R (2017) Natural flavonoid-functionalized silk fiber presenting antibacterial, antioxidant, and UV protection performance. ACS Sustain Chem Eng 5(11):10518–10526
38. Yetisen AK, Qu H, Manbachi A, Butt H, Dokmeci MR, Hinestroza JP, Yun SH (2016) Nanotechnology in textiles. ACS Nano 10(3):3042–3068
39. Pereira C, Pereira AM, Freire C, Pinto TV, Costa RS, Teixeira JS (2020) Chapter 21—Nanoengineered textiles: from advanced functional nanomaterials to groundbreaking high-performance clothing. In: Mustansar Hussain C (ed) Handbook of functionalized nanomaterials for industrial applications. Elsevier, pp 611–714
40. Zhang X, Wang S, Xu L, Feng L, Ji Y, Tao L, Li S, Wei Y (2012) Biocompatible polydopamine fluorescent organic nanoparticles: facile preparation and cell imaging. Nanoscale 4(18):5581
41. Yetisen AK, Qu H, Manbachi A, Butt H, Dokmeci MR, Hinestroza JP, Skorobogatiy M, Khademhosseini A, Yun SH (2016) Nanotechnology in textiles. ACS Nano 10(3):3042–3068
42. Fateixa S, Pinheiro P, Nogueira HIS, Trindade T (2019) Gold loaded textile fibres as substrates for SERS detection. J Mol Struct 1185:333–340
43. Shah MA, Pirzada BM, Price G, Shibiru AL, Qurashi A (2022) (2020). Applications of nanotechnology in the smart textile industry: a critical review. J Adv Res 38:55–75

Classification of Nanomaterials (Carbon, Metals, Polymers, Bio-ceramics)

3

Fayu Wan, Atul Thakur, and Preeti Thakur

Abstract

One of the most cutting-edge materials available today is nanomaterials. In recent years, nanomaterials have captured the interest of a majority of researchers due to their nanoscale size and distinctive physical, biological, and chemical attributes when compared to bulk materials. These are categorized according to their size, chemical make-up, shape, and origin. In this chapter we are going to learn about the classification and types of nanomaterials based on morphology, dimensionality, size, agglomeration state, and composition. The objective is to understand all the types and classification of nanomaterials at a single place. Nanomaterials have several types, like organic-based, inorganic-based, as well as composite-based. The requirements for several industrial uses have led to the mass production of numerous nanomaterials. These are mostly formed from two types of materials:

F. Wan (✉)
School of Electronics and Information Engineering, Nanjing University of Information Science & Technology, Nanjing, China
e-mail: fayu.wan@nuist.edu.cn

A. Thakur
School of Electronics and Information Engineering, Nanjing University of Information Science & Technology, Nanjing, China

Centre for Nanotechnology, Amity University Haryana, Gurugram, India

Nanotechnology Wing, Innovative Science Research Society, Shimla, India

P. Thakur
Centre for Nanotechnology, Amity University Haryana, Gurugram, India

Nanotechnology Wing, Innovative Science Research Society, Shimla, India

NanoLatticeX, Amity University Haryana, Gurugram, Haryana, India

D. Suhag et al. (eds.), *Integrated Nanomaterials and their Applications*,
https://doi.org/10.1007/978-981-99-6105-4_3

synthetic sources and naturally occurring nanoparticles. We go over to the different kinds and subcategories of nanomaterials in this chapter in detail.

Keywords

Nanomaterials · Types · Classification · Sources

3.1 Introduction

Nanomaterials are becoming increasingly significant as a result of technological developments coupled with their superiority over their bulk equivalents in terms of physical, chemical, and biological properties. A rationalization of their key traits is necessary due to the vast variety of nanomaterials that are now used in industry and in research publications. The size, form, aggregate &/or agglomeration structure can be used to distinguish the nano-objects that make up the nanomaterials [1, 2]. The following are the key divisions used to classify them:

(i) According to the dimensions of nanomaterials, which range from 1 to 100 nm in size,
(ii) Classification of nanomaterials based on kind, taking into account the primary constituents [3],
(iii) According to their shape, which includes core-shell nanoparticles, nanomembranes, hollow nanomaterials, nanosheets, and complete nanomaterials,
(iv) By their source of origin or source of production [4].

The study of nanostructures, in recent years, has become most common for many disciplines, for which scientists and engineers are tailoring the materials and the matters at the atomic levels to obtain materials with significantly improved properties [5] (Fig. 3.1).

3.2 Classification of Nanomaterials

Numerous nanomaterials are created every day as a result of the constantly expanding field of nanotechnology, and in order to distinguish one nanomaterial from another, each one must be identified based on its structure, form, size, and chemical synthesis. Following are some general categories into which nanomaterials can be divided, including size, shape, materials, sources, etc.

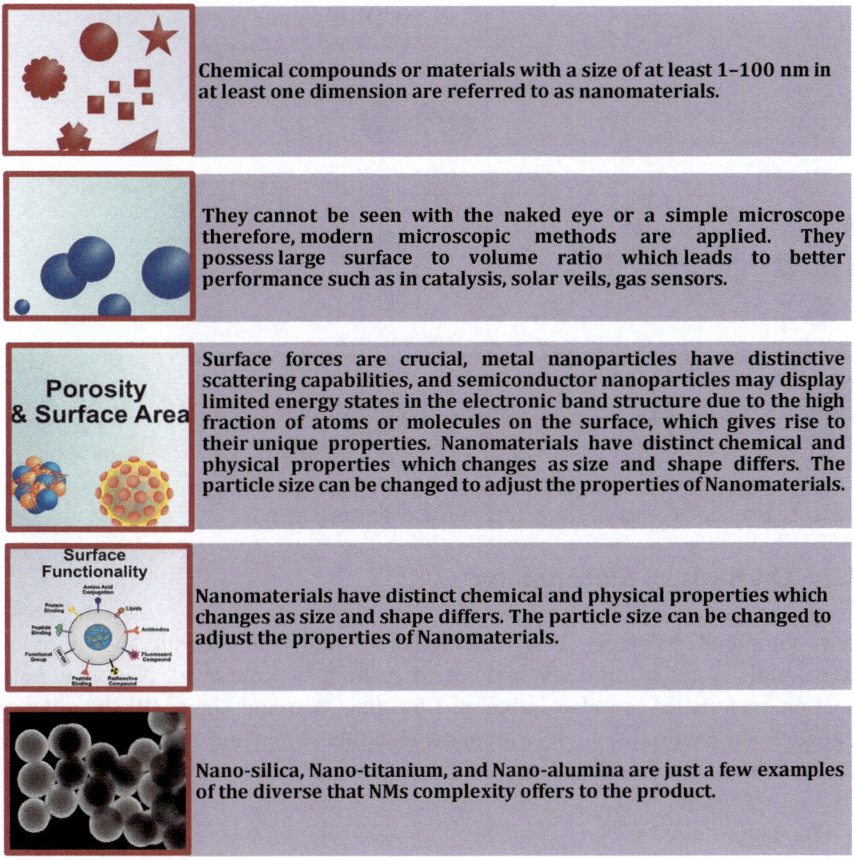

Fig. 3.1 Properties of nanomaterials

3.2.1 Dimensions and Sizes

As there are several types of nanomaterials being produced by researchers for various applications, it becomes necessary to categorize them as per their applications. Since these nanomaterials are merely solid particles, it is simple to determine their dimensions and size using various techniques. Researchers are actively developing diverse nanomaterials for various applications, making it essential to categorize them according to their intended uses. These nanomaterials consist of solid particles whose size and dimensions can be readily assessed through various measurement techniques. This allows for their classification based on crystalline structures and chemical compositions [6, 7]. Moreover, nanomaterials are categorized according to their dimensional properties, including 0-dimensional, 1-dimensional, 2-dimensional, and 3-dimensional variants [8] (Fig. 3.2).

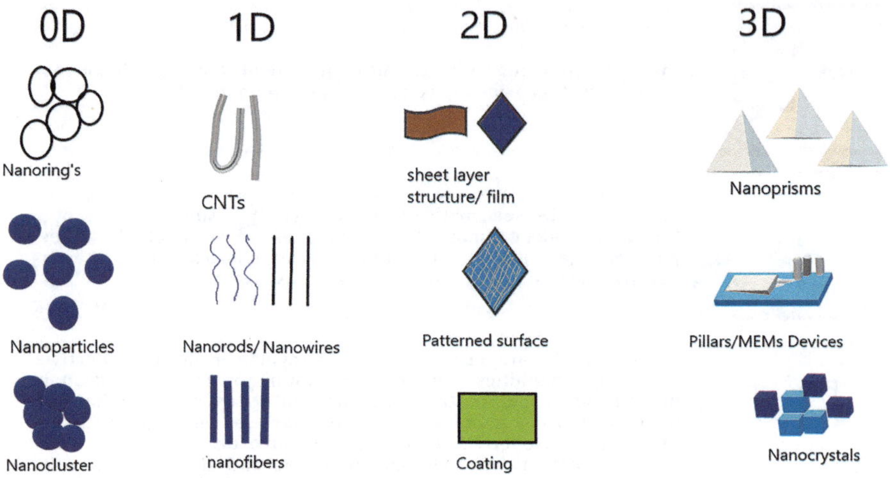

Fig. 3.2 Classification of nanoparticles according to dimensionality

3.2.2 On Basis of Morphology

When compared to their bulk counterparts, the characteristic behavior of nanomaterials is much improved and varied at the nanoscale level. Nanomaterials' optical characteristics are also influenced by their size and shape [9]. In order to categorize nanomaterials and determine which applications they are most suited for, understanding of the following morphology is crucial:

(i) Flatness
(ii) Sphericity
(iii) Aspect ratio
(iv) **High aspect ratio**
  (a) Nanowires, Nanohelix
  (b) Nano zigzags
  (c) Nanopillars, nanotubes
  (d) Nanofibers, nanobelts
(v) **Low aspect ratio**
  (a) Nanospheres
  (b) Nanopyramids
  (c) Nano-cubes

3.3 Carbon Nanomaterials

Fullerene, graphene, nanodiamond, and carbon nanotubes (CNTs) are a few examples of the diverse class of nanomaterials known as carbon-based nanomaterials [5] (Fig. 3.3).

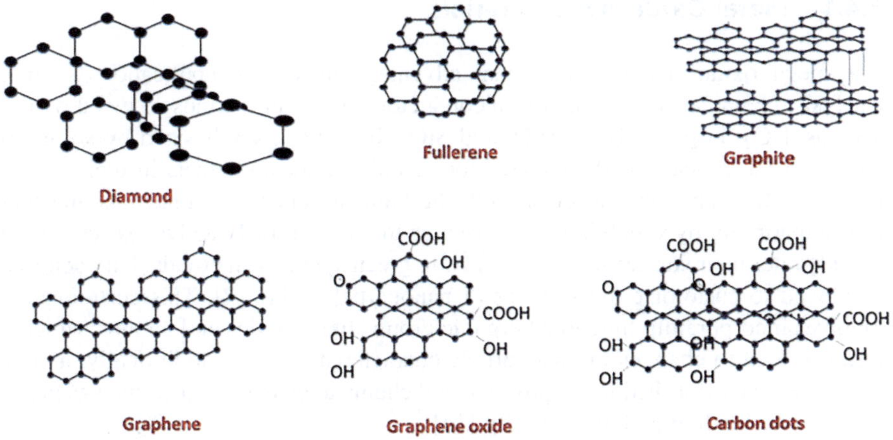

Fig. 3.3 Different structures formed of carbon nanomaterials

3.4 Metal Nanoparticles

In addition, a number of metals, such as gold (Au) and silver (Ag), are utilized to create NMs, and also known as metal nanomaterials or inorganic nanomaterials. Their manufacturing such as lowering particle size to a specified nano range determines their application and qualities. Owing to the growing separation between the atomic coordinates and the unsaturated sites, when size decreases by a nanometer, the atoms on the surface materials become more active [7]. The active surface area of metal nanoparticles, which is used in catalysis and adsorption processes, determines a number of critical features. In general, metal NPs like Pt, Pd, Ni, and Ru have both intra-band transitions like Al, Ag, Au, and Cu and can absorb light via inter-band transition. According to several studies, light irradiation can increase the catalytic activity of metal nanoparticles (NPs) [9] (Fig. 3.4).

Fig. 3.4 Different types of metallic nanomaterials

3.4.1 Metal Oxide Nanomaterials

The metal oxides have a significant advantage in several applications such as catalysis, chemical sensors, and semiconductors [10]. Many kinds of metal oxides, such as TiO_2, Fe_2O_3, Al_2O_3, ZnO, and SiO_2, have been synthesized by sol-gel or hydrothermal processes. Particularly for metallic oxides, changes in their surface characteristics have an impact on both the bandgap energy of materials and their biocompatibility by supplying a surface area that is extremely active. Reactions can quickly alter their surface properties [11]. In green synthesis methods, fatty acids are employed to customize the surfaces of nanoparticles [11, 12]. This customization involves incorporating functional organic groups like epoxies and amines, which in turn can lead to changes in nanoparticle characteristics [13]. Additionally, anionic molecules can offer distinctive physical and chemical characteristics, such as particular analyte binding and high density [14].

3.4.2 Bimetallic Nanomaterials

The bimetallic NPs, as their name suggests, are made up of two metal components that exhibit special qualities such as chemical stability and reactivity [15]. Additionally, their size-dependent electrical and optical characteristics are being used to identify them. The content, shape, and size distribution all have an impact on these characteristics. Bimetallic catalysts are created through the straightforward grafting of binaphthyl moiety as stabilizers on the metallic surface and applications for effective catalytic processes [16]. The preparation technique has a big impact on the size distributions of NPs and their recycleability as catalyst materials. Bimetallic materials like Fe-Cu [17] and Ag-Au [18] among others showed that two metals combined can exhibit varied NP dispersion and stability as well as surface activity in addition to optical and magnetic properties. The miscibility of the two metals and the circumstances of preparation are just two examples of the many variables that have a significant impact on the nanostructure of bimetallic compounds. Bimetallic NPs can generally be categorized into four classes, namely Crown jewel structure, hollow structure, core-shell structure, and alloyed structure.

3.5 Organic Nanomaterials

The category of carbon-based nanoscale materials known as "organic nanomaterials" is suitable with biological applications due to their covalent connections. Noncovalent bonds are weak and quickly broken [19]. To create various shapes of nanomaterials like liposomes, dendrimers, micelles, and polymers, these organic materials can be easily tailored (Fig. 3.5).

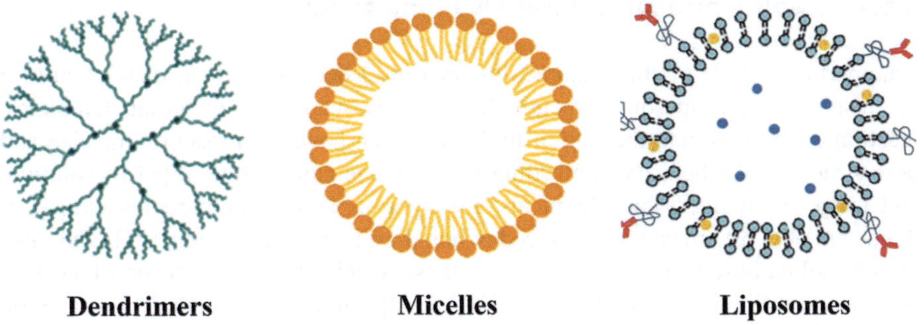

Fig. 3.5 Types of organic nanomaterials [20, 21]

3.6 Nanocomposites

Nanocomposites are made up of two different kinds of nanomaterials. These nanomaterials are produced by joining together nanowires, nanofibers, or materials of a larger size [22]. Any combination of bulk ceramic, metal, or polymer materials, as well as metal, carbon, or organic nanowires and nanofibers, can be used to create these nanocomposites. These composites are solid materials that consist of numerous phases, such as nanoscale repetition spacing between phases or dimensions of less than 100 nm [23]. Physical dimensions in the nanometer size range are always used while creating composite structures. A composite is created when numerous materials are blended to produce a variety of qualities, such as wear and gloss retention, flexural strength, water sorption, optical properties, etc. [24]. In accordance with the Box-Behnken design, chitosan-tripolyphosphate/TiO_2 nanocomposites can be created with reduced energy and chemical adsorption of reactive orange dye [25]. Through a variety of interactions, composite materials enable enhancement of the material's adsorption capability (Fig. 3.6).

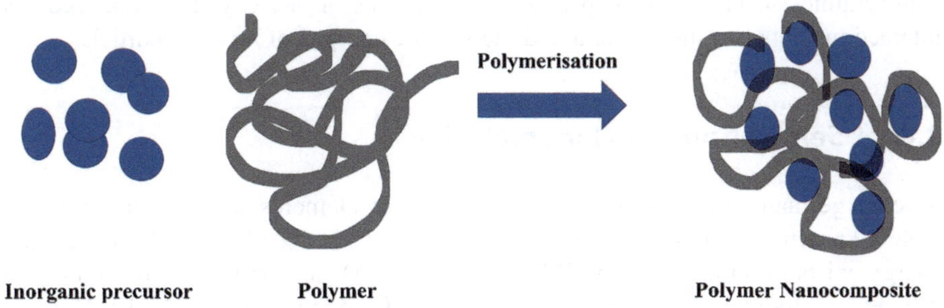

Fig. 3.6 Formation of polymer nanocomposite

3.6.1 Zeolite and Silica-Based Nanomaterial

Nanoparticles made of silica and zeolite have uses because of their distinct surface chemistry and mesoporous structure [26]. A few variables, including templates, reaction conditions, and the rate of hydrolysis, can have an impact on the size and pore structure of silica. By controlling the template concentration, pH of solution, and usage of hydrophobic chemicals, mesoporous silica nanoparticles with pore diameters of 2–50 nm can be produced in both basic and acidic media [27]. The most popular silica-based products are nano-clays, which are nanoparticles of layered silicates with Si atoms bonded to $Al(OH)^3$ or $Mg(OH)^2$ in tetrahedral to octahedral shapes. The van der Waals force connecting these clay layers is typically weak and is easily breakable by the intercalating polymer chain. Since most polymers have different surface energies, they are generally incompatible with the nano-clay structure [28]. The nano-clays can be altered by using a variety of inorganic chemical complexes, such as Pd complex (Hal-Pd) (a nano-clay supported), which is employed as an active heterogeneous catalyst. CNTs can also be added to select few other silica-based nanomaterials to improve their physical and chemical characteristics [29, 30].

3.7 Ceramic Nanomaterials

Ceramics make up a particular sort of nanomaterial. This is further divided into groups based on characteristics and compositions such as heat-resistant, inorganic, and non-metallic solids [31]. Bio-ceramics are composite forms of calcium phosphate material with tiny pores that have been utilized to cover metal joint implants or employed as unloading space fillers for bone ingrowth, and they are made for biological drug administration. These materials have dimensions less than 100 nm [32–34]. These materials showed improved structural ferromagnetic, ferroelectric, electro-optical, and superconductive capabilities. By altering the doping concentration, which could result in the structural distortion of materials by eliminating oxygen vacancies, one can also alter the structural and physical properties of nanoceramics [31]. Their capacity for antibacterial activity is influenced by interactions between the bacteria and the surface charges of the nanoparticles.

3.8 Semiconductor Nanomaterials

Silicon, germanium, gallium arsenide, and other elements close to the periodic table's known metalloid staircase are examples of semiconductors. Low bandgap energy exists in semiconductor NMs [35]. These NMs are created from a variety of different compounds, including those in the II–VI group (ZnO) and the III–V group (GaAs). Due to the quantum size effect or by expanding the surface area, materials' chemical and physical properties can be changed by changing their structure to make them nanoscale objects. The resulting nanostructure determines the materials' high electrical conductivity [36]. There are broadly two categories of semiconductor NMs: depending on the presence or absence of impurities, or dopants in the material.

3.8.1 Intrinsic Semiconductors

Nanomaterials made of intrinsic semiconductors are unexposed, undoped semiconductors that are electrically neutral because they contain an equal quantity of electrons and holes. Diamond, germanium, and silicon are some examples of inherent semiconductor nanomaterials. Having a narrow spectrum of electrical conductivity, intrinsic semiconductor nanoparticles are not particularly helpful on their own for electronic applications.

3.8.2 Extrinsic Semiconductors

Contrarily, extrinsic semiconductor nanoparticles have been purposefully doped with impurities to change their electrical characteristics. Dopants improve or reduce a material's electrical conductivity by introducing extra electrons or holes. Electronic components like transistors and solar cells frequently make use of extrinsic semiconductor nanoparticles.

Extrinsic semiconductor nanoparticles come in both n-type and p-type varieties. Nanomaterials that are N-type semiconductors are doped with impurities that provide extra electrons to the material, making it negatively charged. Silicon doped with phosphorus or arsenic is an example of an n-type semiconductor nanomaterial. Impurities doped into P-type semiconductor nanomaterials give more holes to the material, giving it a positive charge. Silicon doped with boron or aluminum is an example of a p-type semiconductor nanomaterial. On the other hand, silicon doped with boron or aluminum represents an example of p-type semiconductor nanomaterials. By employing both n-type and p-type semiconductors, various electronic applications can be enabled.

3.9 Polymeric Nanomaterials

Solid nanosized polymeric NMs are composed of either synthetic or natural polymers. These materials are frequently employed as medication release controllers that detect the body in pharmaceutical and medical applications [37]. Nanomaterials madeup of polymers includes the following: (i) polymeric micelles produced when amphiphilic block copolymers self-assemble in a specific solvent. Chitosan polymeric micelles can be used to deliver drugs because of their special properties, including their nanosize, stability, biocompatibility, micellar association, and low toxicity; (ii) polymeric NPs, which typically consist of biocompatible and biodegradable polymers and have an average size of 10–1000 nm. To deliver medications

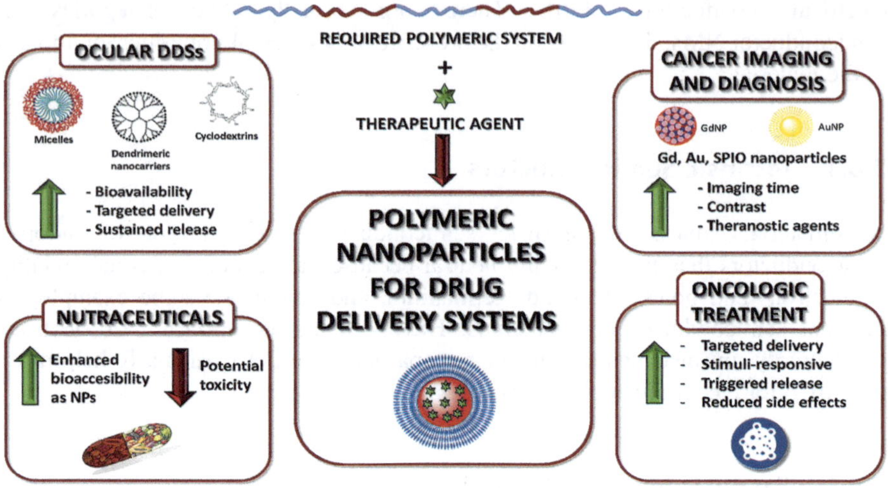

Fig. 3.7 Polymeric nanoparticles for drug delivery systems [39]

to certain targets, these materials are frequently utilized [37]. The specific properties of the material generated are influenced by the procedures used to manufacture the polymer NPs [36]. In general, preparation techniques are divided into three categories: polymerization of monomers, ionic gelation of hydrophilic polymers, and dispersion of polymers; (iii) dendrimers, which are macromolecules with a size of less than 15 nm and 3D shapes. These materials are a new class of polymeric NMs that find extensive use in pharmaceutical and medical applications because of traits like their structure, size, and multivalence [37]; (iii) polymeric nanocomposites, which combine other nanofillers and polymers to produce unique properties and characteristics; and (iv) pharmaceutical and medical applications [38, 39] (Fig. 3.7).

3.9.1 Lipid-Based Nanomaterials

Nanomaterials such as solid lipid nanoparticles, nanostructure lipid carriers, and liposomes are employed in drug delivery due to their ability to transport both hydrophilic and hydrophobic substances, exhibit minimal toxicity, and enable precise control over drug release at specific targets within the human body [37]. These nanoparticles are identified because of their certain beneficial properties, such as site-specific targeting, low cost, lack of toxicity, physical and chemical stability, and the possible control of both hydrophilic and hydrophobic molecules [40]. Lipid-based nanomaterials exhibited negative properties, such as a limited capacity for drug loading and drug expulsion as the materials. They can crystallize during the storage process [41]. They are designed to be used as drug release control materials with an increased stability as compared to SLNs. Liposomes have a size range of 50–100 nm and are composed of cholesterol and phospholipids compounds. These are suitable for drug delivery, especially for cytotoxic drugs, due to their reduced toxicity and enhanced bioavailability [42].

3.9.2 Metal-Organic Nanoparticles

These are hybrid substances made up of organic ligands and inorganic metal ions. A high surface area, a well-organized structure and arrangement, ease of surface modification, and high porosity are only a few of its distinguishing characteristics. They have been widely employed as support materials for things like Fe-BTC (benzene-1,3,5-tricarboxylic acid) due to their strong surface reactivity. These substances serve as supports for the immobilization of enzymes. They use the following synthesis techniques:

(i) Coordination modulation is a technique for creating nanoparticles by chemically regulating the connections between metal and ligand.
(ii) Rapid nucleation is a synthesis approach that uses a regulated concentration and molar ratio of reactants along with accelerated heating techniques like conventional, microwave, and ultrasound heating [43].
(iii) Nanoreactor confinement: A technique for controlling size by isolating nucleation sites under constrained physical circumstances. The nanoreactor method controls particle size by isolating nucleation sites within specific physical constraints. The choice of a non-polar solvent for emulsion formation from uniform nano-scale droplets influences the resulting particle size [44].

3.10 Core-Shell Nanomaterials

Core-shell NPs are made up of a shell material surrounding a core. They are further divided into the subsequent classes:

(i) Depending on the core-shell components, organic–inorganic, inorganic–organic, and inorganic–inorganic types [45].
(ii) If particles with a core-shell have a chemical reaction, they will be coated physically and chemically.

The shell is employed to keep the core's chemical stability and increase surface activity. Coated core particles are intended to have particular qualities, such as magnetic, optical, and catalytic qualities [46]. These nanoparticles are created by building up a number of layers (Fig. 3.8):

(i) The material's core, which is the NPs' center region.
(ii) The top layer, which may branch with various tiny moieties, surfactants, metal ions, and polymeric branches.
(iii) The substance of the shell layer differs from that of the core.

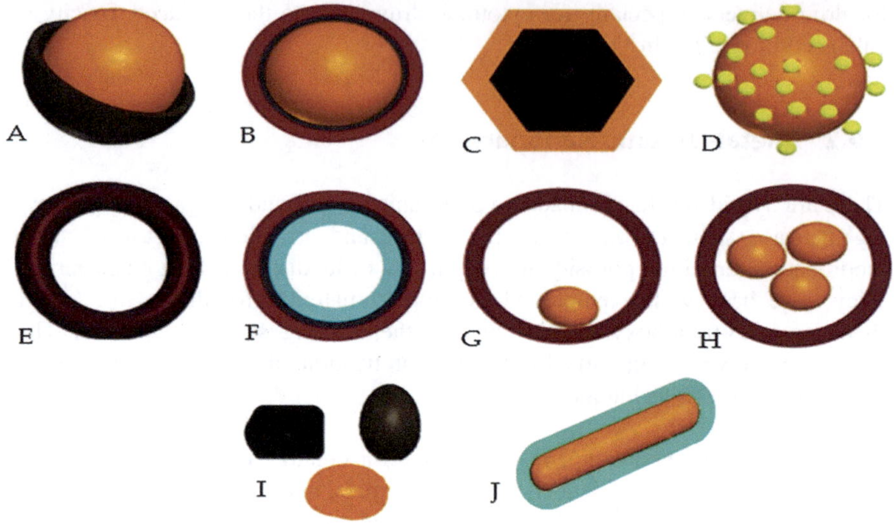

Fig. 3.8 Schematic (**a**–**j**) different structure of core-shell nanoparticles: (**a**) core–shell nanoparticles; (**b**) core double-shell particles, (**c**) polyhedral core/shell nanoparticles, (**d**) core porous-shell nanoparticles, (**e**) hollow-core shell nanoparticles, (**f**) hollow-core double-shell nanoparticles, (**g**) moveable-core shell nanoparticles, (**h**) multi-core shell nanoparticles, (**i**) irregular shape core-shell nanoparticles; and (**j**) rod core-shell nanoparticles [47]

Core-shell nanostructures come in a variety of forms, including the following:

(i) A single or group of small spheres having a metal core and a shell composed of a separate metal. This shell could be a continuous layer, a smaller sphere attached to the core sphere, or a collection of core spheres.
(ii) a nonmetal shell surrounding a metal core,
(iii) a polymer shell surrounding a metal core,
(iv) a nonmetal shell surrounding a nonmetal core,
(v) a nonmetal shell surrounding a polymer-core,
(vi) a polymer-core with numerous shells or a shell constructed of various polymers [48].

3.11 Classification of Nanomaterials on Basis of Source of Origin

Based on the materials used as sources to create them, the nanomaterials are categorized according to where they came from. They fall into two categories: Synthetic nanomaterials and naturally derived nanomaterials.

3.11.1 Synthetic Nanomaterials

The most popular approach for creating nanomaterials is the synthetic method, which enables the creation of nanomaterials by biological, physical, chemical, or hybrid processes [49]. This technique of manufacture has a number of benefits, one of which is the ability to generate vast quantities of nanomaterials in bulk at once in a variety of shapes and sizes. The ability to precisely and accurately link various chemicals or reagents with nanomaterials is one crucial feature of the synthetic technique. But the main issue with artificially created nanomaterials is the lack of knowledge [50]. Additionally, they pose problems to the environment that differs from that of natural nanomaterials. Nanomaterials are already created from a variety of sources for a range of biological purposes. Two different sorts of techniques are often employed for the creation of nanoparticles (Fig. 3.9).

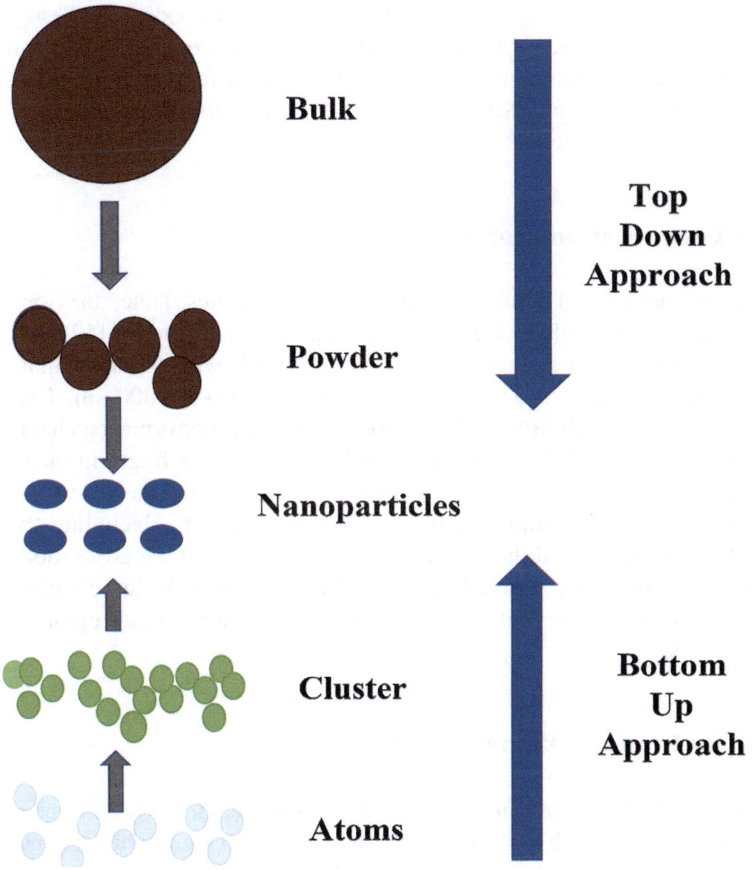

Fig. 3.9 Synthesis approach for nanomaterials

3.11.2 Natural Nanomaterials

Natural nanomaterials are those created from naturally occurring resources. They can be created by biological species like microorganisms or plants as well as by human activity [15]. Because they exist in the hydrosphere, atmosphere, lithosphere, and biosphere, creating them is simple. Natural nanomaterials are present throughout the entire world. There are numerous ways to create nanomaterials, including volcanic eruptions, forest fires, and tree burning. Additionally, some naturally occurring nanomaterials can be found in the shed skin and hair of both plants and animals [51]. The manufacture of nanoparticles was also influenced by anthropogenic activities including transportation and industrial processes. Additionally, reliable sources of nanomaterials are dusts like cosmic dust. The production of nanomaterials in space is typically caused by remarkable temperature changes, pressure gradients, physical collisions, and shock waves under the influence of electromagnetic radiation. According to research, more than half of the aerosol particles in the atmosphere are between 100 and 200 nm in size. Furthermore, we can assert that human body organs including bones, antibodies, and enzymes are predominantly composed of nanomaterials. The body uses these nanomaterials to carry out typical physiological processes. If we take into account that the only components of DNA or RNA, the genetic materials found in animal or human cells, are nanoscale materials [52].

3.11.3 Green Nanomaterials

Green plants are a great source of cellular bio-composites, hence they are used to create nanomaterials, which are a sort of synthetic nanomaterial [52–54]. Natural fibers found in plants are composed of cellulose fibrils with segments that are both crystalline and amorphous and range in length from 100 to 1000 nm. The primary structures benefit greatly from the superior strength and performance characteristics of natural fibers provided by the nano-fibrillar components found in plants. Plants have numerous structural characteristics. For instance, their leaves contain nanostructures that are employed for a variety of functions, including the sliding of insects, mechanical stability, enhanced visible light, and the absorption of damaging UV and radiation. According to 54 different research, the circular layer in plants and insects that prevents them from sinking in water is made up of mounds of nanomaterials.

3.11.4 By-Product Nanomaterials

These are a subset of nanoparticles that are created accidentally as a byproduct of industrial processes (such as the combustion of fuel in an automobile engine) and are also known as subsidiary nanomaterials. Additionally, burning trees and forest fires can yield these nanomaterials. These are a host of naturally occurring nanomaterials, though, and they can be found in the bodies of humans, animals, plants, insects, and other organisms. If we pay great attention, there is a distinction between nanomaterials that are intentionally created and those that are naturally occurring or incidental. These are nothing more than naturally occurring molecules consisting of atoms, which are the building blocks of all living and non-living entities. Interestingly, these atoms and molecules undergo multiple natural manipulations to create nanomaterials. Nanomaterials that are created accidentally or spontaneously are continually being released into the environment, underground water, and the atmosphere. However, the variation in structure between accidental and artificial nanoparticles is the main distinction [55] because artificially manufactured nanoparticles' morphologies can be better regulated than those of naturally occurring nanomaterials.

3.11.5 Engineered Nanomaterials

Engineered nanoparticles are those created by the combustion of fuels such as coal, oil, gasoline, diesel, and coal for transportation, making chemicals and other industrial processes, or refining oil. These nanomaterials, which are commonly employed in the manufacturing of toothpaste, cosmetics, and sports equipment, mostly consist of carbon and hydroxyapatites. This is a brand-new class of artificial nanoparticles that have detrimental impacts on both human health and the environment. Transportation by car results in nanoparticles with a size range of 20–130 nm. Numerous uses for these nanoparticles are found in the healthcare sector as well. The majority of the nanoparticles utilized in commercial applications are designed and made utilizing chemical, biological, and physical processes [56–60].

3.12 Conclusion

A range of nanomaterials exists with various dimensionalities (0, 1D, 2D, and 3D), morphologies, states, and chemical compositions, all of which significantly affect the way they behave. It is important to note that the substance can be harmful in some circumstances, particularly when it transforms into other species of nanoparticles, the use of NPs must still be taken into consideration [61, 62]. The physicochemical characteristics of anthropogenic and artificial NMs, such as size, chemical composition, crystal structure, surface morphology, surface charge and energy, and aggregation state, may be closely related to their toxicity. The circumstances of synthesis, processing, chemical makeup, and doses all have an impact on the levels of danger. Therefore, more focused research on the interaction between physicochemical characteristics and the toxicity of nanoparticles needs to be developed in the future [63].

References

1. Pokropivny VV, Skorokhod VV (2007) Classification of nanostructures by dimensionality and concept of surface forms engineering in nanomaterial science. Mater Sci Eng C 27(5–8): 990–993
2. Gleiter H (2000) Nanostructured materials: basic concepts and microstructure. Acta Mater 48(1):1–29
3. Marchesan S, Ballerini L, Prato M (2017) Nanomaterials for stimulating nerve growth. Science 356(6342):1010–1011
4. Sharma VK, Filip J, Zboril R, Varma RS (2015) Natural inorganic nanoparticles--formation, fate, and toxicity in the environment. Chem Soc Rev 44(23):8410–8423
5. Aversa R, Modarres M, Cozzini S et al (2018) The first annotated set of scanning electron microscopy images for nanoscience. Sci Data 5:180172
6. Thakur P, Thakur A (2022) Introduction to nanotechnology. In: Thakur A, Thakur P, Khurana SP (eds) Synthesis and applications of nanoparticles. Springer, Singapore. https://doi.org/10.1007/978-981-16-6819-7_1
7. Asghari F, Jahanshiri Z, Imani M, Shams-Ghahfarokhi M, Razzaghi-Abyaneh M (2016) Chapter 10—antifungal nanomaterials: synthesis, properties, and applications. In: Grumezescu AM (ed) Nanobiomaterials in antimicrobial therapy. William Andrew Publishing, pp 343–383
8. Thakur P, Thakur A (2022) Nanomaterials, their types and properties. In: Thakur A, Thakur P, Khurana SP (eds) Synthesis and applications of nanoparticles. Springer, Singapore. https://doi.org/10.1007/978-981-16-6819-7_2
9. Bleeker E, De Jong WH, Re G, Groenewold M, Heugens EHW, Koers-Jacquemijns M, Van De Meent D, Popma J, Rietveld A, Wijnhoven S, Cassee FR, Oomen AG (2013) Considerations on the EU definition of a nanomaterial: science to support policy making. Regul Toxicol Pharmacol 65(1):119–125
10. Singh A, Wan F, Yadav K, Salvi A, Thakur P, Thakur A (2023) Synergistic effect of ZnO nanoparticles with Cu2+ doping on antibacterial and photocatalytic activity. Inorg Chem Comm 157:11425
11. Thakur A, Thakur P, Kumari S, Singh S, Ravelo B, Wan F (2023) Mg-Zn ferrite nanoparticles: a brief review on synthesis, characterizations, and applications. Chem Bio Eng Rev 10:1–27
12. Khanra A, Sangam S, Shakeel A, Suhag D, Mistry S, Rai MP, Chakrabarti SK, Mukherjee M (2018) Sustainable growth and lipid production from chlorella pyrenoidosa using N-doped carbon nanosheets: unravelling the role of graphitic nitrogen. ACS Sustainable Chemistry & Engineering
13. Saleh TA, Fadillah G (2019) Recent trends in the design of chemical sensors based on graphene–metal oxide nanocomposites for the analysis of toxic species and biomolecules. Trends Anal Chem 120:115660
14. Das L, Das P, Bhowal A, Bhattachariee C (2020) Synthesis of hybrid hydrogel nano-polymer composite using graphene oxide, chitosan and PVA and its application in waste water treatment. Environ Technol Innov 18:100664
15. Qi Y, Ye J, Zhang S, Tian Q, Xu N, Tian P, Ning G (2019) Controllable synthesis of transition metal ion-doped CeO2 micro/nanostructures for improving photocatalytic performance. J Alloys Compd 782:780–788
16. Liu Z, Yao L, Pan X, Liu Q, Huang H (2018) A green and facile approach to the efficient surface modification of alumina nanoparticles with fatty acids. Appl Surf Sci 447:664–672
17. Chu P, Zhang H, Zhao J, Gao F, Guo Y, Dang B, Zhang Z (2017) On the volume resistivity of silica nanoparticle filled epoxy with different surface modifications. Compos Part A Appl Sci Manuf 99:139–148
18. Gaur P, Banerjee S (2019) C-N cross coupling: novel approach towards effective aryl secondary amines modification on nanodiamond surface. Diam Relat Mater 98:107468
19. Xue Y, Lv P, Lin H, Duan H (2016) Underwater superhydrophobicity: stability, design and regulation, and applications. Appl Mech Rev 68:3

20. Barhoum A, García-Betancourt ML, Jeevanandam J, Tawfik E, Mekkawy SA, Mostafa M, Omran MM, Abdalla MS, Bechelany M (2022) Review on natural, incidental, bioinspired, and engineered nanomaterials: history, definitions, classifications, synthesis, properties, market, toxicities, risks, and regulations. Nano 12(2):177
21. Liu P, Chen G, Zhang J (2022) A review of liposomes as a drug delivery system: current status of approved products, regulatory environments, and future perspectives. Molecules 27(4):1372
22. Saha R, Arunprasath D, Sekar G (2019) Surface enriched palladium on palladium-copper bimetallic nanoparticles as catalyst for polycyclic triazoles synthesis. J Catal 377:673–683
23. Xu B, Wang W, Zhang X, Cao W, Wu D, Liu S, Dai H, Chen S, Wang K, Sun XW (2017) Bright and efficient light-emitting diodes based on MA/cs double cation perovskite nanocrystals. J Mater Chem C 5(25):6123–6128
24. Hao H, Li H, Wang S, Cheng Z, Fang Y (2020) Epitaxial growth of Ag-Cu bimetallic nanoparticles via thermal evaporation deposition. Appl Surf Sci 505:143871
25. Abdulhameed AS, Mohammad A, Jawad AH (2019) Application of response surface methodology for enhanced synthesis of chitosan tripolyphosphate/TiO2 nanocomposite and adsorption of reactive orange 16 dye. J Clean Prod 232:43–56
26. Zheng D, Hu C, Gan T, Dang X, Hu S (2010) Sensors Actuators B Chem 148:247–252
27. Lozhkomoev AS, Pervikov AV, Chumaevsky AV, Dvilis E, Paygin VD, Khasanov OL, Lerner M (2019) Fabrication of Fe-Cu composites from electroexplosive bimetallic nanoparticles by spark plasma sintering. Vacuum 170:108980
28. Zhang Q, Gao Q, Qian W, Zhang H, Tian W, Li Z (2019) A C-coated and Sb-doped SnO2 nanocompsite with high surface area and low charge transfer resistance as ultrahigh capacity lithium ion battery anode. Mater Today Energy 13:93–99
29. Kumar S, Malik MM, Purohit R (2015) Synthesis of high surface area mesoporous silica materials using soft templating approach. Mater Today Proc 5(2018):4128–4133
30. Barua S, Gogoi S, Khan R, Karak N (2019) Chapter 8—silicon-based nanomaterials and their polymer nanocomposites. In: Karak N (ed) Nanomaterials and polymer nanocomposites. Elsevier, pp 261–305
31. Himanshi JR, Prakash J, Verma R, Thakur P, Kandwal A, Wan F, Thakur A (2023) Synthesis, characterization, and applications of doped barium hexaferrite: a review. Phys B: Cond Mater 667:415202
32. Suhag D, Bhatia R, Das S, Shakeel A, Ghosh AB, Singh A, Sinha OP, Chakrabarti SK, Mukherjee M (2015) Physically cross-linked pH-responsive hydrogels with tunable formulations for controlled drug delivery. RSC Adv 5(66):53963–53972
33. Chalia S, Bharti MK, Thakur P, Thakur A, Sridhara SN (2021) An overview of ceramic materials and their composites in porous media burner applications. Ceram Int 47(8): 10426–10441
34. Kala D, Gupta S, Kaushal A (2022) Nanotechnology in healthcare. In: Thakur A, Thakur P, Khurana SP (eds) Synthesis and applications of nanoparticles. Springer, Singapore. https://doi.org/10.1007/978-981-16-6819-7_19
35. Saleh TA (2018) Simultaneous adsorptive desulfurization of diesel fuel over bimetallic nanoparticles loaded on activated carbon. J Clean Prod 172:2123–2132
36. Tian Y, Xue F, Fu Q, Zhou L, Wang C, Gou H, Zhang M (2018) Structural and physical properties of Ti-doped BiFeO3 nanoceramics. Ceram Int 44:4287–4291
37. Nie S, Xing Y, Kim GJ, Simons JW (2007) Nanotechnology applications in cancer. Annu Rev Biomed Eng 9:257–288
38. Yan X-W, Joharian M, Naghiloo M, Rasuli R, Hu M-L, Morsali A (2019) Metal–organic framework derived porous 2D semiconductor C/ZnO nanocomposite with high electrical conductivity. Mater Lett 252:325–328
39. Begines B, Ortiz T, Pérez-Aranda M, Martínez G, Merinero M, Argüelles-Arias F, Alcudia A (2020) Polymeric nanoparticles for drug delivery: recent developments and future prospects. Nano 10(7):1403

40. Yang X, Lian K, Tan Y, Zhu Y, Liu X, Zeng Y, Yu T, Meng T, Yuan H, Hu F (2020) Selective uptake of chitosan polymeric micelles by circulating monocytes for enhanced tumor targeting. Carbohydr Polym 229:115435
41. Sur S, Rathore A, Dave V, Reddy KR, Chouhan RS, Sadhu V (2019) Recent developments in functionalized polymer nanoparticles for efficient drug delivery system. Nano-Struct Nano-Objects 20:100397
42. Mandal B, Bhattacharjee H, Mittal N, Sah H, Balabathula P, Thoma LA, Wood GC (2013) Core-shell-type lipid-polymer hybrid nanoparticles as a drug delivery platform. Nanomed Nanotechnol Biol Med 9:474–491
43. Matsuyama T, Hayashi H, Komatsu S, Asoh T-A, Kikuchi A (2019) Preparation of thermoresponsive nanoparticles exhibiting biomolecule recognition ability via atom transfer radical dispersion polymerization. Colloids Surf B: Biointerfaces 183:110370
44. Okrugin BM, Neelov IM, Leermakers FAM, Borisov OV (2017) Structure of asymmetrical peptide dendrimers: insights given by self-consistent field theory. Polymer 125:292–302
45. Fu S, Sun Z, Huang PY, Li Y, Hu N (2019) Some basic aspects of polymer nanocomposites: a critical review. Nano Mater Sci 1(1):2–30
46. Zhong Q, Zhang L (2019) Nanoparticles fabricated from bulk solid lipids: preparation, properties, and potential food applications. Adv Colloid Interf Sci 273:102033
47. Khatami M, Alijani HQ, Nejad MS, Varma RS (2018) Core@shell nanoparticles: greener synthesis using natural plant products. Appl Sci 8(3):411
48. Meikle TG, Strachan JB, Conn CE (2019) Lipid nanomaterials for the delivery of antimicrobial peptides. In: Methods in microbiology. Academic Press, pp 173–195
49. Thakur A, Verma R, Wan F, Ravelo B, Edelman I, Ovchinnikov S, Thakur P (2023b) Investigation of structural, elastic and magnetic properties of Cu2+ ions substituted cobalt nano ferrites. J Magn Magn Mater 581:170980
50. Verma R, Thakur P, Sun AC, Thakur A (2023) Investigation of structural, microstructural and electrical characteristics of hydrothermally synthesized Li0.5-0.5xCoxFe2.5-0.5xO4 (0.0≤x≤0.4) ferrite nanoparticles. Phys B: Cond Mater 661:414926
51. Li C, Adamcik J, Mezzenga R (2012) Biodegradable nanocomposites of amyloid fibrils and graphene with shape-memory and enzyme-sensing properties. Nat Nanotechnol 7:421
52. Jha AK, Prasad KV, Kulkarni AR (2009) Synthesis of TiO2 nanoparticles using microorganisms. Colloids Surf B: Biointerfaces 71(2):226–229
53. Hochella MF, Spencer MG, Jones KL (2015) Nanotechnology: nature's gift or scientists' brainchild? Environ Sci Nano 2(2):114–119
54. Thakur A, Chahar D, Thakur P (2022) Synthesis of nanomaterials by biological route. In: Thakur A, Thakur P, Khurana SP (eds) Synthesis and applications of nanoparticles. Springer, Singapore. https://doi.org/10.1007/978-981-16-6819-7_5
55. Sharma VK, Filip J, Zboril R, Varma RS (2015) Natural inorganic nanoparticles—formation, fate, and toxicity in the environment. Chem Soc Rev 44(23):8410–8423
56. Verma R, Thakur P, Chauhan A, Jasrotia R, Thakur A (2023) A Review on MXene and its composites for electromagnetic interference (EMI) shielding applications. Carbon 208:170–190
57. Krishnia L, Thakur P, Thakur A (2022) Synthesis of nanoparticles by physical route. In: Thakur A, Thakur P, Khurana SP (eds) Synthesis and applications of nanoparticles. Springer, Singapore
58. Taneja S, Punia P, Thakur P, Thakur A (2022) Synthesis of nanomaterials by chemical route. In: Thakur A, Thakur P, Khurana SP (eds) Synthesis and applications of nanoparticles. Springer, Singapore
59. Chahar D, Thakur P, Sun AC, Thakur A (2023) Investigation of structural, electrical and magnetic properties of nickel substituted Co-Zn nanoferrites. J Mater Sci: Mater Electron 34: 901
60. Peterov D, Edelman I, Thakur A, Thakur P, Sukhachev A, Ovchinnikov S (2023) Correlation between magnetic and electric properties in the series of CoxZn1-xFe2O4 nanoparticles. JETP Lett 117:765–768

61. Nath D, Banerjee P (2013) Green nanotechnology—a new hope for medical biology. Environ Toxicol Pharmacol 36(3):997–1014
62. Seil JT, Webster TJ (2012) Antimicrobial applications of nanotechnology: methods and literature. Int J Nanomedicine 2767
63. Saliani M, Jalal R, Goharshadi EK (2015) Effects of pH and temperature on antibacterial activity of zinc oxide nanofluid against E. coliO157:H7 and Staphylococcus aureus. Jundishapur J Microbiol 8:2. https://doi.org/10.5812/jjm.17115

Structural Properties of Nanoparticles

4

Atul Thakur, Preeti Thakur, and Sahbi Baccar

Abstract

In contrast to nanomaterials to other microscopic materials made up of atomic attractions, they have very different features. Materials with varying nanometre sizes have distinct preparation methods, characterisation methods, and applications than bulk materials. Since nanomaterials are very different from their macroscopic, bulk allotropes in many ways, they necessitate specialised scientific and engineering methods. For a long time, various nanomaterials have been employed in batteries, fuel cells, and other electrochemical applications. Because of their structural peculiarities and superior physical and chemical characteristics, carbon nanotubes have been found to be preferable for conventional carbon-based electrodes for use in fuel cells, supercapacitors, and batteries. Nanomaterials made of biomaterials, nanometals, and nanocomposites have all been employed in addition to those based on carbon. In this chapter, several structural characteristics of nanoparticles will be covered.

Keywords

Nanomaterials · Allotropes · Bio-ceramics · Nanometals · Nanocomposites

A. Thakur · P. Thakur
Centre for Nanotechnology, Amity University Haryana, Gurugram, India

NanoLatticeX, Amity University Haryana, Gurugram, Haryana, India

Nanotechnology Wing, Innovative Science Research Society, Shimla, India

S. Baccar (✉)
ESIGELEC, Bureau C1-226, Technopôle du Madrillet Avenue Galilée, Saint-Etienne du Rouvray, France
e-mail: sahbi.baccar@esigelec.fr

D. Suhag et al. (eds.), *Integrated Nanomaterials and their Applications*,
https://doi.org/10.1007/978-981-99-6105-4_4

4.1 Introduction

The form of nanomaterials, in addition to its overall size and size, as well as its porosity, aspect ratio, and surface roughness or surface structure all affect its surface-to-volume ratio and other attributes. One of the ways that nanomaterials are categorised is based on the shape and size ratio, often known as dimensionality or proportion ratio [1]. This feature of the classification of nanomaterials, and how it is restricted to zero, one, two, and three dimensions, has already been discussed in previous literature. One-dimensional (1-D) objects have nanometre size in two directions and are also larger than 0-D, i.e., micrometre length for nanowires, and the most famous 1-D objects are nanotubes and multiwalled and single-walled nanotubes made of carbon. Zero-dimensional (0-D) objects have nanometre features in every direction, which is expansion in every direction if we talk about quantum dots, nanoparticles, and fullerenes; these are all 2-D nanomaterials [2, 3].

Preferably on the scale that we are able to attain by applying external forces, the asymmetry brought about by the forces near to surface atoms in solids alters the bond length of nanomaterials. When it comes to clusters and nanoparticles at the lower end of the nanoscale, typically between 1 and 5 nm (about 50–5000 atoms), the significance of the effect of these changes in bond length grows with the surface atom to volume atom ratio (dispersion) [4]. Although ionic bonds contract significantly less than the other two types of bonds, it is clear that the phenomena occur in all types of bonds. In comparison to their parent bulk materials, nanomaterials have substantially greater bond energies [5, 6].

This chapter's major goal is to compile and present structural characteristics of nanomaterials in a way that will both make them understandable to new readers and serve as a reminder of the diligent effort of scientists and engineers. We will discuss about the intricate structural characteristics of nanoparticles in this chapter, including those of nanocarbons, nanometals, nanocomposites, nanoceramics, and nano-porous materials, as well as their subtypes [1].

4.2 Nanocarbons

Diamond, graphite, and amorphous carbon are three primary allotropes of carbon, distinguished by their crystalline and non-crystalline structures, as well as by the hybridization of carbon atoms within them. Diamond exhibits a cubic crystal structure with carbon atoms arranged in a sp^3 hybridization state. Van der Waals forces hold the parallel hexagonal lattice planes in the layered structure of graphite, which is composed of sp^2 hybridised carbon atoms [7]. The atoms in amorphous carbons are covalently bonded to one another and hybridised with a mixture of sp^2 and sp^3 carbons; they lack long-range crystalline order. We are all aware that diamonds can be sized down to create nanoscale diamonds. Additionally, graphite can be broken down into lower-dimensional carbon allotropes. Graphene is the name given to a single layer of graphite [7]. Graphite, carbon nanotubes, and fullerenes have all been seen using graphene as the fundamental building block. Fullerenes are the zero-dimensional (0-D) allotropes of carbon nanotubes, which are their one-dimensional (1-D) counterpart. Although CNTs and fullerenes are not singular structures, they do represent a separate family of structures that are extensively discussed in this chapter [8].

4.2.1 Graphene

The two-dimensional allotrope of carbon is called graphene. Carbon atoms are arranged in a hexagonal pattern on a single plane to form its structure. By emphasising the close relationship between graphene and its structure and properties, as is the case with other carbon allotropes, it is possible to characterise the atomic and electronic structure of graphene. Graphene demonstrates a wide range of exceptional and distinctive electrical and optical characteristics. By employing electromagnetic fields and nanofabrication, certain characteristics of graphene are changed. Graphene has been suggested for use in a number of mechanical, electrical, and optoelectronic goods. The properties and applications of graphene derivatives like graphene oxides are also interesting [8].

4.2.1.1 Structural Properties of Graphene

Most of the structural and physical characteristics of graphene are similar to those of its parent substance, graphite. Two interpenetrating triangular sublattices make up the two-dimensional hexagonal lattice in which the carbon atoms are organised. The hybridised carbon atoms are sp^2. The carbon–carbon bond measures 1.42 Å in length [9]. The remaining p-orbital is delocalised to produce the valence and conduction in electronic bands and is positioned perpendicular to the plane of the carbon atoms. The distance between the planes is 3.45 Å, or interplanar separation. It has been suggested that two-dimensional structures like graphene are inherently unstable and that long wavelength oscillations can ruin the long-range order of 2-D crystals. Even 2-D crystals set up in 3-D space are prone to folding [10]. Due to inherent tiny undulations, where the surface normal fluctuates by several degrees and is still out of the plane, suspended 2-D graphene sheets are unstable. Weakly linked graphene monolayers on surfaces have also shown periodic waves. A common feature of graphene is the presence of wrinkles, some of which may be several nanometres wide. There has also been evidence of scrolling along the edges of graphene flakes. When graphene is processed in a liquid phase during microfabrication rather than a solid or gas phase, scrolling results. Although not entirely inherent, the ripples in graphene cause changes in the electronic structure that have an impact on many of the material's electrical and chemical properties [11].

4.2.2 Fullerenes

About 25 years ago, buckminsterfullerene was identified. This discovery launched a new sector in materials science where several disciplines such as physics, chemistry, medicine, mathematics, and engineering joined together. It is a clear example of what we now refer to as nanoscience. Thus, it demonstrates a variety of uses for carbon nanotechnology.

4.2.2.1 Structural Properties of Fullerenes

Studying fullerenes' crystals, or fullerites, is a necessary step in working with them. Isolated fullerene molecules engage in chemical interactions with other substances. Fullerene crystallisation produces fullerites. Under normal circumstances, these crystals showed hexagonal close-packing arrangements [12]. Fullerite crystals are typically face-centred cubic (fcc), with some orientational disorder in the molecules. However, as the temperature drops, these molecules become organised without rotation, and the fcc transforms into a simple cubic (SC) phase [13]. The bandgap of the semiconductor fullerite, which has a 1.5 eV bandgap, is altered by pressure. A rhombohedral phase is discovered when the temperature is raised by 800 °C [14]. Sundqvist has gone into great length about the fullerite phase diagram and compression behaviour. Without a doubt, research into crystals made from different fullerenes should continue to find new uses, perhaps in electronics [15].

4.2.3 Single-Walled Carbon Nanotubes

Single-walled and multiwalled carbon nanotubes are the two types. One atomic sheet of carbon atoms is arranged in a honeycomb lattice to form single-walled carbon nanotubes, which are hollow, long cylinders with incredibly high aspect ratios. They have exceptional mechanical, thermal, transport, optical, electrical, vibrational, and structural properties [16]. These one-dimensional nanomaterials are uncommon. Their electrical characteristics, which span a wide variety of energy scales, are highly sensitive to their microscopic atomic configurations and symmetry. In numerous fields, including energy storage, molecular electronics, nanomechanical devices, composites, chemical and biosensing, carbon nanotube research is currently being intensively pursued. Depending on their diameter and chirality, they can be metallic or semiconducting with different bandgaps [17].

4.2.3.1 Structural Properties of Single-Walled Carbon Nanotubes (SWCNT)

In terms of structure, carbon nanotubes resemble rolled-up sheets of single-layer graphene because they are composed of sp^2-bonded carbon atoms, just like graphite. Typically, they have a diameter in the nanometre range. They are classified as 1-D nanostructures since their length frequently exceeds microns and even centi-metres. Due to their diameter-dependent, direct bandgaps, semiconductor nanotubes are particularly used in photonic device applications. While metallic tubes have a wide range of electronic applications, such as power transmission cables and nanocircuit components, they may not always be considered ideal due to specific electrical and material properties [18].

Their other characteristics can be thought of as two half-fullerene caps atop a graphene sheet that has been folded into a cylinder. These species' diameters typically fall between 0.7 and 2.5 nm. They act like 1-D materials due to the extremely severe restriction along their diameter. A monolayer of sp^2-hybridised carbon atoms set up in a honeycomb lattice structure makes up graphene [19]. The primitive vectors specify what type of cell it is. The two atoms of the graphene sheet that are recognised during the creation of the tube are then specified by the chiral vector, Ch, that identifies each SWCNT. Regarding the graphene primitive vectors, the chiral vector is defined as:

$$\mathrm{Ch} \equiv \mathrm{n}\mathrm{a}_1 + \mathrm{m}\mathrm{a}_2 \equiv (\mathrm{n}, \mathrm{m})$$

where n and m are integers such that $0 \leq m \leq n$.

When the angle θ formed between Ch and a_1 is a quantity of great importance for determining the electronic properties of the SWCNT it is called the chiral angle and is defined through

$$\cos\theta = \mathrm{Ch} \cdot \mathrm{a}_1 \mid \mathrm{Ch} \mid\mid \mathrm{a}_1 \mid$$

The nanotube diameter can be computed from the chiral vector.

The names armchair and zigzag nanotubes refer to the cross-sectional geometry along Ch [20]. The contrast between the three of these different structures will be crucial for understanding their electronic structure, phonon dispersion, which will exhibit a variety of physical characteristics.

4.2.4 Multiwalled Carbon Nanotubes

The elongated cylindrical nanoparticles known as multiwalled carbon nanotubes (MWCNTs) are constructed of sp^2 carbon. They have a 3–30 nm diameter. Their aspect ratio can range between 10 and ten million because they can grow to be many centimetres long. Due to their stiffness and multiwalled structure, they can be recognised from single-walled carbon nanotubes. The distinct wall structure, smaller outer diameter, and hollow interior of carbon nanofibers are characteristic features that differentiate them from other carbon-based nanostructures [14].

4.2.4.1 Structural Properties of Multiwalled Nanotubes (MWCNTs)

Successful mounting of semiconducting quantum dots and metallic nanoparticles on MWCNTs is possible. An emerging subject with significant application potential in photocatalysis and sensorics is the anchoring of quantum dots to MWCNTs and the coating of MWCNTs with inorganic nanoparticles. This material type is typically represented by nanocomposites of multiwalled carbon nanotubes with hydroxyapatite, TiO_2, rare-earth fluorides [21], etc., and WS2. Despite the fact that such MWCNT-oxide mixes are fairly simple to synthesise using sol-gel techniques. It is highly challenging to verify that the inorganic particles are indeed adorning the nanotubes rather than generating a different phase [22].

4.2.5 Modified Carbon Nanotubes

It is commonly known that carbon nanotubes have exceptional electrical and mechanical capabilities. If these properties are altered, new traits may also develop. These modified carbon nanotubes must also undergo more research to determine their toxicity and biocompatibility because they differ from their pure counterparts [23]. By utilising a variety of microscopy and spectroscopic techniques, these small alterations may be understood and can significantly alter the electrical and chemical properties of the target molecules.

In 1985, the first fullerene, or nanocarbon cages, was discovered. Nanotubes, another kind of nanocarbon, were discovered shortly after [24]. The graphene nanoribbons were also recently synthesised and fabricated using several methods. As a result, carbon is incredibly flexible and may be formed into any nanomorphology. But graphene and graphene nanoribbons are likewise becoming more and more significant. Particularly, depending on their geometry, carbon nanotubes have intriguing electrical, optical, and mechanical capabilities.

4.2.6 Structural Properties of Carbon Nanofibers

The well-known carbon nanotubes are not the same as carbon nanofibers. Their geometry can be visualised as consisting of regularly spaced, truncated conical or flat layers along the filament length, which distinguishes them from concentric carbon nanotubes that have a whole hollow core. These metal nanoparticles are heated to a high temperature. The term "carbon nanofibers" refers to flexible, sp^2-based linear filaments with a diameter of about 100 nm. Their aspect ratio is more than 100 [25]. Fibre-based materials are extremely valuable in both science and practice. Nanofibers can be used in our daily lives as well as in the creation of durable composites for use in automotive and aerospace applications because of their combination of high specific area, flexibility, and high mechanical strength. Due to their distinctive structure, which confers semiconducting behaviour and chemically active end planes on both the inner and outer surfaces of the nanofibers, these materials are advantageous as catalyst support materials [26]. By properly combining organic polymer electrospinning with heat treatment in an inert atmosphere, carbon nanofibers can be created. Due to their structure and characteristics, carbon nanofibers could be viewed as a one-dimensional (1-D) version of carbon (Fig. 4.1).

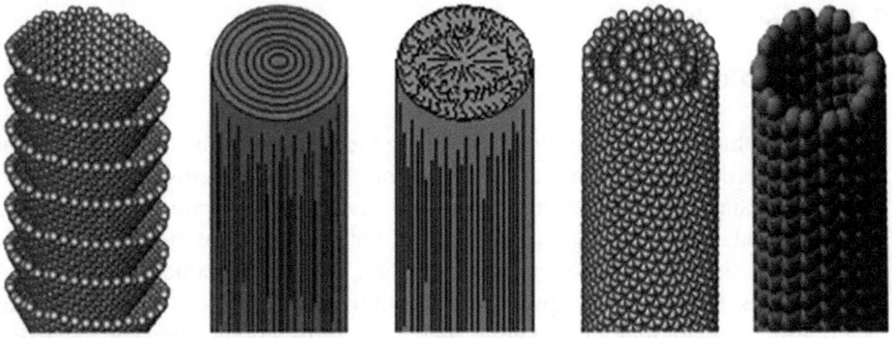

Fig. 4.1 Graphical representation of carbon fibres and carbon nanotubes [26]

4.2.7 Nanodiamonds

Harnessing the energy of explosions, nanostructured diamond particles were initially created in the early 1960s. First, colloidal suspensions of distinct, 4–5 nm-sized nanodiamond particles. Shock wave compressions have been used to create diamond powders from mixtures of catalyst and carbon compounds like graphite and carbon black [27]. The primary grains in the polycrystallinc particles were between 10 and 20 nm in size. The polishing industry used the product extensively. Despite having diameters of 4–5 nm, detonation nanodiamonds (DND) have a tendency to combine. As a result, commercial suspensions of DND frequently contain bigger aggregates, some of which can reach several hundred nanometres in size and tolerate ultrasonic treatment. Nanodiamonds (NDs) can be easily used in a variety of applications due to their chemical reactivity, which enables them to take part in a wide range of chemical reactions and its capacity to change surface chemical groups. The least poisonous carbon nanoparticle known is called ND. With a specific surface area of 428 and 57 m^2/g, respectively, for ND particles with a diameter of 4 and 30 nm, it can absorb and load nanoparticles. The strong optical transition makes it possible to identify single flaws with a microscope. In order to create effective single-photon emitters for quantum information processing or magnetic sensors with nanoscale resolution, the spin state of negatively charged NV centres can be polarised by optical pumping and modified by using electron paramagnetic resonance (EPR). Even at ambient temperature, single-colour centres exhibit remarkably good photostability. Nanodiamond particles synthesised by using various methods are related to the manufacture of particles with impurity defects that are 4–5 nm in size and possess stable luminescence and distinctive spin characteristics [28, 29].

4.3 Nanometals

4.3.1 Noble Metals

In general, noble metal nanoparticles, including their gold and silver analogs, exhibit unique properties and have diverse applications in various fields of science and technology. Since prehistoric times, gold and silver have enthralled mankind, and their nanoscale uses have garnered interest for thousands of years. Nanoscale counterparts are being investigated because of their distinctive functional characteristics that set them apart from the bulk. Properties can be modified to have unusual capabilities by changing their size, form, composition, or local surroundings. Materials' electrical, chemical, optical, and other properties can be accurately controlled by altering their chemical composition at the nanoscale [30].

The physical, optical, and electrical characteristics of anisotropic nanoparticles are shape, size, and surface dependent, which makes them attractive in the context of contemporary materials research. However, creating these materials with exact control over their size and shape is the most crucial step in making use of their features. It takes specialised synthetic techniques to create them with exact control over size and shape. Top-down approaches are used to describe synthesis techniques because they begin with bulk materials. The following strategies are utilised in this process [31]:

(i) High-energy ball milling: By grinding bulk materials, nanoscale powders are created.
(ii) To create conducting nanomaterials, such as metals, wire explosion is used. An explosion results from the supply of a sudden, high current pulse.
(iii) Arc discharge: To evaporate materials, alternating current or direct current arcs are used.
(iv) By condensing evaporating atoms in a matrix, inert-gas condensation produces particle growth.
(v) High-energy lasers used in laser ablation cause evaporation.
(vi) Ion sputtering: Evaporation is caused by the impact of highly energetic ions, typically rare gases.

4.4 Alloys on Nanoscale

It is demonstrated that new physical phenomena, as well as novel or significantly altered features, occur in metal multicomponent mano-systems because of the enhanced freedom associated with composition and chemical ordering.: For instance, new structural concepts may emerge based on the size of the system, the energy and structural properties of the metal components. As a result, multicomponent nanostructured systems exhibit properties in their catalytic activity, mechanical strength, plasmonic and nonlinear optical, as well as magnetic responses that are distinct from and can theoretically be finely controlled in comparison to their pure counterparts [32].

4.4.1 Structural Properties of Nanoalloys

Given the fact that metal atoms can sustain a wide range of coordination numbers (up to 12–14) and that there are not any particularly high-energy barriers between different structures, which allows particles to move between them under practical circumstances, metal nanoparticles and nanoalloys display an impressive range of different morphologies and mixing patterns.

4.4.1.1 Morphology in the Gas Phase

The morphology, or structure, of free nanoparticles is firstly described. These are divided into crystalline, non-crystalline, and those specific to very small clusters.

(i) Crystalline Forms: The atoms of the clusters of this family are situated at the crystalline sites character. The goal of cutting a particle from a bulk crystal is to reduce the surface energy of the resulting cluster as much as possible. One method is to produce cuts by only exposing closely spaced facets, which gives the octahedron shape—two opposed square pyramids sharing a base [32]. Despite the close packing of all octahedron faces, the particles shape, high surface-to-volume ratio prevents it from optimising surface energy. Cutting the vertices of the octahedron, which results in a truncated octahedron, allows for the creation of clusters with more spherical shapes. This structure has five rotational axes in total.

(ii) Non-crystalline: Icosahedral clusters can only have (111) closely spaced facets, reducing the surface energy of the cluster. Since the Mackay icosahedra are severely stressed, it is possible that icosahedra are only the most advantageous forms at small sizes. There are other non-crystalline structures besides icosahedra [32, 33]. The Ino decahedron represents another structure. Two pentagonal pyramids sharing a base make up this building. Although it only has (111) closely spaced facets on its surface, the object is far from spherical in shape. Although it is desirable to truncate the five corners of the pyramids to expose (100)-like facets, this cut rarely results in the best possible decahedron, as illustrated in Fig. 4.2 [33].

4.5 Magnetic Nanostructure and their Properties

The unique magnetic features found at the nanoscale, such as superparamagnetic, enhanced magnetic moment, high saturation field, form anisotropy, etc., make advancements in magnetic nanostructures particularly beneficial. Dots, nanoparticles, nanocrystals, nanowires, nanotubes, and thin films are some of the typical morphologies of magnetic nanostructures. Magnetic nanoparticles of various

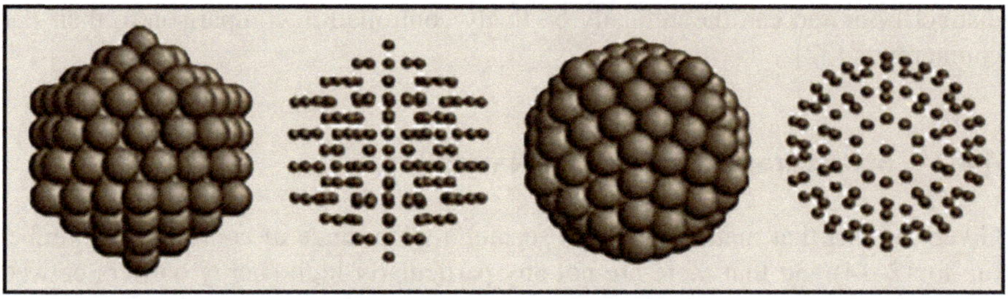

Fig. 4.2 Face-centred cubic cluster decahedron

sizes and shapes have been created, and their structural property connections have been identified, thanks to the new development of new synthesis and fabrication techniques. Remanence enhancement, random anisotropy, and gigantic magnetoresistance effects are seen when the size of magnetic materials approaches those with size-specific magnetic characteristics. Shape and surface anisotropy are induced in nanostructured materials, and thickness dependence of anisotropy is seen in thin films [34]. A variety of geometries, including nanoparticles, nanoparticle arrays, nanowires, nanodots, nano rings, nanotubes, and thin films, can be used to create magnetic nanostructures. A magnetic nanoparticle array can be created by self-assembling magnetic nanoparticles onto a substrate surface after they have been implanted in a liquid to create a composite. Two-dimensional nanoparticle arrays may be used for high-density magnetic recording. The following categories apply to magnetic nanostructures based on the relationship between structural and magnetic properties:

- Materials in bulk that have a nanoscale structure.
- Isolated particles with nanoscale diameters.

At the nanoscale level, thin-film or multilayer magnetic nanostructures have one dimension [35]. To create a heterostructure of thin films, magnetic and nonmagnetic layers are placed on top of one another. Interesting magnetic phenomena such as perpendicular anisotropy and moment changes at surfaces and interfaces can be observed in magnetic thin films [31]. Due to factors such as shape anisotropy, magnetostatic interactions, localized magnetic modes, spin waves, and current-induced magnetization reversal, the two dimensions in a nanowire structure display fascinating magnetic property. By depositing magnetic nanowires onto a surface and leveraging surface anisotropy, one can harness specific properties for various applications in nanotechnology. Between classical and quantum magnets, magnetic molecule clusters exhibit peculiar qualities.

4.6 Nano-ceramics

Because of their optical, magnetic, and electronic characteristics, metal oxide nanoparticles are a significant class of materials with numerous uses, including as catalysts, sensors, optical materials, electrical materials, and magnetic storage. This chapter also covers the use of nano-oxides as sorbents for environmental treatment. Nano-crystalline functional oxide metals are utilized for their unique electrical, magnetic, and optical properties in various materials. These characteristics make nanostructured metal oxides valuable in a variety of applications, including magnetic storage, optical materials, electrical materials, sensors, catalysts, and sensors [36]. Depending on their bandgaps, metal oxides can function as semiconductors or insulators. They behave as semiconductors when the bandgap value is less than 3.5 eV, and as insulators when the bandgap value is greater than 3.5 eV. Iron oxides are coloured because of their narrow bandgap, but zinc oxides are colourless because of their wide bandgap.

4.6.1 Structural Properties of Nano-ceramics

Materials can first be categorised according to their bonding, underlying characteristics or structures, or provenance. Depending on the long or short range of crystallographic ordering, they can either be crystalline or amorphous [36]. They are classified as opaque, transparent, semi-transparent, or reflective materials based on their surface finishing and their dielectric properties. In several high-tech fields, functional materials now have a very significant role. These materials are categorised based on the tasks they can carry out rather than their origin, kind of bonding, or processing methods. Actually, considering metal oxides, metal sulphides, tellurides, arsenides, bimetallic, and of course metals themselves, the prospective novel materials based on nanostructure are even more diverse than the solid components of the Periodic Table. Bandgaps for semiconductors, magnetic moments, specific temperatures, melting points, surface chemistry, and morphology/particle form are among the characteristics that change in the nano regime [37].

4.6.2 Piezoelectric Nano-Ceramics

As it is more suited for the commercial production of ceramics, the unique creation of an unorthodox pressure-less two-step sintering strategy including densification without grain growth is thought to be especially promising [38]. This method made it possible to create high-density pure barium titanate (BT) ceramic samples, which let researchers better understand the size effect in BT ceramic systems at the nanometre scale and address practical commercial applications. BSPT ceramics have been created using a two-step sintering process, with average grain sizes ranging from the micrometre scale down to 10 nm [39].

4.7 Graphite Oxide

Nanometre-thick sheets having infinite lateral dimensions, such as graphene, are commonly referred to as two-dimensional nanomaterials. Applications in sensing, catalysis, and energy storage might benefit from the unusual electrical properties and accentuated surface effects produced by the quantum confinement in the thickness direction. Many new techniques have been developed, including silicon carbide (SiC) vacuum graphitisation, metal-ion intercalation, liquid-phase exfoliation of graphite, chemical vapour deposition (CVD) growth, and metal-ion intercalation. Although each method has merits and demerits of its own, GO was thought to be one of the most promising routes to graphene because it can be processed using wet chemicals and is widely available in monolayer form [40].

4.7.1 Structural Properties of Graphite Oxide

At least six alternative structural models for GO have been proposed based on the characterisations. The first model was altered in 1946 by adding hydroxyl groups to the lattice and corrugating the basal plane, both of which were likewise stoichiometrically regular. For the first time in 1957, the presence of hydroxyl groups in this model explained the hydrogen content of GO, and they were later reinforced by C=C bonds, ketone and enol groups, as well as carboxylic groups at the margins. This model was altered by the stereochemistry into a corrugated carbon layer made up of ribbons of quinoidal structure alternately coupled with opened, chair-conformer cyclohexane rings [40, 41].

4.8 Compound Crystals

Due to the large number of RIM atoms with dangling bonds, graphite nanoparticles are well recognised for being unstable and, under the right circumstances, closing into fullerenes and nanotubes [42]. It was suggested in 1992 that this feature is not unique to carbon-based nanoparticles but also applies to those made of layered inorganic materials. In fact, layered materials and later a variety of other layered compounds were used to create inorganic fullerene-like nanoparticles and inorganic nanotubes [43].

4.9 Nanostructures

Nanoparticles, fullerenes, nanorods, nanobelts, nanowires, and nanotubes are examples of zero- and one-dimensional nanostructures that frequently have distinct properties from single-crystal to bulk materials [44]. Nanostructures have a very high surface-to-volume ratio and a relatively tiny number of atoms in their smallest dimension. As a result, their shape, interfaces, and structure, down to the level of a single atom, may significantly affect their properties. For instance, a particle's electrical properties may be strongly influenced by the interfaces between different phases or compositions inside the particle as well as the atomic configurations of the different areas.

High internal stress affects nanostructures created by the outside-in mechanism. In contrast, identical nanostructures are made by using inside-out, nucleation, and growth synthetic processes appear to have more relaxed mechanical properties. It should be emphasised that both the number of first layers generated and the electron diffraction were used to estimate the structure [45].

4.10 Glasses on Nanoscale

Glasses are said to be homogeneous in their design. However, heterogeneity can occasionally be found at the molecular level. The fabrication of nanostructured glass-derived materials can take advantage of heterogeneities such as amorphous phase separation, incipient crystallisation, and concentration gradients. One of the often-employed spectroscopic techniques for characterising the structure of glassy materials is nuclear magnetic resonance (NMR), which also serves as the foundation for new NMR approaches [46].

Some of the structures that can be formed in glasses include lanthanide-containing nanocrystals, quantum dots, and metal nanoparticles. A final viewpoint for glass waveguides including these nanoscale heterogeneities is offered. Glasses are special so they may be produced in any form [47].

The following are the top three definitions of glasses:

(i) Glasses are amorphous solids because of their lack of translational symmetry and non-crystalline structure [36].
(ii) Glasses are made of amorphous materials that exhibit Tg.
(iii) Glass-transition temperature: not exhibiting sufficient internal stability, as seen in real glasses, to permit the observation of a phase transition of the second order [48].
(iv) Any thermodynamically metastable condition that is frozen comprises all possible metastable phases for the system in question.

However, in some instances, a number of advanced characterisation approaches have shown molecule size heterogeneity. Such heterogeneities would be exemplified by concentration gradients, amorphous phase separation, and nuclei in the early stages of crystallisation. With new, emerging qualities, it is possible to create glass ceramics or nanoceramics with small crystal volume fractions and crystal diameters of a few nanometres [46, 48]. For generations, the creation of glasses and glass ceramics has been based on the manipulation of matter at the nano- and sub-nanoscale. As a result, in the virtually unlimited list of anticipated applications, nanoscale characterisation is important.

4.11 Nanocomposites

4.11.1 Carbon in Polymers

The mechanical, electrical, and other properties of the resulting composites are significantly influenced by the dispersion and orientation of these fillers as well as by the interfacial adhesion between the filler and the matrix material (Fig. 4.3) [49].

Due to their strong sp^2 hybridised carbon–carbon bonds and flawless hexagonal forms, nanotubes and graphene have outstanding mechanical properties. For more than a century, carbon-based nanoparticles like carbon black have been employed as reinforcement for polymers. Despite being primarily a particulate filler, carbon black has structure, and the networks that the particles form provide better reinforcing that would be possible with individual particles [51].

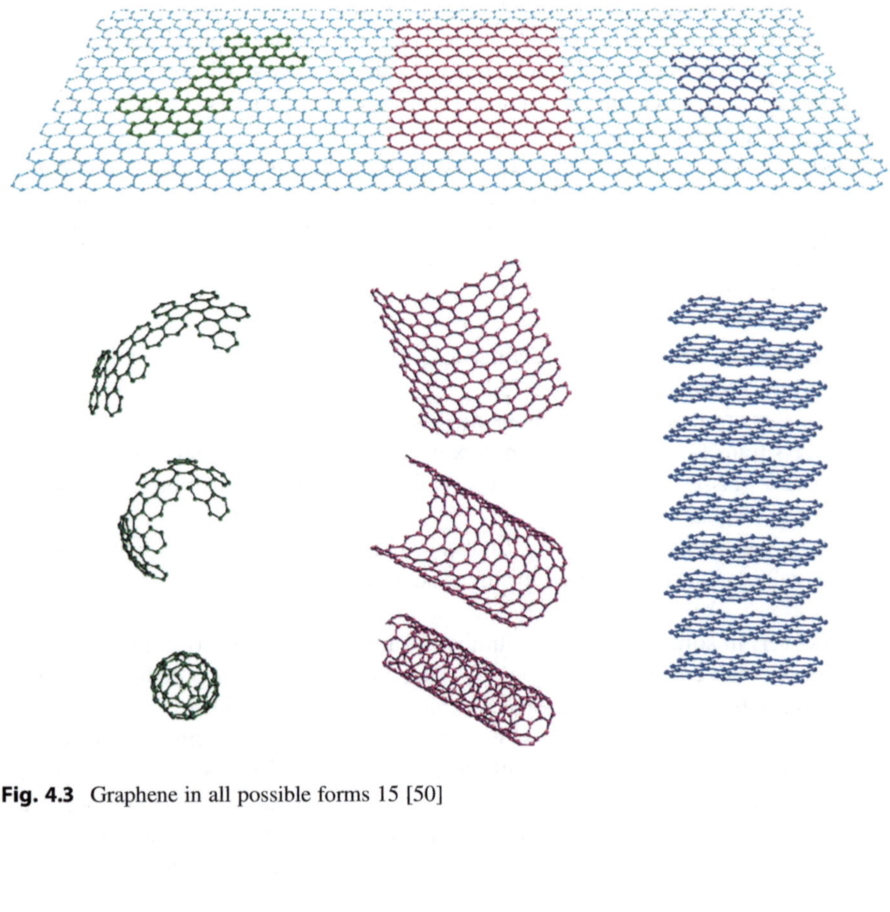

Fig. 4.3 Graphene in all possible forms 15 [50]

4.12 Nanoparticle Dispersion

The oldest cave paintings, made of various kinds of dyestuff such as haematite, charcoal, manganese oxide, and red and yellow ochre, are more than 30,000 years old. These paints, i.e., dispersions were made by grinding the solid followed by mixing in water, grease, and sometimes blood. The basic recipes were quite similar. To adjust colour, milled particles of several different minerals, including calcite, gypsum, lime, haematite, goethite, etc., were used. Lampblack or charcoal water-based inks were created with the addition of gum and sugar. Natural dyes as well as silver and gold powders were used for colouring. With the development of the press in 1450, oil-based inks including carbon black and metal particles (Cu, Pb, Ti) were created that appropriately moistened various lead-based alloy kinds [52]. Similar to the paints used in ancient times, they are still useful now. Traditional paint for wooden houses has always been ochre and seal oil.

4.13 Nano Porous Materials

An intriguing class of nanostructured material with nanosized porosity is known as nano-porous metals. Nano-porous metals can be divided into two groups based on the porosity distribution: those with a regular pore distribution and those with a random porous structure [53]. Dealloyed nano-porous metals have a three-dimensional discontinuous void structure with a length scale of few nanometres to hundreds of nanometres, and thermal annealing and other processes can change the characteristic size to as large as several microns. The nano-porous metals' pore size and microstructure can be precisely controlled by the template technique, but dynamic manipulation of the dominating length scale is almost impossible. Metallic formations that range in length from a few micrometres to more than one centi-metre are different from nano-porous metals. Nano-porous metals have prospective uses in the areas of catalysis, fuel cells, sensors, actuators, etc. [54]

4.13.1 Nano-Porous Metals

Transmetallation, hydrothermal synthesis, filter casting, electrodeposition, surfactant emulsion template, one-step square-wave potential pulse treatment, and wet-chemical approach are only a few of the known ways to make nano-porous metals. However, the dealloying technique is currently the most widely used technique for creating nano-porous metals with random porosity structure [55]. Dealloying is the process of selectively removing one or more alloy components. The development of nano-porous structures during the dealloying or depletion gilding process was unknown at the time.

4.13.1.1 Structural Properties of Nano-Porous Metals

The characteristics of nano-porous metals, particularly the typical length scale of ligaments/channels, are intimately related to their microstructure. Metals with nano-porous microstructures are easily controllable. The precursor microstructure, the type of electrolytes used, and other constituent elements are among the factors that influence the final outcome in this context. Surface diffusion of the more noble element along the alloy/solution contact causes the creation of nano-porous metals during the dealloying process [56]. By modifying the dealloying parameters and the associated electrochemical environment, we may regulate the microstructure of nano-porous metals. As a result, by dealloying, nano-porous metals typically display a ligament-channel structure with a length scale of only a few nanometres. Face-centred cubic (fcc) nanocrystals are randomly arranged within the nano-porous metal [57].

4.14 Zeolites

Zeolites are organic substances that have been a part of our environment since the dawn of human history. They developed into the twentieth century's wonder substances. Their employment in the catalysis of numerous sorts of processes is made possible by the diversity of their three-dimensional channel system, their pore size diversity, which can reach nanometre dimensions, their customisable structural properties, and the confined environment they give [57, 58].

4.14.1 Structural Properties of Zeolites

One of the most popular zeolites is Linde type A zeolite. Either a combination of double 4-ring (4–4) units or a truncated octahedral joined together by double 4-rings can be used to describe it. Truncated octahedra and truncated cuboctahedra are connected to one another through 6-membered rings and other cavities, respectively, to form the pore structure [59]. Zeolite A is most commonly used for ion exchange and as adsorbents. Synthetic faujasite materials are mainly zeolite X and zeolite Y. The sodalite cages in the faujasite structure are arranged in a diamond-like pattern, with 6–6 secondary building units (SBUs) connecting them. The primary cavities of faujasite material, which are in addition to the sodalite cages, measure around 11 in diameter and are joined by 12-membered rings with apertures that are 7.4 in diameter [60].

4.15 Mordenite

Mordenite was first created by Barrer and is a naturally occurring mineral. Its 4- and 5-membered ring structure creates a porous system that is enclosed by 8- and 12-membered rings. The porous system comprises 8- and 12-membered ring-shaped pores in linear channels that are oriented along the crystallographic c-axis (6.77.0) [57]. A supplementary pore system with 8-membered rings connects these channels together. Although initially identified as an 8-membered ring small-pore zeolite, a fault-free structure with adsorption characteristics in line with a 12-membered ring pore system can be synthesised. Mordenite has a larger silica content than zeolite A, X, or Y, which contributes to its superior thermal stability. The mordenite structure includes 5-membered rings in addition to a greater Si/Al ratio, which is expected to improve the material's acid site strength [60, 61].

4.16 Porous Anodic Aluminium Oxide

One of the usual self-organised fine structures, porous anodic aluminium oxide (AAO), has uniformly sized, cylindrical pores that are grouped in a hexagonal pattern. AAO has frequently been utilised as a model for creating nanostructures out of a range of materials [61].

4.16.1 Structural Properties of AAO

AAO with homogeneous cylindrical pores can be used as a mould to create 1-D and nano-heterostructures, two types of nanostructures. The resultant nanostructures can have controlled widths and lengths by varying the pore diameter and the AAO template's thickness [62]. In fact, these nanostructures follow the outline of the AAO pattern. Thus, by carefully determining the geometry of the nanochannels in the AAO template, it is possible to create nanostructures with linear and even more intricate designs (such as branching and step shapes). One of the most popular templates for the synthesis of 1-D nanostructures, such as NWs and NTs, is an AAO with homogeneous and dense pores. In the nanochannels of AAO templates, 1-D nanostructures of different materials, such as metals, semiconductors, and nonconducting and conducting polymers, can be built. The pore size of the template determines the diameters of the 1-D nanostructures [63].

4.17 Porous Silicon and its Properties

By electrochemically etching crystalline silicon, porous silicon (PS) is a nano-porous material created. Porous silicon has historically been studied for its strong visual luminescence and the potential to create effective light emitters made of silicon. It permits practically constant fine-tuning of etched nano-porous structures from the scale of a few nanometres to tens of microns [64].

Since then, PS has had a sharp increase in interest due to the interesting prospects of creating effective silicon-based light-emitting devices. And some fundamentals of Si are still not fully understood. PS morphologies still are not available. Typically, aqueous hydrofluoric acid (HF) solutions are used to etch PS [65]. Surfactants are often included into solutions or mixed with another solvent (such alcohols, dimethyl sulfoxide (DMSO), or dimethylformamide (DMF)). Such additives improve wettability and lower surface energy, allowing solvent to easily fill pores. Additionally, the hydrogen that develops during etching may be removed effectively without bubbles, which would otherwise result in pore inhomogeneities and surface roughness. The typical length scales for PS might range from a few nanometres to tens of micrometres. Low-doped substrates and concentrated HF solutions yield the bigger macropores. Due to tiny current holes that reach their surface and the fact that their growth orientation is not parallel to the carrier flux, side-pores grow at a slower rate. The majority of the carriers are efficiently collected by the primary pore, which expands more quickly. Increased separation between the terminals of pores expanding along lateral ones' termination is the overall result. The specifics of the etching circumstances have a significant impact on the final PS structure. A thin transition layer appears during the initial etching period. In this layer, pores attempt to arrange themselves into a porous structure that is consistent with the etching conditions rather than growing in a steady state regime and with a regular structure [66]. The reduced anisotropy of the etching process in p-type substrates is one of the primary distinctions between p- and n-type silicon. In reality, a p-doped semiconductor's high hole density encourages chemical etching and causes a significant lateral increase of pore size during the etching process. Pores on n-type substrates tend to be more rounded, but on p-type substrates, where there is less anisotropy in the etching process, it is possible to realise complex-shaped structures from a basic geometric pattern [67].

4.18 Organic and Biomaterials

Organic nanomaterials offer the potential benefits of creating building blocks using synthetic chemistry at multiple levels and open the bottom-up route for the synthesis of organic and polymeric nanostructures, thanks to the utility of weak noncovalent interactions for the design and self-assembly of molecules one by one into desired structures. It is still difficult and demanding to assemble precise nanostructures over broad surfaces, to tune noncovalent interactions for effective assembly, and to understand the dynamics and model of the self-assembly process. In reality, despite the significant effort put into modifying their morphologies and functionalities, organic nanomaterials have been predicted to play an active role in optoelectronic devices [68].

4.18.1 Properties of Organic Nanomaterials

Organic nanomaterial morphology control has had mixed results. Modulating the molecular structures in a succession of derivatives is a typical method for controlling organic nanostructures. The morphologies of inorganic nanostructures have been effectively regulated by selecting the right materials and altering the synthesis processes, which is a crucial task for nanoscale science. An n-type semiconductor molecule has its own nanobelt architectures. The nanobelts have a length that varies from ten to a few tens of micrometres. The measurement of electron diffraction and polarised microscope imaging have been used to infer the highly organised molecular packing (uniaxial crystal phase). The molecules are stacked one on top of the other in the observed optical axis [69].

A clear impact of side-chain substitutions on the morphology of perylene diimide molecule self-assembly was also observed with two derivatives that had side-chains that were distinctly different from one another. The two molecules completely differ in morphologies during self-assembly as a result of differing side-chain interference [70].

4.19 Nanocomposites as Bone Implant Materials

Researchers have been working on several man-made materials that may be utilised as a suitable replacement for natural bone and are inexpensive and simple to construct as a result of the growing need for a bone implant material. The potential for achieving the goal is quite high as nanotechnology in the field of medicines continues to progress. Nanocomposites have a great deal of potential for usage as a suitable bone implant material since they are the result of coordinated efforts in tissue engineering and nanobiotechnology to create the optimal orthopaedic implant [71]. Major goals of this chapter are to clarify nanocomposite as a potential next-generation bone implant material, along with reviewing current/conventional bone implant materials and their drawbacks. The goal of this review is to provide brief and pertinent information on different nanofabrication procedures, the characterisation of nanocomposites, and the in vitro and in vivo biocompatibility research of these materials. Although significant advancements are still required to put nanocomposites into practical use, initial research suggests that they are a perfect implant material for orthopaedic applications. Current trends in nanobiotechnology point to a bright future for the use of nanocomposites in orthopaedics. In the initial studies investigating the impact of nanometer-sized ceramics, it was demontrated that Osteoblast adhesion significantly increased on nanophase alumina and titania substrates when compared to conventional micrometer-sized substrates [71, 72].

4.20 Nanofiber Biomaterials

In view of their special qualities and capacity to interact with cells at the same scale as natural extracellular matrix fibrils, nanofibers have drawn interest. There are numerous approaches to control the nanoscale structure and properties of biomaterials, thanks to new fabrication technologies [73].

Compared to current micro- and macroscale biomaterials, nanofiber biomaterials provide a variety of benefits. Due to the small size, high surface area, surface characteristics frequently outweigh those of bulk materials. Extracellular matrix (ECM), which is made up of biopolymer fibrils and fibres that structurally keep cells together, is a component of every tissue in the human body. It acts as a network of insoluble signals for communication between biochemical and biomechanical systems. Designing and utilising nanofiber biomaterials for tissue engineering and medical device applications requires an understanding of the fundamental functions of the ECM. The ECM's fundamental composition and characteristics in relation to how cells interact with nanofibers. Although these nanofibers have a lot in common, there are structural variances as well, depending on the circumstances of manufacturing and the polymers utilised. The nanofibers' predominant structural characteristic is a-helix. When they are shaped into scaffolds, they have an incredibly high-water content, which is advantageous for some tissue engineering applications. The environmental factors and the amino acid sequence of the peptide nanofibers play a significant role in determining their final structure. By controlling the self-assembly, these can create protein nanofiber matrices in a range of planar geometries [74].

4.21 Conclusion

Nanostructured materials have recently garnered significant attention due to their array of remarkable chemical, mechanical, electrical, and optical properties. The general interest in carbon studies has increased dramatically as a result of the discovery of fullerenes. For many years, CNTs and other materials containing carbon have been employed in fuel cells, batteries, and other electrochemical applications. Due to their better physical and chemical properties, carbon nanotubes have been demonstrated in numerous studies to be preferable to conventional carbon-based electrodes for use in fuel cells, supercapacitors, and batteries. Exposure studies are crucial to the precise description and management of their risks, and adequate characterisation will improve the future of produced nanomaterials in terms of toxicity, ecotoxicity, and management of their risks. Ideally, one should measure one or more fundamental metrics, such as diameter, shape, aggregation state, and surface area, on the most relevant samples in media that accurately represent the significant biological or environmental questions. To accurately analyse the physicochemical characteristics of nanoparticles, standardised tests and methodologies for characterisation are required.

References

1. Granqvist CG, Buhrman RA (1976) Ultrafine metal particles. J Appl Phys 47(5):2200–2219
2. Dresselhaus MS, Dresselhaus G, Eklund PC (eds) (1996) Science of fullerenes and carbon nanotubes. In Elsevier eBooks.
3. Thakur P, Thakur A (2022) Nanomaterials, their types and properties. In: Thakur A, Thakur P, Khurana SP (eds) Synthesis and applications of nanoparticles. Springer, Singapore. pp 19–44. https://doi.org/10.1007/978-981-16-6819-7_2
4. Mohamed MM, Volkov V, Link S, El-Sayed MA (2000) The 'lightning' gold nanorods: fluorescence enhancement of over a million compared to the gold metal. Chem Phys Lett 317(6):517–523
5. Thakur P, Thakur A (2022) Introduction to nanotechnology. In: Thakur A, Thakur P, Khurana SP (eds) Synthesis and applications of nanoparticles. Springer, Singapore
6. Almessiere MA, Slimani Y, Thurkanov AV, Baykal A (2022) Structural and morphological characterization of nanomaterials. In: Thakur A, Thakur P, Khurana SP (eds) Synthesis and applications of nanoparticles. Springer, Singapore
7. Sofo JO, Chaudhari AS, Barber GD (2007) Graphane: a two-dimensional hydrocarbon. Phys Rev B Condens Matter Mater Phys 75(15):153401
8. Elias D, Nair R, Mohiuddin TMG, Morozov SV, Blake P, Halsall MP, Ferrari A, Boukhvalov DW, Katsnelson MI, Geim AK, Novoselov KS (2009) Control of Graphene's properties by reversible hydrogenation: evidence for Graphane. Science 323(5914):610–613
9. Nair RR, Ren W, Jalil R, Riaz I, Kravets VG, Britnell L, Blake PG, Schedin F, Mayorov AY, Yuan S, Katsnelson MI, Cheng H, Strupinski W, Bulusheva LG, Okotrub AV, Grigorieva IV, Grigorenko AN, Novoselov KS, Geim AK (2010) Fluorographene: a two-dimensional counterpart of Teflon. Small 6(24):2877–2884
10. Withers F, Dubois M, Savchenko AK (2010) Electron properties of fluorinated single-layer graphene transistors. Phys Rev B 82:7
11. Geringer V, Liebmann M, Echtermeyer T, Runte S, Schmidt MP, Rückamp R, Lemme MC, Morgenstern M (2009) Intrinsic and extrinsic corrugation of monolayer graphene deposited onSiO2. Phys Rev Lett 102:7
12. Ugarte D (1992) Curling and closure of graphitic networks under electron-beam irradiation. Nature 359(6397):707–709
13. Mackay AL, Terrones H (1991) Diamond from graphite. Nature 352(6338):762
14. Iijima S, Ichihashi T (1993) Single-shell carbon nanotubes of 1-nm diameter. Nature 363(6430): 603–605
15. Zhu YS, Hsu W, Terrones H, Grobert N, Chang BH, Terrones M, Wei B, Kroto HW, Walton DJ, Boothroyd C, Kinloch IA, Chen G, Windle AH, Fray DJ (2000) Morphology, structure and growth of WS2 nanotubes. J Mater Chem 10(11):2570–2577
16. Miller GP (2007) Carbon Nanotubes, Properties and Applications Edited by Michael J. O'Connell (Theranos, Inc., Menlo Park, CA). CRC Press and Taylor & Francis Group: Boca Raton, FL. 2006. xviii + 320 pages. $99.95. ISBN 0-8493-2748-2. Journal of the American Chemical Society, 129(19), 6343–6344
17. Carbon nanotubes. (2008). In Topics in applied physics. Springer Science+Business Media
18. Avouris P, Freitag M, Perebeinos V (2008) Carbon-nanotube photonics and optoelectronics. Nat Photonics 2(6):341–350
19. Dresselhaus MS, Jorio A, Hofmann M, Dresselhaus G, Saito R (2010) Perspectives on carbon nanotubes and graphene Raman spectroscopy. Nano Lett 10(3):751–758
20. Neto AHC, Guinea F, Peres NMR, Novoselov KS, Geim AK (2009) The electronic properties of graphene. Rev Mod Phys 81(1):109–162
21. Krätschmer W, Lamb LD, Fostiropoulos K, Huffman DR (1990) Solid C60: a new form of carbon. Nature 347(6291):354–358
22. Lozovik YE, Popov AV (1997) Formation and growth of carbon nanostructures: fullerenes, nanoparticles, nanotubes and cones. Physics-Uspekhi 40(7):717–737

23. Iijima S (1991b) Helical microtubules of graphitic carbon. Nature 354(6348):56–58
24. Oberlin A, Endo M, Koyama T (1976) Filamentous growth of carbon through benzene decomposition. J Cryst Growth 32(3):335–349
25. Yoon SH, Lim S, Song Y, Ota Y, Qiao W, Tanaka A, Mochida I (2004) KOH activation of carbon nanofibers. Carbon 42(8–9):1723–1729
26. Liu Q, Ren W, Chen Z, Yin L, Li F, Cong H, Cheng H (2009) Semiconducting properties of cup-stacked carbon nanotubes. Carbon 47(3):731–736
27. Ōsawa E (2007) Recent progress and perspectives in single-digit nanodiamond. Diam Relat Mater 16(12):2018–2022
28. Balasubramanian G, Chan I, Kolesov R, Al-Hmoud M, Tisler J, Shin CS, Kim C, Wójcik AK, Hemmer PR, Krueger A, Hanke T, Leitenstorfer A, Bratschitsch R, Jelezko F, Wrachtrup J (2008) Nanoscale imaging magnetometry with diamond spins under ambient conditions. Nature 455(7213):648–651
29. Krueger A (2008) Diamond nanoparticles: jewels for chemistry and physics. Adv Mater 20(12): 2445–2449
30. Pérez-Juste J, Pastoriza-Santos I, Liz-Marzán LM, Mulvaney P (2005) Gold nanorods: synthesis, characterization and applications. Coord Chem Rev 249(17–18):1870–1901
31. Guo S, Dong S, Wang E (2009) Rectangular silver nanorods: controlled preparation, liquid–liquid Interface assembly, and application in surface-enhanced Raman scattering. Cryst Growth Des 9(1):372–377
32. Del Fatti N, Christofilos D, Vallée F, Christofilos D (2008) Optical response of a single gold nanoparticle. Gold Bull 41(2):147–158
33. Netzer FP (2010) "Small and beautiful"—the novel structures and phases of nano-oxides. Surf Sci 604(5–6):485–489
34. Löffler JF, Braun H (2007) Magnetization response in bulk nanostructured magnets. Materials Science and Engineering A-structural Materials Properties Microstructure and Processing
35. Thomas S, Al-Harthi S, Kumar DS, Al-Omari IA, Ramanujan RV, Yoshida Y, Anantharaman MR (2008) Microstructure and random magnetic anisotropy in Fe–Ni based nanocrystalline thin films. J Phys D 41(15):155009
36. Jayaraman V, Krishnamurthy D, Gnanasekaran T, Thiruvengadasami ARS, Prasad MNV (2007b) Development of yttria-doped thoria solid electrolyte for use in liquid sodium systems. Ionics 13(5):299–303
37. Sehirlioglu A, Sayir A, Dynys F (2009) High temperature properties of BiScO3–PbTiO3 piezoelectric ceramics. J Appl Phys 106(1):014102
38. Zou T, Wang X, Wang H, Zhong C, Li L, Chen I (2008) Bulk dense fine-grain (1−x)BiScO3–xPbTiO3 ceramics with high piezoelectric coefficient. Appl Phys Lett 93(19):192913
39. Chen S, Dong X, Mao C, Cao F (2006) Thermal stability of (1-x)BiScO3-xPbTiO3Piezoelectric ceramics for high-temperature sensor applications. J Am Ceram Soc 89(10):3270–3272
40. Gilje S, Han S, Wang M, Wang KL, Kaner RB (2007) A chemical route to graphene for device applications. Nano Lett 7(11):3394–3398
41. Stankovich S, Dikin DA, Piner RD, Kohlhaas K, Kleinhammes A, Jia Y, Wu Y, Nguyen ST, Ruoff RS (2007) Synthesis of graphene-based nanosheets via chemical reduction of exfoliated graphite oxide. Carbon 45(7):1558–1565
42. Tahir M, Yella A, Sahoo JK, Annal-Therese H, Zink N, Tremel W (2010) Synthesis and functionalization of chalcogenide nanotubes. Physica Status Solidi B-basic Solid State Phys 247(10):2338–2363
43. Rinke M, Eckert H (2011) The mixed network former effect in glasses: solid state NMR and XPS structural studies of the glass system (Na2O)x(BPO4)1−x. Phys Chem Chem Phys 13(14): 6552
44. Shibuta Y, Suzuki T (2010) Phase transition in substrate-supported molybdenum nanoparticles: a molecular dynamics study. Phys Chem Chem Phys 12(3):731–739

45. Qi W (2005) Comment on: 'The cluster size dependence of thermal stabilities of both molybdenum and tungsten nanoclusters' [Chem. Phys. Lett. 354 (2002) 165]. Chem Phys Lett 402(1–3):279–281
46. Thakur P, Chahar D, Taneja S, Bhalla N, Thakur A (2020) A review on MnZn ferrites: synthesis, characterization and applications. Ceram Int 46(10B):15740–15763
47. Ramgir NS, Ganapathi SK, Kaur M, Datta N, Muthe KP, Aswal DK, Gupta S, Yakhmi JV (2010) Sub-ppm H2S sensing at room temperature using CuO thin films. Sens Actuators B Chem 151(1):90–96
48. Jang BZ, Zhamu A (2008) Processing of nanographene platelets (NGPs) and NGP nanocomposites: a review. J Mater Sci 43(15):5092–5101
49. Kalaitzidou K, Fukushima H, Drzal LT (2007) A new compounding method for exfoliated graphite–polypropylene nanocomposites with enhanced flexural properties and lower percolation threshold. Compos Sci Technol 67(10):2045–2051
50. Kalaitzidou K, Fukushima H, Askeland P, Drzal LT (2008) The nucleating effect of exfoliated graphite nanoplatelets and their influence on the crystal structure and electrical conductivity of polypropylene nanocomposites. J Mater Sci 43(8):2895–2907
51. Pallecchi P, Giachi G, Colombini MP, Modugno F, Ribechini E (2009) The painting of the Etruscan "Tomba della Quadriga Infernale" (4th century BC), in Sarteano (Siena, Italy): technical features. J Archaeol Sci 36(12):2635–2642
52. Hayes JM, Nyce GW, Kuntz JD, Satcher JH, Hamza AV (2007) Synthesis of bi-modal nanoporous Cu, CuO and Cu2O monoliths with tailored porosity. Nanotechnology 18(27): 275602
53. Deng Y, Huang W, Chen X, Li Z (2008) Facile fabrication of nanoporous gold film electrodes. Electrochem Commun 10(5):810–813
54. Lee YC, Park DW, Park JY (2008) Fully packaged nonenzymatic glucose microsensors with nanoporous platinum electrodes for anti-fouling. IEEE Sensors J 8(11):1922–1927
55. Cao L, Zhang B, Lu L (2011) Fabrication of large-area platinum-coated silicon nanowire arrays and its application to enhanced electrocatalytic activity for methanol oxidation. Chin J Anal Chem
56. Joo SH, Kim J (2010) Electrooxidation of glucose at nanoporous gold surfaces: structure dependent electrocatalysis and its application to amperometric detection. Electroanalysis 22(9):939–945
57. De Lara EC, Kahn R, Mezei F (1983) Determination of the intracrystalline diffusion coefficient of methane in A zeolites by means of neutron spin-echo experiments. J Chem Soc 79(8):1911
58. Derouane EG (1990) Relationship between the physicochemical properties of zeolitic systems and their low dimensionality. In Springer eBooks (pp. 373–379)
59. Shulga OA, Zhou D, Demchenko AV, Stine KJ (2008) Detection of free prostate specific antigen (fPSA) on a nanoporous gold platform. Analyst 133(3):319
60. Xu XP, Yu S, Lau SS, Li L, Zhao B (2008) Magnetic and thermal expansion properties of vertically aligned Fe nanotubes fabricated by electrochemical method. J Phys Chem C 112(11): 4168–4171
61. Liu L, Zhou W, Xie S, Song L, Luo S, Liu D, Shen J, Li D, Xiang Y, Liu T, Ren Y, Wang C, Wang G (2008) Highly efficient direct electrodeposition of Co−Cu alloy nanotubes in an anodic alumina template. J Phys Chem C 112(7):2256–2261
62. Seah SYK, Meng G, Xu Q, Han F, Kong M, Zhang L (2008) Electronic transport behavior of bismuth nanotubes with a predesigned wall thickness. J Phys Chem C 112(23):8614–8616
63. Inguanta R, Piazza S, Sunseri C (2007b) Template electrosynthesis of CeO2nanotubes. Nanotechnology 18(48):485605
64. Föll H, Christophersen M, Carstensen J, Hasse G (2002) Formation and application of porous silicon. Mater Sci Eng R 39(4):93–141
65. Arita Y, Sunohara Y (1977) Formation and properties of porous silicon film. J Electrochem Soc 124(2):285–295

66. Dimova-Malinovska D, Sendova-Vassileva M, Tzenov N, Kamenova M (1997) Preparation of thin porous silicon layers by stain etching. Thin Solid Films 297(1–2):9–12
67. Barrett C, Iacopino D, O'Carroll DM, De Marzi G, Tanner DB, Quinn AAJ, Redmond G (2007) Synthesis of pentacene nanotubes by melt-assisted template wetting. Chem Mater 19(3): 338–340
68. Jiang L, Fu Y, Li H, Hu W (2008) Single-crystalline, size, and orientation controllable nanowires and ultralong microwires of organic semiconductor with strong Photoswitching property. J Am Chem Soc 130(12):3937–3941
69. Zhao Y, Di C, Yang W, Yu G, Liu Y, Yao J (2006) Photoluminescence and electroluminescence from Tris(8- hydroxyquinoline) aluminum nanowires prepared by adsorbent-assisted physical vapour deposition. Adv Funct Mater 16(15):1985–1991
70. Zhao G, Raines AL, Wieland M, Schwartz Z, Boyan BD (2007) Requirement for both micron- and submicron scale structure for synergistic responses of osteoblasts to substrate surface energy and topography. Biomaterials 28(18):2821–2829
71. Kunzler TP, Drobek T, Schuler M, Spencer ND (2007) Systematic study of osteoblast and fibroblast response to roughness by means of surface-morphology gradients. Biomaterials 28(13):2175–2182
72. Chandel S, Thakur S, Thakur A (2007) Low leakage current density and improved dielectric behavior of BiFexO3 nano-ceramics. J Alloys Compd 845:156287
73. Hauser CAE, Zhang S (2010) Designer self-assembling peptide nanofiber biological materials. Chem Soc Rev 39(8):2780
74. Paramonov SE, Jun H, Hartgerink JD (2006) Self-assembly of peptide–amphiphile nanofibers: the roles of hydrogen bonding and amphiphilic packing. J Am Chem Soc 128(22):7291–7298

Carbon Nanomaterials in the Field of Theranostics

5

Fayu Wan, Atul Thakur, and Preeti Thakur

Abstract

Numerous unique nanomaterials have been discovered as a result of the rising demand for nanotechnology that will be used in the biomedical and pharmaceutical sciences. Drug delivery systems made of nanomaterials have enormous potential for implementing nanotechnology in medical settings. Graphene, reduced graphene oxides, carbon dots, and fullerenes are just a few examples of the many carbon-based compounds that have attracted a lot of attention in recent years. Because of their excellent mechanical, electrical, thermal, optical, and chemical properties as well as their unique structural dimensions, these materials have attracted a lot of attention in a number of sectors, including biological applications. Recent research has focused on imaging of cells and tissues as well as the delivery of therapeutic molecules for the treatment of disease and tissue restoration. A promising imaging agent for tumor diagnosis, carbon-based

F. Wan (✉)
School of Electronics and Information Engineering, Nanjing University of Information Science & Technology, Nanjing, China
e-mail: fayu.wan@nuist.edu.cn

A. Thakur
School of Electronics and Information Engineering, Nanjing University of Information Science & Technology, Nanjing, China

Centre for Nanotechnology, Amity University Haryana, Gurugram, India

Nanotechnology Wing, Innovative Science Research Society, Shimla, India

P. Thakur
Centre for Nanotechnology, Amity University Haryana, Gurugram, India

Nanotechnology Wing, Innovative Science Research Society, Shimla, India

NanoLatticeX, Amity University Haryana, Gurugram, Haryana, India

D. Suhag et al. (eds.), *Integrated Nanomaterials and their Applications*,
https://doi.org/10.1007/978-981-99-6105-4_5

nanomaterials have a broad-range one-photon property, are biocompatible, and are simple to functionalize. Deep-tissue optical imaging is made possible by the intrinsic two-photon fluorescence property of carbon-based nanomaterials in the long wavelength range (near-infrared II). This chapter explains the potential and promising diagnostic and therapeutic applications of carbon-based nanomaterials for the treatment of many diseases, including cancer, and highlights current developments in one-photon and two-photon imaging.

Keywords

Drug delivery · Carbon-based materials · Graphene · Fullerene · Carbon dots

5.1 Introduction

The term "theranostics" combines the words "therapeutics" and "diagnostics." Using and developing nanomedicine strategies for cutting-edge theranostics is the goal of nano-theranostics. Nano-theranostics includes both nanoscale diagnosis using equipment like MRI, PET-CT scans, etc., and nanoscale therapy using equipment like chemotherapy, radiotherapy, photodynamic therapy, etc. [1, 2]. The process of using one radioactive drug to detect (diagnose) and another radioactive drug to administer therapy to both the primary tumor and any metastatic tumors is known as theranostics. John Funkhouser, who at the time served as the CEO of Pharmanetics, coined the term "theranostics" in 2002. The term, which is effectively a portmanteau of the words "therapeutics" and "diagnostics," describes the Pharmanetics business strategy for creating diagnostic tests to spot cancer cells and administering certain remedies. The theranostic approach is personalized, using both diagnosis and therapy tools as part of the treatment.

Carbon-based nanomaterials (CBNs) have had a significant impact on the biomedical field over the past few decades due to their ability to deliver therapeutic molecules and enable the visualization of cells and tissues, both of which are necessary for the treatment and cure of diseased and damaged tissues [3]. Examples of CBN (CQDs) include fullerenes, carbon nanotubes (CNTs), graphene (G) and its derivatives (graphene oxide (GO)), nanodiamonds (NDs), and carbon-based quantum dots. The graphic suggests that CBN could be employed for photothermal therapy, photodynamic therapy, drug/gene delivery, bioimaging, cell fluorescence labeling, stem cell engineering, and biosensing. CBNs unique optical properties, including their inherent fluorescence, tunable narrow emission spectrum, and great photostability, make them potentially helpful for imaging and diagnosing cells and tissues. Additionally, to enhance their properties, their surfaces can be modified with functional groups (carboxylic acid, hydroxyl, and epoxy). CBN has a high surface area and exceptional optical, mechanical, and electrical properties, making it one of the most sought-after and qualified choices for theranostic applications. The biological safety of CBNs, which is related to their aqueous stability and interactions with cells and tissues, is one of the main difficulties for their real biomedical application. Recent studies have sparked interest in their possible role in anti-inflammatory and anticancer treatments. For instance, CBN induces reactive oxygen species (ROS), which results in DNA and lipid damage as well as cell death, when it is taken up by cancer cells [4]. Additionally, the effects of graphene materials on macrophage metabolism include an increase in ROS levels, disruption of the mitochondrial membrane, and cell death by apoptosis [5]. Conversely, biofunctionalized CBN improves delivery efficiency by lowering clearance and raising body retention. The optical properties of CBN have recently enabled significant progress in one-photon and two-photon imaging applications, among other things. Raman imaging [6, 7], photoacoustic (PA), up-conversion near infrared (NIR), and other optical (even non-optical imaging) modalities have received a lot of attention recently. CBNs are high-performance one-photon and two-photon imaging agents with a variety of multifunctional, multimodal, and multiple functionalities. This is now a widely accepted fact. As a result, they exhibit great potential for use in diagnostics and cell labeling for the treatment of diseases and the regeneration of damaged tissue. As a result, various organizations have recently researched the synthesis of CBN and its novel applications; however, the theranostic applications received little attention. In this chapter, the theranostics applications of CBN are emphasized, along with its pertinent attributes such as optical characteristics, biocompatibility, and in vivo applications. We first discuss the mechanisms that led to the development of the various imaging properties of CBN, then we highlight their promising uses as diagnostic and therapeutic nanomaterials [8] for the treatment of cancer and a number of other diseases, and finally we discuss the opportunities and difficulties associated with developing future usable theranostic platforms.

5.2 Carbon Nanomaterials

Although the presence of fullerenes (C60, C70, and C84), which are carbon-based compounds, was predicted in 1970, C60 was only identified in 1985. Although certain fullerenes, such as C60, have oblong shapes resembling rugby balls, most of them (such as C70) are spheroids. The 1991 discovery of the carbon nanotube (CNT) tremendously benefited further research into carbon-related nanomaterials. The structural properties of a one-nanometer-diameter tube formed from a one-atom thick sheet of graphite can be rolled into a range of qualities depending only on how the nanotubes are rolled [9, 10]. CNTs are very stiff because of their high tensile strength and modulus. Their one-dimensional nanotubular structure also greatly enhances and frequently even surpasses that of conductive metals in terms of electrical and thermal conductivity. CBNs have been researched in a variety of other fields, including composite materials, energy research, catalysis, data storage devices, sensors, and bioelectronics, as well as most recently biomedicine and theranostics. Several important studies on CBN (and their blends and hybrids) have been conducted, especially for medicinal and diagnostic purposes (Figs. 5.1 and 5.2).

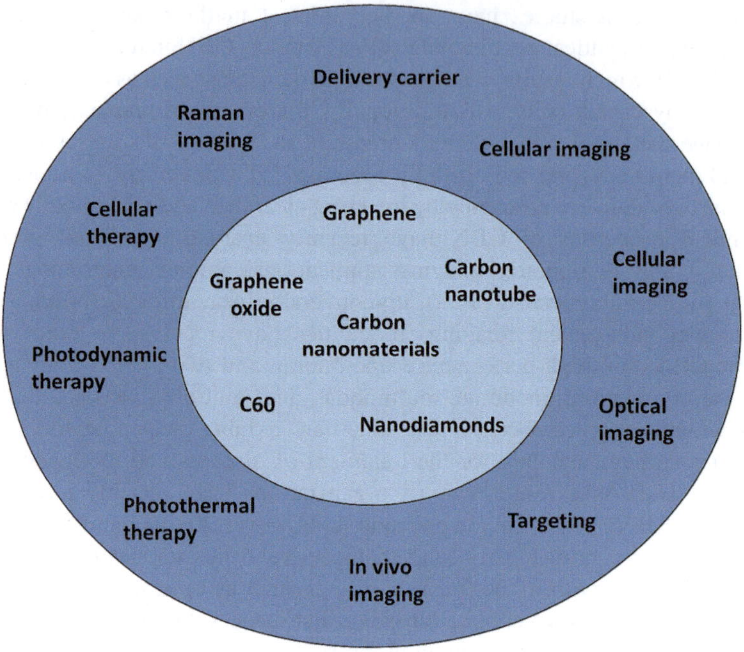

Fig. 5.1 Applications of carbon-based nanomaterials (CBNs) in theranostics [11]

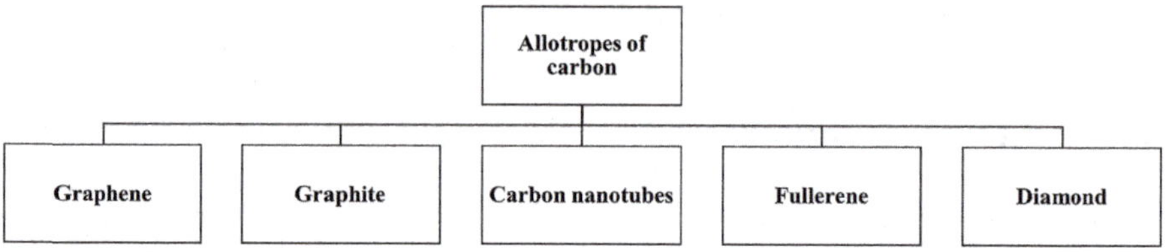

Fig. 5.2 Allotropes of carbon

5.2.1 Fullerenes

On account of its unique photophysical and photochemical properties, buckminsterfullerene (C60), a fullerene discovered in 1985, has generated a lot of interest. Buckminsterfullerene belongs to the class of spherical fullerenes. Because they are key sensitizers for the photoproduction of singlet oxygen (1O2) ROS, fullerenes are employed for blood sterilization and photodynamic cancer therapy [12–14]. However, a significant obstacle to fullerenes' usage in nanomedicine is their dispersibility. The fundamental issue is their limited solubility in many solvents, particularly water, where singlet oxygen has an extensive lifespan. In order to increase fullerenes' solubility in water, a number of techniques have been devised to functionalize them with hydrophilic groups. As a result, the created fullerenes have prospective applications as antibacterial, antiviral, and antioxidant agents. Fullerenes' ability to scavenge free radicals like reactive oxygen species (ROS) and reactive nitrogen species (RNS) and act as an antioxidant has fueled its use in biological applications. Cells are helped by derivatives of glutathione C_{60} to avoid apoptosis brought on by nitric oxide [15]. When human mast cells (hMCs) and peripheral blood basophils were pre-incubated with C_{60}, the IgE-dependent mediator generated by these cells is significantly inhibited, demonstrating the role of fullerenes as a negative regulator of allergic reaction. Fullerenes are another group of prospective photosensitizers. Depending on the polarity of the medium, they can absorb photons in the visible and ultraviolet ranges to create triplet fullerene species that are photoexcited as well as singlet oxygen, or ROS. Furthermore, fullerenes can be equipped with light-harvesting antennae to increase the quantum yield of ROS generation. Therefore, cancer treatment and microbial eradication can be accomplished by the use of fullerenes in photodynamic therapy (PDT).

The cage-like nanoscale structure of fullerenes enables the production of molecular or particulate entities with one or more functional groups covalently bonded to the fullerene cage surface in a geometrically controlled manner. In this case, it is appropriate to use receptor ligands for agonizing or antagonizing cellular and enzymatic functions as well as targeted drug transport across biological membranes. A different method of preparing fullerenes for use in pharmaceutical applications that improves distribution, absorption, and delivery efficiency is liposome formulation. Although fullerene medicine has achieved significant scientific advancements, clinical studies have not been successful due to concerns about the long-term safety and toxicity of fullerenes. In contrast, fullerene-based cosmetics have been examined for long time in clinical settings and utilized in human skincare, suggesting that at least topical fullerene treatment is risk-free [16–18]. Due to the great attraction of the fullerene cage for Gd^{3+} ions, which prevents their release into biological environments, there are less long-term safety issues connected with Gd^{3+} ions. As multivalent drug delivery systems with the potential for diverse targeting features, fullerenes have the capacity to self-assemble into fullerosomes (Fig. 5.3).

5.2.1.1 The Advantages and Disadvantages of Using Fullerene in Medicine

Due to its extensive surface area, fullerenes have the potential to encapsulate a diverse range of atoms and molecules within their structure. New fullerene-derived photosensitizers (PSs) have greater MRI activity. Instead of the conventional PS used for PDT, fullerenes display less photobleaching and are more photostable. Fullerenes can be used in conjunction with light-harvesting antennas, electron transport dyads, and other components to shift absorption wavelengths toward the red end of the spectrum. These coupled antennas could be helpful in facilitating the charge separation and electron transport processes. Additionally, a medicine's dosage and mode of action in biological systems can be modified by modifying the chemical makeup of fullerenes. Both natural and functionalized fullerenes have the ability to kill cancer cells after being cultured. In addition, they have the ability to render viruses, fungi, and bacteria dormant. In vivo studies show that they can kill

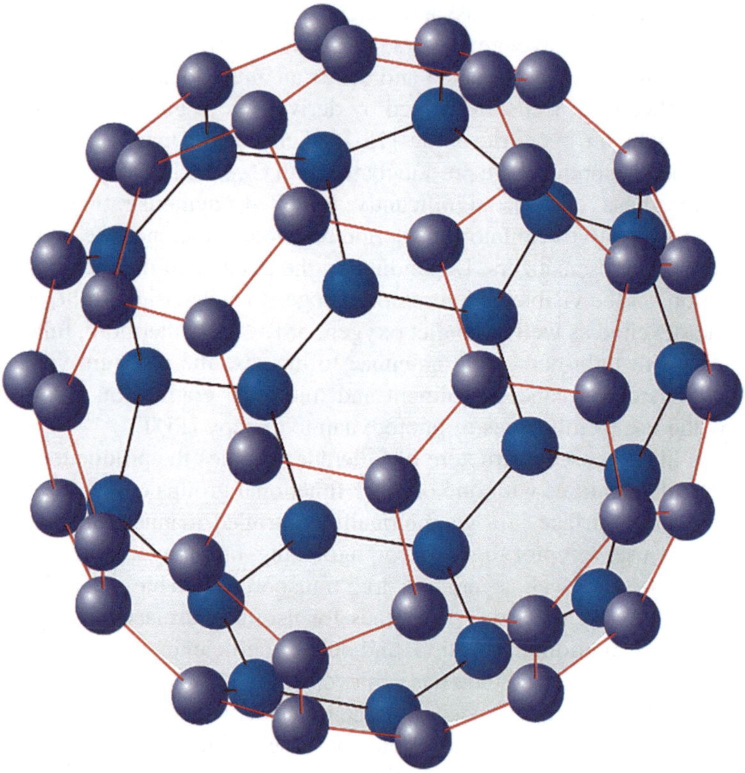

Fig. 5.3 A type of carbon called fullerene (Buckminsterfullerene) is made up of 60 carbon atoms linked together to create an approximately spherical "buckyball" (which looks rather like a soccer ball). The image is from the MDPI article "Carbon-Based Nanomaterials for Delivery of Biologicals and Therapeutics: A Cutting-Edge Technology" which is an open access article [19]

tumors or inhibit their growth in mice. Pure fullerenes are exceedingly hydrophobic and have a tendency to cluster in wet environments. Even when they are functionalized with water-soluble polar groups, this defect might not entirely be rectified. Their harmless, non-toxic molecular structure sets fullerenes apart. Fullerenes may be absorbed by several cell types after entering the body and result in a range of cellular alterations, including modifications to viability, proliferation, inflammatory reactions, and oxidative reactions. The molecular weight of C_{60} is 720 g/mol, which is quite heavy. When numerous additional moieties are conjugated to the cage, the MW might rise considerably.

The primary restriction is the toxicity that results from the in vivo release of metal ions during metabolic activities. The expensive expense of acquiring them is another restriction on their utilization.

5.2.2 Nanotubes Made of Carbon

Carbon nanotubes (CNTs) are seamless cylinders made by rolling up graphene sheets that have special inherent characteristics. CNTs are divided into two categories: single-wall carbon nanotubes (SWCNTs) and multiwall carbon nanotubes depending on how many graphene layers are present in the cylindrical tubes (MWCNTs) [20]. SWCNTs and MWCNTs have diameters that range from 0.4 to 2.5 nm and a few nanometers to 100 nm, respectively. Van der Waals forces govern the interactions between layers in MWCNTs, and a variety of 2D crystal combinations with various electrical, optical, and mechanical characteristics can be used to create multilayered CNTs, which can be used to create a variety of physical phenomena and functional devices [21]. One of the main obstacles to CNTs employment in nanomedicine, however, is their poor dispersibility. Multiple functionalization pathways have been created as a result to spread them out and thereby increase their biocompatibility. Covalent functionalization is possible through the defective carbon atoms on the sidewall or at the end, where carboxylic acid groups or carboxylated fractions are created through oxidation and later chemically altered through amination or esterification. Grafting various polymers, metals, and biological molecules onto the surfaces of carboxylated CNTs has recently become popular. Despite the fact that it is widely acknowledged that when the surface of CNTs is properly functionalized (modified), their cell and tissue compatibility can be significantly improved, the biocompatibility (cell and tissue toxicity) of CNTs has been raised as a practical concern, and numerous in-depth studies are still in progress [22, 23].

Biosensing, illness diagnosis, and treatment are just a few of the many biomedical applications where functionalized CNTs have opened up vast possibilities. They are used to deliver therapeutic compounds such as medications and genes as well as to detect a variety of biological targets and enable biomedical imaging. Their inherent spectroscopic characteristics, including photoluminescence and Raman scattering, can offer useful tools for monitoring, identifying, and imaging illnesses. They can also aid in tracking the status of in vivo therapy, pharmacodynamic behavior, and drug delivery effectiveness.

5.2.2.1 Medical Applications of Carbon Nanotubes: Benefits and Drawbacks

The biodistribution and biodegradability of CNTs is a critical topic from the viewpoint of their potential usage in living things. Hydrophilic nanotubes and other nanomaterials [24] that are soluble in human fluids do not have any harmful or deadly effects. It is demonstrated that nanotubes have good biocompatibility, comparable to that of the polysulfone frequently used in medicine. The beneficial electrical properties of single-walled nanotubes are one of the primary distinctions between them and multi-walled nanotubes. The number of layers affects the properties of multi-walled nanotubes. Tensile strength, flexibility, and high heat conductivity are the key benefits of nanotubes in terms of their mechanical, thermal, kinetic, and electric properties (Fig. 5.4).

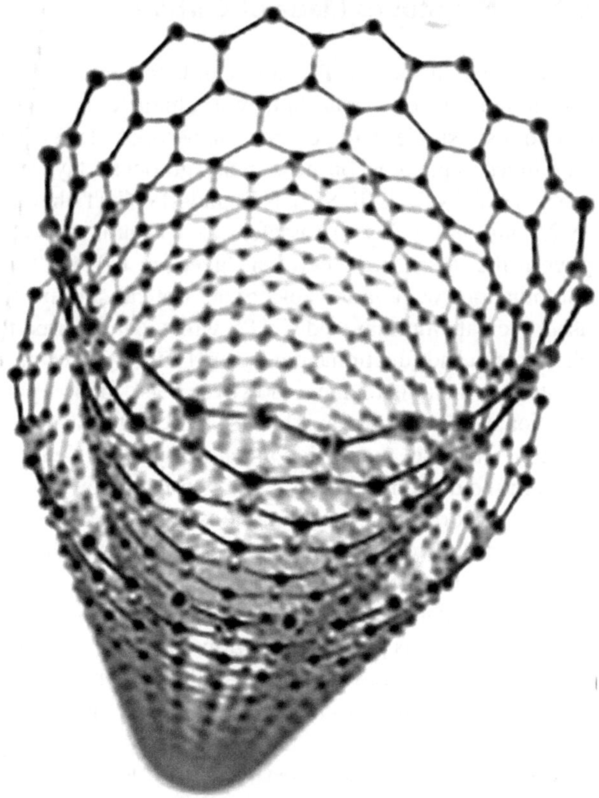

Fig. 5.4 Applications of carbon nanomaterials for delivery of biologics and therapeutics [19]

Conversely, due to this nanomaterial's physical resemblance to asbestos fibers, there are worries over potential cytotoxicity. Cytotoxicity in CNTs is concentration-dependent [25, 26]. The quality of CNTs is harmful to the cells. SWCNTs that are not fully purified are typically more cytotoxic [27]. Additionally, the toxicity of CNTs may differ depending on their length. It has been demonstrated that the stronger the cytotoxic influence on cells or tissues, the longer the fibers are [28]. It is important to keep in mind that the toxicity of CNTs is influenced by the degree of surface functionalization and the type of functional groups attached.

5.2.3 Graphene and its Byproducts

Another class of sp^2 nanocarbon materials, graphene is a single or few-layered two-dimensional sp^2 bonded carbon sheet that exhibits a wide range of remarkable physics and chemistry features. Graphene has been extensively researched in a variety of sectors ever since it was discovered in 2004. By leveraging the interesting optical, electrical, and chemical capabilities of graphene, numerous graphene-based biosensors have been developed to detect biomolecules with high sensitivity

[29]. For use in drug delivery, graphene's poly-aromatic surface structure with its extremely high surface area allows for the efficient loading of aromatic drug molecules via p-p stacking. It has mechanical rigidity of 1060 GPa and a thermal conductivity of up to 3000 W m^{-1} K^{-1} [30]. Individual graphene sheets exhibit exceptional electrical transport capabilities, according to recent findings. Incorporating graphene sheets in a nanocomposite substance is one approach that might be taken to take advantage of these characteristics for biomedical purposes. Additionally, the perfect nanometer-sized graphene fragments known as graphene quantum dots (GQDs) have unique features that make them desirable candidates for a wide range of innovative biomedical applications. These qualities include being effective electro-catalysts and attractive fluorophores because of their photostability, water-solubility, biocompatibility, nontoxicity, and profitability [31].

Since virgin graphene, despite having excellent electrical conductivity, is weakly soluble in water, derivatives of graphene have been developed, including graphene oxide, reduced graphene oxide, few-layer graphene oxide, and chemically altered graphene. GO and rGO are good choices for the synthesis of biocompatible nanocomposites and are substantially more well-suited to sol-gel chemistry. The Hummers method can be used to produce large surface area single layers of GO sheets, which are the oxygenated equivalents of one-atom thick graphene sheets. A wide range of biotechnologies, including drug delivery, nanoprobes, cellular imaging, and biosensors, have made use of GO. The functional groups at the edges and on the surface of GO cooperate to block electron transfer. Because of this, the electrical conductivity of GO is lower than that of its reduced form (rGO). GO-based nanocarriers have received a lot of attention recently for anticancer medication delivery and imaging due to their high drug loading and effective delivery capability. They have a specific surface area that is more than twice as large as the majority of nanomaterials, at roughly 2600 m^2 g^{-1}. In contrast to pure graphene, GO also exhibits excellent water dispersibility and a pH-dependent negative surface charge to maintain strong colloidal stability. However, salt media like phosphate buffered saline and cell culture media that are high in protein might cause GO to aggregate. One more intriguing quality of many aromatic medicinal compounds, including doxorubicin, a strong anticancer medication, can be loaded using GO, which is physisorption via p-p stacking [32, 33]. GO is a potential material for biological applications because of its tiny size, inherent optical characteristics, high specific surface area, low cost, and advantageous non-covalent interactions. Additionally, GO can release medication molecules in response to stimuli like NIR light, which strengthens the case for using it as a delivery vehicle. To fully comprehend the in vivo features of GO, including its blood circulation, inflammatory reactions, and clearance mechanism, more thorough research will be required.

5.2.3.1 The Benefits and Drawbacks of Graphene Use in Medicine

The numerous properties of graphene and materials made from it can be easily changed. Particular characteristics including biocompatibility, rigidity, flexibility, and thermal properties have drawn a lot of attention. Theranostics and controlled medication delivery systems both use nanoparticles because of their precise timing and increased permeability. Researchers have developed graphene-based sensors for detecting biological molecules, thanks to graphene's capacity to hold pharmacological compounds with aromatic rings, interaction with nucleic acids, and potential for attaching diverse chemically active groups to the fundamental lattice structure (and ultimately selected cells). In addition to being employed as surfaces for enzyme immobilization and medication packaging, graphene derivatives are also used as vectors in gene treatments [34, 35].

Since carbon-based nanomaterials cannot biodegrade, potential toxicity and environmental risks are worries. However, Kotchey et al.'s research was able to demonstrate that GO was susceptible to biodegradation by using an oxidative attack with hydrogen peroxide and horseradish peroxide. Unfortunately, this has not yet been done in the instance of the R-GO. This research could aid designers in developing safer biodegradable graphene-based products, lowering the dangers to both human health and the environment. There are various applications for pristine graphene in biomedical research that are not feasible because of its hydrophobicity and unstable homogenous dispersion [36]. The detrimental biological effect (such as apoptosis, increased cell death, cytoskeleton abnormalities, or inflammation) appears to be reduced by thorough cleaning, proper reducing agent or surfactant selection, and adequate material preparation. Additionally, the discovered results can be used to cutting-edge therapeutics, such as novel anticancer medications that cause cancer cells to undergo regulated apoptosis.

5.2.4 Atomic Carbon Dots

Carbon quantum dots (CQDs) are tiny carbon nanoparticles with a size of less than 10 nm. The first study on quantum-sized dazzling and colorful photoluminescence CQDs, using surface passivation and laser ablation of a carbon target, was published by Sun et al. in 2007 [37]. Recently, intensive study has been conducted on 32 CQDs to produce high fluorescence quantum yields (QY) using straightforward synthesis methods. Synthesis of organic materials, such as natural polymers (from sources like chitosan, gelatin, and other sources), has frequently been used to make CQDs. Additionally, vegetables, grape peel, apple juice, and amino acids have been used to make CQDs. Several simple and affordable processes have also been developed for the synthesis of CQD, such as laser ablation, electrochemical oxidation, combustion/thermal microwave heating, aided synthesis, chemical oxidation, hydrothermal carbonization, and pyrolysis. Utilizing a one-step solvothermal approach (referred to as DMEU), Nitrogen-doped carbon quantum dots (CQDs) were successfully synthesized with uniform characteristics. This synthesis method employed nitrogen-rich solvents like N-methyl-2-pyrrolidone (NMP) and dimethyl-imidazolidinone. Recently, a straightforward chemical procedure was developed to produce -C(O)OH-modified CQDs, which is expected to be a unique method for attacking CNTs with acid. In order to modify the fluorescence properties of the

CQDs, additional surface defects have been added, their size has been altered, and chemical changes have been made.

5.2.4.1 Medical Applications of Carbon Quantum Dots: Benefits and Drawbacks

CQDs can be produced in a variety of ways, offering full control over the synthesis procedure and associated costs. They successfully inhibit a number of cancer types, including MCF-7 and MDA-MB-231 (human breast cancer cells), it was also discovered. The simplicity of adding alterations and modifying them for absorption in accordance with cell needs is another benefit of this carrier. A further indication that CQDs are suited for serving as photosensitizers is the possibility that they concentrate selectively within tumors, have a high tumor-to-background fluorescence contrast, and have low fluorescence levels in other tissues and organs. An important advantage of this carbon nanomaterial is its great solubility.

Depending on the type of surface functional group that may have an impact, nonspecific cellular binding is one of the main downsides. Additionally, nonspecific protein adsorption of 66 kDa serum protein albumin from conventional blocking solutions may occur on the surface of quantum dots [38].

5.2.5 Nanodiamonds

The fact that NDs are composed of tetrahedrally bonded carbon atoms arranged in a three-dimensional (3D) cubic lattice gives them the properties of diamond and an onion-shaped carbon shell with a layer of functional groups on their surface.

Bonds between sp^2 and sp^3 exist. NDs are extremely flexible and can adopt two different geometrical shapes. A stretched diamond face, for instance, may behave like a graphene plane, while a puckered graphene sheet may mimic a diamond surface. The unique characteristics of NDs are incredibly intriguing, and they get better as they go smaller. According to theoretical study, NDs with a band gap less than 2 nm will have quantum confinement implications. The size and properties of NDs are impacted by the synthetic process, as was previously mentioned (Fig. 5.5).

It is easy to functionalize NDs with a variety of ligand molecules that act as platforms for the conjugation of different biological molecules, chemicals, and medicines. Due to their extensive surface area, straightforward functionalization, and simplicity of doping, NDs have been researched as theranostic agents. Nitrogen-vacancy (NV) defect centers, which are NDs that include an atom of nitrogen next to a vacancy, are what give NDs their optical characteristics and make it possible for them to be employed as photoluminescent probes. By blasting NDs with high energy particles like electrons, protons, and helium ions, vacuum annealing at 600–800 °C is utilized to create NV centers. Both of these particles produce vacancies that migrate and are eventually captured by nitrogen atoms found in diamonds. The NV centers also emit a strong glow between 550 and 800 nm. NDs are a useful fluorescent probe for single-particle tracking in heterogeneous situations due to their

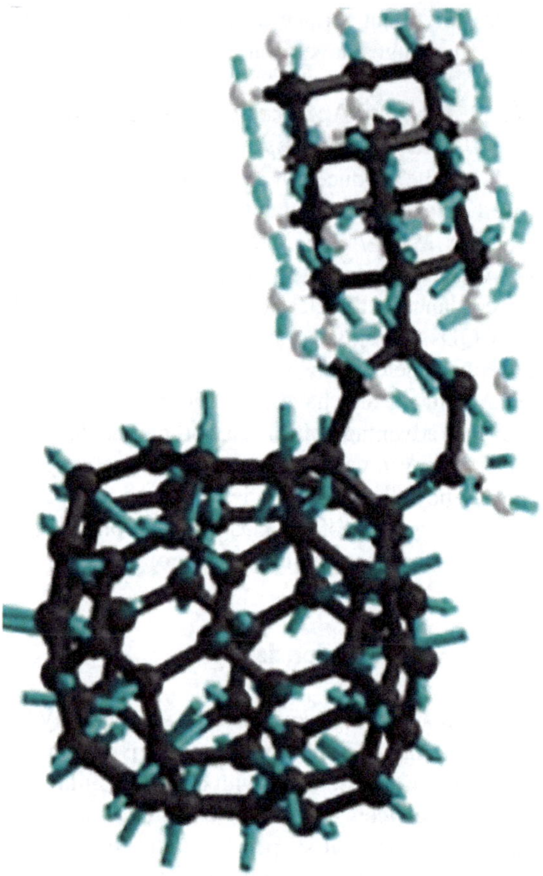

Fig. 5.5 As a complement to fullerenes and carbon nanotubes, diamondoids (nanometer-sized, hydrogen-terminated diamond hydrocarbons, or nanodiamonds) are now being developed [38]

high emission quality and low cytotoxicity. When functionalized, their biocompatibility is known to be superior to that of carbon black and single- and multi-walled CNTs (Table 5.1) [32].

5.3 Why Imaging and Delivery Are Necessary

The potential of CBN as platforms for therapeutic and one-/two-photon imaging has recently come to light. The strong chemical inertness, straightforward functionalization, high photobleaching resistance, biocompatibility, large surface area, and high two-photon absorption cross-section of CBNs over QDs and conventional fluorescence dyes are just a few of their many advantages. These features are especially helpful for use as theranostic nanoparticles. One of their optical modes, two-photon fluorescence microscopy (TPFM), offers a number of benefits over traditional fluorescence imaging methods, including superior 3D spatial resolution,

Table 5.1 Different types of carbon nanostructure, as well as their variations, derivatives, and theranostic uses

	Graphene family nanomaterials	Fullerenes	Carbon nanotubes	Carbon quantum dots
Derivatives and members	Mono-, bi-, multilayer graphene, graphene oxide (GO), reduced (GO)	C60, C70, and other endohedral metallofullerenes	Single-walled carbon nanotubes (SWCNTs), multi-walled carbon nanotubes (MWCNTs)	Graphene quantum dots, carbon quantum dots
Modification	Oxidation: PEG, folic acid, and chitosan to -COOH Ether-carboxyl cycloaddition to -OH Non-covalent surface coating through sorption Ni/au Aptamers Therapeutic substances Polymers (PEI) (PEI) Protein, BSA, and Fe_3O_4	Poly-hydroxylation and carboxylation are examples of oxidation PEGylation Particular covalent functionalization Glycosylation Anions and cations Lanthanide ion chelation and encapsulation, such as Gd^{3+} The combination of cyclodextrin sorption	Ozone oxidation to -COOH: Folic acid, PEG Cycloaddition Non-covalent surface coating through sorption Protein, BSA Glycolipids Surfactants PEGylated phospholipids with Fe_2O_3 DNA Polysaccharides Surface amplification	Covalent bonding of amine-containing agents to oxygen-containing groups. Chelation of lanthanide ions (e.g., Eu^{3+}) Sol–gel technique Doping with N, S, and P elements Nanohybrids with inorganic nanoparticles (e.g., iron oxide, zinc oxide, silica, and titania)
Use of theranostics	Imaging and delivery of drugs/genes A photodynamic treatment Tissue manipulation bacterial resistance Biosensing	Drug/gene delivery Magnetic resonance, photoacoustic imaging Photodynamic, photothermal therapy Enzyme inhibition Antioxidants	Acoustic photothermal treatment Imaging, magnetic resonance Peptide, gene, and drug delivery, biosensing	Drug/gene delivery and imaging Photodynamic therapy Biosensing
References	[39, 40]			

deep-tissue penetration, and little tissue autofluorescence or phototoxicity. To be effective in TPFM, two-photon fluorophore nanomaterials must have a wide two-photon absorption cross-section, high fluorescence quantum yield, low photo-decomposition quantum yield, low toxicity, and low photo-leaching. Over the past few decades, significant effort has been made to design and produce two-photon fluorophores with a variety of compositions, including organic dyes, quantum dots, gold nanoparticles, and carbon-based compounds. A promising nanoplatform for therapies and two-photon imaging is being developed by CBN among them [41]. In order to find the optimal bioimaging and diagnostic applications, it is therefore necessary to investigate the origins of CBNs photo-reactivity as well as a variety of techniques for modifying their optical properties [42].

5.4 Carbon Nanomaterials for Treatment of Cancer (Specifically Used for Breast Cancer)

5.4.1 Drug Release

Chemotherapy and hormone therapy are the two main cancer treatments used to get rid of or control cancer cells all over the body. The term "chemotherapy" refers to the intravenous or oral administration of chemotherapy drugs. Chemotherapeutic drugs can have a number of adverse effects, including hair loss, nausea, vomiting, tiredness, and an increased risk of infection. This is because, despite their extraordinarily high efficacy, these drugs are not selective for cancer cells. Taxanes (paclitaxel and docetaxel) and anthracyclines (doxorubicin and epirubicin) are the most commonly utilized chemotherapy drugs. Between 60 and 70 percent of all breast cancer cases involve hormonal therapy, which is the use of medications like estrogen to block hormones that promote the growth of cancer cells. The efficacy of this kind of treatment depends on the biological characteristics of people with hormone receptor-positive (estrogen-receptor-positive or ER+) condition. Fulvestrant, tamoxifen (TAM), and aromatase inhibitors are typical drugs. TAM is a "antiestrogen," more particularly a selective estrogen receptor modulator (SERM), which binds to the estrogen receptors on breast cancer cells to block the impact of the hormone. Since it has been shown to be successful in both early and advanced stages in women of different ages, TAM is currently utilized to treat the majority of ER+ breast cancers. TAM, though, also has drawbacks [43, 44].

While treating breast cancer, some of the major limitations of standard drugs include their unfavorable pharmacokinetics, restricted biodistribution, lack of selectivity, and tissue damage. These side effects might be prevented if drugs were only given to specific body areas. As a result, the dosage of pharmaceuticals and their concentration in unimportant areas reduced the side effects of chemotherapy and hormone therapy. In this regard, research is being done on the possibility of using carbon nanoparticles to provide medications for the treatment of cancer and breast cancer. These substances are thought to be effective mechanisms for regulating drug release in the body, enhancing the efficacy of treatments, and reducing their toxicity. Carbon nanomaterials, however, have certain limitations due to their chemical functionalization and structural characterization, including their insolubility and slow reproduction rate. Their use in medicine has also been thoroughly tested due to their potential toxicity. Carbon nanoparticles are exceedingly poorly soluble in aqueous or biological environments when they are in their native condition or have not been chemically functionalized. In the latter case, the simple or complex molecule aggregates that give the nanomaterial its special features are not present. Since typical drugs have a propensity to accumulate excessively in biological settings and have a high vulnerability to enzymatic oxidation, which makes it challenging for the body to discard them, insolubility is significant in toxicology research. As a result, fullerenes, CNTs, and graphene are being chemically functionalized to a greater extent, increasing their solubility and reducing their toxicity in living organisms, making them more suitable for the administration and delivery of drugs. CNTs are of great interest to scientists due to their large surface area, high surface area, tubular structure, chemical stability, mechanical resistance, and ability to pass endothelial, vascular, and tumor cell membranes [45, 46].

Due to their surface and strong electrical and thermal conductivity, CNTs can bind to the walls and extremities of molecules including drugs, peptides, and nucleic acids. Therefore, CNTs are considered a possible replacement for the administration and delivery of medication. The very hydrophobic and insoluble fullerene can even be chemically functionalized to become more soluble. For instance, cationic-functionalized fullerenes have a stable solubility and are more effective at binding anionic residues in cancer cells, enabling the targeted and controlled delivery of drugs.

5.4.2 Contrast Agent

Today's medical community values early detection of breast cancer tumors since, in the vast majority of cases, it makes disease eradication possible. The contrast agent technique has received a lot of attention since it allows for the real-time detection of breast tumors. In order to identify antigens expressed on the surface of specific malignancies, this approach employs specific antibodies. These specific antibodies are added to fluorescent molecules or fluorophores, enabling them to absorb electromagnetic radiation (light) at specific wavelengths and emit it at longer ones. Two common contrast agents (CAs) used in magnetic resonance imaging (MRI) are gadolinium complexes and superparamagnetic iron oxides; however, they normally cannot permeate cell membranes. This packaging on or inside of carbon nanostructures allows CAs to be taken up by cells. Although gadolinium, a lanthanide metal, is the most often used contrast agent in MRI, new research has focused on reducing the side effects and dosage requirements of this metal while also enhancing the selectivity of desired organs and tissues. Carbon nanoparticles have been shown to be 1–2 orders of magnitude more effective than regular CAs and to produce dose reductions. Carbon nanostructures also have the capacity to cross cell membranes and combine with pharmaceuticals. Therefore, potent medical imaging together with efficient contrast agents must be created in order to increase the sensitivity and specificity of breast cancer diagnosis.

Carbon nanotubes have been shown to be an appropriate scaffold for contrast agents because they can be utilized to hold imaging agents like fluorescent labels and radionuclides. This aids in the creation of high-quality pictures and the early detection of breast cancer. According to research, CNT can transport imaging chemicals, and functionalized MWCNT can be employed as efficient cell probes for MRI. In vivo MRI on mice showed that CNTs functionalized with Gd (Gd-CNTs) have the ability to serve as a useful contrast agent. Additionally, phospholipid functionalization increases the biocompatibility of CNT, making it more stable while circulating in the reticuloendothelial system. Fullerene and its derivatives have recently been recognized as one of the most promising agents for cancer imaging due to its distinct physicochemical properties. In this regard, polyhydroxy fullerenes (PHF) have shown their potential value for cancer imaging. Additionally, PHF nanoparticles functionalized with amines (PHF-NH2) were discovered to be water soluble and to exhibit green fluorescence with a quantum yield of roughly 17%. Additionally, due to their high surface charge, PHF-NH2 nanoparticles have excellent fluorescence properties and may infiltrate cell lines quickly, including breast cancer cells (MCF-7) in vitro. In addition, endohedral fullerenes, which are enclosed carbon structures resembling cages and containing straightforward or intricate molecules, have been successfully employed as MRI contrast agents [47]. Chelates are commonly employed in MRI studies due to their outstanding contrast capabilities. These chemicals contain gadolinium (Gd + 3) and other heavy metal ions in their structures. The frequent metal loss or transmetalation processes (release of Gd + 3 in the body) that these people undergo have a negative impact on patients with renal or hepatic failure. Recent studies suggest that this can be avoided by introducing Gd + 3 molecules into the cavity of the fullerene cage before functionalization. This makes the chelate contained in the nanomaterial very soluble and non-toxic [48].

One of these fullerenes, the metallofullerenol Gd@C82(OH)22, has a potent immune system-activating effect on T cells when utilized as a contrast agent in MRI. Metallofullerenol successfully suppresses the growth of tumors (paclitaxel) while being less hazardous than several common anticancer drugs used to treat breast cancer, such as taxol. In order to carry chemotherapeutic drugs or imaging agents for the diagnosis and treatment of breast cancer, fullerenes, graphenes, and CNTs can all be functionalized [49, 50]. This provides an added benefit for these systems and makes it possible for them to support the detection of cancer as well as the targeted and careful delivery of medication.

5.4.3 Photodynamic Therapy

Surgery and radiotherapy are two breast cancer treatments that are routinely used to remove, eliminate, or suppress cancer cells in a specific place. Contrary to radiotherapy, which is frequently recommended after surgery to reduce the risk of recurrence, surgery allows for the identification of the extension of the tumor and is frequently used to remove the breast (mastectomy) or the portion of the breast containing the cancerous tumor (breast-conserving surgery). However, since both radiotherapy and surgery require invasive treatments, side effects are frequently reported [51].

A minimally invasive, clinically approved therapy called photodynamic therapy (PDT) has the power to eliminate malignant cells only. PDT requires light, a photosensitizer, and molecular oxygen. The procedure is applying a photosensitizer topically or intravenously to malignant cells, followed by exposure to a certain wavelength of light (within the absorbance band of the photosensitizer). The excited singlet oxygen produced by the electronically excited photosensitizer, which is cytotoxic and damages the tumor microvasculature as well as significantly alters the tumor microenvironment, is created by the transfer of energy from the ground state of molecular oxygen to the excited singlet oxygen. Through apoptosis or necrosis, this leads to the direct destruction of tumor cells [52]. In an effort to replace the invasive thermal procedures now used to eliminate tumors or cancer cells in breast cancer patients, recently, the properties of carbon nanomaterials and their potential use in PDT have been studied extensively. Due to the existence of carbon nanomaterials in PDT, cancer cells develop light-sensitive molecules (carbon nanomaterials), which subsequently absorb infrared radiation from an infrared camera and transform it into heat when oxygen is present. This results in cytotoxic effects that permanently photodamage cancer cells. Carbon nanomaterials have been used to ablate solid tumors in breast. The first study on ZnMCPPc-spermine-SWCNT (zinc mono-carboxy phenoxy phthalocyanine upon conjugation to spermidine, as a targeted molecule, and then adsorbed onto SWCNT) (MCF-7)'s production, photophysical characteristics, and PDT activity was published by Ogbodu et al. [53]. In the absence of light, ZnMCPPc-spermine in SWCNT was found to be relatively safe in an in vitro cytotoxicity assay on cancer cells (MCF-7). The outcomes also demonstrated that PDT is enhanced by the addition of 40 mM spermine, with a 97% reduction in cell viability. Researchers developed a multifunctional system based on fullerene C_{60}, iron oxide nanoparticles, polyethylene glycol, and folic acid that has several uses, such as applications for radiofrequency-assisted heat therapy, PDT, and magnetic targeting. Furthermore, neither in vitro nor in vivo toxicity was discernible in the results of the trials [54]. Nurunnabi et al. [55], however, developed a carboxyl functionalized graphene nanodot that functions as a potential agent for the non-invasive imaging of cancer in deep tissue/organs and efficiently kills breast cancer cells (>70% of MDA-MB231 cell line) through a combination of photodynamic and photothermal effects. Although the results of the aforementioned studies suggest potential uses for carbon nanomaterials in photodynamic treatment, a number of obstacles must yet be overcome before this approach can be widely used to treat breast cancer.

5.4.4 Photothermal Therapy

PTT kills aberrant cells in malignant tissues by causing external light-induced hyperthermia without harming healthy tissue. Through the increased permeability and retention (EPR) effect, carbon nanomaterials like CNTs and graphene selectively concentrate in tumors and exhibit extraordinarily high levels of intrinsic absorbance in the biological transparency windows located within the NIR window

(750–1700 nm) [56]. Additionally, simultaneous imaging and therapy are possible due to the robust fluorescence of carbon nanomaterials, whose emission wavelengths range from visible to NIR-II window. The right surface modified carbon nanomaterials can circulate in vivo with longer half-lives than other nanomaterials like different gold nanostructures, copper sulfide (CuS) nanoparticles, and organic nanoparticles. The first carbon nanomaterial used for PTT was single-walled carbon nanotubes (SWCNT), In a mouse intravenously injected with PEGylated SWCNTs, Choi et al. reported photothermal ablation of human epidermoid mouth carcinoma KB tumor in 2009 [57]. The treated mice had a considerably lower tumor volume than the untreated mice. For effective PTT, Dai's group reported using SWCNTs noncovalently functionalized with PL-PEG and C18-PMH-PEG in 2010 [58]. This method led to a high passive tumor absorption. This was the first study to use SWCNTs to scan tumors in the NIR-II window and the first time these nanomaterials were administered systemically for PTT. Numerous research have since been conducted for PTT in an effort to reduce injection dose and irradiating power density. Due to its significant absorbance in the NIR window, graphene and its derivatives have also been successfully used for PTT in addition to SWCNTs. In 2010, Liu et al. used a fluorescent labeling and imaging approach to examine PEGylated nano-GO behaviors in mice following intravenous (i.v.) injection and discovered the EPR impact of malignant tumors [59]. Because of the strong uptake into the tumor, the photothermal agent nGOPEG was able to effectively remove all tumors when exposed to an 808 nm laser at a power density of 2 W cm^2 for 5 min. Although the injected doses (20 mg kg^1) and NIR laser densities (2 W cm^2) are comparable to those of other photothermal agents like gold nanoparticles, the results showed that nGOPEG looked to be an excellent in vivo tumor NIR PTT agent without demonstrating obvious toxicity. It is widely known that a decrease in GO may result in significantly increased optical absorbance in the NIR range. Dai et al. revealed that ultra-small reduced nanographene oxide (nRGO) could be synthesized by the reduction of PEGylated GO, demonstrating a considerable increase in absorbance in the visible to NIR window. This could minimize the required dosage and laser power density for PTT. The nRGO-PEG could function as a highly effective targeted photothermal agent for in vitro selective cancer cell ablation by further coating with PEGylated phospholipid and conjugating with an RGD-targeted peptide.

5.4.5 Biosensors

It is imperative to create biosensors that can recognize cancerous tumors, particularly for breast cancer, whose early detection is essential for its eradication. Biosensors are capable of detecting tumor markers at such low quantities (traces), while other techniques such as imaging investigations may fall short. Further study has been conducted on the use of carbon nanoparticles as biosensors as a result. Engineers from the University of California, San Diego have developed a graphene sensor that may detect mutations and early-stage breast cancer [60]. According to research from the University of Manchester, graphene oxide is also non-toxic to healthy cells and can be utilized to cure cancer by specifically targeting cancer stem cells. Despite their youth, these techniques have shown positive results in both the prevention and treatment of breast cancer [61].

5.4.6 Molecular Biology/Gene Delivery and Carbon Structures

Gene therapy is the technique of treating disease with therapeutic nucleic acids, to put it simply. Despite being one of the leading causes of death worldwide, chemotherapy is the main treatment for cancer. However, chemotherapy drugs also attack normal tissues in addition to cancerous ones. It follows that the search for safe substitutes for traditional medication to treat cancer and other disorders like infectious diseases, single-gene diseases, and cardiovascular diseases is not surprising [62]. Nucleic acids are often delivered using one of the two techniques: Ex vivo therapy includes injecting the therapeutic nucleic acids into the target cells, letting them grow in a dish, and then injecting them back into the patient. In contrast, nucleic acids can be supplied in vivo by encapsulating them in nanocarriers and then being administered directly to the patient's cells [63]. Genes attached to delivery vectors will improve cell transfection because naked DNA plasmids are prone to degradation and have brief half-lives. Therefore, the vector needs to be secure, deliver nucleic acid to the cytoplasm or nucleus of the target cell, be nuclease-resistant, and be economically feasible.

The transfection rate in viral delivery of nucleic acids (typically adeno-associated viruses and retroviruses) is quite high, despite the fact that nonviral vectors can get over the main issues with viral delivery of genetic material, such as immunogenicity and carcinogenicity [64]. The costs of preparation and the negative consequences of using viral vectors are reduced with the use of nonviral vectors. In addition, protein/nucleic acid complexes like Cas9 (with positive charges)-synthetic single-guide RNA (sgRNA; with negative charges) present more transport difficulties than DNA or mRNA due to their mixed charges. Organic and inorganic nanomaterials are the two main types of nonviral vectors. Examples of organic substances include polyethyleneimine (PEI) and its derivatives, cationic polypeptides, dendrimers, chitosan, polyurethane, and cyclodextrin and its derivatives. Inorganic materials include things like silica nanoparticles, upconverted nanoparticles, carbon-based nanostructures, and gold nanoparticles (AuNPs) [65]. Due to their biocompatibility, high surface-to-volume ratios, ease of functionalization, and acceptable optical characteristics, carbon-based nanoparticles generally have a lot to offer for gene therapy. In general, because to their biocompatibility, high surface-to-volume ratios, ease of functionalization, and acceptable optical characteristics, carbon-based nanoparticles have a lot to offer for gene therapy [66]. Small DNA molecules (iC9 suicide gene, antisense oligonucleotides), large DNA molecules (plasmid DNA), and ribonucleic acid (RNA) molecules (such as small interfering RNA [siRNA], short hairpin [shRNA], micro-RNA [miRNA], and antisense mRNA) can all be delivered to the desired target by carbon nanostructures when their surfaces are properly surface-functionalized.

Currently, 33 gene-therapy drugs or products have FDA approval for use in clinical settings, while more than 3500 clinical trials are still active globally. Obstacles to the transfer of genes using nonviral vectors include blood flow, opsonization, endocytosis by the mononuclear phagocytic system, tissue pressure, endosomal escape, gene transfer effectiveness, specificity, length of gene expression, and safety [67]. As a result, the carbon-based nucleic acid nanocarrier can disintegrate or be intercepted before getting to the desired cell. One way to get around the problems is surface functionalization. Additional challenges include the size of the nanoparticles involved in cellular uptake and the mechanism of cellular uptake. After entering the cells, nano-vectors present a new issue since they start to break down in the lysosome and endosome environments, which are acidic and full of enzymes.

5.5 Toxicity/Concerns Related to Carbon-Based Compounds

The drawbacks of using carbon-based nanostructures for drug/gene delivery have been documented in a number of studies. These drawbacks include issues with autophagy, severe acute inflammatory responses in mice when using carbon dots, heart rate reduction and disruption of embryonic development in wild-type zebrafish when using graphene quantum dots, and an increase in reactive oxygen species (ROS) [68].

The toxicity profile of a carbon-based nanoparticle, however, may differ depending on its physical properties, including size, surface area, dosage, length to diameter ratio, duration, purity, and chemical functional groups on its surface [69]. As a result, it is crucial to consider the differences between each carbon nanoparticle's toxicity profile in great detail. Size, shape, and functionalization have a significant impact on the cytotoxicity profiles of nanoparticles. Small MWCNTs with a mean length of 100–600 nm demonstrated lower levels of toxicity against human umbilical vein endothelial cells in contrast to longer MWCNTs with a length of 200–2000 nm. Even while other research supports the same result that cytotoxicity is reduced with shorter length and diameter of CNTs, several studies does not support the argument [70]. It is a hotly debated topic because the vast majority of studies in the literature fail to adequately address material purity, the manner in which nanoparticles are administered to animal models, and other elements that affect the toxicity profile of carbon-based nanoparticles. Furthermore, GO functionalized with NH2 is more harmful than GO that has not been altered. To better understand the toxicological consequences of CbNMs in a specific human organ, tissue, or cell, a complete understanding of the cellular and molecular interactions with the target tissue or organ is required. The physicochemical characteristics of these nanostructures, how they interact with cells, and how much of them accumulate in a given tissue or organ all have a bearing on these interactions.

Carbon nanostructures can be used to resolve worries about the immunogenicity and carcinogenicity of viral vectors of DNA fragments. To solve their insufficient hydrophilicity, surface functionalization should be applied. Thanks to hybridization and surface functionalization, carbon-based nanostructures can be rendered more soluble in aqueous solutions, have higher transfection rates in the cells they are meant to reach, and have their genetic contents protected against degradation in circulation. For use in clinical studies, further work needs to be done to increase the carbon-based genetic delivery systems' transfection rates as well as their toxicological and pharmacokinetic characteristics. Because relying solely on in vivo studies is insufficient for the successful and efficient clinical translation of these CbNMs because in vivo models do not accurately recapitulate the microenvironment of the human body systems, it is unavoidable to expose a large human population to these nanomaterials in order to monitor the degree of biocompatibility and cytotoxicity to host cells.

5.6 Conclusion

Research, design, manufacture, synthesis, manipulation, and usage of carbon nanomaterials are emerging technologies with potential applications in the field of biomedicine, particularly for the detection, diagnosis, and treatment of breast cancer. But as of right now, there is not a carbon nanomaterial that possesses all the necessary qualities for use in human medicine, including nontoxicity, good solubility, high specificity, ease of functionalization, targeted and controlled delivery system, etc. In order to avoid potential harmful effects, additional research is also necessary. Even while research on the toxicological properties and biological applications of carbon nanomaterials is expanding, there is still little proof of their relevance to human features. The research findings mentioned here are part of human endeavors that promote the future application of nanotechnology-assisted therapies for breast cancer, the prevention of breast cancer's spread and recurrence, and the reduction of its risky side effects. The development of the medical sciences depends on our ability to make use of the information and resources available to us. Previous studies have shown that carbon nanostructures, their derivatives, and their altered forms have theranostic potential. A well-designed nano-system can be used for both therapeutic and diagnostic purposes. Carbon nanomaterials give us the ability to enhance and customize products based on their capabilities. The investigations have already demonstrated that their optical, physical-chemical, and interfacial properties allow us to use them to imagine multimodal and selective non-invasive tumor eradication. Despite the fact that each type of nano-system seems to have a unique advantage, platforms can be built that have even superior capabilities for tumor imaging, monitoring, and detection.

References

1. Suhag D, Chauhan M, Shakeel A, Das S (2020) Emerging trends in nanotheranostics. NanoBioMedicine:335–368
2. Suhag D, Shahdeo D, Gandhi S (2019) Emerging trends and advances in targeted nanotheranostics. SMC Bulletin 10:17–22
3. Thakur P, Thakur A (2022) Nanomaterials, their types and properties. In: Thakur A, Thakur P, Khurana SP (eds) Synthesis and applications of nanoparticles. Springer, Singapore, pp 19–44. https://doi.org/10.1007/978-981-16-6819-7_2
4. Pinto AM, Gonçalves IC, Pestana M (2013) Graphene-based materials biocompatibility: a review. Colloids Surf B: Biointerfaces 111:188–202
5. Yang P, Gai S, Lin J (2012) Functionalized mesoporous silica materials for controlled drug delivery. Chem Soc Rev 41(9):3679
6. Thakur P, Sharma P, Mattei JL et al (2018) Influence of cobalt substitution on structural, optical, electrical and magnetic properties of nanosized lithium ferrite. J Mater Sci Mater Electron 29: 16507–16515
7. Pathania A, Rana K, Bhalla N et al (2017) Raman and Mössbauer spectroscopic studies of tungsten doped Ni–Zn nano ferrite. J Mater Sci Mater Electron 28:679–685
8. Dhanda N, Thakur P, Sun ACA, Thakur A (2023) Structural, optical and magnetic properties along with antifungal activity of Ag-doped Ni-Co nanoferrites synthesized by eco-friendly route. J Magn Magn Mater 572
9. Ruoff RS, Lorents DC (1995) Mechanical and thermal properties of carbon nanotubes. Carbon 33(7):925–930
10. Collins PG, Bradley K, Ishigami M, Zettl A (2000) Science 287:1801–1804
11. Patel KD, Singh R, Kim H (2019) Carbon-based nanomaterials as an emerging platform for theranostics. Mater Horiz 6(3):434–469
12. Margadonna S, Prassides K (2001) Fullerenes. In Elsevier eBooks (pp. 3379–3383)
13. Delgado JL, Filippone S, Giacalone F, Herranz M, Illescas BM, Pérez EM, Martín N (2013) Buckyballs. In: Topics in current chemistry. Springer, pp 1–64
14. Accorsi G, Armaroli N (2010) Taking advantage of the electronic excited states of [60]-fullerenes. J Phys Chem C 114(3):1385–1403
15. Birkett PR (2001) Fullerene chemistry. In Elsevier eBooks (pp. 3367–3371)
16. Benn TM, Westerhoff P, Herckes P (2011) Detection of fullerenes (C60 and C70) in commercial cosmetics. Environ Pollut 159(5):1334–1342
17. Thakur A, Verma R, Wan F, Ravelo B, Edelman I, Ovchinnikov S, Thakur P (2023) Investigation of structural, elastic and magnetic properties of Cu2+ ions substituted cobalt nanoferrites. J Magn Magn Mater 581:170980
18. Singh A, Wan F, Yadav K, Salvi A, Thakur P, Thakur A (2023) Synergistic effect of ZnO nanoparticles with Cu2+ doping on antibacterial and photocatalytic activity. Inorg Chem Comm 157:11425
19. Mahor A, Singh PP, Bharadwaj P, Sharma N, Yadav S, Rosenholm JM, Bansal KK (2021) Carbon-based nanomaterials for delivery of biologicals and therapeutics: a cutting-edge technology. C, 7(1), Article 19. http://creativecommons.org/licenses/by/4.0/
20. Pavelyev V, Tripathi N (2022) Optical characterization of nanomaterials-II. In: Thakur A, Thakur P, Khurana SP (eds) Synthesis and applications of nanoparticles. Springer, Singapore, pp 151–176
21. Avouris P, Hertel T, Martel R, Schmidt TOB, Shea H, Walkup RE (1999) Carbon nanotubes: nanomechanics, manipulation, and electronic devices. Appl Surf Sci 141(3–4):201–209
22. Meng L, Fu C, Lu Q (2009) Advanced technology for functionalization of carbon nanotubes. Prog Nat Sci Mater Int 19(7):801–810
23. Lacerda L, Bianco A, Prato M, Kostarelos K (2006) Carbon nanotubes as nanomedicines: from toxicology to pharmacology☆. Adv Drug Deliv Rev 58(14):1460–1470

24. Kala D, Sharma TK, Gupta S, Verma V, Thakur A, Kaushal A, Trukhanov AV, Trukhanov SV (2021) Graphene oxide nanoparticles modified paper electrode as a biosensing platform for detection of the htrA gene of O. tsutsugamushi. Sensors 21:4366
25. Tsai P, Kuo H, Chiu W, Wu J (2013) Purification and functionalization of single-walled carbon nanotubes through different treatment procedures. J Nanomater 2013:1–9
26. Cataldo F, Da Ros T (2008) Medicinal chemistry and pharmacological potential of fullerenes and carbon nanotubes. In Carbon materials. Springer Nature (Netherlands)
27. Heller DA, Baik S, Eurell TE, Strano MS (2005) Single-walled carbon nanotube spectroscopy in live cells: towards long-term labels and optical sensors. Adv Mater 17(23):2793–2799
28. Lacerda L, Russier J, Pastorin G, Vázquez E, Venturelli E, Dumortier H, Lehto V, Prato M, Kostarelos K, Bianco A (2012) Translocation mechanisms of chemically functionalized carbon nanotubes across plasma membranes. Biomaterials 33(11):3334–3343
29. Almessiere MA, Slimani Y, Thurkanov AV, Baykal A (2022) Structural and morphological characterization of nanomaterials. In: Thakur A, Thakur P, Khurana SP (eds) Synthesis and applications of nanoparticles. Springer, Singapore. https://doi.org/10.1007/978-981-16-6819-7_6
30. Dresselhaus MS, Dresselhaus G (1981) Intercalation compounds of graphite. Adv Phys 51(1): 1–186
31. Sangam S, Garg P, Sanyal T, Pahari S, Khurana SMP, Mukherjee M (2022) Graphene quantum dots and their hybrid hydrogels: a multifaceted platform for theranostic applications. In: Thakur A, Thakur P, Khurana SP (eds) Synthesis and applications of nanoparticles. Springer, Singapore. https://doi.org/10.1007/978-981-16-6819-7_20
32. Mochalin V, Pentecost A, Li X, Neitzel I, Nelson MN, Wei C, Chen Y, Guo F, Gogotsi Y (2013) Adsorption of drugs on nanodiamond: toward development of a drug delivery platform. Mol Pharm 10(10):3728–3735
33. Xu B, Mu Y, Mao Z, Xie Z, Wu H, Zhang Y, Jin C, Chi Z, Liu S, Xu J, Wu Y, Lu P, Lien A, Bryce MR (2016) Achieving remarkable mechanochromism and white-light emission with thermally activated delayed fluorescence through the molecular heredity principle. Chem Sci 7(3):2201–2206
34. Priyadarsini S, Mohanty S, Mukherjee S, Basu S, Mishra M (2018) Graphene and graphene oxide as nanomaterials for medicine and biology application. J Nanostruct Chem 8(2):123–137
35. Hom C, Lu J, Liong M, Luo H, Li Z, Zink JI, Tamanoi F (2010) Mesoporous silica nanoparticles facilitate delivery of siRNA to shutdown Signaling pathways in mammalian cells. Small 6(11):1185–1190
36. Banerjee AN (2018) Graphene and its derivatives as biomedical materials: future prospects and challenges. Interface Focus 8(3):20170056
37. Sun Y, Zhou B, Lin Y, Wang W, Fernando KAS, Pathak P, Meziani MJ, Harruff BA, Wang X, Wang H, Luo PG, Yang HG, Köse ME, Chen B, Veca LM, Xie S (2006) Quantum-sized carbon dots for bright and colorful photoluminescence. J Am Chem Soc 128(24):7756–7757
38. Randel JC, Niestemski F, Botello-Méndez AR, Mar W, Ndabashimiye G, Melinte S, Dahl JJ, Carlson RH, Butova ED, Fokin AA, Schreiner PR, Charlier J, Manoharan HC (2014) Unconventional molecule-resolved current rectification in diamondoid–fullerene hybrids. Nat Commun 5:1
39. Bitounis D, Ali-Boucetta H, Hong BH, Min D, Kostarelos K (2013) Prospects and challenges of graphene in biomedical applications. Adv Mater 25(16):2258–2268
40. Wang Y, Zhang S, Xu T, Zhang T, Yuru M, Liu J, Yan L, Xing F (2018) Ultra-sensitive and ultra-fast detection of whole unlabeled living cancer cell responses to paclitaxel with a graphene-based biosensor. Sens Actuators B Chem 263:417–425
41. Shen Y, Shuhendler AJ, Ye D, Xu J, Chen H (2016) Two-photon excitation nanoparticles for photodynamic therapy. Chem Soc Rev 45(24):6725–6741
42. Bhalla N, Taneja S, Thakur P, Sharma PK, Mariotti D, Maddi C, Ivanova O, Petrov D, Sukhachev A, Edelman IS, Thakur A (2021) Doping independent work function and stable

band gap of spinel ferrites with Tunable plasmonic and magnetic properties. Nano Lett 21(22): 9780–9788
43. Davies C, Godwin JD, Gray R, Clarke M, Cutter DJ, Darby SC, McGale P, Pan H, Taylor CB, Wang Y, Dowsett M, Ingle JN, Peto R (2011) Relevance of breast cancer hormone receptors and other factors to the efficacy of adjuvant tamoxifen: patient-level meta-analysis of randomized trials. Lancet 378(9793):771–784
44. Clarke M, Collins RW, Darby SC, Davies C, Elphinstone P, Evans V, Godwin JD, Gray R, Hicks C, James SJ, Mackinnon E, McGale P, McHugh TD, Peto R, Taylor CB, Wang Y (2005) Effects of radiotherapy and of differences in the extent of surgery for early breast cancer on local recurrence and 15-year survival: an overview of the randomized trials. Lancet 366(9503): 2087–2106
45. Son KH, Hong JH, Lee JY (2016) Carbon nanotubes as cancer therapeutic carriers and mediators. Int J Nanomedicine 11:5163–5185
46. Verma R, Thakur P, Sun AC, Thakur A (2023) Investigation of structural, microstructural and electrical characteristics of hydrothermally synthesized Li0.5-0.5xCoxFe2.5-0.5xO4 ferrite nanoparticles. Phys B: Cond Mater 661:414926
47. Meng J, Liang X, Chen X, Zhao Y (2013) Biological characterizations of [Gd@C82(OH)22]n nanoparticles as fullerene derivatives for cancer therapy. Integr Biol 5(1):43–47
48. Liu Y, Chen C, Qian P, Lu X, Sun B, Zhang X, Wang L, Gao X, Li H, Chen Z, Tang J, Zhang W, Dong J, Bai R, Lobie PE, Wu Q, Liu S, Zhang H, Zhao F et al (2015) Gd-metallofullerenol nanomaterial as non-toxic breast cancer stem cell-specific inhibitor. Nat Commun 6:1
49. Faraj AA, Shaik AS, Halwani R, Alfuraih A (2016) Magnetic targeting and delivery of drug-loaded SWCNTs theranostic nanoprobes to lung metastasis in breast cancer animal model: noninvasive monitoring using magnetic resonance imaging. Mol Imaging Biol 18(3):315–324
50. Chen H, Ma YX, Li Z, Shi Q, Zheng W, Liu Y, Wang P (2012) Functionalization of single-walled carbon nanotubes enables efficient intracellular delivery of siRNA targeting MDM2 to inhibit breast cancer cells growth. Biomed Pharmacother 66(5):334–338
51. Girish C, Vijayalakshmi P, Mentham R, Rao CB, Nama S (2014) A review on breast cancer. Int J Pharm Bio Sci 4(2):47–54
52. Dolmans DEJGJ, Fukumura D, Jain RK (2003) Photodynamic therapy for cancer. Nat Rev Cancer 3(5):380–387
53. Ogbodu RO, Limson J, Prinsloo E, Nyokong T (2015) Photophysical properties and photodynamic therapy effect of zinc phthalocyanine-spermine-single walled carbon nanotube conjugate on MCF-7 breast cancer cell line. Synth Met 204:122–132
54. Shi J, Wang L, Gao J, Liu Y, Zhang J, Ma R, Liu R, Zhang Z (2014) A fullerene-based multifunctional nanoplatform for cancer theranostic applications. Biomaterials 35(22):5771–5784
55. Nurunnabi, Khatun Z, Reeck GR, Lee DH, Lee Y (2014) Photoluminescent graphene nanoparticles for cancer phototherapy and imaging. ACS Appl Mater Interfaces 6(15): 12413–12421
56. Chen Y, Wang L, Shi J (2016) Two-dimensional non-carbonaceous materials-enabled efficient photothermal cancer therapy. Nano Today 11(3):292–308
57. Moon HK, Lee SH, Choi HC (2009) In vivo near-infrared mediated tumor destruction by photothermal effect of carbon nanotubes. ACS Nano 3(11):3707–3713
58. Robinson JT, Welsher K, Tabakman SM et al (2010) High performance in vivo near-IR (>1 mum) imaging and photothermal cancer therapy with carbon nanotubes. Nano Res 3(11): 779–793
59. Yang K, Zhang S, Zhang G et al (2010) Graphene in mice: ultrahigh in vivo tumor uptake and efficient photothermal therapy. Nano Lett 10(9):3318–3323
60. Hwang MT, Landon PB, Lee J et al (2016) Highly specific SNP detection using 2D graphene electronics and DNA strand displacement. Proc Natl Acad Sci USA 113(26):7088–7093

61. Fiorillo M, Verre AF, Iliut M et al (2015) Graphene oxide selectively targets cancer stem cells, across multiple tumor types: implications for non-toxic cancer treatment, via differentiation-based nano-therapy. Oncotarget 6(6):3553–3562
62. Taghavi S, Abnous K, Taghdisi SM, Ramezani M, Alibolandi M (2020) Hybrid carbon-based materials for gene delivery in cancer therapy. J Control Release 318:158–175
63. Yu C, Li L, Hu P, Yang Y, Wei W, Deng X, Wang L, Tay FR, Ma J (2021) Recent advances in stimulus-responsive nanocarriers for gene therapy. Adv Sci 8(14):2100540
64. Rabiee N, Bagherzadeh M, Ghadiri AM, Kiani M, Ahmadi S, Jajarmi V, Fatahi Y, Aldhaher A, Tahriri M, Webster TJ, Mostafavi E (2021) Calcium-based nanomaterials and their interrelation with chitosan: optimization for pCRISPR delivery. J Nanostructure Chem 12(5):919–932
65. Mostafavi E, Soltantabar P, Webster TJ (2019) Nanotechnology and pico technology. In Elsevier eBooks (pp. 191–212)
66. Riley PR, Narayan RJ (2021) Recent advances in carbon nanomaterials for biomedical applications: a review. Curr Opin Biomed Eng 17:100262
67. Zu H, Gao D (2021) Non-viral vectors in gene therapy: recent development, challenges, and prospects. AAPS J 23:4
68. Dominguez GA, Torelli MD, Buchman JT, Haynes CL, Hamers RJ, Klaper RD (2018) Size dependent oxidative stress response of the gut of Daphnia magna to functionalized nanodiamond particles. Environ Res 167:267–275
69. Thakur P, Verma Y, Thakur A (2022) Toxicity of nanomaterials: an overview. In: Thakur A, Thakur P, Khurana SP (eds) Synthesis and applications of nanoparticles. Springer, Singapore, pp 535–544. https://doi.org/10.1007/978-981-16-6819-7_25
70. Ema M, Gamo M, Honda K (2016) A review of toxicity studies of single-walled carbon nanotubes in laboratory animals. Regul Toxicol Pharmacol 74:42–63

MXene Nanomaterial for Medical Application

6

Anand Salvi, Neetu Dhanda, Saarthak Kharbanda, Abhilash Pathania, Preeti Thakur, and Atul Thakur

Abstract

Biomaterials technology has advanced significantly in recent years and 2-D nanostructures have played a key role in this advancement. A new ceramic 2-D nanomaterial, MXene, made up of transition metal carbides, carbonitrides, and nitrides with a planar layout that was produced by etching away "A" from a ceramic phase called "MAX", has emerged to overcome the shortcomings of traditional biomaterials. MXene is used inadequately in biomedical applications due to its weak stability in physiological conditions, lack of prolonged and low biodegradability, and self-controlled drug release. This chapter focuses on cutting-edge examples such as polymer functionalized composites MXene for the developing field of biomedical applications. These applications consist of precise and prolonged antimicrobial activity, biosensing, therapeutics, drug delivery, contrast-enhanced diagnostic imaging, tissue engineering, flexible electronics, and bone regeneration.

Keywords

MXene · Drug delivery · Biosensors · Cancer theranostics and diagnosis · Bio-imaging · Antimicrobial · Tissue engineering

---

A. Salvi · N. Dhanda · S. Kharbanda · A. Pathania (✉) · P. Thakur · A. Thakur
Centre for Nanotechnology, Amity University Haryana, Gurugram, India
e-mail: apathania@ggn.amity.edu

D. Suhag et al. (eds.), *Integrated Nanomaterials and their Applications*,
https://doi.org/10.1007/978-981-99-6105-4_6

6.1 Introduction

In the last few years, lots of development have been observed in the field of biomaterials. A recent report states that between 2019 and 2024, the biomedical market will expand from USD 105 billion to USD 207 billion resenting a 14.5% compound yearly growth rate [1]. According to the American National Institutes of Health, biomaterials are substances that could be used in the short term for analysing, healing, or enhancing the function of the body [2]. The biomaterials are consistent with a body having good power and damage after a course of time [3]. Biomaterials may be produced from the environment or particularly created utilizing ceramic, polymer, composite material, or metal to have varied mechanical, medicinal, and physical qualities [4]. For biomedical and industrial uses, researchers have investigated a variety of synthetic and natural biomaterials in recent times, and their consequent developments must help lead to particular uses based on nanomedicine applications. Due to their exceptional qualities and ultrathin structure, two-dimensional (2-D) nanomaterials have become a popular topic of interdisciplinary research. Researchers have thoroughly investigated carbon based nanodots (CNDs) or several materials of 2-D, including graphene and its products, for their potential use in applications of biomedical [5]. In addition to nanomaterials of 2-D, based ceramics like MXenes are now being studied for a variety of applications. The term "MXenes" was first used by Naguib et al., to describe a class of transition metal nitrides, carbonitrides, and carbides in 2011 [6]. These materials are created by selectively etching "A" and further delamination from their matching parent MAX phase. They have a common principle of $M_{n+1}AX_n$, where "M" stands for an initial conversion metal, 13 or 14 are "A" group of elements, carbon or nitrogen are "X", and "n" ranges from 1 to 3 to make MXenes with the forms M_2X, M_3X_2, and M_4X_3 individually. Following synthesis, several surface terminations, including fluorine (-F), hydroxyl (-OH), and oxygen (-O) are observed. These terminations are characterized by the letter "T" in the simplified formula, $M_{n+1}X_nT_x$. Terminations of the surface allow for additional alterations of the surface [5, 7]. Since the discovery of the first MXene, 2-D titanium carbide (Ti_3C_2), which is formed from Ti_3AlC_2, approximately 70 other MAX phases and 20 MXenes have been successfully identified [8]. MXenes are similarly based on ceramic like 2-D nanomaterial, and their possible use in nanomedicine has been highlighted by a number of their noteworthy characteristics, such as their hydrophilicity, large surface area, nontoxic nature, size tunability, metallic conductivity, near-infrared absorption, biocompatibility, and drug loading capability [9]. Some of the problems of MXenes bio-applications include weak stability in physiological forms, a lack of controlled and prolonged drug release, harm to strong tissues and cells due to weak objective selectivity, etc. By appending surface changes to MXenes, biodegradability, biocompatibility, (700–1300 nm) absorption NIR I and NIR II and photothermal-transformation efficacy can be improved. Polymers involving polyvinyl alcohol (PVA), polyvinylpyrrolidone (PVP), cellulose, soybean phospholipid (SP), polyethylene glycol (PEG), etc., as well as inorganic nanoparticles like polyoxometalates (POMs), mesoporous silica nanoparticles (MSNs), etc., can be used to engineer surfaces [10]. These materials with various surface terminations can detect a variety of analytes due to the charge transfer which results in a change in MXene conductivity throughout the following attachment of the molecular. The Mxene-based nanosheets are crucial to ranostic nano medicine, which combines diagnostics and theranostic Hyperthermia-based photothermal therapy. It is made possible by MXene's extreme photothermal conversion efficiency and mass of extinction coefficient in the bio window of NIR [11]. They introduced photothermal ablation of cancerous cells that are the result of multi-modal imaging and PTT/chemotherapy synergy [2]. Fibrous mats and MXene-modified polymer membranes are also used to promote bone repair and have antibacterial qualities [12]. The composition, physicochemical, and structural characteristics are the only factors that influence the production of nanostructures with these capabilities. Experimental studies of MXenes have proven that Ta,Nb, and Ti are nontoxic, biocompatible, and readily biodegradable due to the presence of nitrogen and carbon, which are vital components of living things [13]. When it comes to biomedical applications such as photothermal therapy, tissue regeneration, antimicrobial activity bio-imaging, biosensing, and drug transport, MXenes outperform graphene and its derivatives. Recent reports on biological uses such as bone regeneration and antibacterial medical bandaging have also been covered. The mechanisms underlying various biomedical applications, such as the detection of hydrogen peroxide (H_2O_2), glucose, and NO_2 photothermal tumour ablation, computed tomography (CT), photoacoustic therapy (contrast-enhanced), stimuli-responsive drug release, radical scavenging, antimicrobial activity, and magnetic resonance imaging (MRI), have been thoroughly discussed and illustrated with diagrams [14–16]. Figure 6.1 shows

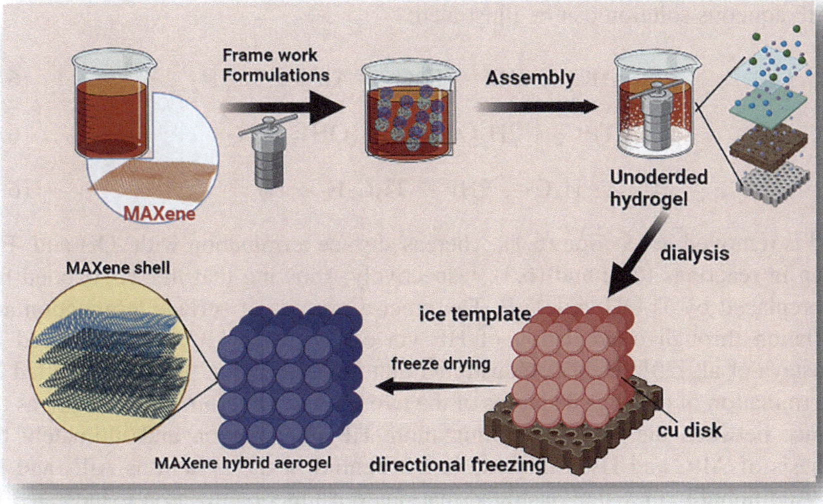

Fig. 6.1 Fabrication of Ti_3C_2, 2-D ultrathin nanosheets with its crystal structure and by delamination

how 2-D ultrathin Ti_3C_2 nanosheets were made, together with the crystal structure, via delamination, surface modification and etching phases are well-known and established due to their unusual combination of characteristics and adaptability, but this family of materials has gone through multiple ups and downs during its brief history, including being referred to under various names.

6.2 Surface Modification/Functionalization and Synthesis of MXenes

6.2.1 Experimental Process

Up to twenty kinds of MXene have been synthesized so far, while up to 70 distinct MAX phases have been throughout experimental prepared. MXene name was given to this novel group of materials of 2-D since the principles underlying behind the process of nitrides and carbides. In this both involves the removal of from the MAX phases of a parent [17]. The M-A bond in the MAX phase is relatively weaker than the metallic-covalent M-X bond and is easily broken, as was previously indicated. Other 2-D nanomaterials with weak Van der Waals interactions between the layers include graphene, phosphorene, and MoS_2 and the exfoliation process in these materials has been extensively researched [18]. In contrast, phases of MAX are strongly composed through robust extra problematic Van der Waals Interactions than the 2-D materials and making the exfoliation process more challenging. The first was hydrofluoric acid (HF) etchant which is used to eliminate layer A; however, due to its extremely damaging chemistry, other synthetic methods have been devised recently [19]. The following chemical interactions between the Ti_3AlC_2 phase and the HF aqueous solution can be illustrated:

$$Ti_3AlC_2 + 3HF \rightarrow Ti_3C_2 + AlF_3 + 3/2\ H_2 \tag{6.1}$$

$$Ti_3C_2 + 2H_2O \rightarrow Ti_3C_2(OH)_2 + H_2 \tag{6.2}$$

$$Ti_3C_2 + 2HF \rightarrow Ti_3C_2F_2 + H_2 \tag{6.3}$$

Al is removed in reaction (6.1), whereas surface termination with -OH and -F is shown in reactions (6.2) and (6.3), respectively, showing that the Ti-Al bond has been replaced by Ti-OH and Ti-F. The precise process of surface termination and exfoliation through intercalation of HF via edges of Ti_3AlC_2 was proposed by Srivastava et al. [20]. They demonstrated that the exfoliation process is started by the termination of the edge Ti atoms of the two layers. HF termination lengthens the distance between the layers, enabling more HF intercalation and ultimately the synthesis of AlF_3 and H_2. F-MXene is left behind in the system as AlF_3 and H_2 exit. Additionally, they demonstrated that random functionalization is indicated by the negative Gibbs free energies (G) of F-MXene and mixed F/OH-MXene [21]. Following the etching procedure, the closely bound MAX phase is transformed into a lightly bound building resembling before the particles of solids are improved by centrifugation or purification. Due to their simplicity and adaptability, fluoride-based etching techniques are widely used for the synthesis of many MXenes. The scalable process is constrained by these artificial methodologies, which limits its usefulness for practical applications. Additionally, Mxenes made with these techniques have a multilayer structure, necessitating additional processing to achieve monolayer morphology [22]. As a result, several Mxenes, such as $Ti_3C_2T_x$, Mo_2CT_x, and Ti_3CNT_x, have been reported using this approach. This approach nevertheless generates traces of HF as seen by the following reaction, although being less dangerous and produces fewer faults than the HF method [23].

$$LiF(aq) + HCl(aq) \rightarrow HF(aq) + LiCl(aq) \tag{6.4}$$

It is interesting to note that the wet etching technique also affects surface characteristics. For instance, the HF etching method results in F-functionalized MXene, whereas the HCl/LiF method results in O-functionalized MXene. By employing the NH_4HF_2 etchant, Halim et al. presented a safer alternative approach for Ti_3C_2MXene. In this method, the delaminated structure is created by NH_4^+ ions intercalating between MXene interlayers [24]. The synthesis of Ti_3C_2 from Ti_3AlC_2 was also studied using bifluoride salts, such as $8KHF_2$ and $NaHF_2$. For MXenes based on carbides and carbonitrides, HF-based procedures work well; however, they did not work for MXenes based on nitrides. Because of this, a novel technique is created to extract Al from the Ti_4AlN_3 phase using molten salts of LiF, NaF, and KF. Li and coworkers created a new, practical, fluoride-free, environmentally friendly approach intended for the production of ($Ti_3C_2Cl_2$ and Ti_2CCl_2)Cl terminated MXene [25]. In this technique, component Zn after molten $ZnCl_2$ is substituted for the component-Al from phases-MAX (Ti_2AlC, Ti_3AlC_2, V_2AlC, and Ti_2AlN), resulting in the production of unique phases-MAX (Ti_2AlC, Ti_3AlC_2, V_2AlC, and Ti_2AlN). As a result of the consequent exfoliation of additional phases-MAX (Ti_2ZnC, Ti_3ZnC_2, etc.) the occurrence of additional $ZnCl_2$, chlorine-terminated MXenes ($Ti_3C_2Cl_2$ and Ti_2CCl_2) were produced. This technique creates new avenues for the synthesis of MXene compositions with novel terminations, which have applications crosswise an extensive variety of technology and knowledge. Moreover, *Yang* et al. produced a high yield (90%) corrosion-anodic of MAX-$Ti_3AlC_2$of double based on an aqueous electrolyte of electrochemical technique for the large-scale manufacture of $Ti_3C_2T_x$ [26]. Therefore, Al-containing MAX phases had been successfully used to synthesize the majority of MXene. A new technique for synthesizing MXene on a wide scale was recently proposed by Gogotsi et al. and involves selectively etching from silicon that contains phases-MAX [27]. From this study, the Si layer was removed from Ti_3SiC_2 MAX, and Ti_3C_2 was produced using a variety of HF oxidants, including HF/HNO_3, HF/H_2O_2, $HF/KMnO_4$, $HF/(NH_4)_2S_2O_8$, and $HF/FeCl_3$.

6.2.2 Modification and Surface Functionalisation

In many electrochemical applications, the interlayer distance in the MXene structure is critical. Thus, a number of strategies were put out to elongate the space between layers and create structures with one or a few layers [28]. The multilayer structure of MXene produced by etching is held together by hydrogen bonds and Van der Waal. The multilayer structure that exfoliates or delaminates can create a single layer MXene. The type of etchant used in the initial step affects the exfoliation method. Li^+ ions intercalate between the interlayers, for example, when MXene is synthesized by using the LiF/HCl technique. These interlayers could be exfoliated through sonication or shaking [8]. Furthermore, trembling, or continuing ultrasonication yields tiny MXenes sheets, more flaws, and low yields. Delamination using intercalating agents (IAs) is undertaken which diminishes interlayer interaction and increases interlayer distance, resulting in a single or few layered structures [29]. While employing the HF approach, exfoliation is done by using ions and organic molecules, that are intercalated between the lengthe and also in the interlayers space between them [30]. It was observed that polar organic molecules and ionic materials were efficient IAs for the delamination of different MXenes. In contrast to other dimethyl sulfoxide (DMSO), MXenes were originating to be effective for $Ti_3C_2T_x$. For the successful oxidation of Nb_2CT_x, $Nb_4C_3T_x$, and $Ti_3C_2T_x$, isopropyl amine was used. Based on the chemical makeup of MXenes, tetramethylammonium hydroxide (TMAOH), urea, and tetrabutyl ammonium hydroxide (TBAOH) also were studied. By using Na^+ intercalation and aryl diazonium salts to modify the surface simple delamination of $Ti_3C_2$was achieved [31]. The fundamental process included the production of Ti_3C_2 using the HF method, followed by intercalation with Na^+ ions, surface modification using diazonium salts, which cause swelling, and moderate sonication required to weaken the interlayer bonds. This is an alternate technique for large-scale multilayer MXene delamination and surface modification using aryl groups that can be researched for other uses. The majority of the synthetic techniques that have been previously mentioned work from the top down, converting bulk precursors into single sheets-like structures, which results in limited control over morphology, size, and structure. Conversely, a bottom-up technique entails constructing smaller nanostructures from atoms and molecules. The bottom-up method is superior to the top-down method in several ways, including fewer flaws, and uniform size or shape homogeneous chemical composition [32]. As a result, it is recommended to synthesize MXenes via bottom-up techniques. Many sensing technologies, such as electrochemical (bio) sensors, piezoresistive wearable sensors, gas adsorptive sensors, and optical sensors have recently been developed using MXene nanostructures [32].

6.3 Structure of MXene

In the MAX phase, The X atoms were in the resulting octahedral places as the "A" element layer interlaced the M layers. Having both metallic and covalent bonding, the M-A bonds are weak and therefore chemically active, causing them to be removed during etching [33]. The subsequently created MXenes with hydroxyl terminations were stabilized by hydrogen bonding and Van der Waals forces.

6.4 Biomedical Applications

These polymers functionalized due to their exceptional qualities and 2-D planar structure. Figure 6.2 shows that MXenes are employed in the medical industry for applications such as antibacterial, photothermal treatment (PTT), medication administration, diagnostic imaging, biosensors, and bone regeneration [34].

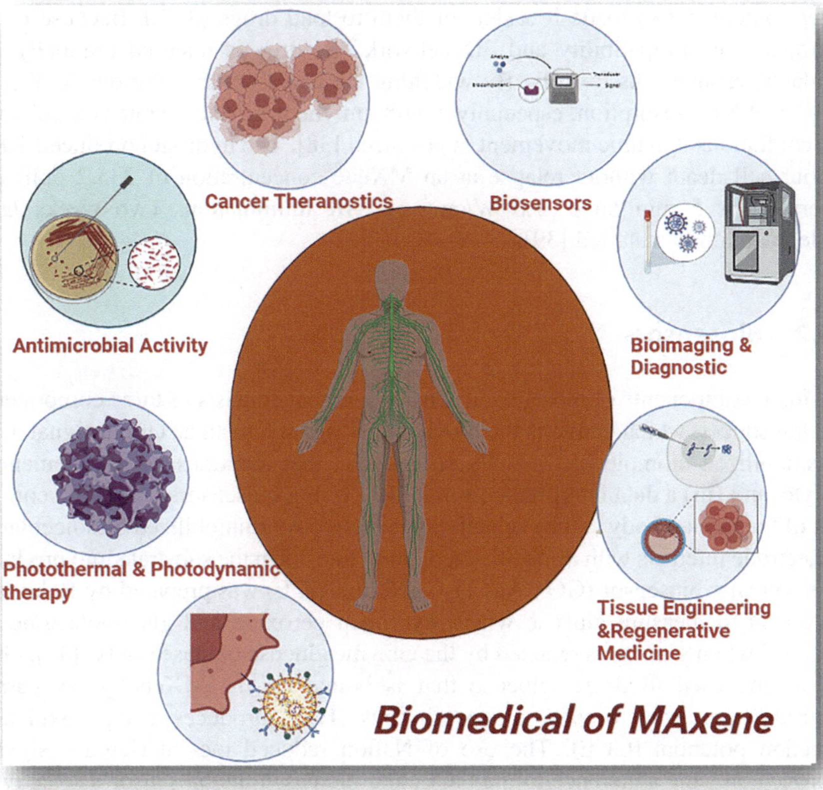

Fig. 6.2 MXenes that have interesting physical and chemical properties could be good options for biomedical research

6.4.1 Drug Delivery

Non-malignant cells also experience negative effects from the traditional cancer treatment methods of chemotherapy and radiation tumour cells have a lower pH [35] than healthy cells do, the creation of stimuli-responsive materials has provided a solution to the problem with current anticancer therapies. The MXene surface contains negatively charged hydroxyl or fluorine groups, which facilitate simple electrostatic interaction with positively charged medicinal molecules. These nanocomposites' multilayer polymeric (negatively charged) coating safeguarded the drugs while they moved through the bloodstream [36]. MXenes with temperature and pH sensitivity is required to keep an eye on the release of chemotherapy drugs modification. Targeted drug release and photothermal ablation work together effectively, thanks to MXene's intrinsic high photothermal conversion and pH sensitivity. The Ti_3C_2MXene surface-modified SP was created by Han and his coworkers for the chemotherapeutic photothermal ablation of cancer cells [9]. A Ti_3C_2MXene-cellulose hydrogel that supports dual anticancer therapy (photothermal/chemo) was synthesized by Xing and coworkers [37]. The nanoplatforms' large pores and high water content (98%) made it easier for them to load drugs (84%). Because of the hydrogel's biocompatibility and 3D network, the drug is released gradually and regulated, reducing its toxicity. By widening the pores in the water due to Ti_3C_2's significant NIR absorption, especially at 808 nm, the drug release rate is accelerated and continuous dynamic movement is produced [38]. The hydrogel produced 100% tumour cell death without relapse at an MXene concentration of 235.2 ppm was observed for 5 min under 1.0 W/cm^2 of NIR illumination. Two weeks later, biodegradation had started [39].

6.4.2 Biosensors

The main components of biosensor a sensing element consists of three components: (i) a transducer, which converts the biochemical signal into an electrical signal, (ii) a transmobilized biomolecule, usually an enzyme, that identifies its complementary analyte, and (iii) a data interpretation unit [40]. Using biosensors, a specific component of the human body can be selectively detected. An immobilized bimolecular on an electrode interacts with a specific element to transform the substrate into products. An enzymatic biosensor (GOx/Au/ Ti_3C_2/Nafion/GCE) was prepared by Rakhi and coworkers to measure glucose where hydrogen peroxide and gluconolactone are produced when glucose is reduced by the enzyme glucose oxidase (GOx) [41]. GO_x, an enzyme used to detect glucose that is bound on the MXene/glassy carbon electrode, is utilized to sense glucose because H_2O_2 produces a large oxidation/reduction potential (GCE). The use of Nafion reduced the interference signals, improved enzyme adherence to the GCE, and increased the biosensor's selectivity. In order to keep the enzyme active during adsorption, lessen its insulating effect for direct electron transfer, and thus increase the electrical conductivity between GO_x and GCE, gold nanoparticles (Au) were utilized. Large-surface area MXenes promote charge accumulation, which improves the sensor's sensitivity and linearity [42]. Figure 6.2 shows the working mechanism of the GO_x which is a dimeric protein that requires the coenzyme flavin adenine dinucleotide (FAD) to function. According to the earlier report [92], the mechanism for sensing glucose is as follows:

$$\mathrm{GOx(FAD)} + \mathrm{Glucose} \rightarrow \mathrm{GOx(FAD_2)} + \mathrm{Glucono-d-lactone} \tag{6.5}$$

$$\mathrm{GOx(FADH_2)} + \mathrm{O_2} \rightarrow \mathrm{GOx(FAD)} = \mathrm{H_2O_2} \tag{6.6}$$

$$\mathrm{Glucose} + \mathrm{O_2} \rightarrow \mathrm{Glucono-d-lactone} + \mathrm{H_2O_2} \tag{6.7}$$

A GOx/Au/Ti_3C_2/Nafion/GCE biosensor has a wide detection range (0.1–18 mM), the sensitivity of 4.2 AmM^{-1} cm^{-2}, and low detection limit (5.9 M), and it delivers exceptional storage stability, repeatability, and reproducibility. In order to build a biosensor (Nafion/Hb/TiO_2- Ti_3C_2/GCE) for the detection of H_2O_2 in 2015, haemoglobin (Hb) was immobilized on TiO_2-Ti_3C_2-GCE [43]. TiO_2 also increased Ti_3C_2's biocompatibility. The use of Nafion reduces the interference signals, improves enzyme adherence to the GCE, and increases the biosensor's selectivity. When TiO_2 nanoparticles are added, proteins on Ti_3C_2 become more stable and bioactive, and a substantial effective area for protein absorption is created [43].

6.4.3 Cancer Theranostic Applications

2-D MXenes have a variety of features that are advantageous for biomedical applications, thanks to their distinctively transformable 2-D in-layer nanostructure and variable chemical compositions. These 2-D multifunctional MXenes and their composites have so far been created for theranostic uses such as photothermal therapy (PTT), photothermal/photodynamic/chemo synergistic treatment, and diagnostic imaging. This section introduces several therapeutic uses for various MXenes 2-D materials [1].

6.4.3.1 Hyperthermia

One of the most common antineoplastic techniques is chemotherapy, although it suffers from non-specificity, significant drug toxicity to healthy cells and tissues, and disappointing therapeutic results [44]. Other physical therapies have been further developed for treating cancer, including focused ultrasound therapy, magnetocaloric therapy, photothermal therapy, photodynamic therapy, and son-dynamic therapy, provided that they are combined with diagnostic imaging guidance, such as magnetic resonance imaging (MRI), photoacoustic tomography (PAT), computed tomography (CT), and ultrasound imaging, among others [45]. Enhancing the tumour specific nanocatalytic efficiency for tumour treatment based on the conventional Fenton-based catalytic reaction is both very desirable and essential. Nanoparticle hyperthermia, a method for using nanoparticles in cancer heat therapy, involves heating tumours to kill cancer cells. Cancerous tumours can be removed using this form of treatment with little harm to the body [46–48].

6.4.3.2 Photothermal Therapy

PTT typically uses photothermal agents that have accumulated in disease areas as energy absorbers to produce local heat when exposed to NIR light, which causes aberrant cells to undergo apoptosis or necrosis [48]. PTT is a helpful modality for treating a variety of diseases because of its advantages of simple operation, little invasiveness, short therapy duration, quick recovery, minimized side effects, and low recurrence. It is commonly acknowledged that two fundamental elements, photothermal agents and outside light, affect the therapeutic effectiveness of PTT [17]. Two essential factors mostly influence the PTT result for photothermal malicious agents. The extinction coefficient (e), which reveals the ability to absorb light, and the photothermal conversion efficiency (g), which reveals the ability to transform light into heat, are the two capabilities. Because of their substantial absorption in the NIR range, developing 2-D MXenes has demonstrated comparatively high photothermal conversion efficiency compared to most traditional photothermal agents, indicating their immense potential as effective photothermal agents.

6.4.3.3 Synergistic Therapy

It is widely recognized that the performance of therapeutic modalities accessible in clinics is substantially hampered by the complexity, diversity, and heterogeneity of malignancies. Due to its synergistic anticancer effects, decreased individual drug/therapeutic modality-related toxicity, and suppression of multidrug resistance via several modes of action, combination therapy has generated a great deal of attention in the scientific community [43] (Fig. 6.3). Due to their exceptional qualities of strong NIR absorption, effective photothermal conversion, simple surface alterations, and multifunctionalities, 2-D MXene-based synergistic therapy has become a highly effective method for the removal of tumours in recent years. The majority of recent research on MXene-based synergistic therapy for cancer mentions PTT/PDT or PTT/chemotherapy. For instance, multifunctional MXene was recently developed by Bai et al. to achieve PTT/PDT/chemotherapy against cancer [49]. To create MXene-based nanocomposites, they allowed the absorption of forming and synthetic polysaccharides onto the surface of Ti_3C_2MXene nanosheets.

6.4.4 Cancer Diagnosis

MXenes are a strong candidate for electrochemical sensing due to their high electrical conductivity, hydrophilicity, the potential for high-density incorporation of several functional groups, low diffusion barrier, guaranteed cytocompatibility, large active surface area, ultrathin 2-D film morphology, and good ion intercalation properties. MXene embedded biosensors are frequently used to detect nitrite, pesticides, phenol, H_2O_2, and, more intriguingly, cancer biomarkers [44]. Due to an exceptional biocompatibility, MXenes can effectively be used as transducers by

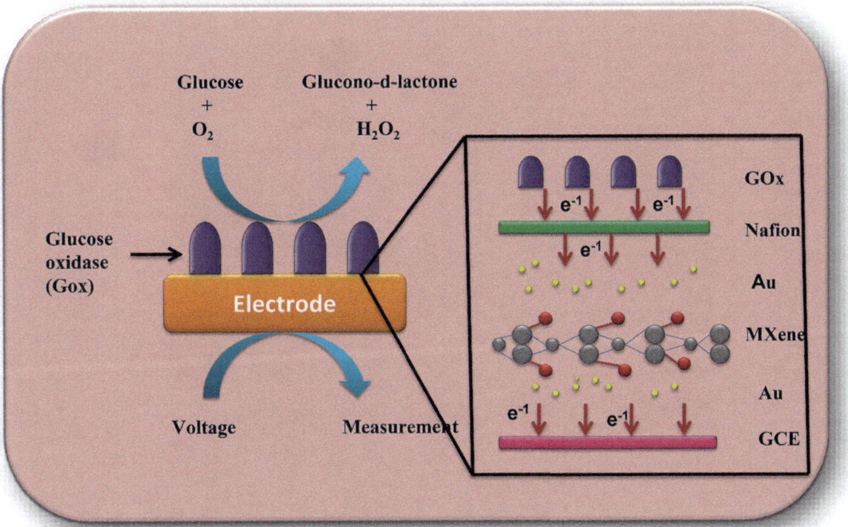

Fig. 6.3 Working mechanism of a glucose biosensor

immobilizing biological receptors on their surface. It is crucial to emphasize that immobilizing enzymes and proteins in other 2-D structures appear to be challenging. However, MXenes' improved electrode kinetic properties, a larger range of detection limits, linear sensitivity, and good selectivity make it simple to achieve. Anti-CEA functionalized Ti_3C_2 MXenes have been synthesized using the delamination process, which is intriguing because it demonstrates their potential for sensing cancer biomarkers. The addition of the MXene interface's surface also increased the total catalytic action [50]. Specifically, Ti_3C_2 nanosheets with $Mn_3(PO_4)_2$ added have been created to mimic the behaviour of biological enzymes for sensing superoxide anions(O_2^-) from HepG2 cells. It is important to note that the addition of an ATP molecule caused $Mn_3(PO_4)_2$ to form a porous ordered nanostructure, which facilitated improved electron transfer between the O2 electrode and electrode, resulting in a faster response time, a sensing limit of 0.5 nM, and an analytical range of 2.5 nM to 14 M. It may also be able to detect O_2 leakage from living cells as well. As a result of the MXene-based electrochemical sensor's success in identifying cancer biomarkers, researchers have decided to use them in MXene-based fluorescence nanoprobes for simultaneous sensing (Fig. 6.4).

MXenes have a lot of potential as contrast agents in X-ray, CT, and MRI imaging in addition to PA imaging. MnOx/Ti_3C_2 has recently been shown to be effective for tumour imaging using T1-weighted MRI. Additionally, because MnO bonds are rapidly broken up in mildly acidic tumour micro-environments, combined MnO_x components demonstrated the unique pH responsive T1-weighted MR imaging potential for tumour imaging [9]. This behaviour can release Mn^{2+} ions, which

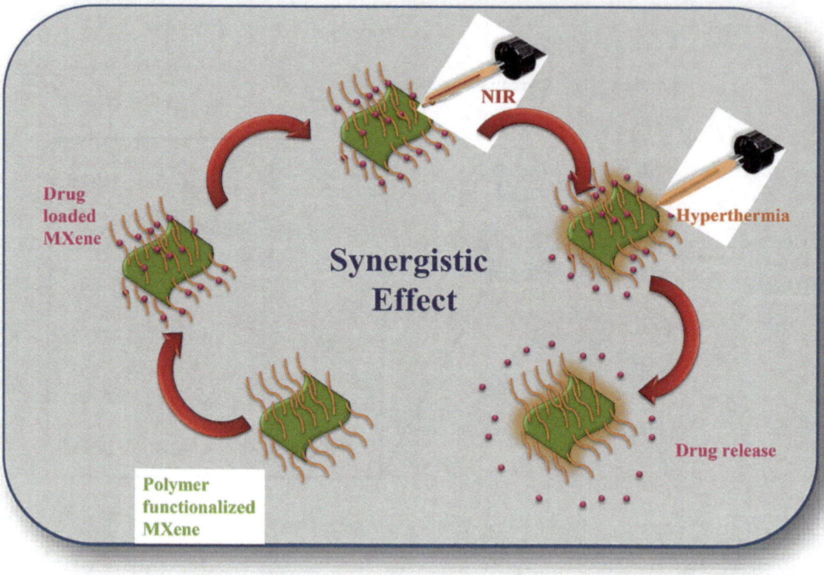

Fig. 6.4 Stages during synergistic effect of photothermal treatment and chemotherapy working

increase the likelihood of interactions between paramagnetic Mn centres and water molecules and improve the performance of T1-weighted MRI. As seen in Ta_4C_3 and Ti_3C_2MXenes, the high atomic number and conductivity of MXenes are also thought to play a significant role in the magic of X-ray/CT imaging.

6.4.5 Bio-Imaging

Although MXenes have a significant amount of potential for biosensing device their direct use in biological and optical applications is severely constrained by their typical low photoluminescence (PL) response in aqueous solution [51]. This is because certain biological applications for nanomaterials require strong PL, small size, and optical absorption [52]. Due to quantum confinement and edge effects when made atomically thin, theoretical and experimental research have demonstrated that creating quantum dots (QDs) is an efficient technique to reveal unique physical features. QDs are better suited for bio-imaging than nanosheets due to their ultrasmall size and luminescent characteristics [38, 41, 42]. In the domains of optical, biological, and cellular imaging, luminescent Ti_3C_2MXene QDs (MQDs) would be potential candidate materials [53]. Although Ti_3C_2-based MXenes have been thoroughly investigated for a variety of applications, their use in biomedicine, theranostic medicinal applications etc. The focus of these articles is T1-weighted MRI-guided efficient PTT against tumour [19].

Monitoring the development of malignancies depends on accumulating multilayer data on several biomarkers with various spatial distributions at the cellular level. Here, a dual signal-tagged titanium carbide MXenes nano probe (cDNA-Ti_3C_2) that responds to biomarkers in many biological sites, including the cytoplasm and plasma membrane, was developed [19].

6.4.6 Flexible Electronics

There is substantial research and development in the miniaturization of devices, especially flexible wearable devices. These next generation flexible wearable electronics are devices that can be mated with the body and be worn on the skin of a human being or any living organism to monitor physiological as well as biological changes. It is a hot research topic among researchers to manipulate all the properties associated with the change in the electrical as well as physiological parameters [54–56]. Flexible and wearable soft devices, such as sensors and energy storage systems, are essential for the development of quick artificial intelligence and interface in human–machine [56]. The creation of flexible electrodes with high sensitivity, quick response times, and broad operating ranges is one of the core aspects of these devices. Also, the development of self-powered and flexible wearable's devices, that are capable of generation of currents and voltages by harvesting the electrical signals, generated from the movement of muscles in human beings. Keywords such as electronic skins, flexible touchable displays are connected with such devices which utilize some of the sensing mechanisms such as piezoresistive, triboelectric, and capacitive effects. Many remarkable nanomaterials have been exploited to prepare substrates and materials for wearable applications [57].

6.4.7 Antimicrobial Applications

Antimicrobial applications of MXene are possible due to the high surface area to volume ratio and simplicity of surface functionalization [45]. Due to their high membrane permeability, biocompatibility, and potential for numerous antibacterial activities, nanoparticles often do not cause bacterial resistance [58]. Before MXenes, the antibacterial effects of graphene-based nanomaterials were primarily mediated via the generation of reactive oxygen species (ROS) and direct interaction with bacterial membranes [59, 60]. The first Ti_3C_2MXene colloidal solution requires a high concentration dose of 200 g mL^{-1} to offer positive inhibition for its antibacterial capabilities. But modified Ti_3C_2 that has been treated with polyvinylidene fluoride (PVDF) significantly enhances the antibacterial properties. Another method of producing Ti_3C_2/chitosan composite nanofibers using electrospinning shows a greater than 62% reduction in microorganisms [61]. Only Ti_3C_2 exhibits antibacterial characteristics when Ti_2C and Ti_3C_2MXenes are compared to *E. coli,* and a closer look at these revealed that smaller nanosheets had stronger antibacterial effects [62].

According to a different study, the distinctive properties of MXene nanosheets, which have sharp edges and are small in size, can penetrate bacterial cell walls, releasing bacterial DNA and finally causing bacteria to disperse [63]. *Trichoderma reesei* (*T. reesei*) was the subject of the sole research of MXenes antifungal activities to far [64–66]. The findings of this investigation showed that the growth of *T. reesei* hyphae was inhibited. Since the MXene nanosheets have sharp lamellar edges, they might also prevent spore germination. As a result, this research demonstrated the ability of Ti_3C_2MXene to interrupt the life cycle of fungi and highlighted the enormous potential of MXenes as strong antifungal drugs. This will significantly impact how effective they are at fighting off microbes. MXenes should ideally be stable for a long enough time to develop bactericidal characteristics while yet remaining biocompatible. Therefore, more studies should be done to boost MXenes stability under physiological settings in order to expand its use as an antibacterial agent.

6.4.8 Tissue Engineering

Due to their ultrathin structural characteristics and unique morphologies with a variety of outstanding physicochemical, electrical, optical, and biological capabilities, MXenes have recently demonstrated great promise in biomedical applications, especially in the applications of tissue engineering [67]. A fascinating biomaterial stage for the varied treatment of various diseases and malignancies, such as cancers and tumours, has been made possible by the development of multipurpose scaffolds with distinctive properties of well-regulated discharge behaviour, highly proficient photothermal translation, and stimulatory tissue regeneration [68].

Future multifaceted biological remedies based on the special advantages of MXene might thus be imagined. MXenes have demonstrated remarkable possibilities in material sciences and stem cell-based tissue therapies [69]. One study created smart biomaterials with titanium carbide (Ti_3C_2) MXene nanofibers for cell culture and tissue engineering [70]. Due to the presence of functional hydrophilic groups, electrospinning and doping might be used to create the resulting hydrophilic composite nanofibers [71]. Bone marrow-derived mesenchymal stem cells (BMSCs) were used to examine the biochemical representative properties; the resulting MXene composite nanofibers showed excellent biocompatibility and noticeably increased cellular activity in addition to promoting BMSC development to osteoblasts. For tissue engineering and repair, smart biomaterials with distinctive multiple biologic features are best suited [72].

6.5 Conclusion and Future Perspectives

2-D MXene composites, according to the prevalent studies have proved themselves as a viable and compatible nanomaterial in biomedical applications due to their excellent properties such as absorption in near-infrared (NIR) region, hydrophilicity, tunable size, electrical conductivity, biocompatibility, minimum or no cytotoxicity, large surface area, and unmatched drug loading capacity. Current issues with these nanomaterials, including mass production, storage, precise structural composition, in vivo retention, and biosafety over an extended length of time, among others, provide barriers to their widespread use in nanomedicine and must be overcome. Deep tissue penetration was made possible by NIR II radiation, although very few photothermal agents have been observed in this biological window. By examining its level of genotoxicity, immune toxicity, and reproductive toxicity, MXenes can be made more effective in immune pathology, temporary implants, fertility treatment, and tissue regeneration. Although NIR II radiation allowed for deep tissue penetration, there are very few reports of photothermal agents in this biological window. MXenes and MXene-based (nano)structures have a lot of functional groups on their surfaces, providing flexible chances for modification and functionalization. They are highly stable, hydrophilic, include entire metal atomic layers, and have customizable compositions, all of which make them desirable candidates for clinical and biological applications. For instance, MXenes and graphene structures have been used to create smart nanoscale systems with the potential for cancer therapy, diagnosis, and medication delivery.

References

1. Manisha H, Priya Swetha PD, Shim YB, Prasad KS (2019) Revisiting fluorescent carbon nanodots for environmental, biomedical applications and puzzle about fluorophore impurities. Nano-Struct Nano-Objects 20:100391. https://doi.org/10.1016/j.nanoso.2019.100391
2. Szuplewska A et al (2022) The 10th anniversary of MXenes: challenges and prospects for their surface modification toward future biotechnological applications. Adv Drug Deliv Rev 182. https://doi.org/10.1016/j.addr.2021.114099
3. Rafatullah M, Sulaiman O, Hashim R, Ahmad A (2010) Adsorption of methylene blue on low-cost adsorbents: a review. J Hazard Mater 177(1–3):70–80. https://doi.org/10.1016/j.jhazmat.2009.12.047
4. George SM, Kandasubramanian B (2020) Advancements in MXene-polymer composites for various biomedical applications. Ceram Int 46(7):8522–8535. https://doi.org/10.1016/j.ceramint.2019.12.257
5. Verma R, Thakur P, Chauhan A, Jasrotia R, Thakur A (2023) A review on MXene and its' composites for electromagnetic interference (EMI) shielding applications. Carbon 208:170–190
6. Ishaq K, Saka AA, Kamardeen AO, Ahmed A, Alhassan MIH, Abdullahi H (2017) Characterization and antibacterial activity of nickel ferrite doped α-alumina nanoparticle. Eng Sci Technol Int J 20(2):563–569. https://doi.org/10.1016/j.jestch.2016.12.008
7. Thakur P, Gahlawat N, Punia P, Kharbanda S, Ravelo B, and Thakur A (2022) *Cobalt nanoferrites: a review on synthesis, characterization , and applications*, no. 0123456789. Springer US
8. Naguib M, Mochalin VN, Barsoum MW, Gogotsi Y (2014) 25th anniversary article: MXenes: a new family of two-dimensional materials. Adv Mater 26(7):992–1005. https://doi.org/10.1002/adma.201304138
9. Wu L et al (2018) 2-D transition metal carbide MXene as a robust biosensing platform for enzyme immobilization and ultrasensitive detection of phenol. Biosens Bioelectron 107 (February):69–75. https://doi.org/10.1016/j.bios.2018.02.021

10. Han X, Huang J, Lin H, Wang Z, Li P, Chen Y (2018) 2-D ultrathin MXene-based drug-delivery nanoplatform for synergistic photothermal ablation and chemotherapy of cancer. Adv Healthc Mater 7(9):1–13. https://doi.org/10.1002/adhm.201701394
11. Hatami M (2017) Toxicity assessment of multi-walled carbon nanotubes on Cucurbita pepo L. under well-watered and water-stressed conditions. Ecotoxicol Environ Saf 142 (April):274–283. https://doi.org/10.1016/j.ecoenv.2017.04.018
12. Fantechi E et al (2014) A smart platform for hyperthermia application in cancer treatment: cobalt-doped ferrite nanoparticles mineralized in human ferritin cages. ACS Nano 8(5): 4705–4719. https://doi.org/10.1021/nn500454n
13. Thakur A, Punia P, Aggarwal RK, Kumar R, Dhar R, Thakur P (2022) Adsorption of Cd and Cr ions from industrial wastewater using Ca doped Ni–Zn nanoferrites: synthesis, characterization and isotherm analysis. Ceram Int 48:18048–18056
14. Thakur A, Mathur P, Singh M (2007) Controlling the properties of manganese-zinc ferrites by substituting In3+ and Al3+ ions, 221,837–845. https://doi.org/10.1524/zpch.2007.221.6.837
15. Bárcena C, Sra AK, Gao J (2009) Applications of magnetic nanoparticles in biomedicine. Nanoscale Magn Mater Appl 167:591–626. https://doi.org/10.1007/978-0-387-85600-1_20
16. Nelson DF et al (1986) A randomized comparison of misonidazole sensitized radiotherapy plus BCNU and radiotherapy plus BCNU for treatment of malignant glioma after surgery: final report of an RTOG study. Int J Radiat Oncol Biol Phys 12(10):1793–1800. https://doi.org/10.1016/0360-3016(86)90321-4
17. Balasubramanian K, Yadav R, Prajith P (2015) Antibacterial nanofibers of polyoxymethylene/gold for pro-hygiene applications. Int J Plast Technol 19(2):363–367. https://doi.org/10.1007/s12588-015-9127-y
18. Heidarian AR, Kord H, Mostafavi K, Lak AP, Mashhadi FA (2011) Investigating Fe and Zn foliar application on yield and its components of soybean (Glycine max (L) Merr.) at different growth stages. J Agri Biotechnol Sustain Develop 3:189–197
19. Zheng J, Wang B, Jin Y, Weng B, Chen J (2019) Nanostructured MXene-based biomimetic enzymes for amperometric detection of superoxide anions from HepG2 cells. Microchim Acta 186(2):8. https://doi.org/10.1007/s00604-018-3220-9
20. Kumari S, Dhanda N, Thakur A, Singh S, Thakur P (2023) Investigation of calcium substitution on magnetic and dielectric properties of Mg–Zn nano ferrites. Mater Chem Phys 297:127394., ISSN 0254-0584. https://doi.org/10.1016/j.matchemphys.2023.127394
21. Zhang Y, Jiang X, Zhang J, Zhang H, Li Y (2019) Simultaneous voltammetric determination of acetaminophen and isoniazid using MXene modified screen-printed electrode. Biosens Bioelectron 130:315–321. https://doi.org/10.1016/j.bios.2019.01.043
22. Yang AS, Zhang P, Wang F, Gaetano A, Lohe MR, and Blom PWM (2020) Angewandte Chemie. https://doi.org/10.1002/ange.201809662
23. Chaudhary S, Umar A, Bhasin KK, Baskoutas S (2018) Chemical sensing applications of ZnO nanomaterials. Materials (Basel) 11(2):1–38. https://doi.org/10.3390/ma11020287
24. Zhu X *et al.* (2017) *Alkaline intercalation of Ti3C2 MXene for simultaneous electrochemical detection of Cd(II), Pb(II), Cu(II) and Hg(II)*, vol. 248, no. Ii. Elsevier Ltd.
25. Thakur P, Chahar D, Taneja S, Bhalla N, Thakur A (2020) A review on MnZn ferrites: synthesis, characterization and applications. Ceram Int 46:15740–15763. https://doi.org/10.1016/j.ceramint.2020.03.287
26. Wang H et al (2016) Surface modified MXene Ti3C2 multilayers by aryl diazonium salts leading to large-scale delamination. Appl Surf Sci. https://doi.org/10.1016/j.apsusc.2016.05.060
27. Arruebo M, Fernández-Pacheco R, Ibarra MR, Santamaría J (2007) Magnetic nanoparticles for drug delivery. Nano Today 2(3):22–32
28. Kumar S, Lei Y, Alshareef NH, Salama KN (2018) Biofunctionalized two-dimensional Ti 3 C 2 MXenes for ultrasensitive detection of cancer biomarker. Biosens Bioelectron. https://doi.org/10.1016/j.bios.2018.08.076

29. Lei Z et al (2007) Effects of nanoanatase TiO2 on photosynthesis of spinach chloroplasts under different light illumination. Biol Trace Elem Res 119(1):68–76. https://doi.org/10.1007/s12011-007-0047-3
30. Zheng J, Wang B (2018) Synthesis of MXene / DNA / pd / pt nanocomposite for sensitive detection of dopamine. J Electroanal Chem 816(January):189–194. https://doi.org/10.1016/j.jelechem.2018.03.056
31. Dhanda N, Thakur P, Thakur A, Green synthesis of cobalt ferrite: a study of structural and optical properties, materials today: proceedings , Volume 73, Part 2,2023, Pages 237-240., ISSN 2214-7853, https://doi.org/10.1016/j.matpr.2022.07.202
32. Ambekar RS, Kandasubramanian B (2019) Progress in the advancement of porous biopolymer scaffold: tissue engineering application. Ind Eng Chem Res 58(16):6163–6194. https://doi.org/10.1021/acs.iecr.8b05334
33. You Q, Zhuang L, Chang Z, Ge M, Mei Q, Yang L (2022) Biosensors and bioelectronics hierarchical Au nanoarrays functionalized 2-D Ti 2 CT x MXene membranes for the detection of exosomes isolated from human lung carcinoma cells. Biosens Bioelectron 216 (August):114647. https://doi.org/10.1016/j.bios.2022.114647
34. Wu H et al (2011) Solvothermal synthesis of cobalt ferrite nanoparticles loaded on multiwalled carbon nanotubes for magnetic resonance imaging and drug delivery. Acta Biomater 7(9): 3496–3504. https://doi.org/10.1016/j.actbio.2011.05.031
35. Dhanda N, Kumari S, Kumar R, Kumar D, An-Cheng S, Thakur P, Thakur A (2023) Influence of Ni over magnetically benign co ferrite system and study of its structural, optical, and magnetic behaviour. Inorg Chem Commun 151:110569., ISSN 1387-7003. https://doi.org/10.1016/j.inoche.2023.110569
36. Tong L, Chuang CC, Wu S, Zuo L (2015) Reactive oxygen species in redox cancer therapy. Cancer Lett 367(1):18–25. https://doi.org/10.1016/j.canlet.2015.07.008
37. Jiang Y et al (2018) Silver nanoparticles modified two-dimensional transition metal carbides as nanocarriers to fabricate acetylcholinesterase-based electrochemical biosensor. Chem Eng J 339:547–556. https://doi.org/10.1016/j.cej.2018.01.111
38. Alwarappan S, Nesakumar N, Sun D, Hu TY, Li CZ (2022) 2-D metal carbides and nitrides (MXenes) for sensors and biosensors. Biosens Bioelectron 205:113943. https://doi.org/10.1016/j.bios.2021.113943
39. Dhanda N, Thakur P, Kumar R, Fatima T, Hameed S, Slimani Y, Sun A-CA, Thakur A (2023) Appl Organomet Chem:e7110. https://doi.org/10.1002/aoc.7110
40. Wang B, Li Y, Hu H, Shu W, Yang L, Zhang J (2020) Acetylcholinesterase electrochemical biosensors with graphene-transition metal carbides nanocomposites modified for detection of organophosphate pesticides. PLoS One 15(4):1–16. https://doi.org/10.1371/journal.pone.0231981
41. Harper A, Anderson MR (2010) Electrochemical glucose sensors-developments using electrostatic assembly and carbon nanotubes for biosensor construction. Sensors 10(9):8248–8274. https://doi.org/10.3390/s100908248
42. Huang H, Dong C, Feng W, Wang Y, Huang B, Chen Y (2022) Biomedical engineering of two-dimensional MXenes. Adv Drug Deliv Rev 184:114178. https://doi.org/10.1016/j.addr.2022.114178
43. Zheng Y et al (2022) Titanium carbide MXene-based hybrid hydrogel for chemo-photothermal combinational treatment of localized bacterial infection. Acta Biomater 142:113–123. https://doi.org/10.1016/j.actbio.2022.02.019
44. Mathur P, Thakur A, Singh M (2008) Effect of nanoparticles on the magnetic properties of Mn-Zn soft ferrite. J Magn Magn Mater 320:1364–1369. https://doi.org/10.1016/j.jmmm.2007.11.008
45. Nam S, Kim J, Oh S, Kim J, Ahn CW, Oh I (2020) Ti 3 C 2 T x MXene for wearable energy devices: supercapacitors and triboelectric nanogenerators Ti 3 C 2 T x MXene for wearable energy devices : supercapacitors and triboelectric nanogenerators. 110701. https://doi.org/10.1063/5.0028628

46. Fartas FM, Abdullah J, Yusof NA, Sulaiman Y, Saiman MI (2017) Biosensor based on tyrosinase immobilized on graphene-decorated gold nanoparticle/chitosan for phenolic detection in aqueous. Sensors (Switzerland) 17:5. https://doi.org/10.3390/s17051132
47. Kharbanda S, Dhanda N, Sun A, Thakur A, Thakur P (2023) Multiferroic perovskite bismuth ferrite nanostructures: a review on synthesis and applications. J Magn Magn Mater 572: 170569., ISSN 0304-8853. https://doi.org/10.1016/j.jmmm.2023.170569
48. Mathur P, Thakur A, Singh M (2008) Low temperature synthesis of Mn0.4Zn0.6In0.5Fe1.5O4 nanoferrite for high-frequency applications. J Phys Chem Solids 69:187–192. https://doi.org/10.1016/j.jpcs.2007.08.014
49. Zhu J et al (2017) Recent advance in MXenes: a promising 2-D material for catalysis , sensor and chemical adsorption. Coord Chem Rev 352:306–327. https://doi.org/10.1016/j.ccr.2017.09.012
50. Gore PM, Naebe M, Wang X, Kandasubramanian B (2019) Progress in silk materials for integrated water treatments: fabrication, modification and applications. Chem Eng J 374 (May):437–470. https://doi.org/10.1016/j.cej.2019.05.163
51. Sundaram A et al (2020) Engineering of 2-D transition metal carbides and nitrides MXenes for cancer therapeutics and diagnostics. J Mater Chem B 8(23):4990–5013. https://doi.org/10.1039/d0tb00251h
52. Thakur A, Thakur P, Hsu JH (2011) Smart magnetodielectric nano-materials for the very high frequency applications. J Alloys Compd 509:5315–5319. https://doi.org/10.1016/j.jallcom.2011.02.021
53. Han X, Huang J, Lin H, Wang Z, Li P, Chen Y (2018) 2-D ultrathin MXene-based drug-delivery nanoplatform for synergistic photothermal ablation and chemotherapy of cancer. Adv Healthc Mater 7(9):1–13. https://doi.org/10.1002/adhm.201701394.D
54. Lei D et al (2020) Research progress of MXenes-based wearable pressure sensors research progress of MXenes-based wearable pressure sensors. 110702(August). https://doi.org/10.1063/5.0026984
55. Rakhi RB, Nayuk P, Xia C, Alshareef HN (2016) Novel amperometric glucose biosensor based on MXene nanocomposite. Nat Publ Group (November):1–9. https://doi.org/10.1038/srep36422
56. Guo Y, Zhong M, Fang Z, Wan P, Yu G (2019) A wearable transient pressure sensor made with MXene nanosheets for sensitive broad-range human—machine interfacing. https://doi.org/10.1021/acs.nanolett.8b04514
57. Xin M, Li J, Ma Z, Pan L, Shi Y (2020) MXenes and their applications in wearable sensors. 8 (April):1–14. https://doi.org/10.3389/fchem.2020.00297
58. Huang J, Li Z, Mao Y, Li Z (2021) Progress and biomedical applications of MXenes. Nano Select 2(8):1480–1508. https://doi.org/10.1002/NANO.202000309
59. Sivasankarapillai VS, Somakumar AK, Joseph J, Nikazar S, Rahdar A, Kyzas GZ (2020) Cancer theranostic applications of MXene nanomaterials: recent updates. Nano-Struct Nano-Objects 22:100457. https://doi.org/10.1016/J.NANOSO.2020.100457
60. Bu F, Zagho MM, Ibrahim Y, Ma B, Elzatahry A, Zhao D (2020) Porous MXenes: synthesis, structures, and applications. Nano Today 30:100803. https://doi.org/10.1016/J.NANTOD.2019.100803
61. Rafieerad A, Yan W, Amiri A, Dhingra S (2020) Bioactive and trackable MXene quantum dots for subcellular nanomedicine applications. Mater Des 196:109091. https://doi.org/10.1016/J.MATDES.2020.109091
62. Mei L et al (2020) Two-dimensional nanomaterials beyond graphene for antibacterial applications: current progress and future perspectives. Theranostics 10(2):757. https://doi.org/10.7150/THNO.39701
63. Li Y, Zhang W, Niu J, Chen Y (2012) Mechanism of photogenerated reactive oxygen species and correlation with the antibacterial properties of engineered metal-oxide nanoparticles. ACS Nano 6(6):5164–5173. https://doi.org/10.1021/NN300934K

64. Lemire JA, Harrison JJ, Turner RJ (2013) Antimicrobial activity of metals: mechanisms, molecular targets and applications. Nat Rev Microbiol 11(6):371–384. https://doi.org/10.1038/NRMICRO3028
65. Dhanda N, Thakur P, Sun A-C, Thakur A (2023) Structural, optical and magnetic properties along with antifungal activity of Ag-doped Ni-Co nanoferrites synthesized by eco-friendly route. J Magn Magn Mater 572:170598., ISSN 0304-8853. https://doi.org/10.1016/j.jmmm.2023.170598
66. Chernousova S, Epple M (2013) Silver as antibacterial agent: ion, nanoparticle, and metal. Angew Chem Int Ed 52(6):1636–1653. https://doi.org/10.1002/ANIE.201205923
67. Akhavan O, Ghaderi E (2010) Toxicity of graphene and graphene oxide nanowalls against bacteria. ACS Nano 4(10):5731–5736. https://doi.org/10.1021/NN101390X
68. Li J et al (2014) Antibacterial activity of large-area monolayer graphene film manipulated by charge transfer. https://doi.org/10.1038/srep04359
69. Rasool K, Mahmoud KA, Johnson DJ, Helal M, Berdiyorov GR, Gogotsi Y (2017) Efficient antibacterial membrane based on two-dimensional Ti3C2Tx (MXene) nanosheets. Sci Rep 7(1): 1–11. https://doi.org/10.1038/s41598-017-01714-3
70. Liu Y et al (2020) Two-dimensional MXene/cobalt nanowire heterojunction for controlled drug delivery and chemo-photothermal therapy. Mater Sci Eng C 116. https://doi.org/10.1016/j.msec.2020.111212
71. Mayerberger EA, Street RM, McDaniel RM, Barsoum MW, Schauer CL (2018) Antibacterial properties of electrospun Ti3C2Tz (MXene)/chitosan nanofibers. RSC Adv 8(62): 35386–35394. https://doi.org/10.1039/C8RA06274A
72. Jastrzębska AM et al (2018) The atomic structure of Ti2C and Ti3C2 MXenes is responsible for their antibacterial activity toward E. coli bacteria. J Mater Eng Perform 28(3):1272–1277. https://doi.org/10.1007/S11665-018-3223-Z

Metal Nanoparticles in the Field of Medicine and Pharmacology

7

Muhammad Ehsan, Deepa Suhag, Raksha Rathore, Atul Thakur, and Preeti Thakur

Abstract

One of the cutting-edge technologies that hold considerable promise for both medical and non-medical applications is metal nanoparticles. Metal nanoparticles are extremely active and have optical, electrical, and chemical properties that are different from those of their macroscale counterparts due to a high surface area to volume ratio. As a result, metal nanoparticles interact with microbes more frequently than larger particles. Because of their distinct physiochemical characteristics, metal nanoparticles have been suggested as a transporter and a therapeutic agent in the biomedical sector. These physicochemical characteristics make them useful in a variety of biomedical fields. Regarding this, magnetic iron nanoparticles can be employed for imaging and tumor cell hyperthermia, whereas plasmonic nanoparticles can be used for tumor cell detection and photothermal death. Particle form and interparticle interaction are both crucial for plasmon resonance absorption, which gives metal nanoparticles their distinctive color. The most efficient of all metal nanoparticles, silver nanoparticles (AgNPs) are utilized extensively in a wide range of fields, including medicine, biotechnology, optics, microelectronics, catalysis, and energy conversion. They are nontoxic and safe. These nanomaterials exhibit super para-magnetism, ferromagnetism, and other magnetic characteristics. Biomedical applications of MNPs include the separation

M. Ehsan (✉)
Departement Informatique, CY Cergy-Paris Université, École CY-Tech, Pau, France
e-mail: Muhammad.ehsan@cyu.fr

D. Suhag · R. Rathore
Centre for Nanotechnology, Amity University Haryana, Gurugram, India

A. Thakur · P. Thakur
Centre for Nanotechnology, Amity University Haryana, Gurugram, India

NanoLatticeX, Amity University Haryana, Gurugram, Haryana, India

D. Suhag et al. (eds.), *Integrated Nanomaterials and their Applications*,
https://doi.org/10.1007/978-981-99-6105-4_7

of proteins, DNA, and bacteria, the development of biosensors, and hyperthermia therapy. Particularly when paired with polymers, metal nanoparticles are in great demand for drug delivery. This is true both for cellular and subcellular distribution as well as blood–brain barrier transport. Surface modification, maximal drug loading, targeted administration, and conjugation of biomolecules are all made possible by polymer coupling (antigen, antibody, enzymes, proteins, DNA, aptamer, etc.

Keywords

Metal nanoparticles · Nontoxic · Drug delivery · Polymer coupling

7.1 Introduction

Solid, colloidal particles with diameters ranging from 10 to 1000 nm are generally referred to as nanoparticles. However, European and other international organizations limit the scope of the aforementioned definition to structures with three-dimensional lengths of at most 100 nm [1]. They are providing benefits that larger particles do not have, like a higher surface-to-volume ratio or better magnetic characteristics.

Nanotechnology is the study, development, and application of materials, whose fundamental unit of measurement is the nanoscale. This category comprises, among other things, physics, advanced materials, chemistry, and the biological and pharmaceutical sciences. Because of their special, size-dependent features, nanoparticles are frequently utilized in many different industrial sectors. These nanomaterials are particularly fascinating for medicine and pharmacology since it is possible to modify the characteristics of nanoparticles [2]. The use of nanoparticles in medicine is connected to the development of particular nanostructures that serve as cutting-edge therapeutic and diagnostic tools. Nanoparticles have a wide range of uses, including as medication delivery systems, radiosensitizers in radiation or proton therapy, in bioimaging, and as fungicides, bactericides and in nanoremediation [3–6]. The purpose of this work is to describe the properties of noble metal-based nanoparticles, with a focus on their uses in medicine and related fields. High biocompatibility, stability, and the potential for large-scale synthesis while avoiding organic solvents are benefits of noble metal-based nanoparticles that are crucial for medical applications. This has a favorable impact on biological systems. Different medications may be able to release under control thanks to these nanoparticles. Additionally, nanoparticles can be freeze-dried to provide a powder formulation. Comparatively, magnetic nanoparticles like those made of iron or cobalt frequently need to be covered to increase their biocompatibility and guard against aggregation, oxidation, or corrosion [7]. There are numerous clinically authorized nanoparticle-based treatments; the most popular ones are based on polymeric (Copaxone) or liposomal (Abelcet, AmBisome, DaunoXome, Myocet) platforms. There are numerous ways to make nanoparticles based on noble metals. One of the most widely used techniques involves chemically reducing the precursor of a noble metal (such as silver nitrate or chloroplatinic acid to produce silver or platinum nanoparticles, respectively) using a suitable reducing agent that also serves as a stabilizer (in the case of nanogold synthesis, sodium citrate) to prevent the agglomeration of the nanoparticles. Additionally, it is possible to vary reaction parameters such as temperature, amount of precursor and reductant, and reaction time. It is possible to generate nanoparticles of various shapes and sizes. Green chemistry techniques that can reduce the use of environmentally harmful compounds are also gaining popularity recently. For this purpose, reducing agents from environmentally friendly sources such as lactic acid, citrus fruits, and coffee seeds are used. Sonochemical, microwave-assisted, and electrochemical techniques are more advanced methods for producing these nanoparticles. Successful treatment of many diseases, including malignancies, depends on early detection. However, conventional diagnostic and imaging methods have limited ability to distinguish between benign and malignant lesions and are unable to detect cancer in its early stages of development. Novel nanoparticle imaging agents, on the other hand, may enable more sensitive and focused imaging of cancer and other diseased tissues. Furthermore, nanoparticles have the potential to be the best vehicle for delivering anticancer drugs and other therapeutics to affected areas without causing collateral tissue damage. The theme issue on Nanoparticles in Medicine provides an in-depth analysis of the latest developments in the biological and medical uses of nanomaterials. This issue is divided into sections dedicated to inorganic (metal and metal oxide) nanoparticles, liposomes, organic nanoparticles, and hybrid nanoparticles. It covers special features of nanoparticles for in vitro detection, in vivo diagnostics, multimodal imaging, chemotherapy, phototherapy, gene therapy, immunotherapy, theranostics, and their clinical implementation.

7.2 History of Nanoparticles

Nanomedicine is an important research field in the twenty-first century. Although there have been many methods of making and using nanoscale particles since antiquity and even hundreds of years ago, the field of nanomedicine as we know it today only emerged in his 1990s. The development of various submicroscopic tools and the study of cellular, molecular, and ultimately atomic structure in twentieth century biology, chemistry, and physics formed the basis of this new science. The approach of nanotechnology proposed by Richard P. Feynman in the 1950s paved the way for the emergence of nanomedicine as an important area of research and application in medicine [8]. Since its inception, nanomedicine has progressed at an incredible speed due to major advances in technology. Throughout history, it has evolved and expanded into various medical applications (such as tissue engineering). Norio Taniguchi first used the term "nanotechnology" in his 1974. The term is still used today to describe the processing of materials by atoms or molecules to separate, solidify or transform. Richard P. Feynman, a Nobel Prize-winning physicist, already foresaw the development of nanotechnology and the opportunities associated with it in his 1959 paper "There's Plenty of Room at the Bottom." It invites us to explore new areas of physics. Although the word "nano" is never used, the work is considered a precursor to nanotechnology. In the early twentieth century, Paul Ehrlich attempted to develop a "miracle drug" containing drugs that could fight disease and eliminate all bacteria with just one treatment, and chemotherapy was born. The development of increasingly complex "magic bullets" became achievable through the wealth of information gathered throughout the 20th century regarding cells, their constituents, intracellular and intercellular processes, and cell signaling, along with advancements in the fields of biochemistry and biotechnology. Became. In the late 1960s, Peter Paul Speiser developed the first nanoparticles that could be used for targeted drug therapy, and Georges Jean Franz Köhler and César Milstein successfully developed monoclonal antibodies in his 1970s. Since then, extensive research has been conducted into the synthesis, applications, and physicochemical functionalization of various carrier systems. Nanoparticles were originally designed in the early 1990s to carry DNA and gene fragments, which were then delivered intracellularly using antibodies. Biocompatible polymers, liposomes, and micelles are currently being investigated as drug, vaccination, and gene carriers. Due to their small size (often less than 200 nm), nanomaterials can circulate within the body and cannot be filtered out of circulation before reaching their intended destination. As a "Trojan horse," the active ingredient may be modified with its hollow interior and surface to allow it to cross cell membranes and other natural barriers. Biosensors (such as antibodies) can be used to identify specific cells or tissues, bind to them, and deliver active agents to targets over long periods of time. Over the last three decades, nanotechnology activity has increased worldwide and has become a major interdisciplinary research topic. The integration of nanotechnology into medicine has greatly spurred this expansion, as nanostructured materials have significant medical effects.

In addition to creating novel medicinal effects, controlling materials in the nanometric range involves cutting-edge, technically difficult chemistry, and production procedures. Because they were not specifically produced at the nanoscale to have therapeutic effects that corresponded to their nanoscale size, conventional small-molecule medicines are omitted from this group. Nanoparticle surfaces contain a variety of functional moieties that allow for multivalent conjugation for therapeutic agent delivery, targeting, imaging, and diagnostic applications.

7.3 Pharmaceutical Nanoparticles

You observe the enormous clouds that surround you as they stretch out like fluffy mountains from your window seat on the aircraft. Or perhaps you'd prefer to hop on a short train and enjoy the scenery as it passes. You've probably seen how things are delivered to their destinations, whether you've been in one of these big cars or are more used to driving yourself to school. Similar to how we use automobiles, planes, and trains to transport products around the world, scientists can build tiny vehicles that travel inside the body. These minuscule vehicles are made out of nanoparticles, which are exceedingly small bits of material. Scientists are creating these nanoparticles to deliver medication to incredibly precise parts of the body.

Why Are Nanoparticles Important for Scientific research? Currently, it might be challenging to treat many illnesses using dietary or alcoholic drugs. This occurs for some ailments, such as diabetes and brain damage, because a barrier keeps the drug from getting to the right places. In diabetes, the walls of the digestive tract serve as a barrier. Insulin, which is necessary for people with diabetes but cannot be digested in the intestines way most food can, just flows through the body. Insulin needs to be supplied; it cannot simply be absorbed by the body. Additionally, using medications to treat the brain might be difficult. Humans have a blood–brain barrier, which keeps chemicals from the bloodstream from reaching the brain. The fluid around the brain can occasionally be injected with drugs by medical personnel. Sometimes the prescription does not work, even though it can address the underlying cause of the problem. When treating many malignancies, this happens. Because cancer cells that affect the body can seem very similar to healthy cells, even if a patient gives medication to the proper area of the body, it may kill both healthy and malignant cells. As a result, patients may have illness and hair loss. The body can occasionally overprotect itself, which is another problem. When medicine enters the body, the immune system could recognize it as a threat and eliminate it before it has a chance to operate. To overcome all of these concerns, medicine can be encapsulated into nanoparticle carriers before being given to patients. Nanoparticles can transport drugs to the appropriate place by overcoming barriers including the blood–brain barrier and digestive system walls. Additionally, they are able to block the immune system from attacking the drug they deliver.

7.4 Exactly How Do Nanoparticles Reach the Correct Location in the Body?

The location of a nanoparticle depends on the size and kind of sensors that are present on its surface. These qualities can be altered to target different diseases or ailments. Their ability to move through the body is determined by the size of the nanoparticles. Since smaller nanoparticles do not get trapped in tiny spaces, the body can tolerate them. Additionally, because nanoparticles must release their drug inside of cells in order to operate, they are easier for cells to absorb. Smaller nanoparticles may also enable more rapid medication release. Although it would seem like a good idea to create nanoparticles as tiny as possible, this is not achievable. Too-small nanoparticles will group together. Nanoparticles that are clumsy can no longer perform as intended. Because of this, it is essential to create nanoparticles that fall within a particular size range. Nanoparticles may go everywhere in the body due to their small size, but they still need to find their intended organs or cell types. Scientists may program nanoparticles to detect certain cell types from various organs, systems, or even malignancies due to the fact that each type of cell has a distinct protein on its surface. A cell uses this to express to which tissue it belongs. Additionally, cancer cells have a unique assortment of proteins. On their surface, nanoparticles include sensors called antibodies that can attach to certain proteins that are shown. When a nanoparticle is attached to a cell, the cell usually absorbs the nanoparticle.

What would happen if a nanoparticle bearing a medicine made contact with a cell? The cell would eat it up along with any drugs within. Therefore, getting nanoparticles to adhere to the appropriate cells is all that has to be done by scientists. It is possible to provide antibodies specific to the target cell type's surface proteins to the nanoparticles to make them attach to that cell type. There are a few more techniques for getting nanoparticles to the right places, depending on the tissue being targeted. If nanoparticles are electrically charged, they will be attracted to particular tissues, like the lungs. As an alternative, the nanoparticles might be made from the fatty molecules that are already delivered to the liver. There are also a few nanoparticles composed of substances that bind to bone molecules.

7.5 Making Nanoparticles Unnoticeable to the Immune System

The job of the immune system is to protect you from foreign invaders, such as dangerous microbes, that can harm you. Immune cells wander through the bloodstream in search of substances that are not normally present. They devour and dissect anything they find alien to the human body. They do this by identifying chemicals and structures normally found in cells and other body substances and attacking those they don't recognize. Since these are rarely produced by the human body, taking them alone will result in immune cells consuming a large number of drugs. To circumvent this problem, scientists have developed nanoparticles that contain or consist of chemicals commonly found in living organisms. These nanoparticles come from outside the body, but they can trick the immune system into believing they are part of the body.

Multifunctional nanoparticles have essentially been developed with (a) stealth-like characteristics to evade the immune system and block photochemical processes, (b) protective layers to prevent proliferation degradation of biological goods (e.g., proteins, DNA), (c) targeting fragments to improve tumor specificity and accumulation, (d) membrane permeation fractions to improve uptake of cells, (e) imaging agent to assess administration and dosage, (f) target-dependent endosperm escape mechanism. Nanoparticles are important scientific components used in pharmacology and biotechnology. They serve as a link between molecular structures and bulk materials. The physical properties of the bulk material are not affected by its size, while the physical and chemical properties of the nanoparticle depend on its size [9]. Therefore, as the size of a material reaches the nanometer scale and the ratio of atoms to its surface increases, its properties change. Because of their distinct qualities, such as large surface area, structural composition, and prolonged blood circulation time compared to small molecules, nanoparticles have emerged as attractive prospects for improving personalized medical therapy. Ability to transform undesirable physicochemical properties of bioactive molecules into desirable bio pharmacological profiles, improving delivery of therapeutic agents across biological barriers and compartments, control the release of biologically active agents, increase therapeutic efficacy through selective delivery of drugs to biological targets, and perform therapeutic functions by combining imaging techniques. Multimodal imaging are all potential advantages of modified therapeutic nanoparticles. Evidence-specific studies of nanoparticles in cell culture and small animal models for medical applications are currently underway. A small number of nanomaterials and nanoparticles are being studied in clinical trials or have been approved by the US Food and Drug Administration (FDA) for use in humans.

7.6 Nanoparticle Types

Depending on their overall shape, nanoparticles are categorized as 0-Dimensional (D), 1D, 2D, or 3D (Fig. 7.1).

1D Nanomaterials 1D nanomaterials are employed in the circuitry of computer chips, as well as for anti-reflective qualities and hard coatings on eyeglasses. They have thin films or surface coatings. These have been employed in engineering, chemistry, and electronics.

2D Nanomaterials Nanostructures in 2D nanomaterials are permanent, lengthy, and have thick membranes. They are utilized to manufacture nanopore filters for filtration and separation of small particles. An illustration of a 2D nanoparticle is asbestos fiber.

3D Nanomaterials 3D nanomaterials are small, fixed nanostructures made of thin films that have atomic-scale porosity, colloids, and free nanoparticles of different morphologies. The free long aspect ratio nanowires and nanoparticles found in 2D and 3D nanomaterials, in contrast to 1D nanomaterials, are harmful to human health [9].

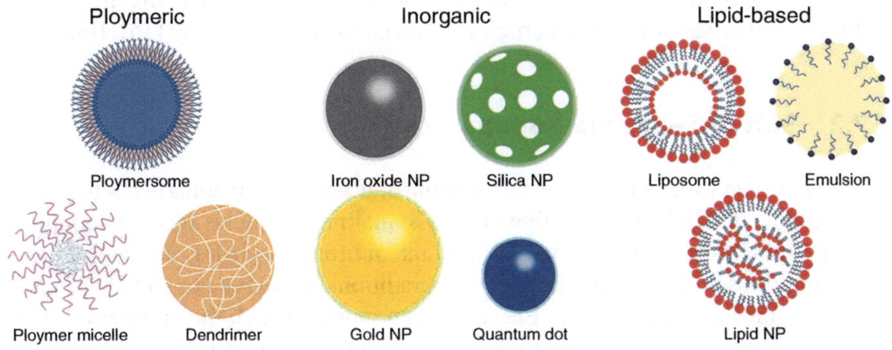

Fig. 7.1 Types of nanoparticles [10]

7.7 Properties of Nanoparticles

In 2008, the International Organization for Standardization (ISO) established a precise definition for nanoparticle, characterizing it as a distinct nano-object with all three Cartesian dimensions measuring less than 100 nanometers. Within the ISO standard, two-dimensional nano objects, such as nano-discs and nanoplates, are categorized as one-dimensional nano objects. These nanoparticle structures exhibit considerable diversity, primarily attributed to variations in their size, shape, and material properties. Some classifications place dendrimers, liposomes, and polymeric nanoparticles within the organic category, while fullerenes, quantum dots, and gold nanoparticles are classified as inorganic nanoparticles. Furthermore, nanoparticles can be categorized based on their composition, which includes carbon, ceramic, semiconducting, or polymeric materials.

7.7.1 Nanoparticles' Physical Properties

The surface layer, the shell, and the core are the three layers that make up the nanoparticle. Surfactants, polymers, metal ions, and others are examples of surface layer chemicals. Nanoparticles can be called a single material or a combination of materials. Nanoparticles can appear as suspensions, colloids or dispersed aerosols, depending on their chemical and electromagnetic properties [11]. The size of the nanoparticles affects their properties [12]. For example, copper nanoparticles less than 50 nm in size are extremely brittle materials that lack the ductility and ductility of solid copper. Additional changes caused by nanoparticle size include superparamagnetism in magnetic materials, quantum confinement in semiconductor Q particles, and surface plasmon resonances in some metal particles.

Absorption of solar radiation in photovoltaic cells in nanoparticles is also significantly higher than in thin films of continuous bulk materials [13]. This is because the nanoparticles are smaller and can absorb more sunlight. The nanoparticles have a high surface-to-volume ratio, which improves their diffusivity at high temperatures [14]. The ability of nanoparticles to sinter at lower temperatures than larger particles make this possible. Although the density of the product is not affected by the diffusion properties of the nanoparticles, it can cause agglomeration [15]. (Fig. 7.2).

7.7.2 Soft or Semisolid Nanoparticles

Liposomes stand out among the many semisolid and soft nanoparticles that have been produced. Clinical applications of liposomal nanoparticles include serving as delivery vehicles for vaccines, antibiotics, and antifungals [16]. Long before nanotechnology was widely used, micelles were traditionally the type of soft nanosystem used in therapeutic applications. Because it can move through the oil interior of the micelle, an organic substance that is normally insoluble in water can be "dissolved" in the surfactant solution by the process of creating the micelle. The drug was

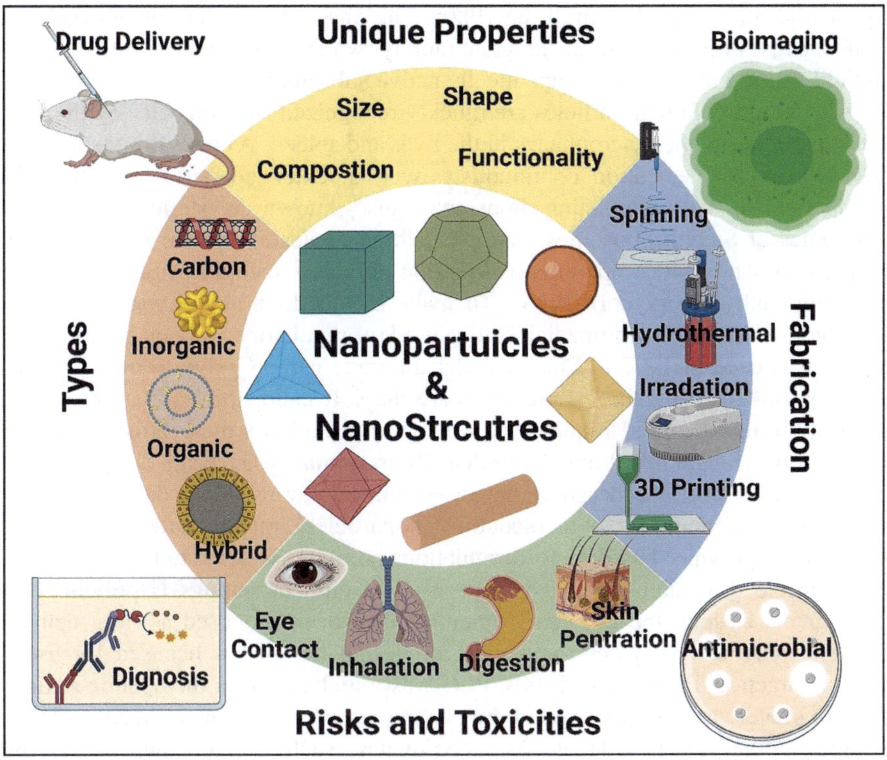

Fig. 7.2 Nanoparticles and their properties, types, applications, risks and toxicities, and fabrication techniques [13]

dissolved through this dissolution mechanism. In some typical examples, soap is commonly used to dissolve phenolic compounds to produce clear solutions widely used for disinfection. Non-ionic surfactants effectively liquefy iodine. These "iodophors" are systems of iodide surfactants that are more stable than iodide systems. However, the limited water solubility of steroids makes it difficult to prepare them for ophthalmic use. Oil solutions or suspensions cannot be used because optical clarity is required. Non-ionic surfactants can be used to produce clear solutions with good sterilization capacity. The solubility of water-insoluble vitamins and essential oils is also achieved using this type of surfactant. Lipid vesicles, subsequently termed liposomes, represent one of the earliest instances of nanotechnology-based drug delivery systems. Advance with nanotechnology. In the middle of the 1960s, their first description of them was made. Initially, they found utility in the exploration of biological membranes. However, during the 1970s, their practical applications began to diversify, notably in the realm of drug delivery. It's worth noting that during this period, Donald Tomalia pioneered dendrimers, inventing, naming, and patenting them, although their application as pharmaceutical agents did not materialize until the 1980s. The phagocyte system of macrophages (MPS) causes liposomes to disintegrate rapidly, which is a serious issue when they are utilized as carriers for therapeutically active substances.

All foreign colloidal particles are quickly recognized as "non-self" by the MPS cells, particularly the macrophages in the liver and spleen. As a result, it is impossible to provide medication continuously over an extended period of time. The production of long-circulating liposomes, also known as sterically stabilized liposomes or Stealth liposomes, involves adding cholesterol to the bilayer as well as predominantly biocompatible, hydrophilic polymers with a flexible main chain, like poly (ethylene glycol) (PEG). To make Doxil, doxorubicin was placed into liposomes. The first nanomedicine to get FDA regulatory clearance was Doxil, which was used in 1997 in Europe under the trade name Caelyx to treat Kaposi's sarcoma linked with AIDS. Liposomes are the soft nanoparticles that are used in clinical settings the most frequently nowadays, especially for the treatment of cancer and systemic fungal infections. More than 20 liposomal products are either available on the market or are undergoing advanced clinical testing. With the exception of liposomes, the majority of therapeutic nanoparticles employed today in clinical settings are polymer-based nano-formulations. Another widely studied method for nanoparticle drug delivery that is now used in clinical settings is polymer-drug conjugates. Despite the fact that numerous have been proposed as drug delivery carriers, only few polymers having linear design have been licensed for use in clinical practice. In the early 1990s, PEG was initially used in therapeutic settings. Approximately a dozen PEGylated drug examples are now being utilized in clinical settings. Other macromolecule-drug conjugates, such as Abraxane, a 130 nm albumin-bound paclitaxel that was authorized by the FDA in 2005 as a second-line therapy for patients with breast cancer, have also been created as drug carriers [17].

The incredible therapeutic potential of biodegradable polymeric micelles, which have a size range of 10–200 nm, as drug delivery nanocarriers has attracted a lot of attention. Polymeric micelles are produced by self-assembling block copolymers consisting of two or more polymeric chains with different hydrophobicities. These copolymers spontaneously coalesce into a core–shell micellar structure in an aqueous environment to lower the Gibbs energy. Long-chain and short-chain phospholipids were combined to form bicelles, which resemble liposomes and micelles. They were initially utilized as model membranes in the 1990s due to their appropriateness for magnetic resonance studies of membrane protein structure, and they have since been used in the bulk of NMR structural examinations of transmembrane proteins. The use of bicollates as a biomedical technique to improve medication absorption through the corneum layer of the skin has been proposed. Dendrimers have developed into another ground-breaking class of drug delivery soft nanoparticle platform due to its distinctive characteristics and well-defined design. Dendrimers have a special chemical structure that allows them to carry a wide range of drugs via covalent conjugation or electrostatic adsorption to their multivalent surfaces. As an alternative, dendrimers can bind medications through chemical bonds, hydrogen bonding, or hydrophobic interaction in the voids in their cores.

Although nanoparticles' medicinal uses are currently restricted, as was previously said, there are intriguing potential. One of them is soft polymer nanoparticles, which offer a lot of potential as nanoscale drug carriers. An excellent drug carrier should have both the targeting property and the stimulus responsiveness to improve the medicine's absorption and reduce adverse effects. It is therefore very desirable to create stimuli-responsive nanoparticles for drug delivery that release the medication in a controlled manner when they reach the target area. Soft polymer nanoparticles can vary in volume or size in response to environmental or stimulus changes. They are referred to as environment-sensitive nanoparticles or stimulus-responsive nanoparticles. These nanoparticles are also known as micro/nanogels. Temperature, electric and magnetic fields, light intensity, pressure, pH, ionic strength, specific (bio)molecules or enzymes, and ultrasound are just a few examples of the stimuli that soft nanoparticles can respond to. The kind of stimulus (physical or chemical) is used to classify these nanoparticles [18, 19]. They cannot be made into soft stimulus-responsive nanoparticles for biomedical applications without the stimulus-sensitive polymer that gives them their structure. Advanced polymerization processes have been developed in order to produce new stimuli-responsive nanoparticles that are sensitive to additional signals such as microwave, redox, and different chemical substances in addition to those already mentioned. Despite current interest, practical clinical applications for the creation of novel stimuli-responsive soft nanoparticles have been restricted, especially in the case of dual/multi-stimulus-responsive varieties, with the exception of very sensitive thermo- and pH-sensitive nanoparticles (Table 7.1).

Table 7.1 Therapeutics based on nanoparticles approved for clinical use

Composition	Trade name	Company	Indication
Liposomal platforms			
Liposomal amphotericin B	Abelcet	Enzon	Fungal infections
Liposomal amphotericin B	AmBisome	Gilead Sciences	Fungal and protozoal infections
Liposomal cytarabine	DepoCyt	SkyePharma	Malignant lymphomatous meningitis
Polymeric platforms			
PEG–adenosine deaminase	Adagen	Enzon	Severe combined immunodeficiency
PEG–GCSF	Neulasta	Amgen	Neutropenia associated with cancer Chemotherapy
PEG–HGF	Somavert	Nektar, Pfizer	Acromegaly

7.8 Therapeutics Based on Nanoparticles in Clinical Trials

The use of nanoparticles in medicine is developing in popularity, with an increasing number of nanoparticle-based therapies now under clinical development. Drug-encapsulated liposomes and polymer-drug conjugates, such as PEGylated medications, predominate in clinical studies. One drawback of employing liposomes is that the therapeutic index is lowered as a result of the speedy clearance of liposomes from the circulation by phagocytic cells of the reticuloendothelial system. Many methods have been developed to minimize this problem. The most common technique is to encapsulate the liposome with biocompatible and inert polymers like PEG to produce long-circulating liposomes. The polymer coating protects the liposome surface, preventing opsonins from identifying the liposomes and reticuloendothelial system clearance as a result [20]. Another strategy is increased liposome accumulation in target cells, tissues, and organs. Specific liposomes have been developed for the administration of various medications, and targeting ligands, such as antibodies, are linked to the surface of the liposome together with small substances (such as folate and transferrin) and peptides [21]. One of these is Onco TCS, an aggressive non-lymphoma Hodgkin's treatment developed by INEX Burnaby, British Columbia, Canada-based pharmaceutical company. Utilizing liposome-based transmembrane carrier systems from INEX. Onco TCSs have the ability to deliver vincristine to intracellular targets. Onco TCS has greater tumor accumulation, a longer blood circulation half-life, and longer drug release properties than free vincristine, according to clinical research results from its phase I and phase II. Liposomal vincristine may thereby lessen the drug's unpleasant side effects while enhancing its efficacy. Additionally, it has been demonstrated that controlled drug release increases the therapeutic effectiveness of liposomal medications. By including a small amount of pH-sensitive phosphatidylethanolamine, dimethyldioctadecylammonium bromide, or oleyl alcohol in the liposome membrane, smart liposomes for preferential intracellular drug delivery have been developed [22]. Under endosomal pH, these liposomes undergo a phase change but are normally stable in blood. PEG has frequently been used to enhance the pharmacokinetics of several nanoparticle compositions. PEG is a highly hydrated flexible polymer chain that increases plasma protein adsorption and biofouling of nanoparticles, extending the half-life of the medication in circulation while decreasing renal clearance of relatively smaller drug molecules. PEG is the best substance to employ in therapeutic settings since it is nontoxic and immune-suppressive. NKTR-118 (PEG-naloxol) in phase I for treating opioid-induced constipation, Hepacid (PEG-arginine deaminase) in phase II for treating hepatocellular carcinoma, and Puricase (PEG-uricase) in phase III for treating hyperuricemia are just a few of the new PEGylated products currently undergoing clinical evaluation due to these favorable properties. Other nanoparticle platforms, such as nano-emulsions, dendrimers, and inorganic nanoparticles, have also demonstrated therapeutic potential (Table 7.2).

These include drug-encapsulated liposomes and polymer-drug conjugates. This technology has considerably increased the number of therapeutic nanoparticles and showed innovative methods for medicinal use. One intriguing instance is NB-001, a medicinal product based on nano-emulsions that recently began its phase II trial as

Table 7.2 Therapeutics based on nanoparticles in clinical trials

Composition	Trade name	Company	Indication
Liposomal platforms			
Liposomal annamycin	L-Annamycin	Callisto	Acute lymphocytic leukemia, acute myeloid leukemia
Liposomal cisplatin	SLIT Cisplatin	Transave	Progressive osteogenic sarcoma metastatic to the lung
Liposomal doxorubicin	Sarcodoxome	GP-Pharm	Soft tissue sarcoma
Polymeric platforms			
PEG–arginine deaminase	Hepacid	Phoenix	Hepatocellular carcinoma
PEG–camptothecin	Prothecan	Enzon	Various cancers
PEG–naloxol	NKTR-118	Nektar	Opioid-induced constipation

a topical therapy for genital herpes infection. Another illustration is Viva Gel, a medication based on poly-L-lysine dendrimers that is currently undergoing phase I testing as an easy, inexpensive, and safe way for women to prevent genital herpes and HIV infection.

7.9 Applications of Nanoparticles in Medicine

Nanoparticles due to their tiny size are very helpful in medicine since they may enter cells or be engineered to attach to specific cells in addition to being able to circulate broadly throughout the body [23]. These qualities have enabled the development of cutting-edge methods for better imaging of organs, tumors, and other diseased tissues in the body. Additionally, they have contributed in the creation of cutting-edge therapeutic delivery methods, such as local heating (hyperthermia), preventing blood flow to tumors and injured tissues, and delivering drug payloads. Magnetic nanoparticles have taken the place of radioactive technetium in the research of cancer lymph node metastasis. The function of the nanoparticles in magnetic resonance imaging (MRI) depends on the contrast change brought about by tiny superparamagnetic iron oxide particles. These particles can also be used to remove malignancies using hyperthermia, which is the localized heating and tissue destruction of such particles by an alternating magnetic field. To enhance the outcomes of ultrasonography, positron emission tomography (PET), or fluorescence imaging, nanoparticles can be produced. These methods frequently need the nanoparticle to be able to recognize a particular cell type or disease state. The same targeting technique might conceivably assist in the exact delivery of a drug to a particular spot of sickness. A liposome or porous nano-sponge structure might be used to administer the drug; these structures would then be held in place by bonds at the targeted location and provide a delayed release of the medication. The development of nanoparticles to assist in medication delivery to the brain through inhalation has

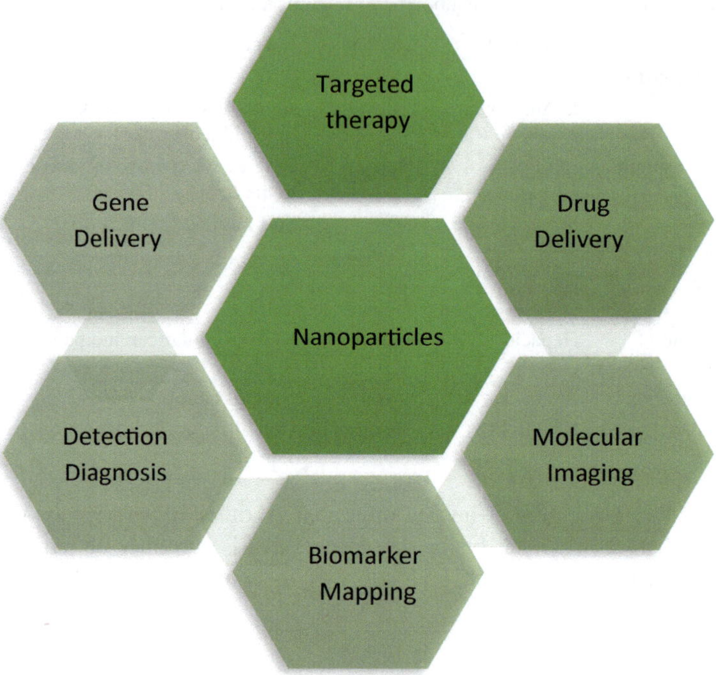

Fig. 7.3 Applications of metal nanoparticles in the field of medicine

significant potential for the treatment of neurological diseases such as Parkinson's, Alzheimer's, and multiple sclerosis.

Nanoparticles and nanofibers are essential ingredients in the development of novel scaffold materials for bone and tissue restoration. These scaffolds include nanoparticles that are biocompatible. Future tissue-repair treatments may, for example, include calcium hydroxyapatite nanoparticles, a natural component of bone, with collagen or collagen replacements. Additionally, nanoparticles have been used in the production of products with a medicinal connection. For instance, the University of Oxford developed the sunscreen Optisol in the 1990s with the goal of developing a secure sunscreen that is transparent to visible light but yet has ultraviolet-blocking action on the skin. In the past, substances such as large particles of zinc oxide, titanium dioxide, or an organic compound that absorbed sunlight were used to create sunscreens. Although potent photocatalysts, zinc oxide and titanium dioxide fall short of expectations because they can damage DNA and skin cells when exposed to water and sunlight (deoxyribonucleic acid). As a result, those compounds did not live up to expectations (Fig. 7.3).

The next step was to make titanium oxide nanoparticles using a little amount of manganese. Studies have shown that nanoparticle-based sunscreen is safer than sunscreen composed of traditional materials. The improvement in safety was attributed to the inclusion of manganese, which changed the compound's semiconducting qualities from n-type to p-type, moved its Fermi level, or oxidation-reduction properties, and reduced the chance of the creation of free radicals.

7.9.1 Nanoparticles for Drug Delivery

Drug targeting of certain organs and tissues has become a significant problem in recent years. This is because conventional medication delivery methods have trouble using the desired dose to reach the target within the allotted time. By utilizing contemporary drug delivery systems (DDS), such as the brain–blood barrier, it is now feasible to solve the issue of biochemical barriers in the body [24]. Medication solubility issues can be resolved by the innovative systems, which can also shield the medication against photodegradation and pH variations [25]. Liposomes, liquid crystals, dendrimers, cyclodextrins, hydrogels [26], polymersomes [27], micelles [28], and nanoparticles (nanospheres and nanocapsules) are the most frequently utilized drug delivery methods [25, 26]. Of all noble metal-based nanoparticles, gold nanoparticles (Au NPs) have the greatest potential for application as drug delivery vehicles. Au NPs in a range of sizes (1–100 nm) and forms (spherical, rod-like, cage-like, etc.) may be synthesized reasonably easily (Fig. 7.4).

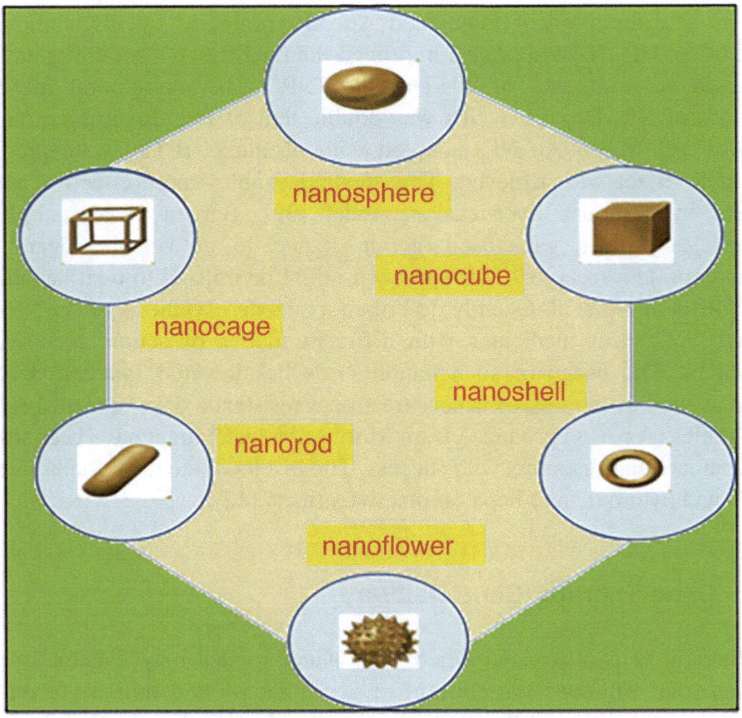

Fig. 7.4 Au NPs with various morphologies [29]

NPs may also be quickly functionalized with a wide range of compounds. Important characteristics of NPs include stability, the macroscopic quantum tunneling effect, and the existence of surface plasmon resonance (SPR) bands [30]. Unfortunately, surface modification has been demonstrated to alter drug toxicity, biodistribution, and pharmacokinetics in Au NPs, which are not biodegradable. Due to their special physicochemical and optical characteristics, gold nanoparticles (AuNPs) can serve as possible probes for the attachment of peptides since they are highly suitable for conjugation with minute biomolecules as DNA, proteins, enzymes, and amino acids [31–33]. For example, Gu et al. found that Au NPs coupled with PEG and 3-mercaptopropionic acid do not produce cytotoxicity in HeLa cervical cancer cells [34]. Using the MTT (3-(4,5-dimethylthiazol-2-yl)-2,5-diphenyltetrazolium bromide) assay, Wójcik and colleagues proved that glutathione-stabilized gold nanoparticles modified with non-covalently bound doxorubicin are more active against feline fibrosarcoma cell lines than unmodified Au NPs) [35]. Gold nanoparticles interact with folic acid PEG-amine, allowing medications to be delivered to tumors via ligation to the folate receptor located on the surface of certain cancer cells, according to Bhattacharya and colleagues [36]. Methotrexate-conjugated gold nanoparticles with diameters of 3 and 20 nm (MTX-Au NPs) were synthesized via a one-pot synthesis. Methotrexate (MTX) is a cytotoxic and folic acid antagonist. On human choriocarcinoma cell lines, researchers observed that 3 nm MTX-Au NPs are more cytotoxic than free MTX [37]. Other studies have produced new doxorubicin-conjugated Au NPs protected by PEGylation on the surface of Au NPs. This approach improved the medication's solubility and release in an acidic environment (Au NPs-Dox-PEG NPs). The intratumoral drug concentration of Au NPs-Dox-PEG NPs was double that of free doxorubicin using this approach [38]. When Au NPs coupled with paclitaxel (PTX) were produced, a comparable effect was achieved. This method enables more effective anticancer treatment in a mouse liver cancer model [39]. Bergen et al. coupled gold nanoparticles with a galactose-targeting ligand to deliver the ligand to the asialoglycoprotein receptor. This approach might be utilized to treat hepatocellular cancer [40]. Faroog et al. recently described a one-step synthesis of Au NPs loaded with two anticancer medicines with different modes of action: bleomycin and doxorubicin. The usage of such nanohybrids has lowered systemic medication toxicity as well as the risk of cancer treatment resistance development [41]. Silver nanoparticles (Ag NPs) produced from Morus alba leaf extract have been utilized to treat hepatocellular disorders with success. The results of the rat model revealed that Ag NPs had cytotoxic and hepatoprotective effects [42].

7.9.2 Gene Therapy/Gene Delivery

Gene therapy is another area where soft nanoparticles may eventually play an important role. Although the concept of gene therapy was initially formalized in the 1970s, it was not until the 1980s that white blood cells extracted from a girl's body were utilized to show in vitro gene transfer for the first time. Before being restored to the girl's body, certain types of genes were put into these cells. This person's immune system has improved, it was discovered. Since then, trials for various conditions have continued and are currently underway. In the approaching years, gene therapy will become a more essential subject for nanomedicine. An in-depth investigation is being conducted to establish whether and how the various targeted transport routes of active chemicals may be used to deliver nucleic acids, DNA fragments, and individual genes into tissue and cells using non-viral nanoparticles. A cutting-edge animal study using rats successfully proved the use of biodegradable, polymeric gene delivery nanoparticles to eradicate glioma cells in the brain and prolong animal longevity. Magnetofection is another option, in which, like active MDT, positively charged iron-containing nanoparticles are loaded with negatively charged nucleic acids and transported to target cells using a magnetic field before being sluiced into them by virtue of their suitably prepared surface structure. The nucleic acids are freed, and the particles are directed to the cell's iron metabolism. This technology, also known as "magnetic gene targeting" (MGT), is tempting for the treatment of cancer, neurological diseases, and myocardial infarction since the target cells are not affected. This notion is shown with a unique method of gene transfer that might be employed in stent angioplasty and could assist to bridge the gap between research and practical use. In a rat model, stents are used as a platform for magnetically targeted gene delivery, in which genes are transported to cells near artery lesions without damaging adjacent organs. The transferred genes are shielded by new magnetic nanoparticles, which facilitate their active distribution to their designated location. By attempting to replace and replenish insufficient DNA, nano-based gene therapy actively interferes with cellular activities at the molecular level. Because the capacity to repair DNA eventually leads to the ability to edit DNA with accuracy, situations involving ethical and societal dilemmas cannot be avoided. Is it permissible to improve human sensory and physical capacities without a medicinal justification, or simply to repair damaged or flawed DNA? And who will have access to the capability of genetic modification, under what conditions, and through what channels? How will we solve the body's weaknesses in the future? Gene therapy is a treatment or prevention strategy that uses exogenous DNA or RNA to cure or prevent illness. Popular viral vectors typically stimulate host immune systems, decreasing the efficiency of gene therapy. Non-viral solutions, such as metallic nanoparticles, can be employed to address the aforementioned issues. Recent study indicates that Au NPs of various forms (e.g., nanospheres or nanorods) preserve nucleic acid by blocking nuclease degradation of DNA or RNA [43, 44]. Au NPs coupled to oligonucleotides have unique characteristics that make them suitable gene regulation agents. Noncovalent (Au NPs may be functionalized with thiolated oligonucleotides) and covalent (Au NPs can be functionalized with thiolated oligonucleotides) carriers can be distinguished [45]. Covalent Au NPs, for example, can activate immune-related genes in peripheral blood mononuclear cells but not in immortalized, lineage-specific cell lines. This discovery suggests that such conjugates might be used in the development of gene delivery devices [46]. Lactoferrin-derived peptides coated with Au NPs were produced by Peng and colleagues. The conjugates created are capable of successfully delivering genes encoding vascular endothelial growth factor (VEGF), which promotes blood vessel development [47].

7.9.3 Nanoparticles for Protein Delivery

An increasing amount of research suggests that nanoparticles can be used as protein carriers. Organothiol, a molecular probe, might be utilized to study the structure and morphology of proteins bound to Au NPs. Joshi et al. developed insulin-functionalized Au NPs that were proven to be effective in transmucosal medication delivery for diabetes therapy in rat models [48]. Covering Au NPs with chitosan, a nontoxic biopolymer that significantly adsorbs insulin on its surface, can improve insulin delivery efficiency. Rathinaraj et al. immobilized anti-HER-2/neu monoclonal antibody herceptin on 29 nm Au NPs, which increased the drug's interaction with the relevant receptors on the surface of breast cancer cells (SK-BR3) [49]. Protein carriers have also been utilized in conjunction with Ag NPs. Di Pietro and colleagues functionalized Ag NPs with a peptide sequence of arginine, glycine, and aspartic acid (RGD), allowing Ag NPs to penetrate leukemia and neuroblastoma cells successfully [50]. Numerous uses of noble metal-based nanoparticles as carriers of biologically active chemicals enhance the prospect of more effective cancer and other contemporary disease therapy.

7.9.4 Radiation-Based Anticancer Therapy Using Nanoparticles

Radiation therapy (RT), in addition to chemotherapy and surgery, is a commonly used technique in the treatment of various malignancies. To boost the efficiency of radiation, several small substances (oxygen and its analogues, gemcitabine, and capecitabine), macromolecules (microRNAs, some proteins and peptides), and nanoparticles can be used [51]. The main goal of radiation treatment or proton therapy is to deliver a deadly dose of radiation to cancer cells while sparing the healthy tissue around them. To achieve this aim, two ways can be used: enhancing the radiation sensitivity of the cancer cells or tailoring the dosage to the tumor volume. Because gamma and X-rays display exponential dose deposition with tissue depth, a portion of the radiation dosage is supplied in front of or behind the tumor [52]. X-rays and other types of high-energy ionizing radiation are commonly employed to ionize water or cellular organelles. The primary target of ionizing radiation is water, which makes up the majority of a cell. As a result of this contact, water molecules are lysed. This mechanism, known as radiolysis, produces charged water species as well as free radicals such as hydrogen and hydroxyl radicals. The interaction of free radicals with DNA and biological components causes apoptosis. Several studies have found that hydroxyl ions are one of the primary sources of cellular damage induced by lipid peroxidation [53]. High atomic number nanoparticles, such as Au NPs, might promote the production of secondary electrons or reactive oxygen species (ROS). Nanoparticles can use this strategy to reduce total radiation exposure while boosting the dosage given locally to the tumor. Furthermore, the harmful consequences can be mitigated [54]. The main physical interactions between X-ray beams and metal-based nanoparticles are the photoelectric and Compton effects. The total or partial absorption of a photon by a nanoparticle causes the loss of one electron from its surface. Ions are created when ionizing radiation has enough energy to remove at least one electron from a nanoparticle. Charged particles are ionized fast because they may make direct contact with atomic electrons via Coulomb interactions and pass some of their kinetic energy [55]. The size of radio-sensitizing nanoparticles influences how they interact with radiation and biological systems. As a result, it is critical to prevent Au NP from accumulating in organs such as the liver and heart. Negative effects may ensue if this is not done. The size of nanoparticles is important in their radiation interactions. When Au NPs grow in size, they face additional ionizing events when they come into contact with secondary electrons and radiation. As a result, the dosage deposited around the Au NPs reduces [56]. According to a modest body of research on the charge of nanoparticles, positively charged nanoparticles interact with negatively charged cell membranes to significantly increase cell uptake. Furthermore, because some cancer cells contain strongly negatively charged glycocalyx on their surface, positively charged nanoparticles can precisely target cancer cells [57]. Silver nanoparticles also exhibit radio-sensitizing characteristics. Their anticancer mechanism is based on the activation of oxidative stress, the encouragement of apoptosis, and the improvement of cell membrane fluidity. The radiosensitization process of Ag NPs is most likely associated with the release of an Ag + cation from the silver-based structures within the cells. This cation may receive an electron and operate as an oxidizing agent, resulting in the production of ROS [58, 59].

Noble metal-based nanoparticles hold great promise for improving cancer therapy. It would be intriguing to see if treating tumor cells with two (or more) distinct types of nanoparticles at the same time might have a synergistic impact. Furthermore, the development of nanocomplexes based on noble metals, such as gold and platinum, seems promising. The examination of the combined effects of noble metal-based nanoparticles with a different type of radio-sensitizing drug (e.g., gemcitabine) may potentially provide excellent findings.

7.9.5 Diagnostic Molecular Bioimaging

Many magnetic nanoparticles (MNPs) have been employed in biomedicine, most notably as MRI contrast agents [57]. Because of advances in nanotechnology, molecular diagnostics now offer higher sensitivity, specificity, and multiplexing. In high atomic number nanoparticles, intense plasmon resonance driven absorption and scattering features are common. X-ray computer tomography (CT) is a frequently utilized diagnostic tool that is both simple to use and reasonably priced. This imaging technique is based on the premise that healthy and ill tissues have different densities, resulting in a contrast. By using contrasting chemicals, the difference between cancer cells and normal cells can be improved. UM-A9 antibodies coupled to gold nanorods have been employed by some. These gold nanoprobes have shown tumor specificity, which improves CT contrast. The aforementioned devices can help in the identification of cellular cancer. Researchers used small Au NPs (1.9 nm) as an X-ray contrast agent to detect cancers in mice [60]. After 24 hours, there were no Au NPs in the blood, but they had already accumulated extensively in the kidney (10% of the injected dose/g, ID/g), tumor (4% ID/g), liver (3% ID/g), and muscle (1%) after just 15 min. The Au NPs were most likely eliminated by the kidneys due to their tiny size and lack of liver concentration [61]. Ag NPs can also be employed in place of iodine and other traditional contrast agents. Karunamuni et al. employed Ag NPs effectively for dual-energy X-ray breast imaging, and their findings are promising. HER81 monoclonal antibodies, which bind to Her2 receptors on the surface of an SKBR3 breast cancer cell line with extraordinary effectiveness, have recently been linked to gold nanorods [62]. Although these probes seem promising, it is uncertain if they will successfully penetrate tumor tissue in vivo. Gold nanoparticles combined with anti-epidermal growth factor receptor (anti-EGFR) monoclonal antibodies had a six-fold greater affinity for binding tumors. Au NPs and Gd NPs were used in magnetic resonance (MR) to distinguish transplanted human brain stem cells due to their higher T1 relaxivity and cell uptake. The resulting conjugates allowed for the detection of 70% of the transplanted cells.

Blythe and Willets have looked at high-resolution fluorescence imaging as a technique for observing fluorescently labeled ligands on the surface of Au NPs. This technique made it possible to reconstitute the orientation and geometry of the substrates for the Au NPs and showed how unevenly DNA binds to the surface of the Au NPs. Nanoparticles can be used to track a target's position or specific intracellular activities. The techniques, including two-photon-induced photoluminescence (TPL), make it possible to identify Au NPs in hippocampal neurons of various shapes (spherical, rod-shaped, and urchin-shaped). A significant discovery is that cells may favor some nanoparticle forms over others [63]. Using Au NPs (nanospheres, nanorods, and nanocages), Nanospheres are the ideal form for cellular internalization. Therefore, nanorods and nanocages are more appropriate for imaging. A prospective technique for observing the accumulation of such nanoparticles in diverse cell areas has been investigated: gold nanoparticles help confocal Raman imaging. Oligonucleotide-based probes that reveal intracellular RNA are essential tools for assessing biological activity in various systems. The molecules of the probe in this case appear to have trouble entering the cell, and the usage of oligonucleotide-based probes is restricted by their susceptibility to nucleases. To get over this limitation, researchers have developed nano-flares consisting of gold nanoparticles functionalized with nucleic acid aptamers with affinity for adenosine triphosphate (ATP). Flow cytometry can distinguish between different cell types based on their gene expression patterns thanks to these structures [64]. To detect cancer cells, unique tags based on the surface enhanced Raman spectroscopy (SERS) method have been created. Manufactured gold dimers connected by biphenyl-4,4′-dithiol (DBDT) generated a greater signal and more cytotoxic activity when compared to normal Au NPs. This implies that imaging based on SERS could be more sensitive than imaging based on traditional fluorescence. Wu et al. developed the first sensor of this type to identify metals in live cells in a distinct study by conjugating DNAzyme onto 13 nm Au NPs [65, 66]. Additionally, nanoparticles have theranostic properties that make it easier to diagnose and treat medical conditions [67]. Even gold nano-stars outperform gold nano-shells in photothermal treatment and imaging (SERS, CT, and TPL imaging) of animal models of sarcomas. There are reports on the theranostic usage of Ag or Pt nanoparticles [68–72].

Noble metal based nanoparticles may easily have their surfaces functionalized, they are non-cytotoxic, and they have optical properties that make them valuable for diagnostics. However, several issues need to be addressed, such as the potential for aggregation, toxicity, or non-specific binding. Another important hurdle is the potential for nanoparticles to become opsonized once they have entered the circulation. Furthermore, the use of nanoparticles for diagnostic purposes may be hampered or rendered impossible by the presence of biological barriers. Therefore, finding creative solutions to these problems is essential.

7.10 Antibacterial, Antiviral, and Antifungal Substances

Silver nanoparticles (NPs) are the most often used antibacterial substances among noble metal-based nanoparticles. Silver salts were frequently employed as medicines as early as the eighteenth century to treat bacterial infections, such as ulcers, burns, open sores, and conjunctivitis. Ag NPs are currently applied as bactericides more frequently since several bacteria are developing antibiotic resistance. Everyone on the planet is at risk from the serious health danger known as AMR, which is expected to cause ten million yearly deaths by the year 2050. Methicillin-resistant Staphylococcus aureus and carbapenemase-producing Enterbacteriaceae are two deadly examples of AMR. Several theories explain how Ag NPs are bactericidal. Yamanaka et al. have shown that Ag NPs connect to the cell walls of bacteria and fungus, causing damage to the cell membrane's structure, enabling internal materials to leak out, and eventually causing cell death [69]. Additionally, silver ions can cause oxidative stress by producing free radicals. Another strategy is based on the harmful interactions of Ag NPs with genomic DNA that prevent normal replication. The silver ions that may be released by Ag NPs might decrease the transcriptional activity of enzymes and other proteins. Ag NPs that are more efficient against Gram-negative bacteria were developed. Ag NPs' function in the dephosphorylation of putative peptide substrates' tyrosine residues, which are essential for cell survival and proliferation and result in bacterial death, was attributed to this activity.

Ag NPs were produced using green chemistry and a chamomile extract. It has been discovered that these two bactericidal substances cooperate to kill the four bacterial strains S. aureus, P. aeruginosa, B. subtilis, and E. coli. In a different study, Ag NPs from clove extract were synthesized and showed that they have stronger antibacterial and antifungal effects than pure clove extract [70]. Despite positive results, it is unquestionably less usual to utilize Au NPs as bactericidal agents. Chemically produced Au NPs have the potential to be used as antibacterial agents, notably against Gram-negative bacteria. Similar to their size, Au NPs' toxicity varied depending on the sort of bacteria they were exposed to. Au NPs accumulate as agglomerates after entering the bacterial cell wall. Zebrafish were used as an animal model to demonstrate Pt NPs' bactericidal activity against E. coli and A. hydrophila [71].

Ag NPs have fungicidal effects that are similar to their bactericidal effects but less effective. It is unclear how Ag NPs interact with fungi. For instance, Trichosporon asahii, a fungus, recently exhibited antibiotic resistance. Novel treatments based, for instance, on silver nanoparticles are urgently needed since trichosporosis can be lethal. Gold nano-discs had the most potent antifungal efficacy. Noble metal-based nanoparticles' antiviral properties have also been supported by an increasing number of research, however almost all of them concentrate on Ag NPs. There hasn't been much research done into how Ag NPs interact with viruses up to this time. The scientists postulate that the antiviral mode of action involves interactions between Ag NPs and the double-stranded DNA of the virus. Speshock et al. obtained 10 nm Ag NPs and tested them as antiviral agents against the Tacaribe virus. These Ag NPs can halt viral replication in its early stages, according to their research [72]. Due to their increasing pharmacological resistance, bacteria require the development of novel, effective techniques of eradication. The usage of noble metal-based nanoparticle carriers is one of the solutions, as shown in the aforementioned cases. The bactericidal effect of such carriers may be enhanced by the addition of particular antibiotics.

7.11 Conclusion

The application of nanotechnology has already had a considerable influence on medicine and the medical sciences. Noble metal-based nanoparticles, in particular Au and Ag NPs, have a wide range of applications, which suggests that further study in this field is important. These structures may be utilized for a variety of things, including as drug delivery systems, components that improve radiation therapy for the treatment of cancer, molecular imaging support, and compounds with bactericidal, fungicidal, and antiviral properties. Because they can be an excellent tool for the detection and treatment of a number of illnesses, with an emphasis on cancer, in light of the aforementioned, it is vital to use nanoparticle-based therapies in clinical trials. However, given the characteristics of particular nanoparticles, such as their biodegradability or porosity, there are a number of challenges that must be overcome. Furthermore, nothing is known about the toxicity of these nanoparticles and how they interact with live, healthy cells. In order for the generated nanoparticles to reach the target site safely and with high selectivity, their pharmacokinetic qualities must be enhanced. This can now be accomplished by giving the nanoparticles the appropriate ligands. Furthermore, a comprehensive approach for sensitizing cancer cells employing a range of noble metal-based nanoparticles can result in superior treatment results. Without a doubt, the development of nanotechnology has made treatment and diagnosis feasible. The currently authorized nanoparticle systems have, in certain circumstances, improved the therapeutic index of medicines by reducing medication toxicity or raising medicinal effectiveness. The next generation of nanoparticle systems may contain targeting ligands like antibodies, peptides, or aptamers, which might improve their effectiveness or reduce their toxicity. More complex systems, including multifunctional nanoparticles that can be utilized simultaneously for treatment, imaging, and targeting, will be the subject of future study. The functionality of nanoparticles becomes increasingly complex as a result of the addition of targeted ligands, necessitating the careful construction of nanoparticles with the physicochemical and biological properties to perform each required function. It is obvious that this has been the fundamental barrier to the adoption of targeted particles in clinical practice since the first targeted liposome was disclosed more than 20 years ago. Few systems have ever made it to clinical trials, and none have ever received a usage license. Thanks to the creation of safer nanomaterials and cutting-edge engineering methods that result in nanoparticles that are appropriately engineered, we believe that more multifunctional nanoparticles will be used in clinical settings in the future.

References

1. Taneja S, Thakur P, Ravelo B, Thakur A (2022) Nanocrystalline samarium doped nickel-zinc-bismuth ferrites: investigation of structural, electrical and dielectric properties. Mater Res Bull 154:111937
2. Lin W (2015) Introduction: nanoparticles in medicine. Chem Rev 115(19):10407–10409
3. Punia P, Bharti MK, Chalia S, Dhar R, Ravelo B, Thakur P, Thakur A (2021) Recent advances in synthesis, characterization, and applications of nanoparticles for contaminated water treatment-a review. Ceram Int 47(2):1526–1550
4. Singh R, Thakur P, Thakur A, Kumar H, Chawla P, Rohit JV, Kaushik R, Kumar N (2021) Colorimetric sensing approaches of surface-modified gold and silver nanoparticles for detection of residual pesticides: a review. Int J Environ Anal Chem 101(15):3006–3022
5. Bharti MK, Gupta S, Chalia S, Garg I, Thakur P, Thakur A (2020) Potential of magnetic nanoferrites in removal of heavy metals from contaminated water: mini review. J Supercond Nov Magn 33:3651–3665
6. Sharma P, Sharma A, Sharma M, Bhalla N, Estrela P, Jain A, Thakur P, Thakur A (2017) Nanomaterial fungicides: in vitro and in vivo antimycotic activity of cobalt and nickel Nanoferrites on phytopathogenic fungi. Global Chall 1(9):1700041
7. Punia P, Bharti MK, Dhar R, Thakur P, Thakur A (2022) Recent advances in detection and removal of heavy metals from contaminated water. ChemBioEng Rev 9(4):351–369
8. Mg K, Krenn HF, Wagner W, Resch R (2015) History and possible uses of nanomedicine based on nanoparticles and nanotechnological Progress. J Nanomed Nanotechnol 06:06
9. Thakur P, Thakur A (2022) Nanomaterials, their types and properties. In: Thakur A, Thakur P, Khurana SP (eds) Synthesis and applications of nanoparticles. Springer, Singapore
10. Yang F, Xue J, Wang G, Diao Q (2022) Nanoparticle-based drug delivery systems for the treatment of cardiovascular diseases. Front Pharmacol 13
11. Mathur P, Thakur A, Lee JH, Singh M (2010) Sustained electromagnetic properties of Ni-Zn-Co nano-ferrites for the high frequency applications. Mater Lett 64(24):2738–2741
12. Dhanda N, Thakur P, Sun AC, Thakur A (2023) Structural, optical and magnetic properties along with antifungal activity of Ag-doped Ni-Co nanoferrites synthesized by eco-friendly route. J Magn Magn Mater 572:170598
13. Nair SS, Krishnia L, Trukhanov A, Thakur P, Thakur A (2022) Prospect of double perovskite over conventional perovskite in photovoltaic applications. Ceram Int 48(23):34128–34147

14. Thakur P, Chahar D, Thakur A (2022) Visible light assisted photocatalytic degradation of methylene blue dye using Ni doped co-Zn nanoferrites. Adv Nano Res 12(4):415–426
15. Rana K, Thakur P, Sharma P, Tomar M, Gupta V, Thakur A (2015) Improved structural and magnetic properties of cobalt nanoferrites: influence of sintering temperature. Ceram Int 41(3): 4492–4497
16. Harish V, Tewari D, Gaur M, Yadav AB, Swaroop S, Bechelany M, Barhoum A (2022) Review on nanoparticles and nanostructured materials: bioimaging, biosensing, drug delivery, tissue engineering, antimicrobial, and agro-food applications. Nano 12(3):457
17. Zhang L, Gu F, Chan JC, Wang SA, Langer R, Farokhzad OC (2008) Nanoparticles in medicine: therapeutic applications and developments. Clin Pharmacol Ther 83(5):761–769
18. Dhanda N, Kumari S, Kumar R, Kumar D, Sun AC, Thakur P, Thakur A (2023) Influence of Ni over magnetically benign Co ferrite system and study of its structural, optical, and magnetic behaviour. Inorg Chem Commun 151:110569
19. Chahar D, Taneja S, Bisht S, Kesarwani S, Thakur P, Thakur A, Sharma PB (2021) Photocatalytic activity of cobalt substituted zinc ferrite for the degradation of methylene blue dye under visible light irradiation. J Alloys Compd 851:156878
20. Woodle MC (1998) Controlling liposome blood clearance by surface-grafted polymers. Adv Drug Deliv Rev 32(1–2):139–152
21. Sapra P, Allen TM (2002) Internalizing antibodies are necessary for improved therapeutic efficacy of antibody-targeted liposomal drugs. Cancer Res 62:7190–7194
22. Simões S, Moreira JN, Fonseca CT, Düzgüneş N, De Lima MCP (2004) On the formulation of pH-sensitive liposomes with long circulation times. Adv Drug Deliv Rev 56(7):947–965
23. Krishnia L, Thakur P, Thakur A (2022) Synthesis of nanoparticles by physical route. In: Synthesis and applications of nanoparticles. Springer Nature Singapore, Singapore, pp 45–59
24. Martinho N, Damgé C, Reis CP (2011) Recent advances in drug delivery systems. J Biomater Nanobiotechnol 02(05):510–526
25. Rigon RB, Oyafuso MH, Fujimura AT, Gonçalez ML, Prado AHD, Gremião MPD, Chorilli M (2015) Nanotechnology-based drug delivery systems for melanoma antitumoral therapy: a review. Biomed Res Int 2015:1–22
26. Anajafi T, Mallik S (2015) Polymersome-based drug-delivery strategies for cancer therapeutics. Ther Deliv 6(4):521–534
27. Vashist A, Vashist A, Gupta YM, Ahmad S (2014) Recent advances in hydrogel based drug delivery systems for the human body. J Mater Chem B 2(2):147–166
28. Yokoyama M (2014) Polymeric micelles as drug carriers: their lights and shadows. J Drug Target 22(7):576–583
29. Hu X, Zhang Y, Ding T, Liu J, Zhao H (2020) Multifunctional gold nanoparticles: a novel nanomaterial for various medical applications and biological activities. Front Bioeng Biotechnol 8:990
30. Bhalla N, Taneja S, Thakur P, Sharma PK, Mariotti D, Maddi C, Ivanova O, Petrov D, Sukhachev A, Edelman IS, Thakur A (2021) Doping independent work function and stable band gap of spinel ferrites with Tunable plasmonic and magnetic properties. Nano Lett 21(22): 9780–9788
31. Ajnai G, Chiu AP, Kan T, Cheng C, Tsai T, Chang J (2014) Trends of gold nanoparticle-based drug delivery system in cancer therapy. J Exp Clin Med 6(6):172–178
32. Shahdeo D, Kesarwani V, Suhag D, Ahmed J, Alshehri SM, Gandhi S (2021) Self-assembled chitosan polymer intercalating peptide functionalized gold nanoparticles as nanoprobe for efficient imaging of urokinase plasminogen activator receptor in cancer diagnostics. Carbohydr Polym 266:118138
33. Shahdeo D, Kesarwani V, Suhag D, Ahmed J, Alshehri SM, Gandhi S (2022) Corrigendum to" self-assembled chitosan polymer intercalating peptide functionalized gold nanoparticles as nanoprobe for efficient imaging of urokinase plasminogen activator receptor in cancer diagnostics"[Carbohydrate Polymers, 266 (2021) 118138]. Carbohydr Polym 284:119239

34. Gu Y-J, Cheng J, Lin C-C, Lam YW, Cheng SH, Wong W-T (2009) Nuclear penetration of surface functionalized gold nanoparticles. Toxicol Appl Pharmacol 237(2):196–204
35. Wojcik M, Lewandowski W, Król M, Pawłowski K, Mieczkowski J, Lechowski R, Zabielska K (2015) Enhancing anti-tumor efficacy of doxorubicin by non-covalent conjugation to gold nanoparticles—in vitro studies on feline fibrosarcoma cell lines. PLoS One 10(4):e0124955
36. Bhattacharya R, Patra CR, Earl A, Wang S, Katarya A, Lu L, Kizhakkedathu JN, Yaszemski MJ, Greipp PR, Mukhopadhyay D, Mukherjee P (2007) Attaching folic acid on gold nanoparticles using noncovalent interaction via different polyethylene glycol backbones and targeting of cancer cells. Nanomedicine 3(3):224–238
37. Tran NQ, Wang T, Lin C, Tai Y (2013) Synthesis of methotrexate-conjugated gold nanoparticles with enhanced cancer therapeutic effect. Biochem Eng J 78:175–180
38. Cui T, Liang J, Chen H, Geng D, Jiao L, Yang J, Qian H, Zhang C, Ding Y (2017) Performance of doxorubicin-conjugated gold nanoparticles: regulation of drug location. ACS Appl Mater Interfaces 9(10):8569–8580
39. Ding Y, Zhou Y, Chen H, Geng D, Wu D, Hong JT, Shen WZ, Hang T, Zhang C (2013) The performance of thiol-terminated PEG-paclitaxel-conjugated gold nanoparticles. Biomaterials 34(38):10217–10227
40. Bergen JM, Von Recum HA, Goodman TT, Massey AP, Pun SH (2006) Gold nanoparticles as a versatile platform for optimizing physicochemical parameters for targeted drug delivery. Macromol Biosci 6(7):506–516
41. Farooq M, Novosad V, Rozhkova EA, Wali H, Ali A, Fateh AA, Neogi PB, Neogi A, Wang Z (2022) Retraction note: gold nanoparticles-enabled efficient dual delivery of anticancer therapeutics to HeLa cells. Sci Rep 12:1
42. Singh AL, Dar MJ, Joshi B, Sharma B, Shrivastava S, Shukla S (2018) Phytofabrication of silver nanoparticles: novel drug to overcome hepatocellular ailments. Toxicol Rep 5:333–342
43. Kong FY, Zhang JW, Li RF, Wang ZX, Wang WJ, Wang W (2017) Unique roles of gold nanoparticles in drug delivery, targeting and imaging applications. Molecules 22(9): 1445. PMID: 28858253; PMCID: PMC6151763
44. Riley MA, Vermerris W (2017) Recent advances in nanomaterials for gene delivery—a review. Nano 7(5):94
45. Ding Y, Jiang Z, Saha K, Kim C, Kim SW, Landis RF, Rotello VM (2014) Gold nanoparticles for nucleic acid delivery. Mol Ther 22(6):1075–1083
46. Kim EY, Schulz R, Swantek P, Kunstman KJ, Malim MH, Wolinsky SM (2012) Gold nanoparticle-mediated gene delivery induces widespread changes in the expression of innate immunity genes. Gene Ther 19(3):347–353
47. Peng L, Huang Y, Zhang C, Hu BX, Chen Y, Chu Y, Jiang Z, Gao J, Mao Z (2016) Integration of antimicrobial peptides with gold nanoparticles as unique non-viral vectors for gene delivery to mesenchymal stem cells with antibacterial activity. Biomaterials 103:137–149
48. Joshi HM, Bhumkar DR, Joshi K, Pokharkar V, Sastry M (2006) Gold nanoparticles as carriers for efficient transmucosal insulin delivery. Langmuir 22(1):300–305
49. Rathinaraj P, Al-Jumaily AM, Huh S (2015) Internalization: acute apoptosis of breast cancer cells using herceptin-immobilized gold nanoparticles. Breast Cancer 51
50. Di Pietro P, Zaccaro L, Comegna D, Del Gatto À, Saviano M, Snyders R, Cossement D, Satriano C, Rizzarelli E (2016) Silver nanoparticles functionalized with a fluorescent cyclic RGD peptide: a versatile integrin targeting platform for cells and bacteria. RSC Adv 6(113): 112381–112392
51. Wang H, Mu X, He H, Zhang X (2018) Cancer radiosensitizers. Trends Pharmacol Sci 39(1): 24–48
52. Retif P, Pinel S, Toussaint M, Frochot C, Chouikrat R, Bastogne T, Barberi-Heyob M (2015) Nanoparticles for radiation therapy enhancement: the key parameters. Theranostics 5(9): 1030–1044
53. Yang X, Gao L, Guo Q, Li Y, Ma Y, Yang JD, Gong C, Yi C (2020) Nanomaterials for radiotherapeutics-based multimodal synergistic cancer therapy. Nano Res 13(10):2579–2594

54. Jeynes JCG, Merchant MJ, Spindler A, Wera AC, Kirkby KJ (2014) Investigation of gold nanoparticle radiosensitization mechanisms using a free radical scavenger and protons of different energies. Phys Med Biol 59(21):6431
55. Butterworth KT, McMahon SJ, Taggart LE, Prise KM (2013) Radiosensitization by gold nanoparticles: effective at megavoltage energies and potential role of oxidative stress. Transl Cancer Res 2(4):269–279
56. Saberi A, Shahbazi-Gahrouei D, Abbasian M, Fesharaki M, Baharlouei A, Arab-Bafrani Z (2017) Gold nanoparticles in combination with megavoltage radiation energy increased radiosensitization and apoptosis in colon cancer HT-29 cells. Int J Radiat Biol 93(3):315–323
57. Haume K, Rosa S, Grellet S, Śmiałek MA, Butterworth KT, Solov'yov AV, Prise KM, Golding JP, Mason NJ (2016) Gold nanoparticles for cancer radiotherapy: a review. Cancer. Nanotechnology 7:1
58. Su XY, Liu PD, Wu H, Gu N (2014) Enhancement of radiosensitization by metal-based nanoparticles in cancer radiation therapy. Cancer Biol Med 11(2):86–91
59. Suhag D, Shahdeo D, Gandhi S (2019) Emerging trends and advances in targeted nanotheranostics. SMC Bulletin 10(1)
60. Cormode DP, Naha PC, Fayad ZA (2014) Nanoparticle contrast agents for computed tomography: a focus on micelles. Pubmed 9(1):37–52
61. Popovtzer R, Singh IP, Kotov NA, Popovtzer A, Balter JM, Carey TE, Kopelman R (2008) Targeted gold nanoparticles enable molecular CT imaging of cancer. Nano Lett 8(12): 4593–4596
62. Karunamuni R, Tsourkas A, Maidment ADA (2014) Exploring silver as a contrast agent for contrast-enhanced dual-energy X-ray breast imaging. Br J Radiol 87(1041):20140081
63. Blythe KL, Willets KA (2016) Super-resolution imaging of fluorophore-Labeled DNA bound to gold nanoparticles: a single-molecule, single-particle approach. J Phys Chem C 120(2):803–815
64. Shah N, Dong J, Bischof JC (2011) Cellular uptake and nanoscale localization of gold nanoparticles in cancer using label-free confocal Raman microscopy. Mol Pharm 8(1):176–184
65. Zheng D, Seferos DS, Giljohann DA, Patel PC, Mirkin CA (2009) Aptamer nano-flares for molecular detection in living cells. Nano Lett 9(9):3258–3261
66. Wu P, Hwang KO, Lan T, Lu Y (2013) A DNAzyme-gold nanoparticle probe for uranyl ion in living cells. J Am Chem Soc 135(14):5254–5257
67. Chandra P, Das D, Abdelwahab AA (2010) Gold nanoparticles in molecular diagnostics and therapeutics. Dig J Nanomater Biostruct 5(2):363–367
68. Lutgring JD, Limbago B (2016) The problem of carbapenemase-producing-carbapenem-resistant-enterobacteriaceae detection. J Clin Microbiol 54(3):529–534
69. Yamanaka M, Hara K, Kudo J (2005) Bactericidal actions of a silver ion solution on Escherichia coli, studied by energy-filtering transmission electron microscopy and proteomic analysis. Appl Environ Microbiol 71(11):7589–7593
70. Parlinska-Wojtan M, Kus-Liśkiewicz M, Depciuch J, Sadik OA (2016) Green synthesis and antibacterial effects of aqueous colloidal solutions of silver nanoparticles using camomile terpenoids as a combined reducing and capping agent. Bioprocess Biosyst Eng 39(8): 1213–1223
71. Ahmed KBA, Raman T, Anbazhagan V (2016) Platinum nanoparticles inhibit bacteria proliferation and rescue zebrafish from bacterial infection. RSC Adv 6(50):44415–44424
72. Speshock J, Murdock RC, Braydich-Stolle LK, Schrand AM, Hussain SM (2010) Interaction of silver nanoparticles with Tacaribe virus. J Nanobiotechnol 8:1

Polymers and Polymeric Composites in Nano/Bio-Medicine

8

Aarti Singh and Deepa Suhag

Abstract

The new class of materials that has captured the attention of scientists is polymeric composite nanomaterials. Their special qualities, including high surface-to-volume ratios, simplicity in functionalization, and photothermal conversion, among others, make them extremely adaptable for a wide range of applications, from energy storage to optoelectronics to biological ones. Recent studies have demonstrated the effectiveness of polymers in tissue engineering, biosensing, drug administration, and cancer photothermal treatment (PTT). When combined with hydrogels and scaffolds, these materials can be made more biocompatible and can aid in treating a variety of illnesses and injuries. However, as two-dimensional nanomaterials-based polymeric composites for biomedical applications are a relatively new topic, there is a considerable amount of fragmented information available.

Keywords

Polymeric composite · Biocompatible · Hydrogels · Scaffolds

A. Singh (✉)
Molecular Science and Engineering, Amity Institute of Click Chemistry Research and Studies, Amity University, Noida, Uttar Pradesh, India

D. Suhag
Centre for Nanotechnology, Amity University Haryana, Gurugram, India

NanoLatticeX, Amity University Haryana, Gurugram, Haryana, India

Nanotechnology Wing, Innovative Science Research Society, Shimla, India

D. Suhag et al. (eds.), *Integrated Nanomaterials and their Applications*,
https://doi.org/10.1007/978-981-99-6105-4_8

8.1 Introduction

In 1959 Nobel Prize winner Richard Feynman was first to forecast the rise of a new domain of science dealing with 1–100 nm size structures. Fifty years later, the uncertain outcome is the skill to employ materials with similar minute scale which is applied by nature [1]. The huge impact of nanotechnology is towering in our lives nowadays. The development of systems and devices holding the tendency to revolutionize medical and healthcare discipline has never been recognized earlier.

At nano-metre scale quantum mechanical character of electrons within matter is affected due to atomic interactions. Fundamental features of material such as their colour, charge capacity, magnetic properties, temperature can be tuned without altering its chemical arrangement. Focussing on this potential will lead to the development of high-performance products that were not feasible earlier.

Large surface-to-volume ratio and mechanical properties are remarkable unique feature among polymer/nanomaterials as it offers many exciting areas of research [2]. The stiffness and hardness demonstrated by nanomaterials make them noteworthy in polymer reinforcement. Recent research reports present that minute additions of nanomaterials ~1 wt % may enhance the mechanical properties remarkably, although clear mechanism behind the dramatic increase is not understood [3].

Altogether, the molecular interaction between the polymer matrices and nanomaterials governs the enhancement with the key role played by availability of a large surface area for such interplay.

8.2 Polymer Nanocomposites

Polymer nanocomposites are branches of composite material in which nanofillers or nanoparticles are diffused inside the polymer matrix where the filler has dimension between 1–100 nm scale. Recent reports have revealed that polymer properties such as dimensional stability, resistance to thermal degradation, resistance to chemical attack, toughness, modulus, strength, flammability resistance, and gas impermeability can be enhanced by the addition of small quantity of nanofiller [4].

Nanocomposite can be classified into three categories on the basis of size and shape of dispersed filler. If all the three dimensions are in nanoscale, we can call it iso-dimensional nanoparticle or nanoparticles, examples are, zinc oxide, aluminium oxide, carbon black, silica [5]. One-dimensional nanoparticle in which two dimensions are in nano-metre range and third forms an elongated structure, examples include carbon nanotubes, etc. The third kind of nanofiller exhibit sheet like shape as two of the dimensions are not confined to nanoscale such as layered silicates, graphene, layered graphite flakes are few examples of two-dimensional nanofillers [6]. Appropriate incorporation of these nanofillers will develop nanocomposite with well-suited features [7]. Earlier micron scale fillers were added into composites now it is replaced by nanofillers. The distribution of nanofiller at the molecular level within polymer matrix can lead to tremendous advancements in the eventual characteristics of the nanocomposites [8]. The interfacial synergy between matrix and dispersed phase along with aspect ratio of the filler play significant role in promoting the properties of composite [9]. On that account nanofillers such as carbon nanotubes, clays, graphene, nanofibres with large aspect ratio and uniform distribution in polymer matrix are proving to be stronghold to architectural materials as at minimum volume fraction of filler is required to strengthen polymer properties. Researchers are working on fabrication methods to achieve the desired properties in polymer nanocomposites.

8.3 Polymer as Biomaterial

The benchmark for selecting a polymer as a biomaterial is to develop an alliance between degradation time till the need of application and mechanical strength. For biomedical applications ideally polymer must own the following features;

(i) Should not elicit any toxic/inflammatory response.
(ii) If administered inside the body, it should be metabolized after meeting its purpose without any trace.
(iii) Easy to process as a final product.
(iv) Shelf-life is acceptable.
(v) Easy to sterilize.

Hence, in spanning area of materials science to chemistry, biology, and bioengineering the main challenge is to facilitate generation of nanocomposite materials with notable better properties. The major goal in these approaches is to tailor the constituents and morphology of fabricated biomaterials, in which the nanofillers are being distributed with favourable orientation in the polymer matrices. Owing huge potency of polymers in biomedical applications, several studies have been described regarding the development of nanocomposites [10].

8.4 Nanofillers

One of the common raw materials utilized in material science is filler. A production of 50 million tons per year of filler is recorded. They can exhibit various shapes, such as flakes, rods, discs, or spherical forms. Primarily the properties of filler are resolved by geometry, chemical coating, size, and functionalization. Nanofillers have one magnitude in the nano range and they act as a dopant. The inclusion of nanofiller in the polymer matrix can affect the material characteristics necessarily by improved thermal, mechanical, rheological, and optical properties. The ratio of mixture between polymer matrix and the nanofiller greatly influences the composite material. Nanofillers can be classified as one-dimensional (1D) nanoplates [11].

8.4.1 Chemistry of Nanofillers

The primary properties determined by nanofiller chemistry are related to interaction between nanofiller and polymer. Firstly, the enthalpic interaction of the polymer chain is affected by nanofiller chemistry. The efficient stress transfer over polymer/ nanofiller interface. Such kind of enthalpic synergy can be described by the van der Waals interplay or covalent bonding between the polymer links and nanofiller. The morphology of the synthesized polymer nanocomposite is significantly influenced by the strength of these interactions. The polymer blend theory suggests that polymer and particle interaction could alter the miscibility barrier for given size scale of the polymer and nanofiller [12]. In agreement to this effect, nanoparticles phase separate giving rise to nanofiller rich domains with low polymer content. Inside these domains attractive inter-particle interrelation makes the aggregates as a large particle with filler rather than individual fillers at nanoscale. Whereas weak interaction leads to the deformation of aggregates playing crucial role during loss and storage modulus and later these deformation processes determine mechanical strength of synthesized polymer nanocomposite [13].

8.4.1.1 Shape and Size of Nanofiller

The filler dimension is at the heart of the effect of "nanoscale" on mechanical characteristics in polymer nanocomposites. Generally, two important features, size and shape of filler dictate the overall properties of polymer nanocomposite:

(i) Surface-to-volume ratio.
(ii) Excluded volume interactions.

The driving force for the fabrication of polymer nanocomposite is surface-to-volume ratio of nanofiller. This ratio depicts the amount of interfacial surface in comparison to bulk material in the nanocomposite. Among the various reasons for its significance, the interfacial region commands structural alignment on the molecular level and is responsible for the efficient stress transfer across the constituents of the nanocomposite. Moreover, if the interfacial region amount is increased, it maximizes the potency of material's novel characteristics.

The excluded volume aids in the form that two solid nanofillers cannot take similar space at the same time, hence because of entropy of filler a force of repulsion is generated. A significant conclusion drawn from excluded volume contributions is that large aspect ratio, for instance, long fibres are difficult to distribute iso-tropically as there is increase in aspect ratio. The excluded volume of nanofiller combined with the configurational entropy of polymer play a crucial role in dispersion of nanofillers in polymer. The excluded volume of the nanoscale fillers coupled with the configurational entropy of the polymer chain has been shown to play an important role in the spatial distribution of nanofillers in polymer [14].

8.5 Different Types of Fillers

8.5.1 Nano-Clays /Layered Silicates

Layered silicates constitute platelets, naturally assembled with a huge number of sheets. Clays are categorized on the ground of placement of ions inside geometry mesh, their crystalline make up and quantity. Various types of clay minerals are smectites, bentonites, and hectorite [15]. The most widely used nano-clay in material science currently is montmorillonite which is a class of bentonite. A single sheet of montmorillonite was stated to have 178 GPa–265 Gpa in Young's modulus [16]. This clay nanofiller is intertwined between two octahedral layered silicate of alumina. With 1–5 nm thickness and diameter up to 100–500 nm these alumina silicate platelets have high aspect ratio (>50). The stiffness of the minerals presents in clay displayed enhanced mechanical properties in the nanocomposite blend [17].

8.5.2 Carbon Nanofiller

Carbon nanomaterials has attracted much attention in diverse research field such as zero-dimensional (0D) fullerene, one-dimensional (1D) carbon nanotube, and two-dimensional (2D) graphene sheet [18]. They are one of the important nanomaterials for the development of scaffolds due to their similar dimension; they act as physical analogue with extracellular matrix (collagen). They possess outstanding mechanical properties to reinforce organic/inorganic artificial scaffolds [19]. Moreover carbon nanomaterials have good conductive properties which can be applied for electrical simulation to artificial scaffolds. Carbon nanomaterials have shown function in wound healing. Interesting properties such as obstruction of pathological process during wound healing have been exhibited by fullerenes [20]. These materials showed anti-inflammatory and antioxidant ability which designate them as wound healing agents. Recently for fungal and bacterial wound infection photothermal treatment by utilizing graphene oxide nanosheets have been reported [21]. An alternative to antibiotics as a non-invasive technique the synergistic effect created by graphene oxide and irradiation with near-infrared laser elicit action against many pathogens thus resulting in rapid wound healing.

Carbon nanomaterials have exhibited remarkable biocompatibility due to their physicochemical properties such as surface roughness, functionalization, and reduction state which makes them suitable substrate for growth and differentiation of several mammalian cells. Carbon nanomaterials with 3D framework were generated through electrospinning technology, hydrothermal process, and solvent casting method. The electrical and mechanical property of carbon nanomaterial-based scaffolds was developed due to electrostatic interaction and π–π stacking among the organic/inorganic substance and carbon nanomaterial [22] (Fig. 8.1).

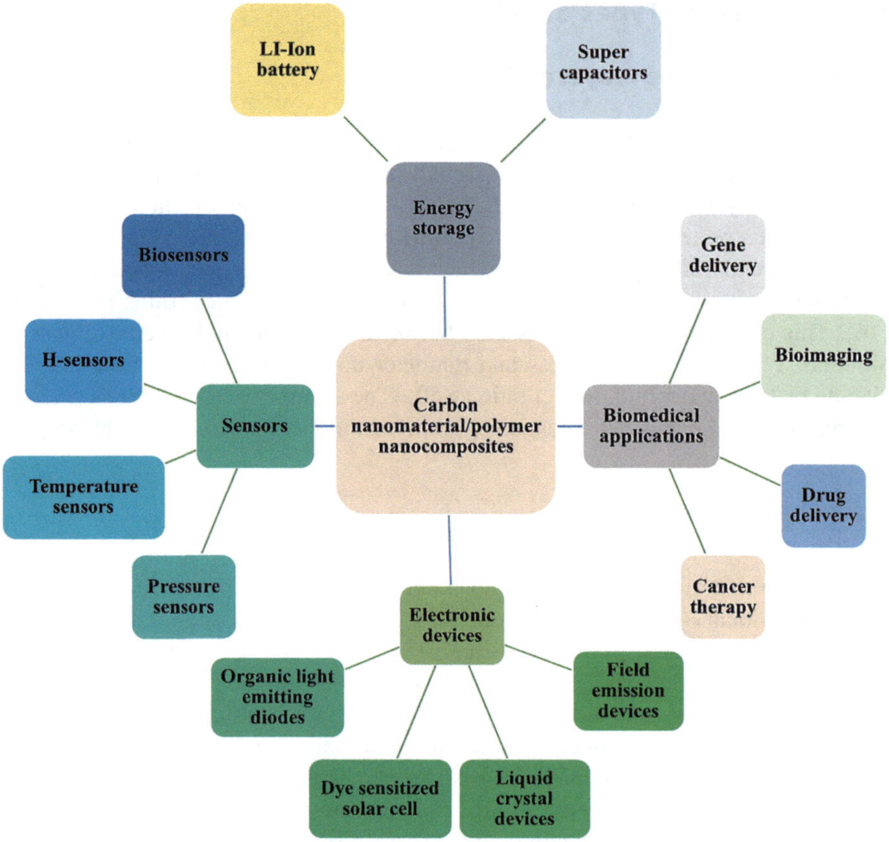

Fig. 8.1 Applications of carbon-based polymer nanocomposite

8.5.3 Metallic and Metal Oxide Nanofiller

Well known for its antibacterial effect, silver acts by blocking the respiratory enzyme pathway and changing microbial DNA and cell wall. Being effective against biofilm forming and multi-resistant bacteria, silver aids in healing wounds. In burn and chronic infections silver sulfadiazine and silver nitrate are applied extensively. However, tissue toxicity was observed in silver sulfadiazine. To avoid this limitation silver nanoparticles were synthesized, which reduced the toxicity of the material due to high surface-to-volume ratio in comparison to conventional compounds of silver [23]. Cytokine modulation during inflammation was observed to reduce by using silver nanoparticles with lighten scar formation. Surface functionalization of silver nanoparticles with thiolated oligonucleotide (5′-HS(CH2)6- TAATGCTGAAGG-3) resulted in the release of silver ions for longer duration thereby reducing cytotoxicity. Nanocrystalline silver can cover large wound surfaces due to its ability to release clusters of silver nanoparticles. The limitation of silver-based nanomaterial lies in

Table 8.1 Properties of nanofillers to polymers

Enhanced characteristics	Limitations
Mechanical properties (Toughness, Tensile, Strength, Stiffness)	Optical issues
Thermal expansion	Dispersion difficulties
Chemical resistance	Sedimentation
Dimensional stability	Viscosity increases which limits processability
Ablation resistance	
Gas Barrier	
Thermal conductivity	
Flame retardant	

reappearance of silver resistant bacteria and blue-grey skin coloration if used for longer time [24].

Zinc oxide nanoparticles have shown better intrinsic antibacterial activity and cytocompatibility than silver nanoparticles in mammalian cell. They are utilized in many hydrogels based wound bandages as it promotes collagen deposition and re-epithelization. In another related study proangiogenic activity was observed with zinc oxide nanoflowers further confirmed by in vivo assay. However, as wound healing dressing material metallic nanoparticles suffers from drawback such as cytotoxicity thus limiting their potential in further treatment [25] (Table 8.1).

8.6 Fabrication of Polymer Nanocomposites

Once the particular polymer and suitable nanofiller is selected for a particular application, the upcoming challenge is to identify the synthesis route for the fabrication of desired polymer nanocomposites. Generally, solid thermoplastic polymers or thermosetting reactive prepolymers with solid nanofillers are processed using the following methods:

(i) Melt intercalation
(ii) Solution intercalation
(iii) Roll Milling

For soft thermosetting prepolymers or thermoplastic polymers reinforced with solid nanoparticles, recommended procedures are as follows:

(i) In-situ polymerization
(ii) Emulsion polymerization
(iii) High-shear mixing

8.6.1 Melt Intercalation

The stacked sheets of nanofiller (layered silicate) are combined with the solid polymer in the liquefied state. If the chosen polymer and the layered nanofiller are compatible to each other, then the polymer can be inserted within the interlayer space to form either an exfoliated or intercalated nanocomposites without the addition of any solvent. Application of high-shear forces helps the nanofillers to disperse uniformly in the matrix and to prevent the chances of agglomeration. This methodology is dependent on the organization of polymer morphology during their molten state. Consequently, the chains present in polymer are free to move and thus mixed properly in the course of melting state of polymer [26]. The shortcoming in this technique is on the incorporation of solid particles in the polymeric phase, mainly when the filler has opposite surface properties to that of polymer matrix. Hence to attain improved compatibility among the two phases surface functionalization is performed. For example, the poor dispersion of graphene sheets into polymeric system is due to strong bonds present between graphene layers and the viscous polymer is overcome by surface functionalization of graphene sheets to favour better component interaction [27].

8.6.2 Solution Intercalation

The solution casting method is mostly used to fabricate homogeneously distributed nanocomposites due to its simplicity. The nanofiller is sonicated in the solvent initially prior to its addition into the polymer matrix. At an ambient temperature the solvent is later evaporated followed by vacuum drying. The resultant nanocomposite is washed properly to remove any undesired impurities [28].

8.6.3 In Situ Polymerisation

In this technique, uniformity of distribution of nanofillers within the polymer matrix is achieved. This method establishes polymers between layers of embedded sheets. When the liquid monomer and layered sheet come into contact with each other the nanofiller gets swollen and hence polymerization is initiated by rise in temperature by adding a suitable initiator. This process allows the formation of interaction between the polymer chain and chemically active surfaces through polymerization (tethering effect) thermoset-clay nanocomposite is produced by utilizing this method [29].

8.7 Polymer Nanocomposite in Biomedical Science

Novel materials with feasible biomedical application have been fabricated to substitute, direct or supplement the normal functioning of living tissues. Materials with biocompatible features must accomplish physical, chemical, biological, and structural (mechanical) affinity with host tissue [30]. Few instances to compile application of polymer nanocomposite are as follows (Fig. 8.2):

(i) Dental implants-surface modified SiO_2 and ZrO_2 filled acrylates.
(ii) Bone plates and screws-nylon, polypropylene, polyethylene reinforced with carbon fibres [30].
(iii) Bone cement: acrylate glass powder.
(iv) Joints replacements: Ultrahigh molecular weight polyethylene with carbon nanocomposites for total hip replacement [31].
(v) Bone fracture repair: Carbon fibre-epoxy composite for external fixator.

8.7.1 Skin Defects and Wound Healing

The human skin essentially have two layers of tissue: an upper layer called as epidermis which provide barrier properties to skin and contains keratinocyte cells and a lower layer known as dermis whose core components are collagen and elastin matrix giving strength and flexibility to skin along with cells which include adipocytes, fibroblasts, and macrophages cells, these layers protect the skin from outer environment by acting as a shield [30]. In order to restore integrity of the tissue after a cutaneous injury, the process of wound healing involves overlaying of

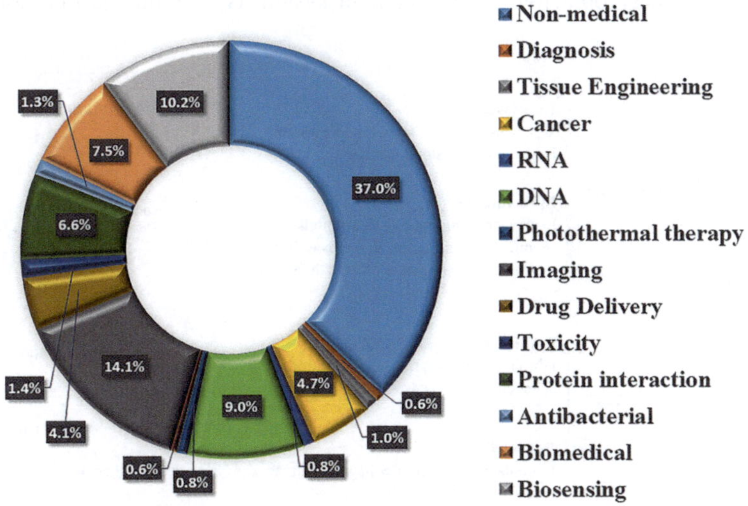

Fig. 8.2 Biomedical application of carbon-based materials

systematic steps which involves haemostasis, inflammation, proliferation, and remodelling. Each stage is crucial for wound healing to progress favourably, hence a well planning to fabricate wound dressings that target these stages is required [32].

8.7.2 Haemostasis

Shortly after an injury, vasoconstriction, which is the narrowing of blood vessels, occurs to halt uncontrolled bleeding [33]. Clotting factors, e.g., polyphosphates and growth factors, e.g., platelet-derived growth factor (PDGF) and transforming growth factor beta (TGF-β) are released due to platelet aggregation which comes in contact with collagen [34]. Coagulation of blood is activated by two pathways: intrinsic pathway which is triggered by the activation of factors XII (FXII) through negatively charged surface with blood [35]. The extrinsic pathway is triggered by tissue factor, a lipoprotein secreted by damaged cell on the extravascular tissue surface which is a chief contributor towards coagulation [36]. A clot is generated as thrombin is formed from prothrombin which promotes platelet activation and crosslinked fibrin formation acting as a seal over wound bed. The clot stimulates complimentary system, recruits inflammatory cells and thereby sets a platform for subsequent stages of healing (Fig. 8.3) [37].

8.7.3 Inflammation

In a period of a day or two, inflammatory phase starts as activated monocytes along with neutrophils are recruited at the site of wound. The complement system gets activated and in response to platelet degranulation immune cells get recruited [38]. The primary objective of complement system is to eliminate microorganism

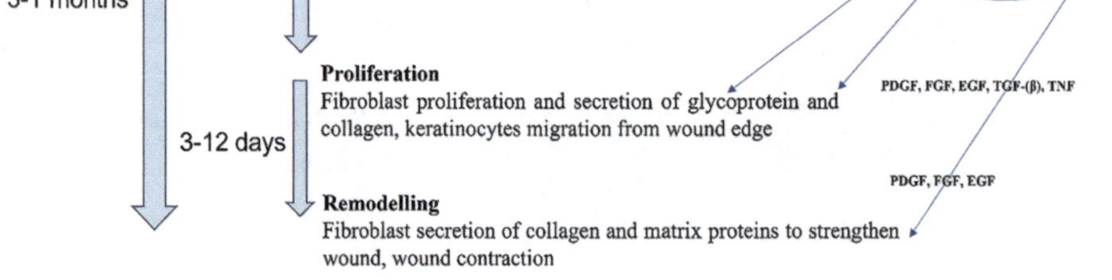

Fig. 8.3 Stages of wound healing

as a part of innate immunity. With the activation of complement system, the main complement protein C3 which cause the release of anaphylatoxin C3a and initiates terminal pathway which ends in forming terminal complement complex (TCC) which is kind of protein which disrupts the membrane of pathogens, causing cell death [39]. The protein formed through complement system stimulates inflammation and phagocytosis of foreign particles. With the presence of activated monocytes and neutrophils at the wound site, the process of cleaning up starts. Phagocytic macrophages which is differentiated from monocytes is primary inflammatory cell responsible for wound healing as it also coordinates angiogenic process. The late stage of inflammatory phase macrophages releases PDGF and TGF-β to initiate the process of proliferation [40]. The transition of macrophages from proinflammatory to anti-inflammatory is believed to play crucial role in the switch to proliferative stage from inflammatory stage [41].

8.7.4 Proliferation

After 3–4 days of injury, the proliferative phase starts which is described by re-epithelization, angiogenesis, new tissue formation, and wound contraction. A new epithelial layer is formed due to the migration of keratinocytes while granulation tissue consisting new blood vessels, macrophages, ECM, and fibroblast covers the wound bed [37]. Fibroblast activation and stimulation is aided by macrophages which provide wound bed with growth factors. New ECM is formed by myofibroblasts with contractile properties [42]. While oxygen and nutrients are supplied to the wound bed through angiogenic sprouting [43]. Contact guidance process mediates re-epithelialization in which keratinocytes attain flatter morphology which allows them to move across the granulation tissue from opposite side of wound bed to re-establish contact [44]. Wound contraction is aided by myofibroblast thus reducing the wound gap [42]. After the formation of keratinocyte basal layer on the wound bed, these cells initiate the development of height, generating cornified, spinous and granular layer of epidermis [45]. This process is Ca^{2+} dependent. Remodelling stage is initiated when the wound bed is covered by granulation tissue and new epidermis.

8.7.5 Remodelling

This final stage of wound healing, where stronger, permanent tissue replaces temporary granulation tissue which can last for weeks or several months. Proliferation of immune cells returns to their usual state. The capillary density reduces as the demand for nutrition decreases to normal range from almost thrice the normal level during the earlier phases of wound healing [46]. Hereby the disappearance of redness of the scar marks the completion of the visible healing. Furthermore, collagen fibres crosslinks and rearrange under the scar to give strength to newly formed tissue.

During this process type I collagen takes over type III collagen with matrix-metalloproteinase-mediated action [45].

8.8 Elements of Chronic Wound

Wounds that have failed to proceed according to normal cascade of wound healing are defined as chronic wounds and thus need treatment in order to heal completely. In developing countries ~2% of the population encounter a chronic wound in their lifetime [35]. Chronic wounds pose a huge threat to healthcare systems globally. The reason behind chronic wounds is due to systemic or local disorder which causes delay in healing. Infections and foreign bodies are local factors which effects the wound healing directly whereas interruptions occurring from health conditions such as age or diseases are categorized as systemic factors [47]. Production of proinflammatory cytokine (interleukin (IL)-1 and tumour necrosis factor alpha (TNF-α) and reactive oxygen species (ROS) increases if there is any microbial infection which induces inflammation leading to breakdown of neighbouring tissues hence hindering the process of healing. Moreover, an extreme inflammation leads to reduction in the level of growth factors which are necessary for healing process [48].

Diabetes is also one of the conditions to cause delay in healing and develop chronic wounds. The impaired ability to heal among diabetic patients is due to prolonged hypoxia around wound due to improper angiogenesis, which increases the oxidative stress level response and early inflammatory reaction leading to destruction of tissue and slow down the process of normal healing [49]. Furthermore, both delayed and excessive inflammation is detrimental during the process of chronic wound healing.

8.9 Wound Dressings

It is important to care for wounds whether it is a major incision or a minor cut. Wound dressings are included in part of this action. Bandages just hold the dressings in its position whereas wound dressings are designed to be in direct contact with the injured tissue. During the nineteenth century antiseptic technique was utilized to treat wound infections and reduce mortality. Modern wound dressings arrived in the twentieth century, occlusive dressings were fabricated to maintain moist environment to wound as growth factors, proteinases, etc. will be exposed to the covered wound which is not in the case of open wound. Wound dressings create hypoxia at wound bed which results in lowering of pH, ultimately reducing wound infection hence promoting re-epithelization, angiogenesis, and collagen synthesis [50]. In 1948 Oscar Gilje interpreted moist chamber effect for treating ulcers. The first modern wound dressings which absorbed fluids and supplied moisture were brought to market in the 1980s, e.g., hydrocolloids, foams. Synthetic wound dressings bloomed during mid-1990s into several forms such as hydrogels, silver/collagen dressings, silicone meshes, vapour-permeable adhesive, etc. [51] (Fig. 8.4).

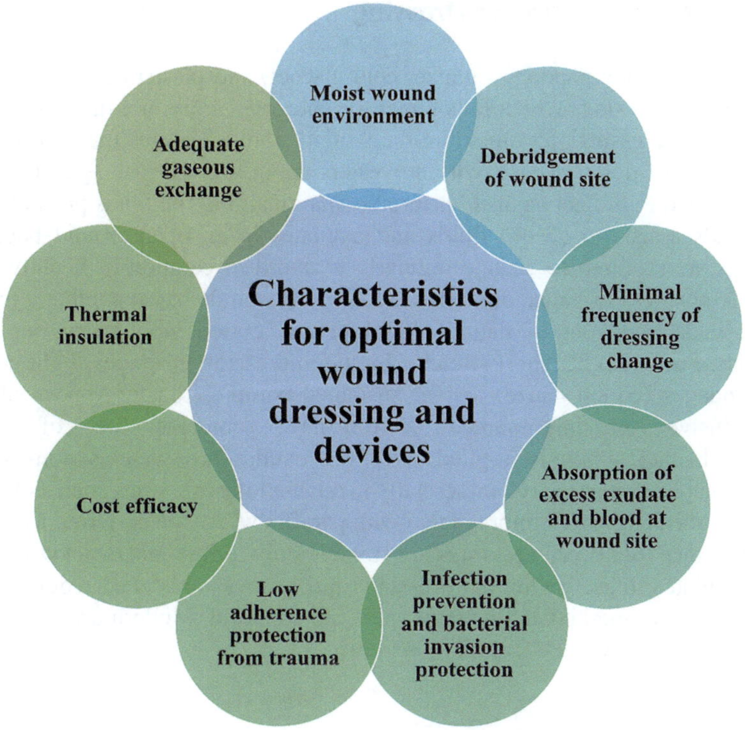

Fig. 8.4 Functions and desirable characteristics of wound dressings and devices related to treatment clinical significance

8.9.1 Components of an Optimal Wound Dressing

Depending on the type of wound, suitable wound dressing must be applied. Following are the selection criteria for a dressing:

(i) Must be non-allergic, sterile, and non-toxic.
(ii) Ability to maintain a moist environment.
(iii) To permit gaseous exchange between environment and injured tissue.
(iv) To improve epidermal migration.
(v) To sustain apt temperature of tissue to enhance blood flow.
(vi) To boost the development of connective tissue and angiogenesis.
(vii) To boost enzyme accumulation and debridement movement to complement epidermal migration.
(viii) Should not adhere to the wound and could be removed easily.
(ix) Able to control formation of bacterial colony at site of wound [52].

8.9.2 Traditional Wound Dressing

Natural or synthetic bandages, gauze, cotton wool, and plasters come under traditional wound dressings which falls under the category of dry dressings and are used as primary or dressings for shielding the wound from infections [53]. Cotton fibres either woven or non-woven, rayon, polyesters are utilized to make gauze dressings have managed to protect against bacterial contamination. The fibres present in these gauze pads assists to absorb fluids and exudates in an open wound bed. These dressings are required to change regularly to avoid maceration of healthy tissues. Cost effectiveness of gauze dressings are low. Polyamide based synthetic materials or dressings made out of natural cellulose and cotton wool carry out distinct functions. For example, short stretch / high compression bandages, light compression bandages (cotton gauze) deliver sustained compression for venous ulcers. A non-occlusive dressing named XeroformTM contains 3% of Bismuth tribromophenate which is applied on slight exuding as well as non-exudating wounds petrolatum based. Commercially available tulle dressings such as Paratulle, Jelonet, and Bactigras are impregnated with paraffin which are preferred for cleaning wounds. Since traditional dressings are suitable only for dry and clean wounds with slight exudates, these traditional dressings fail to treat moistened wounds as they adhere to the wound bed and its removal is quite painful. Modern dressings replace them with more advanced formulations [54].

8.9.3 Modern Wound Dressing

Modern wound dressings have been formulated to keep the wound hydrated rather than just to cover it which will facilitate the activity of the wound for healing. Numerous agents for wound healing are available in the market on the basis of type and cause of wounds making it difficult to select. Synthetic polymers form the base for modern wound dressings, mainly classified as bioactive, passive, and interactive products. Hydrogel, hydrocolloids, foams, and films are semi-occlusive/occlusive interactive dressings. Non-occlusive dressings such as tulle and gauze are categorized under passive dressings [53]. An electrical potential is formed between the drier region of non-wounded area and the moist surface of wounded area due to inhibition of desiccation of wound. Hence epithelial cell migration at wound bed is stimulated. This electrical stimulus also aids in the expression of growth factors on fibroblast cells. Within shorter nursing duration, occlusive dressings have shown effective results despite being costly [55].

8.9.3.1 Alginate Dressing

Calcium and sodium salts consisting of guluronic and mannuronic acid units make alginate dressings. Bacterial infection and wound exudates are minimized due to the formation of hydrophilic gel. Thomas et al. have stated that macrophages are activated by alginates to generate TNF-α which begins inflammatory signals thus accelerating the healing process. A protective film is formed on the wound site due to the exchange of ions in alginate and blood. Alginate dressings are advisable for third degree burns and heavy drainage wounds [56]. As these dressings could dehydrate the wound bed, they are not recommended for dry wounds and secondary dressings are required additionally. Some of the commercially available alginate dressings are Algisite™, Sorbsan™, and Kaltostat™.

8.9.3.2 Hydrocolloid Dressing

Hydrocolloid dressings are used mostly as secondary dressings. They are interactive dressings which contain an outer water-impermeable layer and a colloidal inner layer. They are formed from a combination of pectin, carboxymethylcellulose, and gelatin, with other adhesives and elastomers. Hydrocolloids have features to absorb exudates and debridement along with being impermeable to bacteria. Hydrocolloids provide moist surroundings that absorb exudates and protect granulation tissue. Examples of hydrocolloid dressings available in the market include Comfeel™, Tegasorb™, and Granuflex™. These dressings, however, have limitations when used for highly exuding wounds or neuropathic ulcerst [55].

8.9.3.3 Bioactive Wound Dressings

In the wound healing process biomaterials play a significant role, thus giving rise to bioactive dressings. These dressings are derived from natural sources such as elastin, natural or artificial sources, chitosan, collagen, and hyaluronic acid as these materials are known as non-toxic, biocompatible, and biodegradable. Depending on the type and nature of wound the polymeric forms of these biomaterials can be applied either individually or in combination with antimicrobials or growth factors to improve the process of wound healing [56].

Collagen has similarity with the human skin composition and morphology. Therefore, it is an ideal natural polymer for the development of a wide array of wound dressing compositions. The primary protein present in the extracellular matrix of mammals is collagen. It forms a major portion of dry skin weight ~ 70–80%. It has a tough triple stranded helical structure make up. It provides cellular functions such as cell proliferation, differentiation, and cell migration as well as structural support to the body [57]. The wound healing is stimulated by molecular and cellular cascade which helps in wound debridement and new tissue formation. Collagen dressings are fabricated from porcine, avian, bovine sources and are purified to develop as non-antigenic dressings [58].

Chitosan is derived from chitin, a polysaccharide found in the exoskeletons of crustaceans like crabs and shrimps, through a process of deacetylation, resulting in the formation of Poly N-acetylglucosaminoglycan, also known as deacetylated chitin. It shows antibacterial properties, non-toxic, biodegradable, non-immunogenic, and biocompatible. It can be readily moulded into sponges, gels, films, and scaffolds, sponges can be applied topically as a wound dressing. It forms a necessary platform to initiate collagen synthesis and stimulate fibroblast proliferation on it. Chitosan has features like haemostatic and antibacterial which makes it unique polymer for wound healing ideal for fabrication of materials as wound dressing [59].

Hyaluronic acid comprises a linear polysaccharide, which consists of units of N-acetyl-D-glucosamine and glucuronic acid. Hyaluronic acid can be formulated in the form of mesh, gel, sheath, and film which can be applied widely as a medicament for wounds and severe burns. It can decrease the intensity of chronic injury to acute injury promising its role in skin tissue engineering in clinical studies on animal [60]. Wound without any scar is achieved. The three-dimensional matrix which is formed by hyaluronic acid after mixing with other polymers is equivalent to the human skin biologically as well as structurally.

8.9.3.4 Hydrogel Dressing

Hydrogels are synthesized from synthetic polymers. Hydrogels contains large amount of water which gives a moist environment to epithelium and granulation tissues [61]. Hydrogels dressings possess soft and elastic characteristics for painless administration and removal after healing is complete without damaging the tissue [62]. Hydrogel dressings maintain cooling and soothing effect at wound bed by decreasing temperature. They are widely applied for burn wounds, necrotic wounds, and chronic wounds [63]. Hydrogel dressings are polymeric crosslinked gels that can maintain a moist wound environment. Hydrogels are able to debride the necrotic tissue and clean them. Hydrogels do not adhere at wound sites and can be removed without causing any pain. Hydrogel also has the benefit of delivering therapeutic agents (drug delivery) to the wound [64]. Few examples of hydrogel as wound medicament are Intrasite™, Nu-gel™, Aquaform™ [65].

8.10 Hydrogels

Hydrogels are crosslinked polymer network with three-dimensional structure and are hydrophilic in nature which have capacity to imbibe huge quantity of water as well as biological fluids compared to their actual mass [65]. The polymer network present inside the hydrogels are insoluble due to chemical or physical crosslinking either in the form of crystallites or entanglements which imparts physical stability and structural networks. These hydrogels show compatibility with water which causes them to swell in aqueous medium [66].

8.10.1 Types of Hydrogels

8.10.1.1 Chemically Crosslinked

These hydrogels are synthesized via chain growth polymerization, electron beam polymerization, and condensation polymerization. Chemically crosslinked hydrogels do not disintegrate when come in contact with aqueous phase, due to the expansion of their crosslinked structures. The residual number of crosslinking agents which remain in the polymer product comprises the biocompatibility [67].

8.10.1.2 Physically Crosslinked

Physically crosslinked hydrogels are prepared by hydrogen bonding, ionic interaction, hydrophobized polysaccharides, crystallization, and stereocomplex formation. During physical crosslinking, no harsh chemicals are required and synthesis can be done at room temperature, under mild conditions, and physiological pH. This procedure of crosslinking do not need the presence of ionic groups in the polymer. Mechanically stronger hydrogel can be formed if metallic ions are utilized [68]. The hydrogels formed by this crosslinking are usually biodegradable in nature.

8.10.1.3 pH Responsive Hydrogels

A unique system of pH responsive hydrogels is a class of environment responsive hydrogels which respond to variation in electrolytic concentration or pH change by showing difference in swelling property. These unique change in volume or solubility properties of such pH responsive hydrogels is attained through the exchange of hydroxyl groups or protons which are present in the solution [69]. The pendant functional groups either acidic or basic which are present on the polymer matrix are accountable for such volume change at a particular pH. Moreover, polybasic or polyacidic segments are present in the polymer networks of these pH responsive hydrogels, and when these segments are neutralized, they transform into polyampholytes. These hydrogels have wide applications in the area where external stimulus of pH is required, for example, controlled drug release or gene delivery [69].

8.11 Hydrogel Nanocomposite

The recent advancement in hydrogel technology has guided the progress of hydrogel nanocomposite for application in the biomedical field; currently there is a rise in research in their growth. The fabrication of organic–inorganic structure with soft networks of hydrogel is the need of today's material. In 2002 Haraguchi developed clay hydrogel nanocomposites which exhibited superior properties [46]. The hydrogel nanocomposite displays superior features which includes swelling and mechanical properties when compared with traditional hydrogels. This was a breakthrough study in the domain of hydrogel nanocomposites. Hydrogel nanocomposites are three-dimensional crosslinked networks generated in the existence of nanostructures. These materials are developed either by chemical or physical crosslinking [64]. Hydrophobic interactions, ionic complexations, and hydrogen bonding are significant for physical crosslinkages whereas strong covalent bonds are involved in chemical crosslinking. The addition of nanoparticles into the hydrogel matrix is performed either by absorbing them within the matrix or by homogeneous dispersion in the hydrogel network and also by nanoparticle entrapment in the hydrogel mesh. The presence of nanofiller inside the hydrogel matrix causes improvement in the intrinsic as well as extrinsic properties of hydrogel nanocomposite. The nanofiller

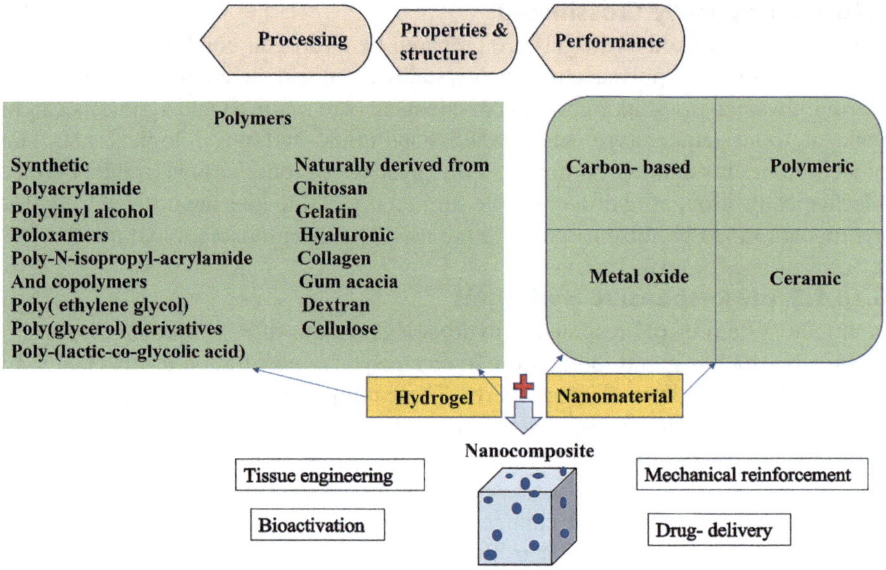

Fig. 8.5 Processing factors influence the final properties and performance of hydrogel nanocomposites, both natural and synthetic polymers can be used to fabricate hydrogel in which nanomaterial can be reinforced to form hydrogel nanocomposite with a variety of applications

contributes towards unique characteristics of the hydrogel matrix and thus promotes them for application as actuators, in microfluidics, catalysis, biomedical, and sensors. The existence of such diversity is due to the ability of these hydrogel nanocomposites to be designed in different forms such as hollow tubes, sheets, thick and thin membranes, and bellows (Fig. 8.5) [30].

8.11.1 Carbon Nanomaterial Reinforced Hydrogel Nanocomposite

Conventional hydrogels are reinforced with carbon-based nanostructures such as graphene and carbon nanotubes (CNTs) to enhance their electrical and mechanical features [70]. In general, both graphene and CNTs based hydrogel nanocomposite are being considered for applications such as biomedical devices, tissue engineering scaffolds, actuators, drug delivery, and biosensors [68]. The two forms of CNTs are single wall and multi wall. The interaction between CNTs and hydrogel networks is limited due to the hydrophobic nature of CNTs as the strong π–π self-interplay causes aggregate formation. To circumvent such condition surface functionalization of CNTs is performed by polar groups such as carboxyls, amines, and hydroxyls [71]. CNTs based hydrogel nanocomposites can be utilized to come up with electrically conductive tissues such as cardiac, nerves, and muscle tissues as CNTs possess high electrical conductivity [72]. Another nanomaterial graphene which has two-dimensional architecture is an excellent heat and electrical conductivity. The solubility of graphene with hydrophilic polymers under physiological environment can be enhanced by its modification into graphene oxide (GO) under acidic medium [73]. In conclusion the conjugation of graphene derivatives and functionalized CNTs with hydrogel networks allows us to procure hydrogel nanocomposite with ameliorated properties. On the other hand, the actual application of carbon reinforced nanocomposite as a platform for tissue engineering or as for controlled drug delivery is yet to be determined due to concern related to their biocompatible characteristics, which has not been analysed completely. Various cell line studies have been performed on carbon nanomaterials, and the obtained results are quite contradictory [74]. For instance, GO could be a foundation for adhesion and proliferation of osteoblasts, human embryonic stem cells, and kidney cells [75], while on the down side, other reports revealed a cytotoxic effects of GO against fibroblasts depending on concentration [75]. These puzzling outcomes have been explained roughly considering structural defects of graphene, preparation technique of GO, charge, and size distribution, and the presence of oxygen moieties has effect on the overall interaction and in vitro and in vivo biocompatibility [76].

8.12 Conclusion

The chapter on polymers and polymeric composites in nano/bio-medicine provides an overview of the use of polymers and composites in various biomedical applications, including drug delivery, tissue engineering, and medical implants. Polymeric materials have several advantages over other materials in these applications, including their biocompatibility, versatility, and tunable properties. The use of polymers and composites in nano/bio-medicine has led to the development of new and improved medical devices, as well as more effective drug delivery systems.

In conclusion, the chapter highlights the potential of polymers and polymeric composites in nano/bio-medicine and their continued importance in the field of biomedical engineering. Ongoing research and development in this area will undoubtedly lead to further advances in medical technology and treatment options for patients. However, the challenges in translating these materials from laboratory research to clinical applications need to be carefully addressed, including issues related to safety, regulatory approval, and manufacturing processes (Fig. 8.6).

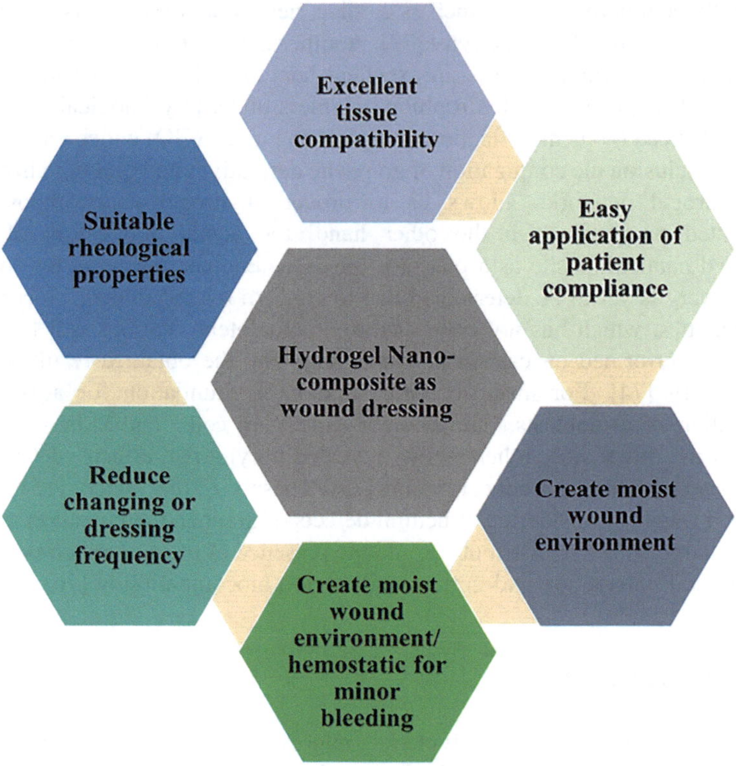

Fig. 8.6 Some advantages of hydrogel nanocomposite as wound dressing

References

1. Kumari S, Dhanda N, Thakur A, Gupta V, Singh S, Kumar R, Hameed S, Thakur A (2023) Nano Ca-Mg-Zn ferrite as tuneable photocatalysts for UV light induced degradation of rhodomine B dye and antimicrobial behaviour for water purification. Ceram Int 49(8): 12469–12480
2. Chahar D, Thakur P, Sun AC, Thakur A (2023) Investigation of structural, electrical and magnetic properties of nickel substituted Co-Zn nanoferrites. J Mater Sci Mater Electron 34:901
3. Dementjev AP, Maslakov KI (2012) Chemical state of carbon atoms on a nanodiamond surface: growth mechanism of detonation nanodiamond. Fuller Nanotub Carbon Nanostructures 20: 594–599
4. Wu X, Wang Y, Liu J, He S, Zhang L (2012) Improved crack growth resistance and its molecular origin of natural rubber/carbon black by nanodispersed clay. Polym Eng Sci 52: 1027–1036
5. Bindu PH, Thomas S (2013) Viscoelastic behavior and reinforcement mechanism in rubber nanocomposites in the vicinity of spherical nanoparticles. J Phys Chem B 117(41): 12632–12648
6. Rao CNR, Sood AK, Subrahmanyam KS, Govindaraj A (2009) Graphene: the new two-dimensional nanomaterial. Angew Chem 48(42):7752–7777

7. Wang H, Xie G, Fang M, Ying Z, Tong Y, Zeng Y (2015) Electrical and mechanical properties of antistatic PVC films containing multi-layer graphene. Compos B Eng 79:444–450
8. Mohan VB, Lau K, Hui D, Bhattacharyya D (2018) Graphene-based materials and their composites: a review on production, applications and product limitations. Compos B: Eng 142:200–220.
9. Chen W, Chen S, Hu W, Li D, Dai Z (2017) The preparation approaches of polymer/graphene nanocomposites and their appilcation research progress as electrochemical sensors. J New Mater Electrochem Syst
10. Ulery BD, Nair LS, Laurencin CT (2011) Biomedical applications of biodegradable polymers. J Polym Sci B 49(12):832–864
11. Cadek M, Coleman JP, Barron V, Hedicke K, Blau WJ (2002) Morphological and mechanical properties of carbon-nanotube-reinforced semicrystalline and amorphous polymer composites. Appl Phys Lett 81(27):5123–5125
12. Erukhimovich I, De La Cruz MO (2007) Phase equilibrium and charge fractionation in polyelectrolyte solutions. J Polym Sci B 45(21):3003–3009
13. Sheng N, Boyce MC, Parks DM, Rutledge GC, Abes J, Cohen RC (2004) Multiscale micromechanical modeling of polymer/clay nanocomposites and the effective clay particle. Polymer 45(2):487–506
14. Zhang Q, Gupta SK, Emrick T, Russell TP (2006) Surface-functionalized CdSe nanorods for assembly in Diblock copolymer templates. J Am Chem Soc 128(12):3898–3899
15. Verma R, Thakur P, Sun AC, Thakur A (2023) Investigation fo structural, microstructural and electrical characteristics of hydrothermally synthesized Li0.5-0.5xCoxFe2.5-0.5xO4 ($0.0 \leq x \leq 0.4$) ferrite nanoparticles. Phys B Condens Matter 661:414926
16. Chen B, Evans JR (2006) Elastic moduli of clay platelets. Scr Mater 54:1581–1585
17. Verma R, Chauhan A, Kumari S, Jasrotia R, Ali A, Gopalakrishnan C, Kumar R, Ghotekar S (2023) Green synthesis of ZnO NPs using Timbur (*Zanthoxylum armatum* DC.) plant extract for antimicrobial and dye degradation applications. Chemical Paper:1–11
18. Novoselov KS, Geim AK, Morozov S, Jiang D, Zhang Y, Dubonos SV, Grigorieva IV, Firsov AA (2004b) Electric field effect in atomically thin carbon films. Science 306(5696):666–669
19. Kala D, Gupta S, Kaushal A (2022) Nanotechnology in healthcare. In: Thakur A, Thakur P, Khurana SP (eds) Synthesis and applications of nanoparticles. Springer, Singapore
20. Zhou Z, Joslin S, Dellinger A, Ehrich M, Brooks B, Ren Q, Rodeck U, Lenk RP, Kepley CL (2010) A novel class of compounds with cutaneous wound healing properties. J Biomed Nanotechnol 6(5):605–611
21. Khan MI, Abdelhamid HN, Wu H (2015) Near infrared (NIR) laser mediated surface activation of graphene oxide nanoflakes for efficient antibacterial, antifungal and wound healing treatment. Colloids Surf B: Biointerfaces 127:281–291
22. Tang Z, Shen S, Zhuang J, Wang X (2010) Noble-metal-promoted three-dimensional macroassembly of single-layered graphene oxide. Angew Chem 49(27):4603–4607
23. Dhanda N, Thakur P, Kumar R, Fatima T, Hameed S, Slimani Y, Sun AD, Thakur A (2023) Green synthesis of Ni-Co nanoferrites using aloe vera extract: structural, optical, magnetic, antimicrobial studies. Appl Organomet Chem 2023:1–13
24. Pastar I, Nusbaum AG, Gil J, Patel S, Chen J, Valdes J, Stojadinovic O, Plano LRW, Tomic-Canic M, Davis SM (2013) Interactions of methicillin resistant Staphylococcus aureus USA300 and Pseudomonas aeruginosa in polymicrobial wound infection. PLoS One 8(2):e56846
25. Peterson LR (2009) Bad bugs, no drugs: no ESCAPE revisited. Clin Infect Dis 49(6):992–993
26. Vaia RA, Ishii HA, Giannelis EP (1993) Synthesis and properties of two-dimensional nanostructures by direct intercalation of polymer melts in layered silicates. Chem Mater 5(12):1694–1696
27. Dhanda N, Thakur P, Sun AC, Thakur A (2023) Structural, optical, and magnetic properties along with antifungal activity of Ag-doped Ni-Co nanoferrites synthesized by eco-friendly route. J Magn Magn Mater 572:170598

28. Vashist A, Kaushik AK, Ghosal A, Bala J, Nikkhah-Moshaie R, Wani AW, Manickam P, Nair MN (2018) Nanocomposite hydrogels: advances in nanofillers used for nanomedicine. Gels 4: 75
29. Huang C, Cheng Q (2017) Learning from nacre: constructing polymer nanocomposites. Compos Sci Technol 150:141–166
30. Sun X, Liu Z, Welsher K, Robinson JA, Goodwin AL, Zaric S, Dai H (2008) Nano-graphene oxide for cellular imaging and drug delivery. Nano Res 1(3):203–212
31. Sahoo NG, Bao H, Pan Y, Pal M, Kakran M, Cheng HS, Li L, Tan LP (2011) Functionalized carbon nanomaterials as nanocarriers for loading and delivery of a poorly water-soluble anticancer drug: a comparative study. Chem Commun 47(18):5235
32. Thakur A, Thakur P (2023) Ferrite nanoparticles for agriculture related activity. In. Applications of nanostructured Ferrites. Woodhead Publishing. 315–350
33. Delavary BM, Van Der Veer WM, Van Egmond M, Niessen F, Beelen RH (2011) Macrophages in skin injury and repair. Immunobiology 216(7):753–762
34. Wu C, Mohammadmoradi S, Chen JZY, Sawada H, Daugherty A, Lu H (2018) Renin-angiotensin system and cardiovascular functions. Arterioscler Thromb Vasc Biol 38:7
35. Siddiqui AM, Bernstein JM (2010) Chronic wound infection: facts and controversies. Clin Dermatol 28(5):519–526
36. Lane DP, Philippou H, Huntington JA (2005) Directing thrombin. Blood 106(8):2605–2612
37. Gurtner GC, Werner S, Barrandon Y, Longaker MT (2008) Wound repair and regeneration. Nature 453(7193):314–321
38. Sinno H, Prakash S (2013) Complements and the wound healing cascade: an updated review. Plast Surg Int 2013:1–7
39. Giang J, Seelen MA, Van Doorn MBA, Rissmann R, Prens EP, Damman J (2018) Complement activation in inflammatory skin diseases. Front Immunol 9:639
40. Lucas T, Waisman A, Ranjan R, Roes J, Krieg T, Müller WE, Roers A, Eming SA (2010) Differential roles of macrophages in diverse phases of skin repair. Pubmed 184(7):3964–3977
41. Porcheray F, Viaud S, Rimaniol A, Léone C, Samah B, Dereuddre-Bosquet N, Dormont D, Gras G (2005) Macrophage activation switching: an asset for the resolution of inflammation. Clin Exp Immunol 142(3):481–489
42. Werner S, Krieg T, Smola H (2007) Keratinocyte–fibroblast interactions in wound healing. J Investig Dermatol 127(5):998–1008
43. Tonnesen MG, Feng X, Clark RF (2000) Angiogenesis in wound healing. J Investig Dermatol Symp Proc 5(1):40–46
44. Santoro M, Gaudino G (2005) Cellular and molecular facets of keratinocyte reepithelization during wound healing. Exp Cell Res 304(1):274–286
45. Keck M, Lumenta DB, Andel H, Kamolz L, Frey M (2009) Burn treatment in the elderly. Burns 35(8):1071–1079
46. Schexnailder PJ, Gaharwar AK, Bartlett RL II, Seal BL, Schmidt G (2010) Tuning cell adhesion by incorporation of charged silicate nanoparticles as cross-linkers to polyethylene oxide. Macromol Biosci 10(12):1416–1423
47. Khodaii Z, Afrasiabi S, Hashemi S, Ardeshirylajimi A, Natanzi M (2019) Accelerated wound healing process in rat by probiotic lactobacillus reuteri derived ointment. J Basic Clin Physiol Pharmacol 30(3):20180150
48. Bjarnsholt T, Kirketerp-Møller K, Jensen PØ, Madsen KG, Phipps R, Krogfelt K, Høiby N, Givskov M (2008) Why chronic wounds will not heal: a novel hypothesis. Wound Repair Regen 16(1):2–10
49. Swift ME, Burns AL, Gray KL, DiPietro LA (2001) Age-related alterations in the inflammatory response to dermal injury. J Invest Dermatol 117(5):1027–1035
50. Miranda O, Srinivasan G (2016) Advanced trends in treatment of wounds. Curr Sci 111:641–647

51. Daunton C, Kothari S, Smith LE, Steele DA (2012) A history of materials and practices for wound management. Wound Pract Res: Journal of the Australian Wound Management Association 20:174
52. Szycher M, Lee SJ (1992) Modern wound dressings: a systematic approach to wound healing. J Biomater Appl 7(2):142–213
53. Chhabra S, Chhabra N, Kaur A, Gupta N (2017) Wound healing concepts in clinical practice of OMFS. J Maxillofac Oral Surg 16(4):403–423
54. Ramos-e-Silva M, Ribeiro de Castro MC (2002) New dressings, including tissue-engineered living skin. Clin Dermatol 20(6):715–723
55. Grada A, Mervis J, Falanga V (2018) Research techniques made simple: animal models of wound healing. J Invest Dermatol 138(10):2095–2105.e1
56. Boyce ST, Lalley AL (2018) Tissue engineering of skin and regenerative medicine for wound care. Burns Trauma 6:4
57. Chahar D, Kumar D, Thakur P, Thakur A (2023) Visible light induced Mg-doped Co-Zn nanoferrites. Mater Res Bull 162:112205
58. Manjubala I, Scheler S, Bössert J, Jandt KD (2006) Mineralisation of chitosan scaffolds with nano-apatite formation by double diffusion technique. Acta Biomater 2(1):75–84
59. Shahdeo D, Kesarwani V, Suhag D, Ahmed J, Alshehri SM, Gandhi S (2021) Self-assembled chitosan polymer intercalating peptide functionalized gold nanoparticles as nanoprobe for efficient imaging of urokinase plasminogen activator receptor in cancer diagnostics. Carbohydr Polym 266:118138
60. Anjum S, Arora A, Alam MS, Gupta B (2016) Development of antimicrobial and scar preventive chitosan hydrogel wound dressings. Int J Pharm 508(1–2):92–101
61. Shakeel A, Singh A, Das S, Suhag D, Sharma AK, Rajput SK, Mukherjee M (2017c) Synthesis and morphological insight of new biocompatible smart hydrogels. J Polym Res 24(7). https://doi.org/10.1007/s10965-017-1267-7
62. Suhag D, Bhatia R, Das S, Shakeel A, Ghosh AB, Singh A, Sinha OP, Chakrabarti SK, Mukherjee M (2015) Physically cross-linked pH-responsive hydrogels with tunable formulations for controlled drug delivery. RSC Adv 5(66):53963–53972
63. Raucci MG, Alvarez-Perez MA, Demitri C, Sannino A, Ambrosio L (2012) Proliferation and osteoblastic differentiation of hMSCs on cellulose-based hydrogels. J Appl Biomater Funct Mater 10(3):302–307
64. Sangam S, Garg P, Sanyal T, Pahari S, Khurana SMP, Mukherjee M (2022) Graphene quantum dots and their hybrid hydrogels: a multifaceted platform for Theranostic applications. In: Thakur A, Thakur P, Khurana SP (eds) Synthesis and applications of nanoparticles. Springer, Singapore
65. Boonkaew B, Kempf M, Kimble RM, Supaphol P, Cuttle L (2014) Antimicrobial efficacy of a novel silver hydrogel dressing compared to two common silver burn wound dressings: ActicoatTM and PolyMem Silver®. Burns 40(1):89–96
66. Merino S, Martín C, Kostarelos K, Prato M, Vázquez E (2015) Nanocomposite hydrogels: 3D polymer–nanoparticle synergies for on-demand drug delivery. ACS Nano 9(5):4686–4697
67. Ishihara M, Ono K, Sato M, Nakanishi K, Saito Y, Yura H, Matsui T, Hattori H, Fujita M, Kikuchi M, Kurita A (2001) Acceleration of wound contraction and healing with a photocrosslinkable chitosan hydrogel. Wound Repair Regen 9(6):513–521
68. Ishihara M, Obara K, Ishizuka T, Fujita M, Sato M, Masuoka K, Saito Y, Yura H, Matsui T, Hattori H, Kikuchi M, Kurita A (2003) Controlled release of fibroblast growth factors and heparin from photocrosslinked chitosan hydrogels and subsequent effect onin vivo vascularization. J Biomed Mater Res 64A(3):551–559
69. Bajpai AK, Shukla SK, Bhanu S, Kankane S (2008) Responsive polymers in controlled drug delivery. Prog Polym Sci 33(11):1088–1118
70. Haraguchi K (2007) Nanocomposite gels: new advanced functional soft materials. Macromol Symp 256(1):120–130

71. Vivaldini SM, Ribeiro RM, Mosimann G Jr, Kr T, Pereira GFM, De Araújo WN (2021) A real-life study of the positive response to DAA-based therapies for hepatitis C in Brazil. Braz J Infect Dis 25(2):101573
72. Bharti MK, Gupta S, Chalia S, Garg I, Thakur P, Thakur A (2020) Potential of magnetic nanoferrites in removal of heavy metals from contaminated water: mini review. J Supercond Nov Magn 33:3657–3665
73. Mottaghitalab F, Farokhi M, Zaminy A, Kokabi M, Soleimani M, Mirahmadi F, Sadeghizadeh M (2013) A biosynthetic nerve guide conduit based on silk/SWNT/fibronectin nanocomposite for peripheral nerve regeneration. PLoS One 8(9):e74417
74. Huang YC, Hsu SH, Kuo WC, Chang-Chien CL, Cheng H, Huang YY (2011) Effects of laminin-coated carbon nanotube/chitosan fibers on guided neurite growth. J Biomed Mater Res A 99(1):86–93
75. Zhang Y, Ali SF, Dervishi E, Xu Y, Li Z, Casciano D, Biris AS (2010) Cytotoxicity effects of graphene and single-wall carbon nanotubes in neural phaeochromocytoma-derived PC12 cells. ACS Nano 4(6):3181–3186
76. Agarwal S, Zhou X, Ye F, He Q, Chen GC, Soo J, Chen P (2010) Interfacing live cells with nanocarbon substrates. Langmuir 26(4):2244–2247

Bio-Ceramics and Bio-Glasses for Orthopedics and Tissue Engineering

9

A. C. Sun, Atul Thakur, and Preeti Thakur

Abstract

Modern medicine depends heavily on biomaterials for body part repair and regeneration, and their influence on current culture is growing. Given the significance of orthopedics (the second most frequently replaced organ after the blood is the bone), bioactive glasses and ceramics serve as crucial points of reference for future technological advancements in this area. Depending on the use, bio-ceramics can directly interact with the tissue in the area, either promoting tissue growth or, in the case of bioactive ceramics, triggering fresh tissue regeneration. As in the case of bioinert ceramics, it can also remain inactive at the application site and serve as a mechanical load carrier. In comparison to other classes of materials, bio-ceramics and bioactive glasses, among others, have been demonstrated to have better biological characteristics. The popularity of hydroxyapatite and other calcium phosphates in recent years in orthopedic and dental surgery is largely attributable to their favorable biocompatibility, osteoconduction, and osseointegration. Nanomedicine and tissue engineering are important fields that can help us develop new methods of treating fractures and damaged tissues. Bio-ceramics, a broad category of materials, is a good choice for various forms of reconstructive and regenerative medicine. Currently available bone graft alternatives include ceramics (including glasses), either alone or in conjunction with another material, in around 60% of the cases. This chapter discusses a variety of bio-ceramics and bio-glasses topics, ranging from the

A. C. Sun (✉)
Department of Chemical Engineering and Materials Science, Yuan Ze University, Taoyuan City, Taiwan
e-mail: acsun@saturn.yzu.edu.tw

A. Thakur · P. Thakur
Centre for Nanotechnology, Amity University Haryana, Gurugram, India

D. Suhag et al. (eds.), *Integrated Nanomaterials and their Applications*,
https://doi.org/10.1007/978-981-99-6105-4_9

fundamentals through applications. Additionally, this chapter highlights current developments in bio-ceramics, including coatings, bone cement, medication delivery, 3D-printed ceramic scaffolds, and associated in vitro and in vivo performances.

Keywords

Orthopedics · Bio-ceramics · Tissue regeneration · Nanomedicine · Scaffolds

9.1 Introduction

According to the definition proposed at the Consensus convention of the European Society for Biomaterials in Chester, England, in March 1986, biomaterials are "any component, except a psychoactive substance, or combination of substances, either artificial or herbal in source, that may be used over any duration of time, as a whole or as a part of a device, which treats, enhances, or replaces any tissue, organ, or functioning of the human frame." [1]. In recent decades, synthetic alternatives have been progressively vital within the repair of aging, deteriorating, or wounded tissues. Keeping in mind the example of bone, the size of the global market for bone grafts and replacements, which was anticipated to be worth USD 2.80 billion in 2022, is expected to increase at a compound annual growth rate (CAGR) of 6.3% between 2023 and 2030. The allograft category, which had the biggest revenue share of 57.3% in the material type segment in 2022, is expected to see significant growth prospects throughout the course of the projection period. Even when the replacement tissue is available, 90–95% of these presently involve extracting bone from the patient or another donor, necessitating two surgical procedures for the grafts [2]. Utilizing a synthetic alternative cuts the number of invasive procedures by 50% greatly lowers the risk of infection or other problems and lowers expenses. (One of the key factors behind the development of bone substitute materials has been this). Functional ceramics such as electrical, photonic, and energy-related ceramics have gradually been developed as ceramic science advances. Even though research is being done to enhance the capabilities of bio-ceramics, the term's definition is still up for debate. Ceramics utilized in medical and dental procedures on the human body are generally categorized as bio-ceramics in this aspect. Alumina, zirconia, bioactive glass, glass-ceramics, hydroxyapatite, and resorbable calcium phosphates are a few examples of materials that fall under the category of bio-ceramics. They have been employed in dentistry as endodontic sealers, bone defect fillers, root repair materials, apical fill materials, perforation sealers, and regeneration aids. They have the capacity to resorb and promote the regeneration of natural tissues or to operate as human tissues.

The desire for implant materials that could connect to live tissues, which were meant to replace inert metal and plastic implants that were poorly tolerated by the body, led to the development of bioactive glasses (BGs) by Larry Hench in the late 1960s. A group of bioactive glasses known as "Bio-glass" are made of silicon dioxide, sodium oxide, calcium oxide, and phosphorous pentoxide. This substance produces a surface pore structure with a pore size of 5–20 nm, excellent surface area, and the capacity to stimulate the production of apatite. The osteogenic potential of the released ions, the strong bone mineral binding affinity, and the regulated disintegration of bioactive glasses make them particularly appealing for bone regeneration.

9.2 Scaffolds and Bio-Ceramics Work Perfectly Together for Tissue Engineering

Through the application of interdisciplinary technologies, tissue engineering produces implants that, once inserted, can fuse and restore a particular functional tissue. A key element of this method is a synthetic scaffold, which serves as the primary two- or three-dimensional (2- or 3-D) framework for both the in vitro and in vivo development of hard and soft tissues. Moving on to orthopedics, bone is one of the few tissues with an innate capacity for restitutio ad integrum (Latin for "return to original condition"). On the other hand, the ability of bone tissue to regenerate is restricted to a certain defect size, beyond which the body is unable to repair the fault, leading to a series of bone tissue ingrowths. For bone lesions larger than 5 cm [3], this supposedly upper limit has already been achieved. The regeneration of bone may also be negatively impacted by a number of patient-related and independent risk factors. In these conditions, fibrous tissue or cartilage in growth is typical, leading to non-union of the margins of the bone defect and, consequently, biomechanical instability. Biocompatible materials, in addition to using autologous, donor-specific, or xenogeneic bone tissue, have been clinically demonstrated to be helpful for the treatment of critical-size bone lesions [2]. For a number of reasons, including their limitless supply, sterilizability, and lack of donor-site morbidity, pathogenic transmission, or immunogenic issues, clinicians prefer the use of synthetic bone transplant materials. This is strengthened by the fact that additive manufacturing techniques can be used to process the majority of synthetic materials, enabling the development of anatomically precise bone implants. Thus, there is a need for novel biomaterials that can repair injured tissues, promote the body's natural regeneration processes, and facilitate tissue healing. It is thought that the creation of three-dimensional tissue requires "scaffolds," which are porous templates. Bio-ceramics, a distinct class of completely, partially, or non-crystalline ceramics developed for the restoration and repair of broken body parts, have a lot of potential [3].

Around 20 years ago, the initial investigation into the use of bio-ceramic materials in orthopedic surgery started. Only biologically inert materials were first investigated, but recently, several scientists have started looking into bioactive ceramics. There has been a marked rise in interest in this field during the past 20 years. The use of prosthetic implants, bone cavity filling, intermaxillary fixation, and other words linked to orthopedic surgery are all being addressed with great effort. Biomaterials (bio-ceramics) used in orthopedic surgery can be classified as either bioactive (can promote the formation of bone tissue and build a link strong enough to withstand physiological stresses) or bioinert (have no effect on tissue). Alumina is a bioinert ceramic with exceptional mechanical resilience and very little wear when given the right structural properties. As a consequence, it is used to create small joint prostheses, cover metal prosthetic shafts, and create prosthetic hip components (acetabulum and head). Hydroxyapatite (HA) is currently the bioactive ceramic with the greatest potential use in orthopedic surgery. It may be used to cover metallic prosthetic substrates and fill in fractured bones in a variety of ways because of the close interfacial connection it establishes with bone tissue [4–6].

9.2.1 Scaffolds

One of the current advances in the field of tissue engineering is the use of scaffolds, which can aid in the proliferation of cells in a particular location to regenerate tissues and eventually biodegrade so that they can be eliminated. These kinds of scaffolds have been developed using a variety of substances, such as polymer composites and calcium phosphate, to aid in the development of bone and tissue cells (Fig. 9.1). For scaffold-related applications, porous biomaterials are typically used. There are two types of scaffold applications: in vivo and in vitro. Delivering the scaffold to a particular location of interest within the body is a common step in in vivo applications so that the cells can grow under control at the site of the disease [4]. These methods frequently entail adding stem cells (or the particular cell type) to the scaffold, whereupon they multiply and connect to the local tissue. The scaffold will break down and be eliminated by the body once it has served its purpose. In vitro scaffolds, in contrast, are a method of external cell multiplication. A big culture of cells will be grown outside the body, and the growth will be directed in one of two ways: either from a small number of stem cells to a large number of stem cells or by boosting the growth and regulating the types of cells created, so they may be employed in a particular location. The patient can then be given the cell cultures.

9.2.2 Bio-Ceramics

Bio-ceramics refers to a vast range of ceramics designed primarily for the regeneration and repair of dysfunctional or injured body parts. Utilizing solid pieces (for instance, in the reconstruction of the tympanic cavity or as load-bearing components of joint prostheses), microparticles and granules for bone filling, coatings on metal joint prostheses, intravenous formulations (bone cement), and porous scaffolds are some of the current clinical applications. Based on how their usage affects the body's tissues, bio-ceramics may be categorized into three major categories: practically inert substances like alumina and zirconia, bioactive substances like bioactive glass, and resorbable substances like tricalcium phosphate (TCP). Inert ceramics are employed for the femoral heads and acetabular cups in almost all dental implants and hip replacements, but these materials are rarely used as scaffolds since their inertness results in the creation of a 1–3-mm thick "protective" fibrous capsule on the

Fig. 9.1 Bio-ceramic Scaffolds [4]

implant's surface [2]. The implant and host tissue do not develop a connection, even if there is no significant foreign body reaction. The ability to create a solid connection with the host tissue is crucial when employing bio-ceramics to create scaffolds. Ceramics that are bioactive and bioresorbable are a great solution in this regard. The later ones also have the added advantage of progressively degrading over time while being replaced by the host tissue's native tissue. As a result, they disappear after their job of serving as templates for new tissue is finished. In the sections that follow, the materials are conceptually categorized into their respective class from a microstructural perspective: crystalline ceramics, bioactive glasses, glass-ceramics, and composites. In addition to pastes, granules, or monoliths, bio-ceramics are added to bioinert implants to increase their ability to connect with bone and prevent fibrous encapsulation. The regeneration efficiency of currently implanted bioactive materials, such as HA or bio-glass, is limited as compared to autografts because of their long-term in vivo stability. The sections that follow focus on the primary bio-ceramic materials that are currently employed to create scaffolds (Figs. 9.2 and 9.3) [4, 7].

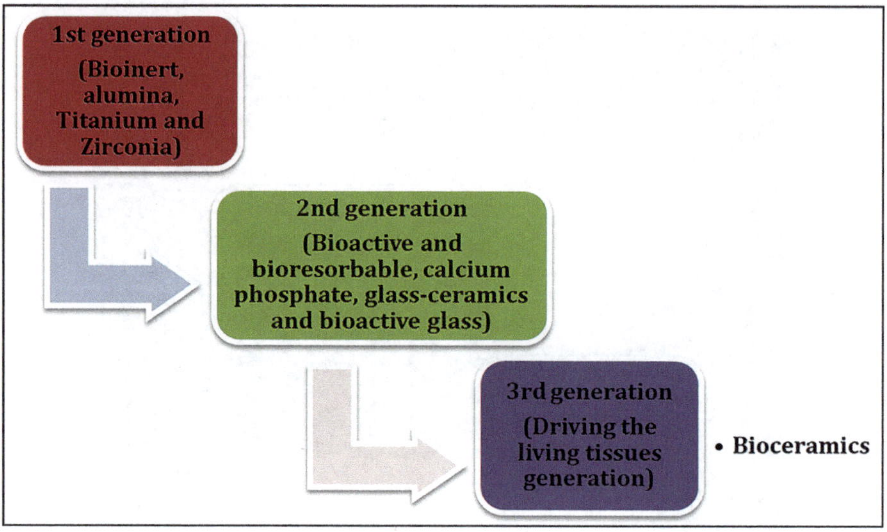

Fig. 9.2 A schematic layout of the three generations of bio-ceramics

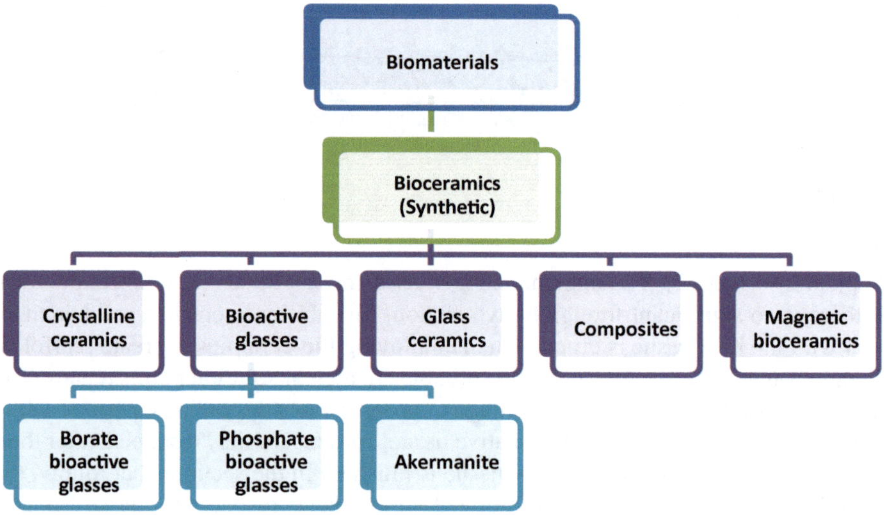

Fig. 9.3 Classification of bio-ceramics

9.2.2.1 Ceramics Made of Crystal

The principal members of this family are calcium phosphates, which are among the most widely used crystalline ceramics for bone tissue regeneration. This is a result of their exceptional properties, which include (i) Osteoconductivity, or the ability to provide a biocompatible interface as bone migrates and subsequently bonds to the host tissue without the formation of scar tissue, and (ii) Similarity to the mineral phase of bone in terms of structure and chemical composition. Synthetic HA ($Ca_{10}(PO_4)_6(OH)_2$), with a stoichiometric calcium-to-phosphate ratio of 1:67, is the calcium phosphate phase that, from the viewpoint of crystallography, most closely mimics genuine bone apatite. Because of its exceptional biocompatibility and osteoconductivity, HA is effectively used as a bone filler in the form of cement or granules as well as coatings on metallic joint prostheses. But its usage as a scaffold material is limited due to its weak mechanical properties and extraordinarily slow rate of resorption [4].

Other calcium phosphates with varying degrees of solubility have developed as a result, with the rate of dissolution increasing as the calcium-to-phosphorus ratio decreases and the crystallographic structure also playing a role in solubility. Specific papers addressing calcium phosphate bio-ceramics are directed to interested readers. All calcium phosphate bio-ceramics manufactured in a porous form have limited mechanical properties (brittleness, low fatigue strength), which essentially limit their clinical application to small load-bearing parts of the skeleton. It is important to remember that sintering, which does not occur in a viscous flow regime and may not completely result in the densification of scaffold struts, is routinely used to strengthen calcium phosphate scaffolds. Some materials, such bioactive glass-ceramics and mixes, seem to be more advantageous for creating robust, long-lasting scaffolds. The thigh bone, acetabular cup, and tibial plate of hip and knee joint replacements are all made of alumina, another well-known example of crystalline ceramic comparable to calcium phosphates. This is mostly due to alumina's high strength, superior wear resistance, and bioinertness (associated with preserving the required physico-chemical and mechanical characteristics over time). In clinical settings, only spherical porous scaffolds constructed of porous alumina are used to create orbital implants for enucleation. These scaffolds should not degrade over time in the anophthalmic sockets of patients and allow fibrovascular ingrowth through their porous network.

9.2.2.2 Bioactive Glasses

Numerous studies have shown that bioactive glasses are capable of attaching securely to live tissues (mostly bone), triggering a range of biological processes like tissue regeneration and angiogenesis, and eventually dissolving. These properties of bioactive glasses are the consequence of a surface modification that evolves over time following contact with a physiological environment [8]. As the interface for interacting with host tissues, the glass surface forms a layer of physiologically active hydroxycarbonate apatite (HCA), and the byproducts of dissolution (such as silica, salt, calcium, phosphate ions, etc.) promote the formation of new tissue in the host tissues. 45S5 Bio-glass®, a component of the $45SiO_2$-$24.5Na_2O$-$24.5CaO$-$6P_2O_5$ (wt.%) system, was created by Prof. Larry Hench and colleagues. It has been used in therapeutic settings since 1985. More silicate, borate, and phosphate glasses have been suggested for use in medical applications over the years. To create bioactive glasses, the sol-gel method or conventional melting-quenching methods are frequently used. Glasses formed from melt can be poured into molds to create rods and bars or cast as components in a variety of sizes and forms. The melt can also be cooled in cold water to form "frit," which are simple-to-powder granules and fragments of various sizes that can be utilized to make porous scaffolds in later processes [9]. Not to mention, the glasses may be spun to produce glass fibers, particularly the phosphate ones, which have attracted attention in recent years for application in soft tissue engineering as guides for muscle or nerve repair as well as for the development of glassy bone scaffolds. The silica content must be at least 60 mol% for melt-derived silicate glasses to bond with bone.

Additionally, hydroxy-carbonated apatite layer production and bone bonding may be accomplished with glasses that contain up to 90% mol.% silica if the glass is made using a sol-gel procedure. Because sol-gel glasses have a much greater surface area per gram that may be used for ion-exchange processes (tens vs. few meter square per gram), they were shown to produce a nanocrystalline hydroxy-carbonated apatite surface layer more quickly than thermal glasses in general. Mesoporous bioactive glasses (MBGs) have advanced over the past 10 years, enabling the development of a single, adaptable biomaterial with excellent bioactive properties (the capacity to form a surface HCA layer after only a few hours of contact with biological fluids) [4].

(i) **Borate bioactive glass:** Researchers have found that borosilicate or borate bioactive glasses encourage tissue infiltration in vivo as well as cell growth and differentiation in vitro. As quickly as 45S5 glasses, borate bioactive glasses also deteriorate quickly. Due to their poor chemical stability, they entirely transform into a substance that resembles HA. By adjusting the glass composition, the rate of degradation can be regulated. In addition, the toxicity of boron discharged into the solution as borate ions $(BO_3)^{3-}$ is a worry.

(ii) **Phosphate Glass with bioactivity:** CaO and Na_2O operate as network modifiers as it generates networks. Due to the ions present in the organic mineral phase of the bone, it exhibits a chemical affinity for bone. This glass' degradability can be managed by changing its composition. It has potential as a resorbable biomaterial for tissue engineering due to the flexibility it has demonstrated.

(iii) **Akermanite:** Due to its controlled mechanical properties and quick rate of decomposition, Akermanite ($Ca_2MgSi_2O_7$) has recently drawn greater interest. When compared to beta-TCP, marrow-derived or adipose-derived stem cells and osteoblasts have shown good osteogenesis and proliferation on akermanite. Recent research suggests that this Mg-containing silicate ceramic may be more suitable for bone regeneration than b-TCP as a bone transplant material. The bioactivity of akermanite is yet unknown, though, through what mechanism. One of the key elements in the proliferation and cell differentiation of different types of cells is the materials chemistry of biomaterials [9]. Biomaterials are essential to tissue engineering. They serve as a 3D template, provide mechanical support, and enable the creation of an artificial extracellular matrix (ECM) to assist the growth of new tissues [10]. This indicates that using just one kind of biomaterial won't affect the engineering of hard and soft tissues. As a result, each type of biomaterial, including metals, ceramics, and polymers, has a specific role to play in the creation of scaffolds for tissue engineering. Consequently, composite materials have become more prevalent.

9.2.2.3 Glass-Ceramics

Glass may be heated to a point where it partially crystallizes and contains a variety of crystalline phases, the size and composition of which can be altered depending on the thermal treatment settings. The resultant glass-ceramic material, in general, has greater mechanical properties than its parent glass, such as a higher elastic modulus, hardness, failure strength, and wear resistance. Sintering, which involves heating glasses over their glass transition temperature to initiate localized flow, is a common method for producing scaffolds. The base glass used to create scaffolds from 45S5 Bio-glass® has a propensity to crystallize before the sintering end (complete densification) is reached, leading to exceptionally brittle glass-ceramic porous products. Additionally, according to Chen et al. (2006), the formation of a sodium-calcium-silicate crystalline phase appears to be partially reducing scaffold bioactivity. Numerous research teams that proposed alternative glass-ceramics in an effort to address these problems have produced some interesting discoveries [4, 11].

9.2.2.4 Composites

For scaffolds to be effective in tissue engineering and regeneration, they must have properties that are appropriate for the tissue in question and the mechanical stresses that it will experience in vivo. Like other ceramic materials, bio-ceramics have the flaw of being brittle, which would prevent its usage in load-bearing applications. In addition, because of their high stiffness, bio-ceramics might not be appropriate for non-ossified applications where sufficient compliance with soft tissues is required. One method to overcome these challenges is to combine bio-ceramics with polymers to create a composite scaffold that makes the most use of both materials. To enhance the mechanical properties of the polymer matrix, i.e., to increase strength and stiffness as well as to successfully induce more bioactivity, bio-ceramics are frequently used as fillers or coatings. We employed a novel technique to produce porous composites, utilizing a porous inorganic matrix (P3HB) to cover a bio-ceramic scaffold (45S5 Bio-glass®) made of bio-ceramic material. In reality, once the scaffold struts started to fail, the P3HB layer acted like a glue, holding the inorganic particles together. Specifically, the polymer was used to support the 45S5 Bio-glass® scaffold structure. This type of composite scaffold has compressive strength (up to at least 1.5 MPa) that was twice as high as bare 45S5 Bio-glass® scaffolds [4]. Both non-degradable and biodegradable polymers have been used to create composite scaffolds, although strong polymers frequently have low biocompatibility since they frequently surround themselves with fibrous medicines after implantation. As a result, several initiatives to develop composites that include biodegradable polymers and bio-ceramics have been made. Nano-bio-ceramic/polymer composites have recently gained attention due to its potential to interact with host tissue and cells. One of the most fascinating challenges in this regard is to develop clever composite biomaterials with nanoscale interaction between the bioactive inorganic phase and the organic one so that the scaffold degrades as a single fabric instead of having choppy rates of deterioration for the glass and polymer levels.

The so-called Big name gels, a specific class of organically altered silicates (ormosils) with an organic center surrounded by flexible arms that can end in alkoxysilane corporations and are capable of forming a silica-like network during the sol-gel technique, merit special consideration. These hybrid materials are intriguing for tissue engineering applications that need adequate long-term fatigue behavior because they exhibit bioactive properties and have fracture toughness that is comparable to cancellous bone and greater than that of sol-gel glasses. Additionally, it has been reported that bio-ceramic/metallic composites may be produced while the ceramic component is used as a coating. Steel materials including stainless steel, titanium, and Co-Cr-Mo alloy have become popular choices for load-bearing prostheses because of their high power, exceptional fatigue resistance, and exceptional machining qualities. Some metallic chemicals, however, might have negative side effects, such as releasing a lot of metallic ions into the tissues, which could lead to concerns including inflammatory and immunological responses. Studies on people living on the go have shown that TiNbZr scaffolds with changed surfaces are preferable to bare metal ones for osteoblast-like cells to attach to and multiply on [4].

9.2.2.5 Magnetic Bio-Ceramics

The creation of magnetic bio-ceramics, which include a ferrimagnetic, ferromagnetic, or superparamagnetic segment integrated into any bio-ceramics, is probably the maximum creative and thrilling area of research because it has such a lot of ability programs, inclusive of the capability to deal with most cancers via hyperthermia or to desire the osteoinduction phenomenon while applying magnetic fields (or no longer). The usage of external magnetic fields, particular nano-systems or nanoparticles may be hired as drug vendors to deliver medicines to precise elements of the body.

9.3 Bone Replacement and Regeneration

Depending on the bone in question and the degree of harm, an expansion of materials can be used in bone regeneration and restoration strategies. Those techniques have so far utilized alumina, metallic bio-glass, bio-glass-metal fiber composites, polymer-carbon fiber composites, calcium phosphate, and hydroxy apatite [11]. These substances have been applied for a selection of functions, which include porous materials to inspire the natural bone cells to renew and as dense substances for direct implants and reconstruction substances (plates, and many others). Bio-ceramics are a collection of materials that can be utilized in reconstructive approaches together with total artificial hip, knee, shoulder, elbow, and wrist replacements, intramedullary nails for fracture repair, Harrington rods for spinal curvature correction, vertebrae spacers and extensors for congenital deformity correction, alveolar bone replacements, mandibular reconstruction, and dental tactics

[12]. With a selection of businesses that offer a biomimetic system manage to simulate organic fluids and lead them to extraordinarily biocompatible solutions, hydroxyapatite may be stabilized and functionalized similarly [13]. In a nutshell the issues brought on by bone abnormalities and injuries may be addressed by bone replacement and regeneration utilizing bio-ceramics. Bio-ceramics are the best material for encouraging bone healing and repair because of their great biocompatibility, bioactivity, and mechanical qualities. Results in bone replacement and regeneration are anticipated to be further improved by ongoing research into bio-ceramic materials, as well as by their applications in bone tissue engineering and drug delivery systems.

Raw materials used for the following:

(i) **Hip joint**

The powder form of alumina (Al_2O_3) is white. The ceramic that arises from shaping, compressing, and heating to a high temperature (sintering) has high density, high strength, great corrosion resistance, good biocompatibility, and high wear resistance. It can also be machined, honed, and polished to produce a high-quality item.

Zirconia (ZrO_2) is a white powder as well. It can be compressed and sintered into a very durable ceramic, just like alumina. Its wear resistance qualities are inferior to those of alumina [14]. A more durable bio-ceramic known as Y-TZP can be created by adding yttrium oxide and trace amounts of magnesium oxide.

(ii) **When mending bones**

For a very long period, calcium phosphates have been used as synthetic bone grafts. Recent studies have shown the efficiency of modifying compositional and textural characteristics, such as nano-, micro-, and macro-porosity, to regulate and synchronize bone development and material resorption.

Biomimetic calcium phosphates have a better capacity to change the material characteristics as compared to conventional high-temperature sintering methods. The structure and chemical makeup of these calcium phosphates closely mirror those of bone minerals.

Calcium phosphate, sometimes referred to as hydroxyapatite or $Ca_{10}(PO_46$ $(OH)_2$, is the primary component of natural bone in the body. Ceramics made from synthetic calcium phosphate may potentially be used in the medical field. The problem is that these ceramics are not as strong as those composed of zirconia or alumina.

(iii) **For teeth**

Alumina, zirconia, bio-active glass, glass-ceramics, hydroxyapatite, and bioabsorbable calcium phosphates are some of the examples of bio-ceramic materials. They have been employed in dentistry as endodontic sealers, bone defect fillers, root repair materials, apical fill materials, perforation sealers, and regeneration aids. They can be generically categorized in endodontics as either calcium phosphate/tricalcium/hydroxyapatite based, calcium silicate based, or calcium phosphate and silicate based mixes [14].

9.4 Manufacturing Methods for Bio-Ceramic Scaffolds

There are several different processing techniques that may be used to create bio-ceramic porous scaffolds for tissue engineering. Foaming, solid freeform fabrication (SFF), starch consolidation, organic phase burning-out, and sponge replication are a few examples of these. These procedures were commonly adapted from other settings or constructed on the spot. Different methods may be used to manage the 3-D structure of tissue engineering structures, and processing has a big influence on a lot of the scaffold's characteristics. In reality, the range of strut thickness and orientation, interconnectivity, and total porosity that may be produced using each approach is where it performs best. In order to meet the needs of the specific type of tissue, the best process must be carefully selected. Additionally, as the overall cost of each fabrication process varies, some are better suited for mass production while others are better for the manufacture of things with added value [4].

9.4.1 Foaming Techniques

Ceramics with holes ranging from 20 nm to 1–2 mm can be made by dispersing a gas in the form of bubbles into a ceramic suspension or colloidal sols, followed by solidification. Using an aqueous solution of H_2O_2, ceramic powder is foamed, and the resultant material is then cast into molds and heated to 60 °C. At this temperature, H_2O_2 decomposes and foams because the oxygen produced prefers to create bubbles in the slurry. The sample is then fused to create crystalline ceramics, bioactive glass, and calcium phosphate scaffolds, depending on the starting powders. To alter the pore size and porosity fraction, alter the heat treatment and amount of H_2O_2 utilized. The inherent flaw in this foaming method is that pores are only connected in a laminar fashion, leading to insufficient connection perpendicular to the laminae.

Sol-gel foaming is a chemical-based wet synthesis technique that involves transforming a solution containing ceramic precursors (sol) into a network of covalently bonded silica. It combines mechanical foaming with sol-gel technology. It is possible to develop a glass or glass-ceramic structure with a hierarchical organization, interconnected macropores for tissue ingrowth, and a mesoporous texture (channels in the 2–50 nm range) that promotes cell adhesion and the adsorption of biological metabolites while accelerating surface reactions in vitro and in vivo. This mesoporous structure can be achieved through the process of evaporation-induced self-assembly (EISA), where a surfactant is introduced as a template to facilitate supramolecular self-assembly within the sol.

By using this technique, it is possible to create highly bioactive 3D scaffolds that resemble bone; however, a drawback is that the porous result is inherently fragile due to its nano-porous texture, which creates serious problems for the device's safe implantation (too low mechanical properties).

9.4.2 Consolidation of Starch

This process creates porous ceramics by using granules of starch extracted from corn, rice, or potatoes as both a pore forming and a binder. The low cost and environmental friendliness of this processing method are its key benefits. The technique involves mixing distilled water, ceramic powder, and starch granules to generate a suspension that is maintained between 60 °C and 80 °C.

The steps in the procedure are:

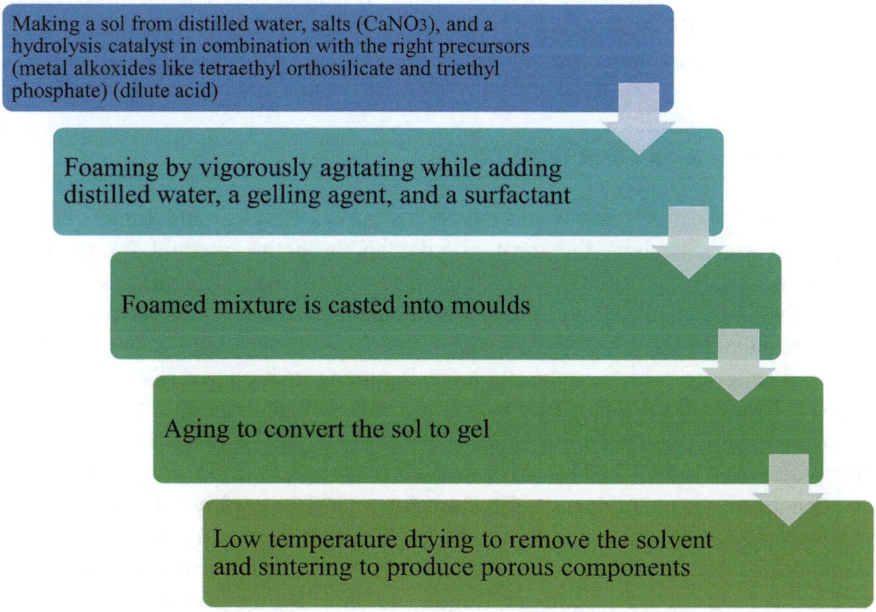

As a result of the water absorbing properties of the starch at these temperatures, a gel-like material is created. This substance is then thermally treated to burn off the organic phase and sinter the ceramic matrix [4]. The final dimensions of the component may be managed more easily after sintering since modest dimensional changes occur during consolidation and drying. Although the resultant glass-ceramic scaffolds' mechanical properties (compressive strength of roughly 6 MPa) were equivalent to cancellous bone's (2–12 MPa) in the past, the porosity was too low (40 vol.%) and the interconnections were not strong enough to support a clinical use. In order to create pores for tissue engineering bioactive glass scaffolds, investigations with several polymer phases (other than starch) have been carried out.

9.4.3 Burning-out of the Organic Phase

The organic phase burning-out (or space holder method) is one such approach for making porous scaffolds. In this method, ceramic particles (like rice husk) are mixed with solid polymeric phases of synthetic (like poly (methyl methacrylate) or polyethylene microbeads) or natural origin. The mixture is then subjected to high-temperature thermal treatment before being compressed to produce a "green body." The polymeric particles fill the component's volume upon heating while the inorganic particles sinter, creating a porous material that exhibits a reverse replication of the original sacrificial template. Because, they feature strong, dense struts, the scaffolds made using this approach can have great mechanical strength that is even comparable to cortical bone.

9.4.4 Using a Sponge Replica

The sponge replication method was created by Schwartzwalder and Somers in 1963, and it has since become the most popular and effective method for producing ceramic scaffolds that mimic foam for tissue engineering. This success is partly attributable to the method's simplicity and versatility, since any ceramic material that can be adequately dispersed into a solution may be employed with it. The reticulated open-cell structure, which can be created using the foam replica method and is made of porous material surrounded by a web of ceramic ligaments, the struts, has been discovered to be the most suitable for bone tissue engineering scaffolds because it closely resembles the 3D trabecular framework of cancellous bone. Another significant benefit of this method is that the first sponge may be readily cut and sculpted to match the size and form of the tissue defect. Ideally, this would make it possible to design scaffolds that are tailored to the patient's particular clinical needs. In a recent application of the sponge replica technique and the EISA methodology, hierarchical porous bioactive glass scaffolds were made using the sponge replica method and a surfactant as co-templates for the macropores and mesopores of the scaffold. These scaffolds are extremely bioactive and have a mesoporous structure, but they are also quite delicate. Some researchers made an attempt to enhance the mechanical properties of these hierarchical porous structures by coating the strut with silk, however their compressive strength (a few hundreds of kilopascal) was still insufficient for a clinical usage to be deemed safe.

9.4.5 Solid Freeform Manufacturing

A group of new mold-free techniques known as rapid prototyping and solid freeform fabrication use layer-wise manufacturing techniques to create scaffolds with customized external shapes and pre-designed internal architectures (strut features, pore arrangements, sizes, and distribution) directly from computer-generated 3-D models. This 3-D reconstruction of the patient-specific tissue defect was produced using the patient's computed tomography or magnetic resonance imaging data. This collection of methods is very useful for creating functionally graded bio-ceramic and composite materials. To produce scaffolds for tissue engineering applications, a variety of SFF techniques have been used [4].

9.5 Bio-Ceramic Scaffolds Can Be Manufactured Using 3D Printing Technology

Thanks to the swift advancement of additive manufacturing technology, or 3D printing, clinical solutions for bio-ceramic scaffolds may now be customized to the particular requirements of patients. Some state-of-the-art 3D printing technologies may manufacture three-dimensional porous scaffolds that are precisely matched to the tissue defect based on clinical data, reducing the time and cost of product design, processing, and application. However, there are still a number of problems with conventional 3D-printed bio-ceramic scaffolds, such as their single use, low rate of bone resorption, and difficulties in inducing the growth of new blood vessels. These problems make 3D printing materials for bio-ceramic scaffolds an uncommon alternative for present treatment options, and it will take some time for them to do so [15]. In order to improve the biological performance of 3D-printed bio-ceramic scaffolds and develop multifunctional scaffolds to increase their applications in the biomedical field, researchers thoroughly investigated the loading activity from the viewpoints of the material's composition and structure [16].

9.6 Advantages of Bio-Ceramics

(i) They offer some benefits, including biocompatibility, non-toxicity, dimensional stability, and most significantly, being bioinert for endodontic applications [17].
(ii) Due to its resemblance to biological hydroxyapatite, they possess excellent biocompatibility qualities. Because of their capability to absorb osteoinductive substances when a bone healing process is present nearby, they have intrinsic osteoinductive capacity.
(iii) Serve as a biodegradable lattice-based regenerative scaffold that offers a framework as the body regenerates tissue. To be able to chemically bond with tooth structure, create a strong hermetic seal, and be radiopaque. Due to in situ precipitation after setting, which is a condition that causes bacterial sequestration, the substance has antibacterial capabilities.
(iv) Bio-ceramics create porous powders with 1–3 nm-diameter nanocrystals to thwart bacterial adherence. Apatite crystals can occasionally include fluoride ions, and the resulting nanomaterial possesses antibacterial characteristics [5].

9.7 Current and Future Clinical Applications of Bio-Ceramic Scaffolds

(i) For low-/non-load-bearing uses or compressive load situations, including bone augmentation and reconstruction, middle ear repair, vertebral replacement, and iliac crest replacement, biologically active glasses, glass-ceramics, and calcium phosphates are usually employed in solid or powder form due to biomechanical limitations. In their surgical procedures, surgeons also apply thermal-sprayed HA coatings to metal joint prostheses. Alumina acetabular cups with bioactive glass-ceramic porous coatings may improve the osteointegration of prosthetic implants, according to recent research [18].

(ii) All applications of bioactive ceramics emphasize bioactivity while limiting the requirement for mechanical qualities, which might present issues with implants that are too porous. Prefabricated porous blocks of various sizes composed of HA, (biphasic) calcium phosphate, and a few bioactive glasses are presently available all over the world and clinically implanted in individuals for the treatment of minor bone abnormalities. The surgeon can alter these implants during surgery to match the size and form of the defect. SFF-derived custom-made HA porous scaffolds are produced when a high level of size precision or complex shapes are required, as in the case of implants for orbital floor replacement. Trabecular bone, a natural bio-ceramic-based composite used in bone banks, is also used as a restorative material. For patients whose eyes have been removed, creating orbital implants is a special non-osseous use where (porous) bio-ceramics are regularly employed. Because they are biocompatible and allow fibro vascularization inside their pore network, porous spherical implants (scaffolds) made of bovine, coralline, synthetic HA, and alumina are routinely implanted during anophthalmic socket surgery. When compared to competing devices on the market, the 45S5 Bio-glass®/polyethylene composite porous orbital implants utilized in early human studies showed enhanced implant fibro vascularization, which may be due to the angiogenic effect of bioactive glass.

(iii) Recently, the usage of bio-ceramic and composite scaffolds has also been proposed for several novel applications in interaction with soft tissues. As an inorganic phase, bioactive glass is frequently used in these scaffolds. Bioactive glasses' angiogenic potential has opened up new possibilities for skin tissue engineering in this field. Larger soft tissue constructions would benefit greatly from the ability of 45S5 Bio-glass® put into PGA meshes to promote scaffold neovascularization, which was initially proven in vitro (using fibroblasts) and in vivo (in rats) [19, 20].

(iv) Blood clotting was hastened by the inclusion of bioactive glass nanoparticles, which also enhanced the scaffold's wettability, surface roughness, and general biocompatibility.

9.8 Bio-Ceramics Based on Magnesium for Orthopedic Use

Magnesium ions have a crucial role in the regulation of ion channels, DNA stability, enzyme activation, and promotion of cell growth and proliferation, to name just a few biological processes in which they are directly engaged. In recent years, this alkaline earth metal has become very popular in orthopedic applications. Magnesium-based bio-ceramics are a broad category of materials that contain magnesium and are used in orthopedic applications like bone cements, bone scaffolds, and implant coatings. Regenerative bone implants should encourage the growth of new bone and decay concurrently with osteogenesis. Clinically proven calcium phosphate bone grafts have a low bioresorption rate despite having great osseointegration and osteoconductive efficiency [19]. Because they are biocompatible, mechanically very stable, and disintegrate considerably more quickly than calcium phosphates under physiological settings, magnesium phosphate (MP) based ceramics represent a possible option [21, 22].

9.9 Description of Importance

Composite ceramic materials made of magnesium oxide and calcium phosphate ($MgO/Ca_3\ (PO_4)^2$) are regarded as a potential class of bioactive materials and are anticipated to be used in artificial bone scaffolds. Although calcium phosphate-based bone substitutes have long been researched, a new trend in the biomaterials industry is magnesium-based bio-ceramics made of magnesium phosphates and silicates. This is because magnesium ions have unique biological significance in processes like enzymatic activation, cell growth and proliferation, and others. Such compositions do not emit hydrogen after breakdown, unlike pure magnesium implants. Similar to calcium-based bio-ceramics, magnesium-based bio-ceramics are used to create a variety of products, including coatings, cements, and macroporous scaffolds [19]. From this angle, we provide a thorough overview of several types of magnesium-based bio-ceramics, their processing procedures for various therapeutic applications, and their behavior both in vitro and in vivo.

9.10 Bioresorption

Ceramics are fragile, therefore implant leftovers are associated with a higher risk of inflammation or fractures of newly formed bone. To overcome these limitations, novel synthetic bone replacements are being developed with the intention of totally replacing them with genuine bone tissue. A fundamental need for this is that the synthetic material degrades with kinetics that correspond to the pace of osteogenesis. Bioresorption, which can include both chemical dissolution and cellular resorption of the material, is often recognized as a critical need for an ideal bone implant. While macrophages and osteoclasts are capable of carrying out cellular resorption, solubility is a chemical characteristic that is intrinsic to a substance and is influenced by environmental variables including pH, ion composition, and temperature. Over the past few years, interest in completely bioresorbable magnesium-based bone implants has significantly increased. This involves both magnesium alloys and magnesium phosphate (MP) ceramics. The elastic moduli of magnesium are similar to those of cortical bone, and it is believed to be biocompatible, osteoinductive, and angiogenic. The in vivo degradation of the metal still presents a challenge, though, since it typically occurs too rapidly and is based on corrosion, which results in the production of hydrogen gas at the peri-implant areas. Despite the dearth of in vivo and in vitro studies, the biocompatible, osteoconductive, and osseointegration properties of MPs have been established since there was no evidence of inflammatory rejection or foreign body responses, and healing occurred without the formation of connective tissue. The limited studies that have been conducted indicate that MP deteriorated over a period of 3–6 months, and that the subsequent replacement of the ceramics by bone resulted in a full conversion of the implant to natural bone tissue [23, 24].

9.11 Bio-Inorganics and Biological Compounds Used in Bio-Ceramics as Drug Delivery Systems

The development of synthetic bone grafts with both osteogenic and antimicrobial properties would be a big step forward. Recent studies have discovered that low temperature biomimetic CaPs are indeed an amazing tool for the inclusion of antibiotics because they enable for such control release kinetics by modifying the textural attributes of the biomaterial, thereby preserving the drug's efficiency. Similar to this, other active ingredients, such as anti-inflammatory or anti-cancer medications, or even growth factors like BMPs, which can improve the bone graft's osteogenic potential, can be added. However, there are still certain challenges, such as regulating the release kinetics over an appropriate time period, watching the release in vivo, or ensuring a repeatable performance in varied habitats, clinical situations, and patient specificities. In addition to the transport of antibiotics or other biomolecules, bio-ceramics may be used as carriers for the local delivery of active ions that can trigger certain biological reactions [25]. Consequently, ions such as copper, strontium, zinc, cobalt, silicon, and boron may be able to encourage angiogenesis and osteogenesis, while copper, zinc, and silver may also have anti-inflammatory and antibacterial characteristics. This technique offers significant financial savings and increased stability.

9.12 Scaffolds Made of Bio-Ceramic with Drug-Induced Antibacterial Function

One of the strategies that is presently most routinely utilized to treat infections associated with bone implants is drug treatment. However, due to general toxicity, pathogenic bacterial resistance, visceral issues, and other significant bad effects, systemic administration of antibacterial drugs or non-targeted drug delivery is not the best course of action. In order to give localized and targeted therapeutic effects, antibiotics or other pharmaceuticals can be added to bionic bone scaffolds. The release of these medications can be regulated by the bio-ceramic materials and structural aspects of the scaffold [26]. Numerous aspects need to be taken into account when mixing a medication with a scaffold, such as the physical and chemical qualities, drug absorption and delivery effectiveness, anticipation, material composition, and scaffold structure. This efficient and less hazardous antibacterial technology focuses on perfectly matching the antibacterial delivery system to the 3d model of the scaffolds in order to achieve the antibacterial effects of bio-ceramic-based scaffolds. Because of its broad-spectrum antibacterial activity, low toxicity, and high effectiveness, tetracycline antibiotics have frequently been used to functionalize bio-ceramic scaffolds for use in orthopedic applications. For instance, doxycycline-loaded Mg-Ca-TiO_2 scaffolds with outstanding drug release capacity, cytocompatibility, bioactivity, and antibacterial activity were produced using the space holder approach. Bio-ceramic-based scaffolds can treat bone implant-associated infections while also boosting bone repair by integrating scaffold production processes with sustained release antibiotic delivery technology. This enhances the controlled-release effect of antibiotics [27].

Bio-glass (BG), a type of ideal bio-ceramics with built-in antibacterial properties, not only has amazing osteoinductive potential but also has a variety of antibacterial behaviors.

It was also demonstrated that the antibacterial effectiveness of novel bio-glass scaffolds built of SiO_2-P_2O_5-CaO-Na_2O-SrO-F or SiO_2-Na_2O-Al_2O_3-CaO-B_2O_3. Due to its exceptional antibacterial property, robust osteogenic activity, and mechanical robustness, forsterite (Mg_2SiO_4) scaffold has also become one of the finest materials for bone regeneration. By increasing the pH and osmotic pressure of the fluid and causing bacterial cell membrane depolarization and death, alkaline ions released from these scaffolds may have had an antibiotic effect. The most well-known antibacterial substance employed in several bio-ceramic-based scaffolds is silver (Ag). However, its antimicrobial effects should be taken into account to prevent a negative impact on osteogenesis, biocompatibility, or bioactivity. Ag could be added to various bio-ceramic-based scaffolds using a variety of techniques, including coating, doping, and mixing, and in a variety of forms, including ions, particles, and oxides.

9.13 Cell-Based Therapies Together with Bio-Ceramics

Bio-ceramics have enormous potential for application in bone tissue engineering. Since they were initially introduced in the 1990s, the use of bio-ceramics as a bone grafting technique has increased. These techniques use hybrid constructions that combine bone marrow cells or MSCs (mesenchymal stem cells) with synthetic scaffolds, which are often made of porous ceramics and molecular signals. This strategy works best in cases of complicated skeletal injuries with insufficient osteogenic potential. Even though this discipline has made tremendous strides recently and a number of clinical investigations are ongoing in both Europe and the USA, more needs to be learned about the functions of the various parts and how they interact [25]. A key development for the effectiveness of this treatment strategy could be the creation of scaffolds, which can assist cell attachment and proliferation while also giving the cells the correct temporary habitat and guiding them in the right direction. It is important to be aware of the wide range of possible outcomes in the field of cell therapy and the lack of published data allowing for the connection of findings. Therefore, in order to facilitate the sharing of knowledge, both researchers and surgeons must immediately establish international clinical trial databases.

9.14 Bio-ceramic's Brittleness, High Young's Modulus, and Poor Workability Are their Main Shortcomings

(i) Brittleness is not a problem when using bio-ceramics in non-stressful environments. However, because bio-ceramics are less reliable than metallic biomaterials when employed in locations with high stress loads, it becomes a significant issue. This brings up the requirement for reliability assurance in the design of bio-ceramics. However, since a significant amount of bone tissue would need to be removed, the invasion of the organs would be exacerbated.
(ii) The bone must be under some weight in order to remain alive; otherwise, it would dissolve away as in zero-gravity. Wolf's law states that when bone tissue is bound to substances with a high Young's modulus, stress shielding is shown because the surrounding bone is resorbed.
(iii) During surgery, the bio-ceramics are frequently shaped by the surgeon. Unfortunately, the workability of bio-ceramics is poor, Due to which molding the bio-ceramics lessens their dependability in terms of mechanical qualities [2].

9.15 Nanoscale Bio-Ceramics Toxicity Risks

Despite all the advantages and superiority of nanomaterials—particularly nanoscale bio-ceramics—over microscale ones that have been highlighted, there are some major health concerns regarding them because of their potential impacts on biological systems and potential for cytotoxicity or genotoxicity. The materials may bind to receptors on the cell surface, interact with extracellular matrix components, and disrupt cell signaling, behavior, and activity at the nanoscale level [28]. The risk of toxicity increases with decreasing size, fiber-like shape, and positive charge. It is important to note that the hazardous effects of nanomaterials are mostly dictated by their size, shape, and charge. It has been proposed to modify surfaces using pico-technology, a major technology and new generation of material engineering, without the possible cytotoxicity issues connected with nanomaterials.

9.16 Need for a more Logical Strategy

It is now obvious that further significant technical advancements require a greater understanding of the bioactivity-inducing processes, namely the fusion of the biomaterial with the biological host. Historically, technical progress has been achieved by trial-and-error techniques in the creation of new biomaterials. In this part, we focus on bio-ceramics and bioactive glasses since, as we have already seen, they share a lot of chemical properties with bone. (It is true that their principal drawback is the broad variety in mechanical attributes). Currently, it is not fully known how ceramics adhere to bone and soft tissue or how biomaterials promote osteoblast activities including adhesion, proliferation, and differentiation. But it is clear that they are connected to the surface structure and chemistry of the biomaterial [1]. This fact is largely highlighted by the discovery that nanophase materials appear to offer higher bioactivity, at least in terms of increased osteoblast adhesion. For instance, it is widely known that both at the macro- and micro-scales, these properties may have a major influence on cell activity. Recent years have seen significant progress in the structural characterization of bioactive glass and ceramics due to significant improvements in theoretical frameworks and experimental methods (like high-energy diffraction, multi-nuclear nuclear magnetic resonance (NMR), and extended X-ray fine structure). Because biomaterials play a key role in both present and future biological applications, it is necessary to utilize this knowledge to understand how they work. It is essential and pertinent to establish the main conclusions from early foundational research on bio-ceramics and bio-glasses in order to guide future work in the field of biomaterials [29]. For instance, it is predicted that thorough understanding of the interactions between surface dissolving, ion release, and the activation of osteogenic cellular processes in bio-glasses would reveal features shared by other biological materials, for which little to no structural information is available. Finally, it should go without saying that bone attachment using metals or polymers has never been effective. On the other hand, there are no ceramic materials that have suitable mechanical properties. Therefore, composites comprised of tough metals, bioactive ceramics, and biodegradable polymers are still an alternative. The biological reaction induced by a biomaterial is ultimately determined by its interaction with cells, which is mediated by proteins adsorbed on the biomaterial surface [30, 31]. As a result, knowing how proteins adsorb to biomaterial surfaces is critical for designing future biomaterials for cutting-edge medical devices. This necessitates an atomic-level description of the surface of both proteins and biomaterials [32–34].

9.17 Conclusion

The bone is composed of a variety of materials, including a 30% flexible matrix and 70% bonded minerals embedded with specialized bone cells. The bone's special makeup enables it to be hard and powerful while also being light. It is crucial to promote bone synthesis above and beyond what the host tissue is capable of when the self-regeneration process fails in certain impaired clinical settings. An effective method to improve the timing of bone formation and material resorption is the creation of biomimetic synthetic materials with composition and structural traits that are more similar to genuine bone. Ideal bone grafts should properly balance bone resorption and bone deposition under a variety of parameters imposed by the patient's age, gender, and social habits (sport, addictions), as well as the particular stressors at the implantation site. To advance in the creation of these materials, it is essential to comprehend the biological mechanism behind the interaction between the graft at the bone site and the patient's unique bio-ceramics. First-generation biomaterials may employ inert bio-ceramics, second-generation biomaterials may use Hench's bioactive glasses, and third-generation biomaterials may use regenerative materials rather than replacement ones. Therefore, it is important to comprehend the processes driving bioactivity and the nature of the biological reaction that causes the biomaterial to interact with the tissue around it. It is now necessary to adopt a more rational approach, one that is based on a fundamental understanding of the relationship between atomic or molecular structure and bioactivity. The vast improvements in bioactive glass and ceramic synthesis in the past have mostly been empirical in nature.

Despite all of the benefits that have already been mentioned, one of the primary drawbacks of nanomaterials is their potential for toxicity, particularly nano-sized bio-ceramics. In addition to the biochemical damage, they can physically interact with biological components, which can result in cell death. Furthermore, nanomaterials pose challenging toxicological hazards because they may boost the generation of reactive oxygen species. Additional research is required to develop the protocols that have been authorized for reducing the cytotoxicity of nanomaterials.

References

1. Thakur P, Thakur A (2022) Introduction to nanotechnology. In: Thakur A, Thakur P, Khurana SP (eds) Synthesis and applications of nanoparticles. Springer, Singapore. https://doi.org/10.1007/978-981-16-6819-7_1
2. Dhanda N, Thakur P, Thakur A (2023) Green synthesis of cobalt ferrite: a study of structural and optical properties, materials today: proceedings. 73(2):237–240., ISSN 2214-7853. https://doi.org/10.1016/j.matpr.2022.07.202
3. Bharti MK, Chalia S, Thakur P, Sridhara SN, Thakur A, Sharma PB (2021) Nanoferrites heterogeneous catalysts for biodiesel production from soybean and canola oil: a review. Environ Chem Lett 19(5):3727–3746
4. Baino F, Novajra G, González-Calbet JM (2015) Bioceramics and scaffolds: a winning combination for tissue engineering. Front Bioeng Biotechnol 3:202
5. Dhanda N, Kumari S, Kumar R, Kumar D, An-Cheng S, Thakur P, Thakur A (2023) Influence of Ni over magnetically benign co ferrite system and study of its structural, optical, and magnetic behaviour. Inorg Chem Commun 151:110569., ISSN 1387-7003. https://doi.org/10.1016/j.inoche.2023.110569
6. Dhanda N, Thakur P, Kumar R, Fatima T, Hameed S, Slimani Y, Sun A-CA, Thakur A (2023) Appl Organomet Chem:e7110. https://doi.org/10.1002/aoc.7110
7. Dhanda N, Thakur P, An-Cheng S, Thakur A (2023) Structural, optical and magnetic properties along with antifungal activity of Ag-doped Ni-Co nanoferrites synthesized by eco-friendly

route. J Magn Magn Mater 572:170598.,ISSN 0304-8853. https://doi.org/10.1016/j.jmmm.2023.170598
8. Mathur P, Thakur A, Singh M (2008) Efect of nanoparticles on the magnetic properties of Mn-Zn soft ferrite. J Magn Magn Mater 320:1364–1369. https://doi.org/10.1016/j.jmmm.2007.11.008
9. Fernandes H, Rebelo A, Brazete D, Stan G (2018) Bioactive glasses and glass-ceramics for healthcare applications in bone regeneration and tissue engineering. Materials 11(12):2530
10. Kaushik S, Gandhi S, Chauhan M, Ma S, Das S, Ghosh D, Chandrasekharan A, Alam B, Parmar AS, Sharma A, Santhoshkumar T, Suhag D (2020a) Water-templated, polysaccharide-rich bioartificial 3D microarchitectures as extra-cellular matrix bioautomatons. ACS Appl Mater Interfaces 12(18):20912–20921
11. Thakur P, Gahlawat N, Punia P et al (2022) Cobalt nanoferrites: a review on synthesis, characterization, and applications. J Supercond Nov Magn 35:2639–2669. https://doi.org/10.1007/s10948-022-06334-1
12. Schlickewei C, Kleinertz H, Thiesen DM, Mader K, Priemel M, Frosch K, Keller J (2019) Current and future concepts for the treatment of impaired fracture healing. Int J Mol Sci 20(22):5805
13. Kumari S, Dhanda N, Thakur A, Singh S, Thakur P (2023) Investigation of calcium substitution on magnetic and dielectric properties of Mg–Zn nano ferrites. Mater Chem Phys 297:127394., ISSN 0254-0584. https://doi.org/10.1016/j.matchemphys.2023.127394
14. AZoM.com. (2019, February 4). Applications of Bio-Ceramics. https://www.azom.com/article.aspx?ArticleID=16479
15. Gul H, Khan MR, Khan AL (2020) Bioceramics. In Elsevier eBooks (pp. 53–83)
16. Eugen G, Claus M, Anna-Maria S, Niklas D, Philipp S, Andrea E, Andrea M, Elke V (2023) Degradation of 3D-printed magnesium phosphate ceramics in vitro and a prognosis on their bone regeneration potential. Bioact Mater 19:376–391
17. Vorndran E, Moseke C, Gbureck U (2015) 3D printing of ceramic implants. MRS Bull 40(2):127–136
18. Kala D, Gupta S, Kaushal A (2022) Nanotechnology in healthcare. In: Thakur A, Thakur P, Khurana SP (eds) Synthesis and applications of nanoparticles. Springer, Singapore. https://doi.org/10.1007/978-981-16-6819-7_19
19. Bioceramics in orthopaedic surgery: state of the art and preliminary results. (1992). PubMed. https://pubmed.ncbi.nlm.nih.gov/1345634/
20. Nabiyouni M, Brückner T, Zhou H, Gbureck U, Bhaduri SB (2018) Magnesium-based bioceramics in orthopedic applications. Acta Biomater 66:23–43
21. Vallet-Regí M (2001) Ceramics for medical applications. RSC 2:97–108
22. Nandhini A, Sudhakar T. Premkumar J (2021) Ceramics and Nanoceramics in Biomedical Applications. In: Hussain CM, Thomas S (eds) Handbook of Polymer and Ceramic Nanotechnology. Springer, Cham
23. Zhou H, Liang B, Jiang H, Deng Z, Yu K (2021) Magnesium-based biomaterials as emerging agents for bone repair and regeneration: from mechanism to application. J Magnes Alloys
24. Davaie S, Hooshmand T, Ansarifard S (2021) Different types of bioceramics as dental pulp capping materials: A systematic review. Ceramics International
25. Kakutani Y, Yamamuro T, Nakamura T, Kotoura YJ (2015) Strengthening of bone-implant interface by the use of granule coatings on alumina ceramics. 3D printing of ceramic implants MRS Bull., 40 (2) (2015), pp. 127-136, https://doi.org/10.1557/mrs.2015.326 Biomed Mater Res. 1989 Jul;23(7):781-808. doi: 10.1002/jbm.820230709.PMID: 2738088
26. Jones JDC (2013) Review of bioactive glass: from hench to hybrids. Acta Biomater 9(1):4457–4486
27. Ginebra MP, Espanol M, Maazouz Y, Bergez V, Pastorino D (2018) Bioceramics and bone healing. EFORT Open Rev 3(5):173–183. PMID: 29951254; PMCID: PMC5994622
28. Negrini NC, Maqbool M, Mouriño VSL, Ciraldo FE, Cresswell M, Jackson PL, Lovell CC, Boccaccini AR (2019) Chitosan/hydroxyapatite composite bone tissue engineering scaffolds

with dual and decoupled therapeutic ion delivery: copper and strontium. J Mater Chem B 7(40): 6109–6124

29. Hayakawa S, Tsuru K, Osaka A (2008) The microstructure of bioceramics and its analysis. In Elsevier eBooks (pp. 53–77)
30. Lyons JB, Plantz MA, Hsu WK, Hsu EL, Minardi S (2020) Nanostructured biomaterials for bone regeneration. Front Bioeng Biotechnol 8:922
31. Oryan A, Alidadi S, Moshiri A, Maffulli N (2014) Bone regenerative medicine: classic options, novel strategies, and future directions. J Orthop Surg Res 9:1
32. Rezwan K, Chen Q, Blaker JJ, Boccaccini AR (2006) Biodegradable and bioactive porous polymer/inorganic composite scaffolds for bone tissue engineering. Biomaterials 27(18): 3413–3431
33. Crowley CP, Wong J, Fisher DS, Khan WS (2013) A systematic review on preclinical and clinical studies on the use of scaffolds for bone repair in skeletal defects. Curr Stem Cell Res Ther 8(3):243–252
34. Chen CH, Woo EM (2018) Polymer science and technology. Chem Phys 15(7):2495–2506

Biocompatibility, Bio-Clearance, and Toxicology

10

Hui-Min David Wang, Preeti Thakur, and Atul Thakur

Abstract

Biomaterials' biocompatibility has been a highly contentious topic for years. The phenomenon's definite significance and underlying mechanisms of biomaterials still need to be discovered. The ability of a biomaterial to support biological function, including molecular signaling systems, without inducing or stimulating local or negative effects to the host is generally referred to as biocompatibility. The selection of a suitable host response is addressed by biocompatibility or safety assessment. Biocompatible materials enable the body to absorb synthetic implants without any unfavorable immune responses, allergic reactions, inflammatory difficulties, or chronic health issues. In addition, they are not carcinogenic. Biocompatibility is highly influenced by the application type also, testing for biocompatibility is difficult because of the new emerging technologies. Understanding, analyzing, and establishing biomaterials toxicity profile is becoming increasingly important in light of the enormous variety and number of prospective uses for nanomaterials as well as their adoption ability. Investigations must be directed immediately at describing the dangers and toxicity these nanostructured biomaterials pose to the environment and/or to human health. The goal of this chapter is to present a thorough overview of data relating to nanostructured biomaterials' biocompatibility, bio-clearance, and

H.-M. D. Wang (✉)
Graduate Institute of Biomedical Engineering, National Chung Hsing University, Taichung, Taiwan
e-mail: davidw@dragon.nchu.edu.tw

P. Thakur
Centre for Nanotechnology, Amity University Haryana, Gurugram, India

A. Thakur
NanoLatticeX, Amity University Haryana, Gurugram, Haryana, India

D. Suhag et al. (eds.), *Integrated Nanomaterials and their Applications*,
https://doi.org/10.1007/978-981-99-6105-4_10

nanotoxicology, pathways at the small level (cell), entry points inside the human body, and their possible effects on human health.

Keywords

Biocompatibility · Biomaterials · Toxicity · Bio-clearance · Human health

10.1 Introduction

Medical devices are a diverse group of products that can vary in many ways, such as their intended use, design, complexity, and technology specially nanotechnology [1]. However, regardless of their differences, all medical devices share essential qualities, such as their intended use for medical purposes and their potential to affect the health and safety of patients and users. To ensure the safety and effectiveness of medical devices, regulatory agencies in different countries have established their laws, guidelines, and standards for the developing, testing, manufacturing, marketing, and post-market surveillance of medical devices [2]. These regulatory requirements aim to minimize the risks associated with medical devices and ensure that they meet the relevant safety, performance, and quality standards [3]. Medical device regulations can vary significantly between countries and regions, depending on factors such as the size and complexity of the market, the level of technological innovation, and the healthcare needs and priorities of the population. Some regulatory authorities have more stringent requirements than others, and some require more extensive clinical testing and documentation before a device can be approved for use. Overall, the regulatory landscape for medical devices is complex and constantly evolving, as new technologies, products, and safety concerns emerge. Nevertheless, the ultimate goal of medical device regulation is to promote the safe and effective use of medical devices and protect the health and well-being of patients and users around the world.

A medical device could be anything from a simple bandage to an advanced stethoscope. Software for diagnostic techniques like computed tomography (CT) scans and magnetic resonance imaging (MRI) [4, 5] may also be included in a medical device. These gadgets come in a variety of sophistication levels and uses [6].

Biomaterial's function is significantly impacted by interactions between the body and the materials. The interaction between the body and the biomaterial is extensive. This definition has three essential elements: first, the biomaterial must function; second, the body's tissues must react to the biomaterial in the best way possible; and third, biocompatibility must always be established for a particular application [7]. The capacity of a biomaterial to produce optimal performance in the host with a suitable surface for connective tissue adhesion without inducing unfavorable local or systemic reactions is known as biocompatibility. Biocompatibility can also be explained as the capacity of a scaffold to enable the proper cellular activities to maximize tissue regeneration without inducing unfavorable systemic and local reactions in the host. Biomaterials based on nanotechnology are the subject of burgeoning research. For their potential uses in medication delivery, diagnosis, therapy, photothermal therapy, photodynamic therapy, bioengineering, and biomedical engineering, nutraceuticals [8] a variety of biomaterials have been produced. To clarify potential toxicological hazards, it is crucial to accurately identify nanostructured biomaterials and gain an understanding of their surface chemistry/reactivity (when they come into contact with biological membranes) [9].

10.2 History

Prior to the seventeenth century, many diseases had more difficult diagnoses and treatments, making it difficult to get the desired results. After the seventeenth century, advances in the identification and treatment of numerous diseases were made thanks to the advent of cutting-edge medical technology. The history of medical technology is marked by key advancements that have revolutionized the way we diagnose and treat diseases. One of the earliest medical devices was the thermometer, which was invented by Galileo in 1603 [10]. Initially used to measure temperature in liquids and gasses, it was later improved by Sanctorius Santorio in the seventeenth century to measure body temperature in humans. Another significant advancement in medical technology was the stethoscope, which was invented by French physician René Laennec in 1819. The stethoscope, a wooden tube with a trumpet shape, enabled doctors to listen to the internal sounds of the body, such as the heart and lungs, and diagnose diseases more accurately. In 1895, German physicist Wilhelm Conrad Röntgen discovered X-rays, a breakthrough that allowed doctors to see inside the body without surgery [11]. This technology revolutionized medical diagnosis and treatment, enabling doctors to identify and treat fractures, tumors, and other conditions.

Other notable advancements in medical technology include the development of the electrocardiogram (ECG) by Willem Einthoven in 1903, which enabled doctors to measure the electrical activity of the heart and diagnose heart conditions, and the invention of the CT scanner by Godfrey Hounsfield and Allan Cormack in 1972, which provided detailed images of the body's internal structures. Throughout history, medical technology has played a crucial role in the advancement of medicine, enabling doctors to diagnose and treat diseases more effectively and improve patient outcomes. By improving the reliability and accuracy of disease diagnosis and treatment, these inventions have made significant advancements in the medical industry and significantly improved patient health and quality of life. According to data from 2015, the United States is the world's leader in the production and export of medical equipment. Germany, the Republic of the Netherlands, Beijing, Brussels, Ireland, the Swiss nation, Mexico, Europe, and Tokyo occupies the following nine positions after the United States. Because medical equipment is used more frequently in metropolitan areas, this industry has profited from constant and rapid innovation [12].

10.3 Categorizing Medical Equipment

The European Union Medical Device Regulation (EU MDR) has established a classification system for medical devices based on their risk level, complexity, and intended use. This classification system ensures that medical devices are safe and effective for use in patients. The three regulatory classes of medical devices are Class I (A), Class II (IIa, IIb, and III), and Class III (D) [13, 14]. Class I (A) devices are considered low risk medical devices, such as tongue depressors and surgical retractors. These devices pose the least amount of risk to patients and require the least amount of regulatory oversight. Class II (IIa, IIb, and III) devices are moderate to high-risk devices and require more regulatory scrutiny and documentation than Class I devices. Hypodermic needles are an example of a Class IIa (B) device. Class IIb (C) and Class III (D) devices are even higher-risk devices, such as implantable pacemakers and artificial heart valves.

The classification of medical devices is important because it determines the level of regulatory scrutiny and documentation required for their approval, as well as the post-market surveillance and reporting requirements. The EU MDR and other regulatory frameworks ensure that medical devices are safe, effective, and meet the relevant quality and performance standards before they are approved for use in patients.

Class III (D) devices have a significant danger to patients, necessitating particular and stringent regulations on these devices by the relevant regulatory bodies to ensure their safe and effective operation. These classifications were produced using a variety of biocompatibility tests carried out in accordance with ISO 10993 rules, whose basic tenet is based on the type and length of bodily contact. This classification was used to establish the type of biocompatibility studies to be conducted prior to clinical trials after product design and development. These medical devices were divided into three categories based on how they made contact with the body: implant devices, surface-contacting devices, and external contacting devices [15] (Figs. 10.1 and 10.2).

10.3.1 Medical Equipment and Biomaterials

The study of biomaterials is a new topic that is important to the advancement of medical device technology. Since the entire business is based on biomaterials, they serve as the fundamental fabric for the market for medical devices. As a result, the global market is enormous [16]. There are, however, few trustworthy sources, and different estimates have been made for the global market for biomaterials, which was ranging more than $130.17 billion by December 2021. The biomaterials industry, in accordance with medical equipment firms, is experiencing constant double-digit yearly growth. Biotechnology, chemistry, materials science and nanotechnology [17, 18] are just a few of the healthcare fields that make use of biomaterials. Biomaterials are substances that are intended to interact with a biological system for purposes of healing, mending, replacing, and diagnosing a range of human

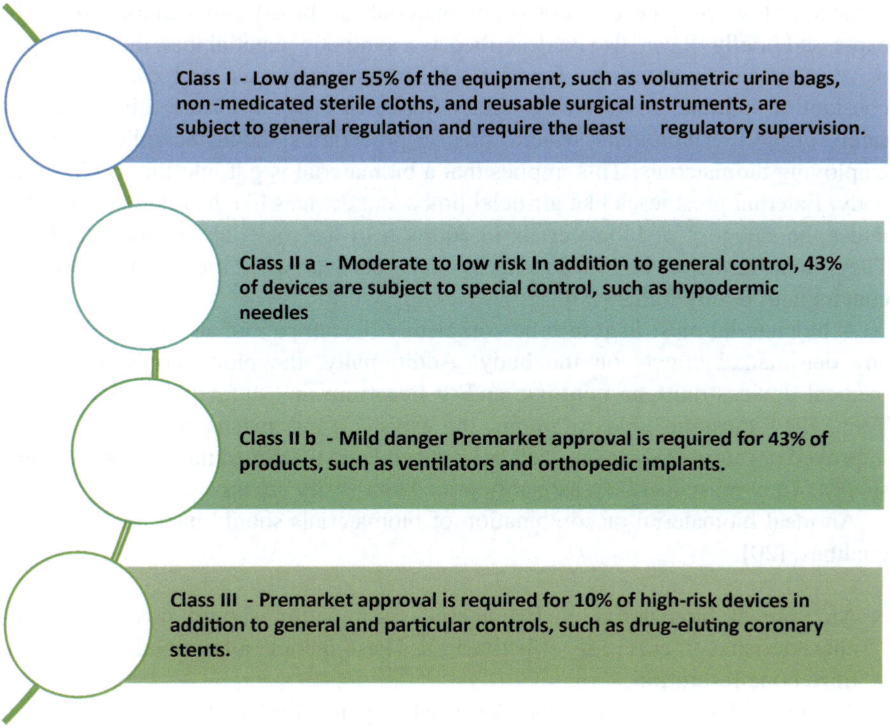

Fig. 10.1 The classification of medical devices

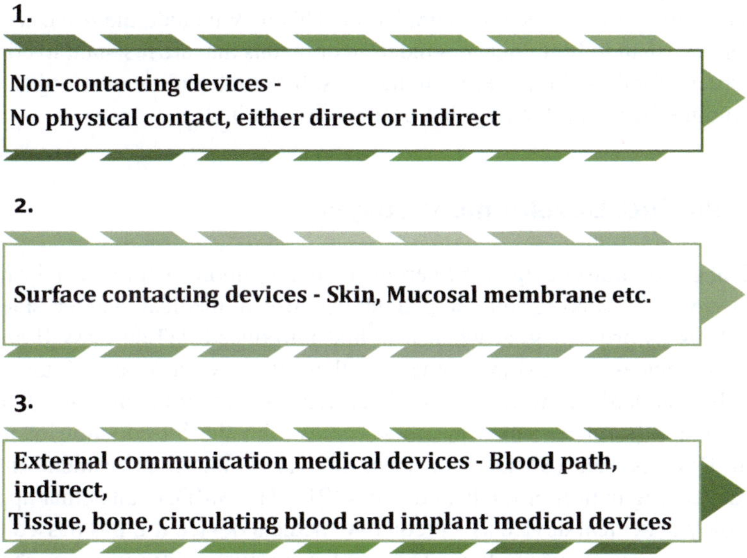

Fig. 10.2 Classification of medical devices based on the type of contact with body

ailments. However, the concept of biomaterials is broad and cannot simply be expressed briefly. It can also be described as a synthetic material that, through repair or replacement, imitates the function of a biological tissue and can come into constant or sporadic contact with bodily fluids. In order to ensure a biomaterial's safety inside the biological system, precise procedures should be followed while employing biomaterials. This implies that a biomaterial is put into the inside of the body. External prostheses like artificial limbs and devices like hearing aids also fall under the category of biomaterials in addition to the materials mentioned above. They can nevertheless be classified as biomaterials even when they do not come into contact with bodily fluids [19].

A biomaterial must therefore be completely biocompatible and should not have any detrimental effects on the body. Additionally, the biomaterials utilized in medical devices must be nontoxic and/or free from any unfavorable side effects. Then, they must be able to mimic the physical and mechanical properties of improved or replaced tissue, as well as be shaped and machined into various, distinct shapes. They must also be reasonably priced and easily accessible.

An ideal biomaterial or combination of biomaterials should have the following qualities: [20].

- An optimal biomaterial or biomaterial combination should possess specific characteristics to ensure its effectiveness. These include a low modulus to minimize bone resorption,
- An elevated resistance to reduce wear debris generation, and
- Appropriate chemical composition to promote biocompatibility and prevent unfavorable tissue interactions.
- To ensure optimal functionality of biomaterials, it is important for them to possess a high degree of resistance to degradation. This may include the use of polymers that are resistant to biological degradation or metals that are resistant to corrosion.
- It is crucial for these biomaterials to have a substantial amount of strength, so they can endure the cyclic loading that occurs within joints.

10.4 Product Development Stages

A medical gadget must navigate numerous regulatory hoops before it can be released onto the market. Making sure a patient can use a medical gadget safely and effectively is its primary goal when it is first introduced. While class II and class III medical devices may exhibit some risk, their efficacy must be maintained at an acceptable standard. Hence, the Medical Device Amendments of 1976 have undergone significant revisions in crucial areas with the introduction of the Safe Medical Devices Act (SMDA) in 1990 to ensure the improvement in safety standards and regulation of medical devices [21]. The SMDA states that applicants may employ a medical device as a predictive medical device if it has passed the 510 (k) process. Prior to now, the FDA's main responsibility was to check the manufacturing and quality assurance records of medical equipment; it had no say

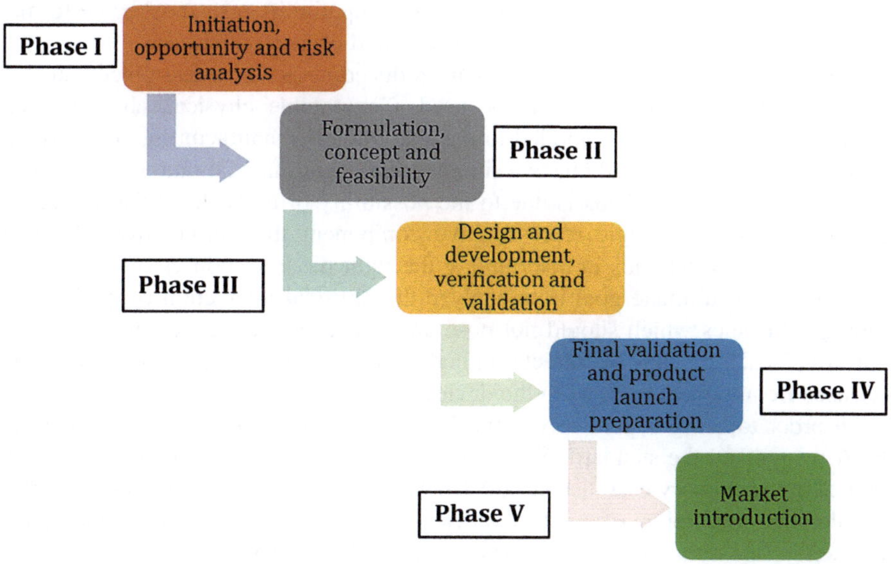

Fig. 10.3 Developmental stages of a biomaterial

over the documentation of research and development. However, the SMDA 1990 granted the FDA the first authority to control the quality system regulation, which went into effect in the middle of 1998, and the medical device design process. According to these regulations, design controls should be implemented for class II and class III medical devices. These controls involve a plan of action that was pertinent for the manufacturing of all new products and an architectural history file where every detail from the stage of development are documented (Fig. 10.3).

The manufacturing of medical devices goes through five stages, including:

- Phase I: Analysis of the situation, the chance, and the risks.
- Phase II: Conceptualization, Feasibility, and Formulation.
- Design and development—verification and validation in Phase III.
- Phase IV: final product validation and launch planning.
- And Phase V Launch of the product and post-launch evaluation.

10.5 Screening for Biocompatibility

Biocompatibility studies must be carried out on all medical equipment, notably class II and class III devices, to guarantee the safety of users/patients. Any potential negative biological reactions resulting from contact with the gadget can be found by doing biocompatibility tests [22]. The device is regarded as safe and approved for usage if the altered/unfavorable biological reactions are small; however, if the increased unacceptable hazards outweigh the benefits, the device will be refused.

To ensure the safety of users and patients, biocompatibility tests must be performed on all medical equipment, particularly class II and class III devices.

For biocompatibility testing, the final produced device or a test object, such as representative components, might be used. Comparable physical, chemical, and surface qualities should be applied to the test article's manufacturing and sterilization processes, and it should have a similar component material ratio to the medical device's finished form. This is due to the possibility of a low-level tissue reaction during testing of a single representative component that differs from the final finished form, which aids in determining the right tissue responses of the selected test piece. The ultimate goal is to analyze the physical interaction of device with biological tissues which should not pose any unacceptable hazards in the form of harmful local or cytotoxic effects, either in the finished product or through the release of component parts from the device.

In order to bring supervisory control over biological compatibility testing, the FDA recognized the standard ISO guidelines that all assays should adhere to. The part of the regulatory established ISO 10993:1 guidelines has the segment heading "Biological evaluation of medical devices—part I." Based on the kind, level, frequency, and duration of device exposure to the body, as well as the physicochemical qualities of typical device materials, the biological reaction to medical equipment is determined. Testing and Evaluation is a Part of a Risk Management Process. The steps involved in performing biocompatibility tests and their results are described in Fig. 10.4:

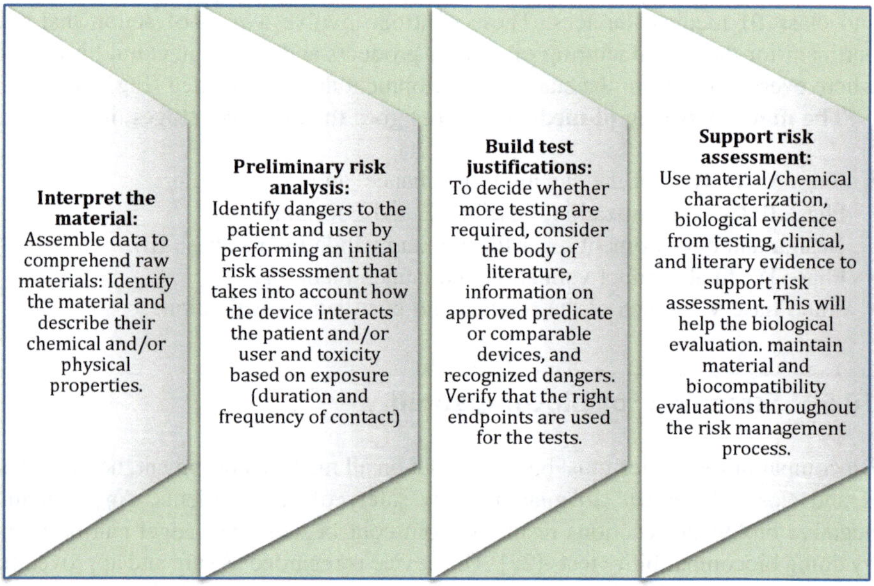

Fig. 10.4 Risk based approach to biocompatibility

10.6 ISO 10993 Biocompatibility Assessment and Risk Management Procedures

When extractants and leachables enter the body of the user or patient from medical devices, they share many of the same chemical properties and differ little from one another. Extractants are described as "chemicals (organic/inorganic) that are extracted from representative equipment components and used to identify and quantify likely leachables" when exposed to specified solvents under particular conditions. Leachables are therefore defined as "chemicals, both organic and inorganic in nature, which transfer representative components to the body/user over time." In contrast to extractants, the release to the patient is neither solvent-dependent nor strictly regulated. The (CDRH), a division of the United States Food and Drug Administration (US-FDA) is in charge of the oversight of medical equipment, regulates the tests for finding extractants and leachables in medical equipment. The conceivable method through which leachables could enter the body of the user or patient served as the foundation for the experimental design. One such way for a medical device to enter the body through leachable material is the medicinal substance contained in syringes, syringe filters, and infusion pumps. The second way that leachable substances enter the body of the user or patient is through direct migration into the tissue in contact with the medical device, such as fixture, prosthesis, stents, a dressing or a gauze, and therapeutic lenses. Some medical devices, such as implants and stents, can be inserted via both medication ingestion and direct touch. The experimental procedure should be created based on the path that leachables from medical devices take to enter the body. However, only medical equipment with the third way of entry is subject to analysis.

The FDA's response to 510(k) calls for two prerequisites. Finding out whether a single medication, a blend of medications, or other biological materials intended for use with the medical device are stable and compatible is the first step. The safety of the drug product and any extractants, leachables, pollutants, or degradation from medical equipment are then assessed. In accordance with the FDA Modernization Act of 1997, the ISO 10993:12 standard entitled Sample Preparation and Leaching Evaluation was created to support the testing of extractants and leachates in medical devices. The Product Quality Research Institute and ISO 10993:12 standards for extractants and leachables in oral and nasal inhalation products are more relevant.

10.7 Study for Biocompatibility Is Intended to Determine

The methodology for study design or Testing is based on the analysis of potentially dangerous components that could be delivered to a patient using a drug or device, can be divided into two sections depending on the route of introduction.

1. **The drug product is the point of entry for extractants and leachables developed for medical devices:**

Prior to learning about the biocompatibility of extractants and leachables from medical devices, FDA-approved drugs designed for device usage were prioritized. The testing device should be checked for the detection of one or more medications. If the device just carries one drug, like an insulin pump, choosing the drug is clear-cut and simple. However, if more than one medicine is present in the device, choosing becomes extremely difficult.

To assess the biocompatibility of extractants and leachables, an experimental method involves two steps. The first step is to prepare the drug delivery system and the second step is to subject it to appropriate testing procedures. The drug delivery system should include commonly used drugs for both the i.v. and epidural routes of infusion, with their compatibility and leachables evaluated. This approach ensures the safety and efficacy of drug delivery systems in clinical settings.

(i) **Step 1:** This entails controlled environment extraction processes for extractables.

The series of activities included in Phase 1 are as follows:

- An appropriate solvent type is chosen, such as polar, semipolar, or nonpolar, depending on the representative medicinal product.
- If the polar buffer(s) and the drug product's chemical characteristics, such as pH and ionic strength, match, a polar solvent should be utilized as the transport medium. If a drug product contains surfactants, ethanol and water should be mixed in a 50:50 ratio; otherwise, IPA should be used for the medicinal product's extraction.
- Which extraction method to use—extensive, heightened, or stimulated—must be determined.
- The exhaustive extraction procedure using the Soxhlet equipment is advised for neat solvents that have not been combined with two or more other solvents.
- The batch extraction procedure with successive mixing or reflux should be maintained for mixed solvents and buffers.
- Choosing an appropriate analytical technique to identify leachables in typical pharmaceutical goods. The (GCMS) technique of evaluation can be used to identify valuable organic extractables.
- The (LCMS) technique of evaluation can be utilized to determine nonvolatile organic extractables.
- LCMS is frequently utilized for extractable inorganic materials.
- Chromatography is frequently used for organic extractables.
- Atomic spectroscopy, which uses techniques for quantitative determination of chemical elements, is used to evaluate organic or inorganic extracts. This includes (ICP) atomic emission spectroscopy to determine very precisely the elemental compositions of samples, and (ICPMS) to measure elements at trace levels in biological fluids. Unknown extractables can be located using mass spectrometry (MS).
- The analytical procedures should be examined for exactitude, recuperation, precision, relevancy, detection limit, quantification limit, linearity

(concentration curve), skepticism interference, and system applicability in accordance with OECD ISO 17025 and GLP requirements.
- Based on the effectiveness of the approach and its intended usage, a standard acceptance criterion should be established.
- To ascertain a drug's stability and biocompatibility with a medical device, analytical assay techniques should be performed in conjunction with the currently accepted standard practices. A new analytical method needs to be developed and validated if these options are not accessible.
- **Step 2** should be carried out after getting all the analysis reports.

(ii) **Step 2:** Here, the amount of leachables that are incorporated into the drug product for medical devices is evaluated, as is the stability of the medications in those devices. The steps of the procedure are as follows:
- The medication should be administered under favorable conditions at a clinically ideal rate for a clinically desirable duration of time (or it should be stored in a device for the therapeutically relevant time), once the drug product has been loaded into the settings of the medical device.
- In concordance with the test substance, a drug product under control is ingested and optimized under the same circumstances as the test substance but is not exposed to medical devices.
- The representative aliquots need to be gathered at the conclusion or midway through. Both the test sample and the control are subjected to an assay.
- A comparison of the two samples is made.
- In addition, the leachables and extractants method is used for the dispensed sample and control analysis.
- The control sample should be omitted if any leachables are discovered.
- Carry out the same steps for every representative medicine.

To pass the United States Pharmacopeia's (USP) approval criteria for an assay, the disparity between the tested substance and the control sample must be the same and without a larger variance. The USP method states that if the medication product value is ±10%, the same ±10% should also be counted for the control. The suggested acceptability criterion for impure and leachable compounds in pharmaceutical products and medical devices is 0.05% [23].

10.8 Toxicological Risk Assessment

The suggested particular techniques are for biocompatibility and toxicology testing. This listing does not suggest that all procedures are required for any specific device or that these are the only tests that are available:

Devices must undergo toxicological risk assessment in order to guarantee the user's or patient's safety. All studies for nonclinical laboratory research must follow FDA 21 CFR, Part 58 GLP. It is important to extract the test substance, the positive

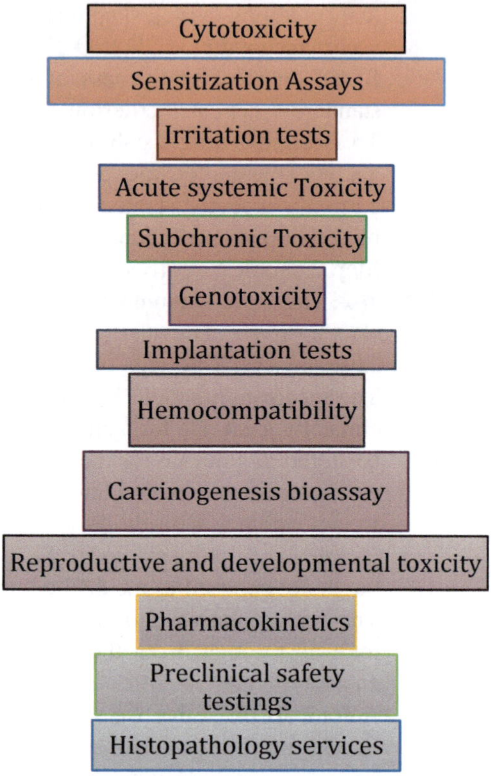

Fig. 10.5 List of toxicological risk assessment tests

control, which evaluates the test validity of the experimental methodology, and the negative control, which is distinguished by the lack of necessary reagents or components (Fig. 10.5).

10.8.1 Cytotoxicity Studies

Biomaterials/medical devices have been evaluated using a range of in vitro cytotoxicity assays, which are detailed in ISO 10993:5. (ISO 10993:52009) [24]. These recommendations cover three methods—an extract test, a direct contact test, and an indirect test—for determining the cytotoxicity of medical devices/biomaterials. Although, a variety of in vitro cytotoxicity assays can be used, and the device components can be grouped according to the assessment criteria and readout system. Cell culture assays use isolated cells in vitro to assess the biocompatibility of a material or extract. These methods can be used to assess the toxicity or irritancy potential of materials and chemicals. They are an excellent method for screening materials prior to in vivo tests.

10.8.1.1 Qualitative Cytotoxicity

There are three commonly used qualitative cytotoxicity tests for medical devices. The direct contact procedure is recommended for low-density materials such as contact lens polymers. In this method, a piece of test material is placed directly onto cells growing in culture medium. The cells are then incubated. During incubation, leachable chemicals in the test material can diffuse into the culture medium can come in contact with the cell layer. Malformation, degeneration, and lysis of cells surrounding the test material indicate the reactivity of the test sample. The Agar Diffusion assay is appropriate for high-density materials such as elastomeric closures. In this method, a thin layer of nutrient-supplemented agar is placed over the cultured cells. After the test material (or an extract of the test material dried on filter paper) is placed on top of the agar layer, the cells are incubated. A zone of malformed, degenerative, or lysed cells under and around the test material indicates cytotoxicity. The MEM Elution assay employs a variety of extracting media and extraction conditions to test devices under realistic or exaggerated conditions. Extract titration can provide a semi-quantitative measurement of cytotoxicity. The extracts are prepared before being applied to a layer of cells and incubated. The cells are examined under a microscope after incubation for malformation, degeneration, and lysis. At least one type of cytotoxicity test should be performed on each component of any device [25].

10.8.1.2 Quantitative Cytotoxicity MTT Assay

For screening purposes, qualitative cytotoxicity tests (direct contact, mem elution, and agar diffusion) are acceptable according to recent regulatory changes on biocompatibility for devices (ANSI/AAMI/ISO 10993-5:2009), however, quantitative evaluation is preferred. The ISO 10993-5:2009 Annex C description of the MTT cytotoxicity assay states that it can reliably measure as few as 950 cells. The MTT is a colorimetric method for measuring the reduction of yellow 3-(4,5-dimethylthiazol-2-yl)-2,5-diphenyl tetrazolium bromide by mitochondrial succinate dehydrogenase. The proportion of living cells in a solution may be measured since only living cells catalyze cellular reduction.

Utilizing the MTT, one may assess the cytotoxicity of extracted components for medical equipment:

- Toxic substances.
- Pollution in the environment and toxins.
- Potential cancer treatments.
- Antibodies to investigate possible growth inhibitors.
- The main benefits of the MTT are its quantitative ability, flexibility in using extracts or direct contact, and the fact that the results are not susceptible to analyst opinion.

Furthermore, the MTT can be carried out on 96-well microplates in a common reader (such a Bio-Tek ELx808), enabling quick screening of several samples. Although the MTT assay is advised, it should be emphasized that it cannot distinguish between different types of cellular death, such as apoptosis and induced cell death. Furthermore, it can underestimate cellular damage and only pick up on death in the very last phases of cellular aging.

10.8.2 Sensitization Tests

For the purpose of assessing potential irritability and delayed-type hypersensitivity of medical devices and their component materials, the ISO 10993:10 guidelines have offered certain risk assessment methodologies [26]. These methods include considerations to be made before running the test, particular actions to follow, and the key variables to consider when interpreting the results. Testing for sensitization is helpful in detecting drugs that, after repeated or prolonged contact, have adverse local or systemic effects. Immunologic processes are involved in these allergic or hypersensitive reactions. Various compounds from the test material, the test material itself, or—most frequently—extracts of the test material may be used in studies to measure sensitization potential. All classes of medical equipment should undergo sensitization testing, according to the Materials Biocompatibility Matrix. The local lymph node assay (LLNA) and the guinea pig maximization test (GPMT) are the two primary delayed-hypersensitivity test methodologies. In addition the Buehler test is a closed patch test that can be used to assess topical medical devices that come into proximity to skin. The three tests listed above all pass the ISO 10993:10 requirements.

(i) The Magnusson–Kligman Method's Guinea Pig Maximization Test is advised for devices that will communicate with or come into contact with the body or body fluids inside. To increase the skin sensitization response in this investigation, the test material is combined with complete Freund's adjuvant (CFA).
(ii) The Closed Patch Test uses repeated topical doses and is advised for devices that will only come into touch with intact skin.
(iii) The quantification of the lymphocyte response to a sensitizer is determined by the murine local lymph node assay (LLNA). When a substance sensitizes the skin, the allergen is transported to the draining lymph nodes by the epidermal Langerhans cells, which in turn stimulates the growth and differentiation of T-lymphocytes. This test is superior to the closed patch test and guinea pig maximization test from the standpoint of animal welfare.

10.8.3 Irritation Test

These studies gauge a substance's propensity for local irritation utilizing locations like the skin or mucous membranes, typically in an animal model. Although the exposure route (skin, eye, or mucosa) and duration of contact should be similar to the device's anticipated clinical use, it is frequently wise to somewhat exaggerate exposure conditions in order to provide patients with a margin of safety. In the intracutaneous test, test material extracts and control materials are administered subcutaneously. Scores are given for erythema and edema at the injection sites (redness and swelling). Devices that will communicate with the body or its fluids inside or externally are advised to follow this protocol. It accurately recognizes the possibility of local discomfort brought on by probable chemical extracted from a biomaterial. Topical devices that come into touch with intact or breached skin on the outside should be subjected to the Primary Skin Irritation test. In this process, the test substance or an extract is directly injected to rabbit skin locations that have been both intact and abraded. The substance is removed after a 24-h exposure, and the locations are scored for erythema and edema. For devices that will come into contact with intact natural channels or tissues and communicate with the outside world, mucous membrane irritation tests are advised. Instead of the original source material, these investigations frequently use extracts. Studies on eye discomfort, cheek pouches, and vaginal irritation are a few frequent techniques.

10.8.4 Acute Systemic Toxicity

The Acute Systemic Toxicity test discovers leachables that cause systemic (as opposed to local) harmful effects by employing extracts of the device or device material. Mice are injected with test material extracts and a negative control blank (intravenously or intraperitoneally, depending on the extracting media). At four additional time periods after the injection, hazardous symptoms in the mice are also monitored. This test is recommended according to the Biocompatibility Matrix of Materials for all products that come into contact with blood. It might be suitable for any other apparatus that makes touch with inside tissues. The material-mediated pyrogen test determines whether a substance has the capacity to raise a person's body temperature after being injected into their blood. The bacterial endotoxin (LAL) test is used for in vitro testing of lot release for pyrogenicity. Each device or substance must have it validated. However, the rabbit pyrogen test is the favored method for determining biocompatibility. The rabbit test is sensitive to material-mediated pyrogens that might be present in test materials or extracts in addition to identifying bacterial endotoxins [27].

10.8.5 Sub-Chronic Toxicity

Sub-chronic toxicity tests are used to evaluate potential negative effects from repeated or protracted exposure to test compounds or extracts for up to 10% of the test animal's entire lifespan (e.g., up to 90 days in rats). Consider the real use circumstances of a medical device when selecting an animal model for sub-chronic toxicity. The best animal models are chosen using case-by-case analysis. Two common sub-chronic testing techniques are suitable for a wide range of devices. Both experiments use mice. One administers medicine intravenously, and the other by intraperitoneal infusion. Implant tests are commonly carried out across a variety of time periods to assess the sub-chronic toxicity of devices and device materials. All permanent devices must undergo sub-chronic testing, and those with extended interaction with interior tissues should be given consideration.

10.8.6 Genotoxicity

Genotoxicity tests can be used to determine whether medical devices can significantly affect DNA structure, which is a developing concern. This test assesses chromosomal, DNA or genotoxicity caused by mutations in medical devices or their components over a period of time. The regulatory requirements for assessing the genotoxicity of devices are provided by ISO 10993:3 [28]. There are a few in vitro and in vivo tests that can be used to determine genotoxicity. A genotoxin is the chemical that causes genotoxicity. A genotoxin is the chemical that causes genotoxicity. Genotoxicity testing may not be required if the product and its extracts have been chemically characterized and adequate safety assurance data are provided. To determine the quantity of genotoxin removed from the device, chemical analysis of the device or its components is necessary. By looking at the device's indication and human exposure, the whole risk–benefit ratio was examined. Genotoxicity testing is taken into consideration for biomedical equipment that are suspected of posing a genotoxic risk although the test results are not yet available for them.

Products are being considered for genotoxicity testing and CDRH [29] requires genotoxicity information when in prolonged (>24 h to 30 days) or persistent (>30 days) contact with tissues such as bone, mucous membranes, blood or any other internal direct contact. They are also taken into consideration if they have not previously been legally marketed in the US market. A single genotoxicity research cannot be used to examine all genotoxins. There are two in vitro tests and an optional in vivo test to assess genotoxicity.

Point mutations along a DNA strand, harm to the DNA's general structure, or harm to the chromosome's structure are three types of genotoxic consequences (which contains the DNA). To ascertain whether harm has taken place at any of these levels, numerous tests have been developed. These tests are run as a battery and complement one another. The Salmonella typhimurium bacteria, which have been chosen for their susceptibility to mutagens, are used in the Ames test, which is the most popular test for mutagenicity, to identify point mutations. Mammalian cells are used frequently in the Mouse Lymphoma and HGPRT assays, which look for point mutations. Using the Mouse Lymphoma assay, clastogenic lesions in genes can also be found (chromosome damage). Unscheduled DNA synthesis (UDS) is a test for DNA damage and repair that can be used both in vitro and in vivo. Chromosome damage can be directly observed using cytogenetic tests. There are both in vitro and in vivo techniques, such as the mouse micronucleus and chromosomal aberration assays.

10.8.7 Implantation Tests

Implant studies are used to assess the biocompatibility of medical devices or biomaterials that come into direct contact with living tissue other than skin (e.g., sutures, surgical ligating clips, implantable devices, etc.). Medical devices that are intended to be implanted for either short- or long-term use in clinical settings can be evaluated using these tests. Using implantation techniques, it is possible to evaluate both absorbable and non-absorbable materials. For a valid evaluation of safety, the implant study must closely reflect the intended clinical use. With the aid of histopathology, it is possible to assess the dynamics of biochemical exchange as well as cellular and immunologic responses during implantation examinations. The histological examination of implant sites greatly increases the amount of data gathered from these studies.

10.8.8 Hemocompatibility

According to ISO 10993:4:2017 guidelines [30], hemocompatibility tests were carried out for medical devices, and the tests were relevant to all equipment that pass directly from the blood regardless of the contact time. For the purpose of risk evaluation, these tests should be performed to assess their hemolytic, complement activation, and thrombogenicity properties. Therefore, for medical devices that come into indirect contact with blood, hemolysis tests are performed regardless of the contact time, and other hemocompatibility tests such as complement activation and in vivo thrombogenicity are not required. If the medical devices are manufactured from new materials that have never been used in devices lawfully marketed in the US for critical uses such as cardiac or vascular devices, or in devices designed to deliver chemicals into the bloodstream, they should be subjected to in vitro thrombogenicity testing. For example, substances that are eliminated from the body and enter the bloodstream can affect blood platelets and the clotting system. However, in vitro thrombogenicity studies have also been performed on medical devices made of advanced materials and intended for indirect contact with the bloodstream. Hemocompatibility tests are not always carried out if the device's risk is identified, so it is advised to present the necessary documentation that demonstrates a solid case for skipping these particular testing. The base materials used in production and operation, material (such as geometry, size, surface imperfections, and roughness) should be considered and compared with equipment in the United States.

10.8.9 Carcinogenesis Bioassay

The occurrence of cancer or second-generation defects that are permanent is one of the most significant and potentially harmful side effects of medical gadgets. Medical equipment that come into direct contact with patients run the danger of promoting the development of cancerous cells. To assure human safety, it is crucial to test the device for mutagenic, carcinogenic, and reproductive risks. For the offered test techniques, there may be some restrictions on test sample size, preparation, and validation. The majority of tests for carcinogenicity are conducted on implantable and externally communicable devices that have prolonged (more than 30 days) interaction with tissues. Testing for carcinogenicity is required for extracorporeal devices, absorbed devices, and permanent implants. However, due to the high cost and testing time, carcinogenicity studies are warranted only if there is information about the risks associated with implants and extracorporeal products. These tests are used to assess the ability of test substances and/or extracts to cause tumors following a single or a series of exposures throughout the course of the test system's lifetime (e.g., 2 years for rat and 7 years for dog). Testing for carcinogenicity in devices is pricy, complicated, and debatable. Almost always, manufacturers can work out an alternative to comprehensive carcinogenicity testing of their products [31].

10.8.10 Reproductive and Developmental Toxicity

These studies evaluate the possible effects of drugs and/or extracts on ovulation, children's health, and early postnatal development. Suitable for products in sustained contact with tissues.

10.8.11 Pharmacokinetics

The metabolic processes of absorption, distribution, biotransformation, and elimination of harmful leachables and possible breakdown products from test materials and/or extracts are studied using pharmacokinetic or ADME. (Absorption/Distribution/Metabolism/Excretion) research, they are particularly suitable for biodegradable substances or for drug device combos.

10.8.11.1 Preclinical Safety Testing

Preclinical safety studies aim to characterize pharmacological and toxicological effects throughout clinical development, not just before the start of human research. Studies conducted in vitro and in vivo can both aid in this characterization [32].

10.8.11.2 Histopathology Tests

The most straightforward method of determining a device's biocompatibility is frequently an implant study. Living tissue is in direct contact with the test substance. The implant site is recovered and checked microscopically for tissue reactivity once the required amount of time has passed. The ability to identify and describe various reactions taking place with tissues and the immune system rests with the histopathologist [33].

Similar to this, different organs and tissues are removed during necropsy and examined microscopically for harmful effects in sub-chronic and chronic research.

Many of these studies also need the clinical chemistry examination of test animal specimens or serum samples.

10.8.12 Studies Concerning Microbiology and Sterility

Microbiological research is necessary to make medical equipment safe and sterile. These studies must comply with ISO 13485:2016 requirements. BSI provides technicians and microbiologists with the ability to verify that test procedures meet sterility, packaging, and microbiological testing standards to ensure medical devices are safe and comply with ISO 13485:2016 [34].

The following special requirements are part of ISO 13485:

- Appropriate manufacturing processes should be used (e.g., clean room and/or controllable and custom setup).
- To guarantee the integrity of the sterile pack, primary packaging procedures must be validated and managed.
- Validation and oversight of medical equipment sterilization procedures.

10.9 Conclusion

The recent development of unique, sophisticated, and sensitive methods to assess the long-term stability and toxicity of novel biomaterial implants has sped up more efficient and biologically pertinent assessment of their biocompatibility. Due to the reactive, degradative, kinetic, and dynamic physiological milieu, rapid screening utilizing in vitro cell culture models alone cannot predict the long-term in vivo response. This eliminates the need to assess many factors, such as genotoxic, epigenetic, carcinogenic, and immunogenic effects, to increase the material's clinical acceptability [27]. The medical device business is a young industry, and testing of these devices is necessary to guarantee the security of users and patients. For its testing, there are numerous regulations accessible. The classification and biocompatibility testing procedures for these devices are regulated by ISO standards, particularly ISO 10993. The medical devices are divided into many sorts based on the kind, length, and proportion of risk experienced.

Medical devices can be thought of as drug device combination goods because they contain pharmaceuticals as well and must be tested according to strict guidelines. A combination product is defined by CFR as a medication or biologic combined with acceptable medical equipment. Regulations for glass syringes used to administer drugs or biological products are provided by ISO 11040:4 [28]. Biocompatibility testing (according to ISO 10993) and risk assessment (according to ISO 14971) are mandatory to ensure user/patient safety. However, the method of manufacture, testing and risk assessment of the device must be approved before use. This section describes the various biocompatibility tests based on ISO 10993 requirements for the examination of biological substances such as extractants, leachates and carcinogenic, mutagenic and antibacterial substances.

References

1. Thakur P, Thakur A (2022) Introduction to nanotechnology. In: Thakur A, Thakur P, Khurana SP (eds) Synthesis and applications of nanoparticles. Springer, Singapore, pp 1–18. https://doi.org/10.1007/978-981-16-6819-7_1
2. Kala D, Gupta S, Kaushal A (2022) In: Thakur A, Thakur P, Khurana SP (eds) Synthesis and applications of nanoparticles. Springer, Singapore, pp 405–416. https://doi.org/10.1007/978-981-16-6819-7_19
3. Jarow JP, Baxley JV (2015) Medical devices: US medical device regulation. Urol Oncol 33(3): 128–132
4. Pathania A, Bhardwaj S, Thakur SS, Mattei JL, Queffelec P, Panina LV, Thakur P, Thakur A (2018) Investigation of structural, optical, magnetic and electrical properties of tungsten doped NiZn nano-ferrites. Phys B Condens Matter 531:45–50
5. Chahar D, Taneja S, Thakur P, Thakur A (2018) Remarkable resistivity and improved dielectric properties of Co–Zn nanoferrites for high frequency applications. J Alloys Compd 843
6. Darrow JJ, Avorn J, Kesselheim AS (2021) FDA regulation and approval of medical devices: 1976-2020. JAMA 326(5):420
7. David Y, Judd TM, Zambuto RP (2004) Introduction to medical technology management practices. In Elsevier eBooks (pp. 166–177)
8. Khanra A. Sangam S, Shakeel A, Suhag D, Mistry S, Rai MP, Chakrabarti S, Mukherjee M (2017) sustainable growth and lipid production from chlorella pyrenoidosa using n-doped carbon nanosheets: unravelling the role of graphitic nitrogen. ACS Sustain Chem Eng
9. Thakur P, Verma Y, Thakur A (2022) Toxicity of nanomaterials: an overview. In: Thakur A, Thakur P, Khurana SP (eds) Synthesis and applications of nanoparticles. Springer, Singapore
10. Rome BN, Kramer DB, Kesselheim AS (2014) Approval of high-risk medical devices in the US: implications for clinical cardiology. Curr Cardiol Rep 16:1–9
11. Challoner DR, Senate US (2011) Medical devices and the public's health: the FDA 510 (k) clearance process at 35 years. Written Statement before the Committee on Health, Education, Labor, and Pensions US Senate, Institute of Medicine of the National Academics, Washington
12. Center for Devices and Radiological Health & Center for Devices and Radiological Health (2019) A history of medical device Regulation & Oversight in the United States. U.S. Food And Drug Administration. https://www.fda.gov/medical-devices/overview-device-regulation/history-medical-device-regulation-oversight-united-states
13. Van Norman GA (2016) Drugs, devices, and the FDA: part 2. JACC Basic Transl Sci 1(4): 277–287
14. Aronson JK, Heneghan C, Ferner RE (2020) Medical devices: definition, classification, and regulatory implications. Drug Saf 43(2):83–93
15. (1993) Medical devices; medical device distributor reporting—FDA. Final rule; notification of status under the Safe Medical Devices Act; confirmation of effective date. Fed Regist 58(168): 46514–46523
16. Park B, Lakes RS (2007) Biomaterials: an introduction, 2nd edn. Plenum Press, New York. 1992. Biomaterials (2007). In Springer eBooks
17. Naagar M, Chalia S, Kharbanda S, Thakur P, Thakur A (2023) Correlating tailored saturation magnetization of normal-inverse spinels with noncollinear magnetic ordering. J Mater Sci Mater Electron 34:473
18. Kharbanda S, Dhanda N, Sun ACA, Thakur A, Thakur P (2023) Multiferroic perovskite bismuth ferrite nanostructures: a review on synthesis and applications. J Magn Magn Mater 572

19. Hanker JS, Giammara BL (1988) Biomaterials and biomedical devices. Science 242(4880): 885–892
20. Jebakumar AZ, Idrees M, Nondo HS (2014) A review on application of biomaterial in medical sciences. J Sci Health Sci 4:390–393
21. Rathi VK, Krumholz HM, Masoudi FA, Ross JS (2015) Characteristics of clinical studies conducted over the total product life cycle of high-risk therapeutic medical devices receiving FDA premarket approval in 2010 and 2011. JAMA 314(6):604
22. Williams D (2003) Revisiting the definition of biocompatibility. Med Device Technol 14(8): 10–13
23. Williams DR (2008) On the mechanisms of biocompatibility. Biomaterials 29(20):2941–2953
24. Thangaraju P, Varthya SB (2022) ISO 10993: biological evaluation of medical devices. Springer eBooks, pp 163–187
25. Gudeppu M, Balasubramanian J, Bakthavachalam P, Lenin NC, Shanmugam PST (2020) Biocompatibility and toxicology. Elsevier eBooks, pp 99–133
26. Cambiaghi A (2018) Biological evaluation of medical devices as an essential part of the risk management process: updates and challenges of ISO 10993-1: 2018. Eurofins Medical Device Testing, Lancaster, PA
27. Strickland J, Clippinger AJ, Brown JR, Allen DT, Jacobs A, Matheson J, Lowit A, Reinke EK, Johnson MH, Quinn MA, Mattie DR, Fitzpatrick S, Ahir S, Kleinstreuer N, Casey W (2018) Status of acute systemic toxicity testing requirements and data uses by U.S. regulatory agencies. Regul Toxicol Pharmacol 94:183–196
28. Boutrand JP (ed) (2019) Biocompatibility and performance of medical devices. Woodhead Publishing
29. De Jong WH, Carraway JW, Geertsma RE (2020) In vivo and in vitro testing for the biological safety evaluation of biomaterials and medical devices. In Biocompatibility and performance of medical devices (pp. 123–166). Woodhead Publishing
30. International Organization for Standardization (2009) ISO 10993-5: 2009-Biological evaluation of medical devices-Part 5: tests for in vitro cytotoxicity
31. Lorenz G, Schüle A (2022) Medical technologies – key learning from two case studies. Wiley, pp 345–355
32. Kumaravel TS, Sathya TN, Balaje R, Pradeepa P, Yogaraj D, Murali MR et al (2022) Genotoxicity evaluation of medical devices: a regulatory perspective. Mut Res Rev Mutat Res 789:108407
33. Thrivikraman G, Madras G, Basu B (2014) In vitro/in vivo assessment and mechanisms of toxicity of bioceramic materials and its wear particulates. RSC Adv 4(25):12763
34. International Organization for Standardization. ISO 11040-4:2015 (2015) Prefilled syringes—part 4: glass barrels for injectables and sterilized sub-assembled syringes ready for filling. www.iso.org/standard/58079.html

Functionalization of Biomaterials

11

Raksha Rathore, Deepa Suhag, Fayu Wan, Ritesh Verma,
Atul Thakur, and Preeti Thakur

Abstract

The rapid development of nanomedicine has raised the likelihood that produced nanoparticles will 1 day come into contact with humans and the environment. A variety of academic fields, including engineering and the health sciences, have taken a keen interest in the development of nanotechnology. Any significant developments in nanomaterial-based applications would depend on the production of functionalized nanoparticles and other nanomaterials, which are anticipated to have application potential in fields like pharmaceutical and biomedical sciences. Early clinical results have demonstrated that functionalizing nanoparticles with specific recognition chemical moieties produces multifunctional nanoparticles with enhanced efficacy while at the same time reducing side effects. These characteristics include targeted localization in tumors and active cellular uptake. Chemical procedures must be developed that reliably bind chemical moieties to nanoparticles. The development of numerous chemical techniques to synthesize functionalized nanoparticles specifically for drug delivery, cancer therapy, diagnostics, tissue engineering, and molecular biology provides extensive information of the structure–function relationship of these functionalized nanoparticles. This chapter's information is essential for the broad

R. Rathore (✉) · D. Suhag · R. Verma · A. Thakur
Amity Centre for Nanotechnology, Amity University Haryana, Gurgaon, Haryana, India
e-mail: raksha.rathore1@s.amity.edu; dsuhag@ggn.amity.edu; athakur1@ggn.amity.edu

F. Wan
School of Electronics and Information Engineering, Nanjing University of Information Science & Technology, Nanjing, China

P. Thakur
Amity School of Applied Sciences, Amity University Haryana, Gurugram, Haryana, India
e-mail: pthakur@ggn.amity.edu

D. Suhag et al. (eds.), *Integrated Nanomaterials and their Applications*,
https://doi.org/10.1007/978-981-99-6105-4_11

and safe use of functionalized nanoparticles and other nanomaterials, particularly in the biomedical sector.

Keywords

Functionalized nanoparticles · Drug delivery · Cancer therapy · Pharmaceutical · Nanomedicine

11.1 Introduction

The study of molecularly sized functional systems, or nanotechnology, is one of the research fields that is swiftly expanding in a variety of scientific disciplines. Beginning in 1959, research in the field of nanotechnology was conducted. The idea of making objects with atomic accuracy was initially brought up by physicist Richard Feynman when he observed, "There is plenty of room at the bottom." Recent advances in the synthesis of NPs have led to the development of various novel applications in a range of sectors, including nanomedicine, biomedical sciences, and engineering. NPs have a rising number of potential applications in biomedicine because of their oxidation resistance, simplicity in production, and optical properties [1]. Organs may also very well survive functionalization if the appropriate ligands are used. Metal NPs make excellent drug and tracer carriers because they are easily functionalized. Although NPs have many advantages, they also have substantial drawbacks, such as the lack of surface features. These drawbacks, however, can be remedied by giving the NPs certain chemical moieties. One benefit of functionalizing NPs is the capability to predictably regulate the required features to suit the specific applications [2–4]. The methods used to create, work with, and employ functionalized nanoparticles (FNPs) give exciting new potential for the creation of brand-new multifunctional instruments for biomedical and nanotechnological applications. Functionalization has also been shown to make NPs compatible in later phases and stop them from clumping together [5, 6]. Functionalization also improves the synergistic physical, chemical, and mechanical properties of NPs [7, 8]. Functionalized nanoparticles (FNPs) exhibit improved pharmacokinetic profiles and transport properties in vivo after systemic injection. They can penetrate deeper into tissues through small capillaries and epithelial lining, enhancing the delivery of medicinal chemicals to specific locations.

When discussing carbon nanoparticles, carbon nanotubes (CNTs) and graphene, two examples of carbon nanostructures with extraordinary mechanical, electrical, and physical characteristics, are widely employed. Due to their exceptional qualities, these CNTs and graphene may be employed as electrode materials, supercapacitors, electrical components (such nanowires, transistors, and switches), catalysts, membranes, sensors, biomedical devices, and structural strengthening agents. The inability of pure materials to dissolve in water or organic solvents and the absence of functionality on native graphene surfaces to modify the material's characteristics make it difficult to employ carbon nanomaterials effectively. Through reactions on the sidewalls of graphene or at the termini of nanotubes, covalent functionalization has grown in popularity as a method for modifying carbon nanomaterials.

11.2 Functionalization/Optimization of Nanoparticles

Following the covalent functionalization of NPs with chemical and biological components, the bioactivity of the resulting material should be closely examined for any adverse changes in intrinsic activity. The physio-chemical properties of the NPs, size, dispersion, surface charge, and type of the FNPs are anticipated to influence the in vivo drug delivery result. It is well knowledge that 20–200 nm particles may be used to systemically administer medications. While tiny particles can pass through the fenestration in the hepatic sinusoidal endothelium and result in hepatic accumulation rather than protracted circulation periods, larger particles are quickly cleared from the circulation by the RES [9]. The ability of NPs to target particular locations for drug delivery is critical when a medication is given systemically [10].

Targeted delivery may decrease any negative impact on healthy tissues by concentrating the medication at the point of action. It is also essential to create NPs with a release profile that can be managed and fits the requirements of the desired application. There will be greater environmental protection and more controllable loading if encapsulation is used rather than adsorption on the NPs. Effective combinations of these traits are anticipated to enhance therapeutic results while reducing negative effects on normal tissues when creating NPs. During functionalization, the surface of the NPs is modified by the addition of a chemical functional group, which encourages self-organization and compatibility. Most often, NPs have been functionalized with thiols, disulfides, amines, nitriles, carboxylic acids, phosphines, and biomolecules. The primary goal of functionalizing NPs is to cover their surface in a molecule that has the required chemical characteristics for the application. When particles are functionalized, their surface characteristics are substantially changed in each situation. The surface chemistry of NPs is already an important part of their synthesis since this feature may be employed to control their size and self-organization throughout development [11]. This may be achieved by complexing groups that form bonds on surfaces, but complexes must first be produced.

11.3 Class of Functionalization

11.3.1 Thiol/Aminothiol

Researchers pioneered the thiol conjugation approach, which is now widely used to functionalize NPs [12]. Guerrero et al. [13] synthesized three different thiol FNP types. Using the molecules dodecanethiol and octanethiol, also known as Au-SC12 and Au-SC8, they first created thiol functionalized gold (Au) clusters, as described by Brust et al. Additionally, Au NPs with the designations Au-SC_8/SC_{11}COOH and caps formed of both octanethiol and thiolated undecanoic acid molecules were synthesized utilizing the Simard et al. synthesis method. Thiol chemistry is one of the most advanced functionalization processes used to make functional NPs. Sulfur compounds with a number of metals, including Ag, Cu, Pt, Hg, Fe, Si, and Au, spontaneously form strong coordination bonds [14]. Sulfur and organosulfur will spontaneously adsorb due to their great affinity for metal surfaces [15]. To make thiol- or disulfide-capped NPs, employ the direct functionalization method, in which the metal precursor and protective ligand are reacted concurrently. The post-functionalization procedure, which is likewise based on thiol chemistry, includes grafting sulfur compounds onto the surfaces of solvent-covered pre-synthesized NPs to replace ligands that contain sulfur.

11.3.2 Bio-Functionalization

NPs functionalized with biomolecules have lately received a lot of attention due to the usefulness of the resultant hybrid materials as drug carriers. Because tiny biological molecules, such as amino acids, may be attached to the surface of the NPs, it is now possible to bind functional biomolecules to bio FNPs [16]. Sousa et al. [17] proposed using aspartic and glutamic acid as chelating agents in the synthesis of NPs. Leucine, arginine, cysteine, and tyrosine were only a few of the amino acids that were directly anchored to the surface of NPs in a two-step synthesis described by Tie et al. [18]. NPs were modified using organosilane and vinyl alcohol/vinyl amine copolymer before being used to conjugate antibodies, enabling them to target particular biological cells and tissues [19].

11.3.3 Polymers in Functionalization

The NPs' adaptability for usage in challenging conditions, such acidic and alkaline liquids, is partly due to the polymer matrix incorporated into them [20]. Polymer NPs are also commonly used in pharmaceutical applications, particularly in the management of cancer and drug targeting. In order to improve the properties of polymeric nanocomposites, such as strength, which is essential for tissue engineering applications, NPs or nanofibers have been used in their production [21–23]. When nanocomposites are created by only mixing, the filler and polymer coupling is weak, resulting in artifactual defects and a reduction in the mechanical characteristics of the nanocomposites [24–27]. This problem will require a well-designed interphase, which will also improve the strength, toughness, and compatibility of the composites. The interfacial interaction of the NPs with the polymer matrix enhances the quality and properties of the nanocomposites. To build a strong connection, stabilize the NPs, and make them compatible with the polymer, a surfactant is required for the functionalization of NPs with polymers. The polymer functionalization enhanced the mechanical characteristics of the NPs. Additionally, these nanocomposites are insoluble in both basic and acidic liquids. By using polymer FNPs made by phage display, Rothenfluh et al. [28] effectively conveyed a hydrophobic medication. Post-polymerization can also boost bioavailability.

11.3.4 Miscellaneous

It has been discovered that Iceland's acidic hot springs contain virus nanoparticles (NPs) with changed chemical groups, which may adapt to a range of challenging conditions, including the acidic environment of the body. Since the stoichiometries were mathematically dictated by geometric constraints, dendritic NPs structures were the first nanomaterials that could be employed for precise stoichiometric functionalization. Stoichiometric mono-functionalization of Au NPs was reported by Hainfeld et al. [29].

11.4 Functionalization of Carbon Nanomaterials

11.4.1 Functionalization of Fullerene

Fullerenes can be functionalized using a variety of different chemical procedures. Here, we will focus on a handful of the more noteworthy responses. Surface functionalization makes it simple to make fullerene soluble in both water and organic solvents [30]. The fullerene surface can be chemically altered using one of two different methods: Through (i) complexation with a solubilizing agent and (ii) covalent functionalization of the fullerene surface using potent acids at high temperatures, oxygen-based functional groups, primarily hydroxyl groups, may be grafted on the surface of fullerenes, the hydrophobic fullerene surface is partially hidden [31]. C60 and C70 fullerenes were mixed with an aqueous solution including nitric and sulfuric acids at a temperature of 85–115 °C. The acid oxidized the fullerene surfaces, which gives grafting of 15 hydroxyl groups on average per fullerene. Another possible method for creating OH functionalized fullerenes (fullerenols) is the addition of strong basic NaOH to C60 benzene solutions with tetrabutylammonium hydroxide acting as a catalyst [32]. There are more grafted hydroxyl groups as a result of the reaction's higher efficiency compared to reactions based on strong acids. In another experiment, a drop of liquid Na/K alloy was added to a suspension of C60 in tetrahydrofuran. The intermediate was then reacted with oxygen and water (Fig. 11.1).

This method is very efficient and produced some of the highest levels of hydrophilicity by extending the hydroxylation of the fullerenes. These procedures are commonly used to make fullerene hydrophilic despite the lack of control over the density of addends. An alternative method for creating highly polyhydroxylated fullerenes uses a suspension of partly hydroxylated fullerene with hydrogen peroxide at a temperature of 60 °C [34]. The final product, fullerenols, has an average of 36–40 hydroxyl groups and 8–9 secondary bonded water molecules, despite the

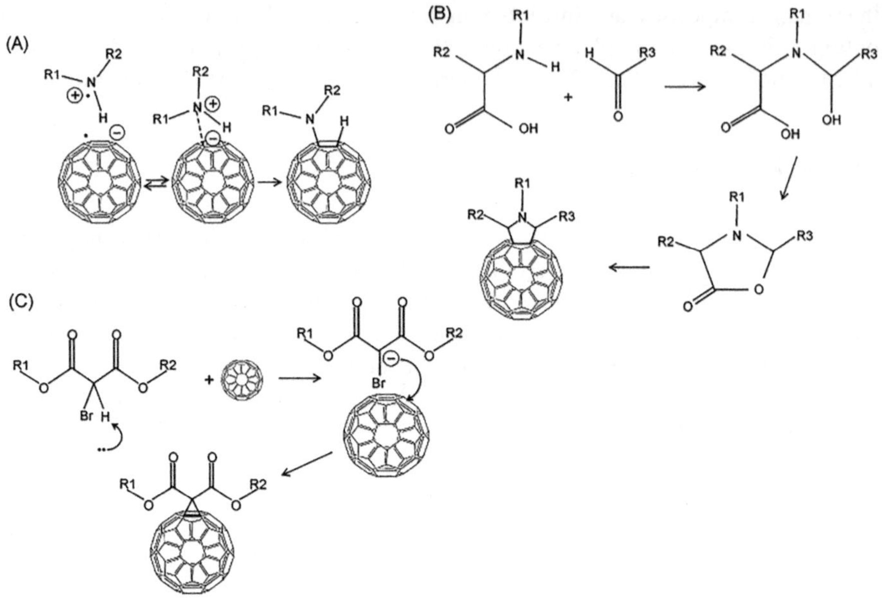

Fig. 11.1 (**a**) The C60 amination mechanism. To create the C60 anion radical, single-electron transfer is the first step. The final product is then produced by proton transfer after radical recombination produces a zwitterion. (**b**) Mechanism of the Prato reaction. (**c**) Bingel reaction mechanism [33]

process taking 4 days. The addition of aliphatic primary amines such n-propylamine, t-butylamine, and dodecylamine to C60 results in the grafting of amine groups onto this compound. In other processes, smaller main or secondary amine chains like methylamine, diethylamine, or ethylenediamine are combined with C60 or C70 fullerenes [35]. A stable product that grafts a pyrollidine ring to C60 is produced by the 1,3-dipolar cycloaddition of an azomethine ylide to a C60 molecule. This is a typical amination procedure. The fullerene moiety may be added to by this reaction in a variety of ways, which is a phenomenon. The Bingel reaction is commonly employed for functionalizing fullerene exohedra. By introducing a -halo ester/ketone group to the cage while the reaction is occurring under strongly basic conditions, a methanofullerene is created [36]. The chemistry of the fullerene surface has been modified several times using the Bingel reaction. For instance, the Bingel reaction was used to produce C60 Hexakis-adducts [37]. One of the several chemical processes for functionalizing fullerenes is the Diels-Alder cycloaddition reaction, which is incredibly flexible and yields a broad variety of cycloadducts. Due to its lack of electrons, fullerene interacts with 1,3-dienes that are abundant in electrons to create adducts that fuse a 6-6 bond to a cyclohexene ring.

11.4.2 Functionalization of Carbon Nano-Onions

Similar to fullerenes, CNOs have different solubilities in polar, non-polar, protic, and aprotic solvents. Because some solvents, like *N-N*-dimethylformamide, are effective proton acceptors, they act differently from other solvents. As a result, CNOs have high solubility and low potential values. Tetrachloroethane, in contrast to chloroform, cannot receive protons from CNOs. A positive -potential and low solubility are the outcomes of this. It is possible to functionalize the CNO surface to improve solubility. Curvature greatly facilitates both covalent and non-covalent functionalization of CNOs. One of the frequent reactions is the oxidation of CNOs produced by oxidative or reductive reagents like HNO_3, H_2SO_4, or KOH. But the use of strong acids or bases can be harmful to the CNOs. These strong acids add functional groups with the hydroxyl, carboxyl, and ether-like chemistries. As an alternative strategy to target the CNOs surface and graft oxygen functional groups, nitric acid or ozone can be employed [38]. Additionally, procedures using azomethine ylide addition (1,3-dipolar cycloaddition) and soluble carboxyl groups were employed to produce CNOs. A direct acid-base reaction, amidation employing in-situ generated acid chloride, and carbodiimide-activated coupling are possible further processes using carboxylated CNOs. As an instance, carboxylated CNOs are amidated directly using polyethylene glycol with a diamine ending. A second possible amidation reaction is produced by treating carboxylated CNOs with 1-Octadecylamine, which is known to be a particularly efficient agent for the amidation of CNTs. Another method for adding NH groups to CNOs suspended in dimethylformamide is microwave irradiation. The most recent method for functionalizing the surface of CNOs is the 1,3-dipolar cycloaddition of azomethine ylide. Organically modifying fullerene C60 with an aldehyde-amino acid condensation process that yields azomethine ylides was a frequent procedure [38]. Another version made use of carbon tetrabromide, 1,8-diazabicyclo [5.4.0] undec-7-ene, and the functionalization of the CNOs by a [2 + 1] Bingel-Hirsch cyclopropanation [39].

Functionalized CNOs are used to image and target cancer cells. In two distinct tumor cell lines, the modified CNOs quickly absorb and show good brightness and photostability. They also display minimal cytotoxicity. It is possible to employ CNOs by attaching polymeric molecules to them. As an example, polypyrrole CNO is used to improve the specific capacitance of supercapacitors [40]. Using chemical polymerization or electrostatic potentiostatic deposition, polypyrrole is added to CNOs as bilayers. The composites created utilizing the two techniques have higher specific capacitances of 800 Fg^{-1} and 1300 Fg^{-1}, respectively, and are more mechanically and electrochemically stable.

11.4.3 Functionalization of Nanodiamonds

The diamond surface chemistry is altered by the ND functionalization procedures, which are mostly well-established. The graphitic shell can also be broken down with $FeSO_4$ and H_2O_2 (Fenton), which swiftly target the non-diamond phases on the surface with OH groups [41]. Diamond surface hydroxylation may also be produced mechanically by milling and ultrasounds in water, both of which result in radical reactions and the -(OH) ending [42]. The ND surface can get oxidized when the diamond core's amorphous/graphitic coating is removed using strongly oxidizing substances such potent acids, singlet oxygen in NaOH, potent ozone, and air treated with a catalyst. Additionally, supercritical water heated to 350–450 °C or ozone are used for surface oxidation. After the ND surface has undergone oxidation, other methods are used to provide the appropriate functional groups. One approach entails non-covalent functionalization, which is carried out using electrostatic or hydrophobic interactions [43]. Polar materials, such as DNA sequences, polymeric complexes, etc., initially bind to the ND surface through electrostatic interaction. The second approach depends on the use of apolar contacts and is based on the molecules' ability to self-assemble on the ND surface. In the creation of composite materials and medication delivery, this chemical functionalization has been widely utilized. Non-covalent functionalization is a simple technique, but because of uncontrolled adsorption process, it has low reproducibility. The covalent functionalization employs carboxylic and hydroxyl groups to graft on additional molecules. For instance, after being activated with thionyl chloride, carboxylic groups can interact with amines. This process also give rise to the production of PEGylated NDs [44]. PEG-chains could contain amino, thiol, or azido groups at the end, which are often used to covalently bond a variety of different biomolecules. Another frequently utilized functionalization is the derivatization of the hydroxyl groups by esterification generated by high temperature -OH carboxylation succinic anhydride. Derivatizing NDs using siloxanes is an alternative approach. The siloxane coating's thickness on the ND core may be controlled to some extent with this method of functionalization, but it also depends on the siloxane precursor used [45]. Covalently anchoring dyes and receptors as well as using siloxane functionalization for drug delivery have all been done effectively [46].

The functionalization of H-terminated NDs with alkenes, hydroxylated NDs with alkyl chlorides, and diazonium salts are further methods [48]. The carboxyl groups on the ND surface can also be converted into azido groups, making them available for other chemical reactions [49]. Some of the typical functional groups used to change the ND surface chemistry is shown in Fig. 11.2.

11.4.4 Functionalization of Carbon Nanotubes

Carbon nanotubes were discovered in 1991 by Iijima. Since then, various effective functionalization methods for CNTs have been devised, and nanoscience has become a distinct field of research within the field of materials science [50]. Pure CNTs have hydrophobic walls due to their graphitic structure. As a result, van der Waals forces produce bundles of the most recent CNT manufacturing. This makes it difficult to separate CNTs and make a stable dispersion in water and most solvents. To solve this problem, the CNT surface can be functionalized. In the 1990s, two separate methods for functionalizing CNTs—covalent and non-covalent—were

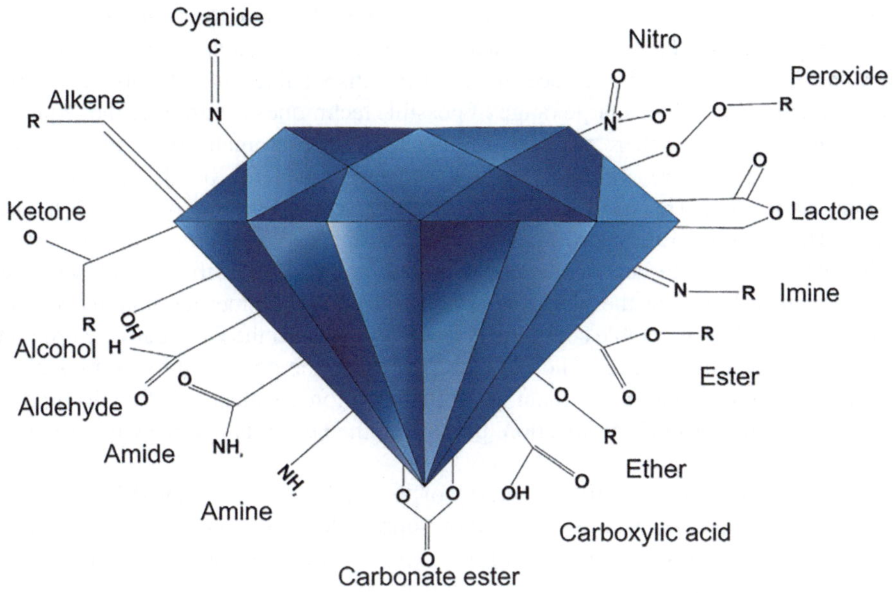

Fig. 11.2 Modified surface chemistry of NDs with various functional groups [47]

devised. Covalent functionalization results in the formation of covalent bonds between a functional group and the carbon atom of a CNT. The covalent functionalization procedure, which comprises a series of chemical processes, results in the formation of covalent chemical bonds between CNTs and the functional entities. Side wall covalent functionalization and oxidation and end/defects functionalization are the two categories into which they may be divided. The defect-functionalization of CNTs continues by using powerful agents to target the flaws in the nanotubes' walls and leave holes functionalized with oxygenated functional groups [51].

Strong acids like HNO_3, H_2SO_4, or their combination, as well as potent oxidants like $KMnO_4$, ozone, or reactive oxygen plasma, are typically used to carry out oxidation. This raises the reaction temperature [52]. These reactions lead to the formation of acidic sites on the CNTs, which may then be used to bind other molecules. Therefore, oxidation of the CNTs, oxygen-containing functional groups including carboxylic acid, ketone, alcohol, and ester groups are created, which may be utilized to affix a wide range of different kinds of chemical moieties. Numerous chemical processes, including as esterification, thiolation, silanation, polymer grafting, and the attachment of specific biomolecules are made possible by these functional groups [53].

In the process of direct covalent sidewall functionalization, the formation of the covalent bond results in a shift in the carbon hybridization from sp^2 to sp^3, as well as a concomitant loss of the p-conjugation typical of the aromatic rings of graphene. The atoms at the ends and in the defects of the CNTs can produce this mechanism by interacting with some compounds that have a high chemical reactivity. One of the most important advantages of functionalizing the CNTs is that the polar or non-polar groups grafted on their surface make them dispersible in solvents. Covalent functionalization offers a large range of possible techniques for grafting the required molecules, however, there are certain downsides. The aromatic rings of the CNT graphitic lattice are broken by the strong covalent bonds utilized for functionalization. The mechanical and electrical characteristics of the CNTs are impacted by this. Every covalent bond function as a scatter point, and every disruption of the surface conjugated network creates faults that might serve as rupture sites, lowering the electrical conductivity [54]. Numerous initiatives have been made to find workable solutions in an effort to lessen the CNT damage brought on by the functionalization. The CNT properties can be preserved by non-covalent functionalization. One of the main advantages of non-covalent functionalization is that it prevents carbon hybridization, preserving the hexagonal lattice with aromatic graphitic rings.

The structural and electrical features of the CNTs are maintained because the non-covalent functionalization is based on supramolecular complexation formed by hydrogen bonds, adsorption based on van der Waals forces, electrostatic forces, or stacking interactions.

11.4.5 Functionalization of Graphene

The chemical functionalization of graphene, which is a critical step in the preparation of the material, determines its final properties and the types of applications it can be used for, including electronics, filler in the synthesis of nanocomposite materials, biomaterials and biomedicine, sensing, energy, and the environment. To transform zero-gap graphene into a semiconductor suitable for electrical applications, it goes through functionalization [55]. With regard to biomaterials, functionalization renders graphene water-soluble, and the appropriate surface chemistry makes graphene a great nanoplatform for drug delivery as well as other uses, such tissue engineering [56]. Graphene is widely utilized as the appropriate electrode material for numerous electrocatalytic applications. When the surface chemistry is changed with the appropriate functional groups, the finished electrode is more efficient than the sum of its parts [57]. In terms of chemistry, graphene is comparable to other sp^2 hybridized carbon allotropes [58]. Similar to CNTs, graphene too exhibits serious flaws. Atoms near the edges of defects are often more reactive than the surface that surrounds them, even though functionalization of graphene may also take place on the surface, as we shall explore later. Similar to CNTs, graphene may be functionalized via covalent or non-covalent methods.

11.5 Covalent Functionalization

The chemical modifications of the graphene surface that include graphene oxide and reduction have drawn the greatest interest. Thus, a wide range of atoms and organic groups may be grafted onto pure graphene, enabling radical additions, electrophilic substitution, and cycloaddition processes. Since Brodie demonstrated in the eighteenth century that graphite could be transformed into graphene oxide using potassium chlorate and fuming nitric acid, studies on graphene oxide (GO) have been debated [59]. In order to reach equivalent levels of graphite oxidation, Hummers and Hoffman proposed an alternative in the 1960s that involved employing permanganate and concentrated sulfuric acid or a mixture of sulfuric and nitric acids. It has been demonstrated, nonetheless, that the degree of oxidation depends not only on the oxidants employed, but also on the quality of the graphite utilized and the environment in which the process takes place [60]. These scalable methods result in the production of hydroxyl and epoxide groups, as well as minor amounts of carboxy, carbonyl, phenol, lactone, and quinone groups. GO is easily dispersible in water due to the large concentration of polar functional groups. GO is a material with many functional groups, however, it can be used as a building block to create particular compounds. Additionally, the hydrophilicity of GO made it possible for consistent thin film deposition, which is essential for applications in electronics [61]. By, adding appropriate functional molecules, oxygen-based functional groups enable further tuning of the surface chemistry. Additionally, GO is functionalized using quantum dots to create platforms for electrochemical applications. Prussian blue and metal hydroxides are other species that are employed to functionalize GO and modify its electrochemical characteristics. The various organic chemicals that may be used to functionalize GO include porphyrins, aromatic dyes, alkylamines, ionic liquids, pyrene and perylene diimide, cyclodextrin, aryl diazonium compounds, and polymers.

Sheets of graphene are also hydrogenated, sp^2 to sp^3 hybrids are converted by hydrogenation. As a result, graphene is transformed from a highly conductive material into an insulating one, significantly altering its electrical characteristics [62]. In a hydrogen environment, graphene was functionalized by RF-CVD. In order to prepare hydrogenated graphene, other research teams looked into wet chemical processes such as the Birch reduction. Doping with halogenated elements can also change the electrical characteristics of graphene [63].

Covalently bonded halogens (graphene halides) significantly alter the characteristics of graphene, just like hydrogen does. For instance, fluorographene is the thinnest insulator and the only stable stoichiometric graphene halide. In contrast to its counterpart graphenes, halogenated graphenes display a wide range of exceptional and intriguing electrical, optical, thermal, electrocatalytic, magnetic, mechanical, and chemical capabilities. In order to create graphene sheets dispersible in common organic solvents, which is essential for the production of nanocomposite materials, this approach has been devised. It is also possible to incorporate organic functional groups. The aryl diazonium salt covalent modification permits the manufacture of functionalized graphene with increased dispersibility in polar organic solvents and water by altering the substituents in the aryl-ring. This method of functionalization can be used to produce high performance graphene-based nanocomposites, and the enhanced polymer host dispersibility of graphene nanosheets led to substantial interlayer cohesive energy and surface inertia. Graphene's electrical properties can be modified covalently using 4-bromobenzene diazonium tetrafluoroborate (4-BBDT).

11.6 Non-covalent Functionalization

Functionalization without Covalent Bonds: Because of the shaky link that exists between the functional group and the graphene during non-covalent functionalization, the essential electrical structure of the material is intact. Non-covalent functionalization retains the inherent chemical and physical characteristics while providing novel selective capabilities through the interaction of the counter molecules with the hexagonal lattice. This kind of functionalization is based on the so-called "-" interaction created between aromatic rings. For non-covalent functionalization between aromatic rings, several chemicals can be used. A number of substances can be used for non-covalent functionalization. Due to its strong affinity for the graphite basal planes, the pyrene moiety has generated a great deal of attention. The functionalization of graphene with 1-pyrenebutyrate improved its water dispersibility. Using non-covalent functionalization to bind pyrenebutanoic acid succinimidyl ester to graphene in order to boost the effectiveness of power conversion in graphene-based photovoltaic systems. More examples of non-covalent functionalization are shown when 1-pyrenecarboxylic acid is used to functionalize graphene to provide unique optical and sensing properties (Fig. 11.3) [64].

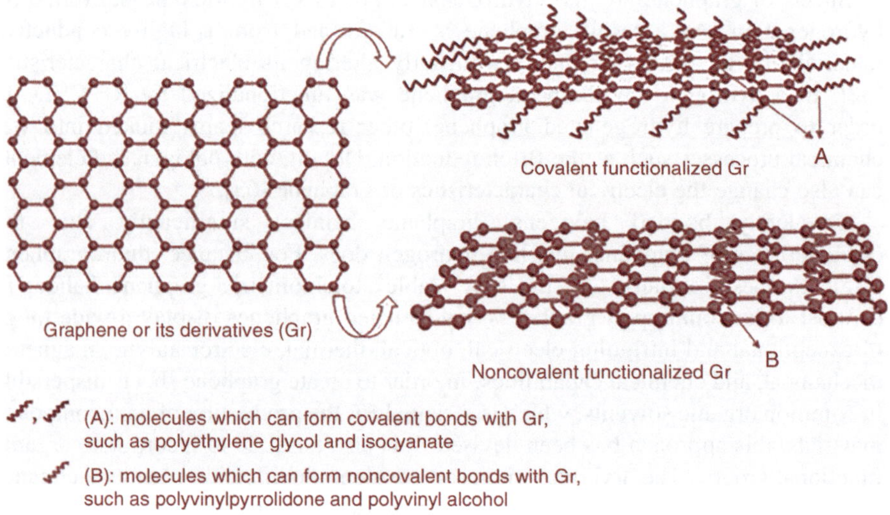

Fig. 11.3 Functionalization of graphene [65]

11.7 FNPS Applications in Biomedical Science

Nanotechnology offer promise for nanomedicine and therapeutic procedures. The development of molecular machines to identify and eradicate host pathogens, in tissue engineering, and the replacement or repair of cells or cellular components in vivo are just a few of the overlapping molecular technologies that nanomedicine depends on. Others include ongoing advances in genomics, proteomics, and nanoengineered microbes. Research into the delivery and targeting of therapeutic and diagnostic agents with NPs is at the forefront of nanomedicine due to some oral and injectable drugs' ineffective formulations, which have novel delivery requirements like optimizing efficacy, minimizing side effects, and improving patient compliance [66]. The use of NPs in formulations can improve drug bioavailability, prolonged release, and more precise targeting to the level of direct intracellular delivery. The blood–brain barrier, the pulmonary system, and the skin's strong epithelial connections may all be avoided by small-sized NPs, which is advantageous for the delivery of targeted medications (Fig. 11.4) [67].

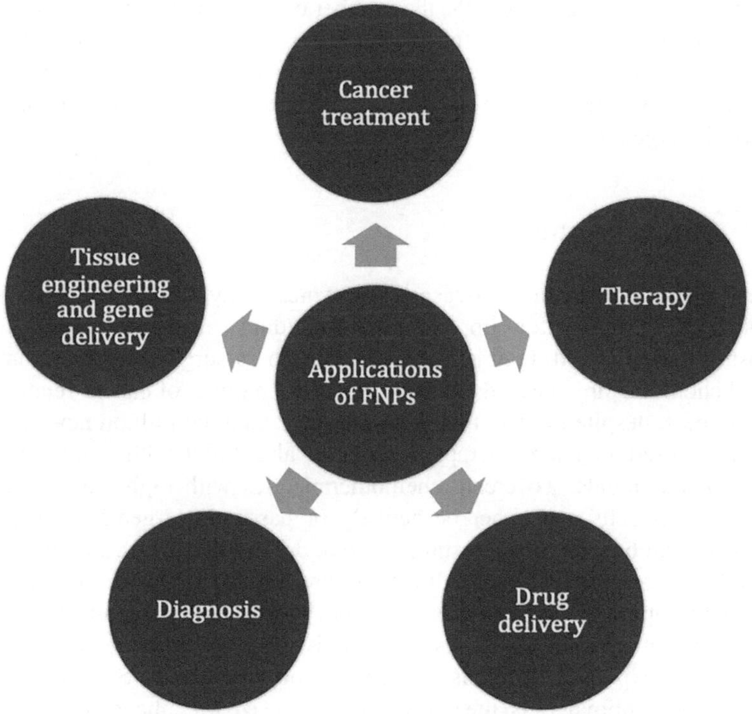

Fig. 11.4 Applications of FNPs

11.7.1 Therapy

By raising medication concentrations in blood or vascularized tissues using polymeric NPs, drug toxicity can be reduced while therapeutic effects are improved. NPs-based drug delivery method for intra-tissue medication release in articular cartilage may be beneficial when utilized in pharmacological and biomolecular treatment for disorders like osteoarthritis. Numerous NP capabilities have been studied, including the ability of NPs to penetrate (endocytosis) or rupture the membranes of dangerous bacteria and tumor cells. Although polymer-based NPs cannot directly target cells, they provide flexible chemistry for the attachment of pharmaceuticals that, in many circumstances, can increase both cell selectivity and uptake. A number of membrane-bound receptors can be targeted via receptor-mediated endocytosis. Folate and other small molecules having a glycosidic moiety allow for precise targeting to certain cell types that express the required receptor protein [68]. The two most well-known examples of cell targeting with functionalized NPs are asialo orosomucoid functionalized with poly-lysine to target the asialoglycoprotein receptor on hepatocytes and NPs functionalized with iron transport protein transferrin. The targeting method's effectiveness depends on a number of factors, including the ligand receptor binding strength, the length of the spacer between the ligand and NPs, the number of targeting ligands per NP, and the conjugation chemistry used for functionalization. Careful adjustment of the many factors that influence cell-specific targeting is necessary because nonspecific electrostatic attachment to the surface of untargeted cells results in ineffective cell-specific targeting [69].

11.7.2 Cancer Treatment

Cancer is one of the deadliest diseases in existence today. The development of new technological systems, for improve understanding of tumor biology, better diagnostics systems, and treatment modalities such as surgical intervention, radiation, and chemotherapy, has led to a decrease in the number of cancer-related deaths in recent years, despite the fact that there are more than ten million new cases each year. Drugs used in chemotherapy frequently also kill healthy cells and cause damage to individuals. To create chemotherapeutics with higher action and less damage, cancer cells can either be actively or passively targeted. A nano-based targeting system began clinical testing in the middle of the 1980s. In the 1990s, the first product based on polymeric liposome and polymer protein conjugates was released. Therapies are created based on the widely used targeting technique of NPs and nanocarrier systems. Drugs and ligands can be combined with nanocarriers to bind to overexpressed antigens or receptors on target cells. Several liposomes, NPs, and other nanomaterials, the optimal threshold size for enhanced extravasations into tumors is 200 nm [70]. One of the main problems of targeting approaches is the inefficient diffusion of nanosystems to tumor cells, which may lead to drug resistance and render many potential treatments inactive. One way to get over this constraint so that it can actively connect to the targeted cells after extravasations is by functionalization. The use of ligand functionalization can be used to provide active tissue targeting using a number of conjugation chemistries. It is essential that drugs target molecules with high selectivity that are only expressed on the cell surface. Because the ligand FNPs may promote receptor-mediated internalization and release the drug within the cells, the target cells should have a surface marker overexpressed in order to maximize selectivity [71]. Most experts agree that higher binding affinities lead to greater targeting efficacy. Targeting can also be improved by using multivalent avidity. Proteins, particularly antibodies and their fragments, nucleic acids, or other receptor ligands including peptides, vitamins, and carbohydrates can all be used for targeting. Furthermore, it is possible to increase binding affinity and selectivity to cell surface targets by creating proteins that recognize a specific conformation of a drug target receptor.

11.7.3 Drug Delivery

One of the primary goals of pharmacology is to optimize therapeutic efficacy of the therapy by giving the correct medicine, in the right dose, to the right patient at the right time. In this sense, drug distribution is one of the most important factors in the field of pharmacology. Nanotechnology could be necessary to accomplish the goals of medicine delivery. The more traditional methods for comprehending medication distribution concentrate on the various routes of drug delivery, such as oral, parenteral, transdermal, and inhalational. Throughout the history of medicine, there has always been a need to develop different methods of giving a patient a medication. Innovative drug delivery techniques' primary goals should be to improve patient compliance, reduce toxicity, boost bioavailability, and achieve precise therapeutic targeting. Drug delivery methods are now widely used in the development and expansion of medicinal potency. Beyond our wildest expectations, the creation of novel pharmacological agents and enhanced delivery systems has the potential to transform disease diagnosis and treatment [72]. Over the past 10 years, nanotechnology has fundamentally altered the field of medication distribution and is currently a key component in the creation of cutting-edge drug delivery systems. Nanotechnology, which may also modify the physical characteristics of medicine particles, can affect the extent and location of medication dispersion. NPs are widely employed in medical procedures and the transdermal delivery of vaccinations [73]. We will go over nanotechnologies once again in relation to the delivery of medications. The application of nanotechnology in the prevention, diagnosis, and treatment of cancer is believed to offer the largest benefit to public health. Osmotic pumps are a good illustration of a microfabrication technique that has been modified by nanotechnology to create secondary nanometer-sized constructs (microspheres) [74]. According to Orringer et al., NPs can target brain malignancies through antigen-dependent selective or nonspecific routes [75]. Targeted delivery can be achieved in two ways: physically induced targeting based on stimuli, such as the magnetic induction between magnetic NPs and a magnet at the site of action, or cell- or tissue-specific targeting based on a strong affinity between a ligand and target receptor, like the RGD peptide to a specific cellular receptor or the mineral affinity of bisphosphonates (BPs) to the mineral phase of bone.

11.7.4 Tissue Engineering and Gene Delivery

More and more nanometer-sized parts are being created and employed in tissue engineering applications. Through the process of tissue engineering, biological constructions that will maintain, improve, and restore tissue function are frequently produced. The appropriate cells can be separated, grown in culture, and then assembled into a three-dimensional (3D) tissue. In vitro genetic modification of this tissue is thus possible before transplantation. Before being implanted into the body, the produced cell tissues can be functionalized with growth agents and enclosed in polymeric 3D matrices. This provides mechanical support for the developing tissues and has the ability to result in the formation of certain tissues. The fundamental difficulties in tissue engineering such as short half-life and stability of growth factor proteins in vivo, as well as the degradation of protein and polymeric matrices upon release from the polymer owing to moisture, temperature, pH, etc. To overcome some of the drawbacks in the existing tissue engineering technologies, gene therapy approaches can be used for tissue regeneration. The practice of treating human illnesses with genetic material injected directly into a patient's individual cells is known as gene therapy. The desired genes can be cloned into an expression plasmid and used as a DNA complex for tissue regeneration by using viral and non-viral gene delivery vectors. In the era of genetic modification and gene delivery, transfection systems on the nanoscale are being specifically built using a range of materials for various uses, notably for patients who suffer from missing, inadequate, or failing tissue and need tissue transplantation. The multidisciplinary approach that has been used to develop this technology (particle mediated gene therapy) will continue to do so. It will combine concepts from both tissue engineering and gene therapy. Polymeric matrices including collagen, PLGA, and alginate hydrogels can be used to transport DNA. Combining these two disciplines enables the functionalization or incorporation of proteins and genes onto nanoparticles (NPs), which can then be implanted or injected to promote tissue regeneration and treat genetic diseases like hemophilia, muscular dystrophy, and cystic fibrosis by replacing errant genes in the affected cell.

Gene therapies are also being developed for cardiovascular, neurological, infectious diseases, wound healing, and cancers by delivering genes to enhance naturally occurring proteins, alter the expression of existing genes, or produce cytotoxic proteins or prodrug activating enzymes to kill tumor cells.

11.7.5 Diagnosis

Computed tomography (CT), magnetic resonance imaging (MRI), and positron emission tomography (PET), among other tomographic imaging techniques, have become indispensable diagnostic tools during the past few decades. While MRI and CT offer intrinsic tissue contrast at high spatial resolution and dynamic imaging, PET provides superior sensitivity in static imaging based on tracer materials. Due to the constraints of CT, which call for the use of ionizing radiation, real-time imaging of 3D volumes with MRI is still difficult. Hybrid organic–inorganic NPs, as they are used in this context, are anticipated to show promise as novel nanocarrier probes for biological applications like MRI, MPI, PET, agents that induce hyperthermia, magnetically controlled medium for the sensitive detection of biomolecules and diseased cells, etc. [76]. Iron oxide is the most promising category of nanoparticles (NPs) for imaging and site-specific delivery of drugs and other bioactive substances. The MPI will improve as an in vivo imaging modality using functionalized magnetic NPs, with the potential to be used as a clinical imaging technique. The ability to monitor in vivo drug delivery to the treated tumors via magnetic resonance imaging, produce a controlled drug release profile, and result in a high uptake of the conjugate in the tumor cells are all advantages that magnetite nanoparticles (NPs), which were first surface-modified with aminopropylsilane (APS) and conjugated with methotrexate, have for treating tumors.

11.8 Conclusion

Because it stands out among the components of the periodic table as a particularly unusual element with specific properties, carbon is the fundamental element of organic chemistry and, ultimately, life. One of the peculiar properties of carbon that makes it possible to develop new biosensors that can detect a range of biological, organic, and inorganic analytes with desired sensitivity levels, low detection limits, and good stability is its capacity to form a variety of structures with peculiar properties. One approach that may be employed to enhance the performance or increase the capabilities of the sensors is the capacity to connect carbon nanostructures to metal or oxide nanoparticles or insert them into other materials like organic polymers or hydrogels. The chemical assembly of carbon-based nanostructures into 2D and 3D structures can also contribute features that improve sensing capabilities. Carbon nanostructures appear especially promising for the fabrication of devices for the multiple detection of analytes for the miniaturization of the sensors allowing a diffuse sensing to control the environment, for example. These devices could include microfluidic chips or microelectronic devices. These advances in biosensor research provide a fascinating glimpse into the development of potential platforms for the monitoring of the environment, air quality, harmful substances, food quality, and biomarkers.

The discovery and development of innovative therapeutic carriers and diagnostic tools depend increasingly on the functionalization of NPs, which provides a unique connection between materials and medicine and back again. Through the use of various types of materials and their functionalization, this technique connects drug targets and other biomedical applications, such as imaging and tissue engineering, to the molecular processes involved in physiology. The addition of functional groups to NPs also enables the creation of multimodality therapeutic approaches that combine tissue engineering, cancer treatment, medication targeting, and diagnostics. As a result of the research that has been done thus far, we now have a better knowledge of FNPs systems and are better able to deliver novel multifunctional nanocarriers, such as targeting molecules, medications, and diagnostic tools, to patients more swiftly.

References

1. Arshady R (1993) Microspheres for biomedical applications: preparation of reactive and labelled microspheres. Biomaterials 14(1):5–15
2. Schulz-Dobrick M, Sarathy KV, Jansen M (2005) Surfactant-free synthesis and functionalization of gold nanoparticles. J Am Chem Soc 127(37):12816–12817
3. Singh R, Thakur P, Thakur A, Kumar H, Chawla P, Rohit JV, Kaushik R, Kumar N (2021) Colorimetric sensing approaches of surface-modified gold and silver nanoparticles for detection of residual pesticides: a review. Int J Environ Anal Chem 101(15):3006–3022
4. Punia P, Dhar R, Ravelo B, Trukhanov AV, Panina LV, Thakur P, Thakur A (2021) Microstructural, optical and magnetic study of Ni–Zn Nanoferrites. J Supercond Nov Magn 34:2131–2140
5. Duguet E, Vasseur S, Mornet S, Devoisselle J (2006) Magnetic nanoparticles and their applications in medicine. Nanomedicine 1(2):157–168
6. Arias JL, Ruiz M, Gallardo V, Delgado ÁV (2008) Tegafur loading and release properties of magnetite/poly(alkylcyanoacrylate) (core/shell) nanoparticles. J Control Release 125(1):50–58
7. Doty RC, Tshikhudo TR, Brust AM, Fernig DG (2005) Extremely stable water-soluble ag nanoparticles. Chem Mater 17(18):4630–4635
8. Moon J, Shul YG, Hong S, Choi YJ, Kim HS (2005) A study on UV-curable adhesives for optical pick-up: II. Silane coupling agent effect. Int J Adhes Adhes 25(6):534–542
9. Moghimi SM, Hunter AC, Murray JC (2001) Long-circulating and target-specific nanoparticles: theory to practice. Pharmacol Rev 53(2):283–318
10. Shahdeo D, Kesarwani V, Suhag D, Ahmed J, Alshehri SM, Gandhi S (2021) Self-assembled chitosan polymer intercalating peptide functionalized gold nanoparticles as nanoprobe for efficient imaging of urokinase plasminogen activator receptors in cancer diagnostics. Carbohydr Polym 266:118138
11. Thakur P, Thakur A (2022) Nanomaterials, their types and properties. In: Thakur A, Thakur P, Khurana SP (eds) Synthesis and applications of nanoparticles. Springer, Singapore. https://doi.org/10.1007/978-981-16-6819-7_2
12. Brust M, Walker M, Bethell D, Schiffrin DJ, Whyman R (1994) Synthesis of thiol-derivatised gold nanoparticles in a two-phase Liquid–Liquid system. J Chem Soc 0(7):801–802
13. Guerrero E, Muñoz-Márquez MA, García ME, Crespo P, Fernández-Pinel E, Hernando A, Fernández AÁ (2008) Surface plasmon resonance and magnetism of thiol-capped gold nanoparticles. Nanotechnology 19(17):175701
14. Dong T, Wu H, Lin M (2006) Superlattice of Octanethiol-protected copper nanoparticles. Langmuir 22(16):6754–6756
15. Gooding JJ, Mearns FJ, Yang W, Liu J (2003) Self-assembled monolayers into the 21st century: recent advances and applications. Electroanalysis 15(2):81–96
16. Shahdeo D, Kesarwani V, Suhag D, Ahmed J, Alshehri SM, Gandhi S (2022) Corrigendum to "Self-assembled chitosan polymer intercalating peptide functionalized gold nanoparticles as

nanoprobe for efficient imaging of urokinase plasminogen activator receptor in cancer diagnostics" [Carbohydrate Polymers, 266 (2021) 118138]. Carbohydr Polym 284:119239
17. Sousa M, Rubim JC, Sobrinho P, Tourinho F (2001) Biocompatible magnetic fluid precursors based on aspartic and glutamic acid modified maghemite nanostructures. J Magn Magn Mater 225(1–2):67–72
18. Tie SL, Lin YQ, Lee H, Bae Y, Lee CH (2006) Amino acid-coated nano-sized magnetite particles prepared by two-step transformation. Colloids and Surfaces A: Physicochemical and Engineering Aspects 273(1–3):75–83
19. von Zur Muhlen C, von Elverfeldt D, Bassler N, Neudorfer I, Steitz B, Petri-Flink A, Hofmann H, Bode C, Peter K (2007) Superparamagnetic iron oxide binding and uptake as imaged by magnetic resonance is mediated by the integrin receptor mac-1 (CD11b/CD18): implications on imaging of atherosclerotic plaques. Atherosclerosis 2007(193):102–111
20. Guo Z, Pereira T, Choi O, Wang Y, Hahn HT (2006) Surface functionalized alumina nanoparticle filled polymeric nanocomposites with enhanced mechanical properties. J Mater Chem 2006(16):2800–2808
21. Sandi G, Kizilel R, Carrado KA, Fernandez-saavedra R, Castagnola N (2005) Effect of the silica precursor on the conductivity of hectorite-derived polymer nanocomposites. Electrochim Acta 2005(50):3891–3896
22. Huang WY, Han CD (2006) Dispersion characteristics and rheology of organoclay nanocomposites based on a segmented main-chain liquid-crystalline polymer having pendant pyridyl group. Macromolecules 2006(39):257–267
23. Mammeri F, Le Bourhis E, Rozes L, Sanchez CJ (2005) Mechanical properties of hybrid organic-inorganic materials. Mater Chem 2005(15):3787–3811
24. Sanchez C, Ribot F (1994) Design of hybrid organic-inorganic materials synthesized via sol-gel chemistry. New J Chem 1994(18):1007–1047
25. Gao S-L, Mader E (2002) Characterization of interphase nanoscale property variations in glass fibre reinforced polypropylene and epoxy resin composites. Compos Part A Appl S 33:559–576
26. Shenhar R, Norsten TB, Rotello VM (2005) Polymer-mediated nanoparticle assembly: structural control and applications. Adv Mater 2005(17):657–669
27. Rana K, Thakur P, Tomar M, Gupta V, Thakur A (2016) Structural and magnetic properties of Ni-Zn doped BaM nanocomposite via citrate precursor. AIP Conf Proc 1731(1):050152
28. Rothenfluh DA, Bermudez H, O'Neil CP, Hubbell JA (2008) Biofunctional polymer nanoparticles for intra-articular targeting and retention in cartilage. Nat Mater 2008(7):248–254
29. Hainfeld JF, Powell RJ (2000) New frontiers in gold labeling. J Histochem Cytochem 48(4): 471–480
30. Almessiere MA, Slimani Y, Thurkanov AV, Baykal A (2022) Structural and morphological characterization of nanomaterials. In: Thakur A, Thakur P, Khurana SP (eds) Synthesis and applications of nanoparticles. Springer, Singapore. https://doi.org/10.1007/978-981-16-6819-7_6
31. Chiang LY, Swirczewski JW, Hsu CS, Chowdhury SR, Cameron S, Creegan KM (1992) Multihydroxy additions onto C60 fullerene molecules. J Chem Soc 24:1791
32. Li J, Takeuchi A, Ozawa M, Li X, Saigo K, Kitazawa K (1993) C60 fullerol formation catalysed by quaternary ammonium hydroxides. J Chem Soc 23:1784
33. Speranza G (2021) Carbon nanomaterials: synthesis, functionalization and sensing applications. Nanomaterials 11(4):967
34. Kokubo K, Matsubayashi K, Tategaki H, Takada H, Oshima T (2008) Facile synthesis of highly water-soluble fullerenes more than half-covered by hydroxyl groups. ACS Nano 2(2):327–333
35. Seshadri R, Govindaraj A, Nagarajan R, Pradeep T, Rao CNR (1992) Addition of amines and halogens to fullerenes C60 and C70. Tetrahedron Lett 33(15):2069–2070
36. Maggini M, Scorrano G, Prato M (1993) Addition of azomethine ylides to C60: synthesis, characterization, and functionalization of fullerene pyrrolidines. J Am Chem Soc 115(21): 9798–9799

37. Li H, Haque SA, Kitaygorodskiy A, Meziani MJ, Torres-Castillo AM, Sun Y (2006) Alternatively modified bingel reaction for efficient syntheses of C60 hexakis-adducts. Org Lett 8(24): 5641–5643
38. Georgakilas V, Guldi DM, Signorini R, Bozio R, Prato M (2003) Organic functionalization and optical properties of carbon onions. J Am Chem Soc 125(47):14268–14269
39. Molina-Ontoria A, Chaur MN, Plonska-Brzezinska ME, Echegoyen L (2013) Preparation and characterization of soluble carbon nano-onions by covalent functionalization, employing a Na–K alloy. Chem Commun 49(24):2406
40. Martín R, Heydorn PC, Alvaro M, García H (2009) General strategy for high-density covalent functionalization of diamond nanoparticles using Fenton chemistry. Chem Mater 21(19): 4505–4514
41. Liang Y, Ozawa M, Krueger A (2009) A general procedure to functionalize agglomerating nanoparticles demonstrated on nanodiamond. ACS Nano 2009(3):2288–2296
42. Shuai C, Li Y, Wang G, Yang W, Peng S, Feng P (2019) Surface modification of nanodiamond: toward the dispersion of reinforced phase in poly-l-lactic acid scaffolds. Int J Biol Macromol 2019(126):1116–1124
43. Dolmatov VY (2018) The influence of detonation synthesis conditions on the yield of condensed carbon and detonation nanodiamond through the example of using TNT-RDX explosive mixture. J Superhard Mater 2018(40):290–294
44. Neburkova J, Vavra J, Cigler P (2017) Coating nanodiamonds with biocompatible shells for applications in biology and medicine. Curr Opin Solid State Mater Sci 2017(21):43–53
45. Whitlow J, Pacelli S, Paul A (2017) Multifunctional nanodiamonds in regenerative medicine: recent advances and future directions. J Control Release 2017(261):62–86
46. Reina G, Zhao L, Bianco A, Komatsu N (2019) Chemical functionalization of nanodiamonds: opportunities and challenges ahead. Angew Chem Int Ed 2019(58):2–14
47. Tinwala H, Wairkar S (2019) Production, surface modification and biomedical applications of nanodiamonds: a sparkling tool for theranostics. Mater Sci Eng C 2019(97):913–931
48. Kennedy ZC, Barrett CA, Warner MG (2017) Direct functionalization of an acid-terminated nanodiamond with Azide: enabling access to 4-Substituted-1,2,3-triazole-functionalized particles. Langmuir 2017(33):2790–2798
49. Chen J, Hamon MA, Hu H, Chen Y, Rao AM, Eklund PC, Haddon RC (1998) Solution properties of single-walled carbon nanotubes. Science 1998(282):95–98
50. Pavelyev V, Tripathi N (2022) Optical characterization of nanomaterials-II. In: Thakur A, Thakur P, Khurana SP (eds) Synthesis and applications of nanoparticles. Springer, Singapore
51. Wang SC, Chang KS, Yuan CJ (2009) Enhancement of electrochemical properties of screen-printed carbon electrodes by oxygen plasma treatment. Electrochim Acta 2009(54):4937–4943
52. Coleman JN, Khan U, Gun'ko Y.K. (2006) Mechanical reinforcement of polymers using carbon nanotubes. Adv Mater 2006(18):689–706
53. Moniruzzaman M, Winey KI (2006) Polymer nanocomposites containing carbon nanotubes. Macromolecules 2006(36):5194–5205
54. Yavari F, Kritzinger C, Gaire C, Song L, Gullapalli H, Borca-Tasciuc T (2010) Tunable bandgap in graphene by the controlled adsorption of water molecules. Small 2010(6): 2535–2538
55. Chung C, Kim Y-K, Shin D, Ryoo S-R, Hong B-H, Min D-H (2013) Biomedical applications of graphene and graphene oxide. Acc Chem Res 2013(46):2211–2224
56. Viswanathan P, Ramaraj R (2019) Graphene-based electrochemical sensors for biomolecules micro and nano technologies. Elsevier, Amsterdam
57. Rodrigues-Perez L, Herranz MA, Martin N (2013) The chemistry of pristine graphene. Chem Commun 2013(49):3721–3735
58. Brodie BC (1859) XIII. On the atomic weight of graphite, vol 149. The Royal Society, pp 249–259
59. Dreyer DR, Park S, Bielawski CW, Ruoff RS (2010) The chemistry of graphene oxide. Chem Soc Rev 2010(39):228–240

60. Miller GP (2006) Reactions between aliphatic amines and [60]fullerene: a review. C R Chim 9(7–8):952–959
61. Xing G, Zhang J, Zhao Y, Tang J, Zhang B, Gao X, Yuan H, Qu L, Cao W, Chai Z (2004) Influences of structural properties on stability of Fullerenols. J Phys Chem B 2004(108): 11473–11479
62. Elias DC (2009) Control of graphene's properties by reversible hydrogenation: evidence for graphene. Science 2009(323):610–613
63. Karlicky F, Datta KKR, Otyepka M, Zboril R (2013) Halogenated Graphenes: rapidly growing family of graphene derivatives. ACS Nano 2013(7):6434–6464
64. Krishnia L, Thakur P, Thakur A (2022) Synthesis of nanoparticles by physical route. In: Synthesis and applications of nanoparticles. Springer, Singapore, pp 45–59
65. Li F, Long L, Weng Y (2020) A review on the contemporary development of composite materials comprising graphene/graphene derivatives. Adv Mater Sci Eng 2020:1–16
66. An X, Butler TW, Washington M, Nayak SK, Kar S (2011) Optical and sensing properties of 1-pyrenecarboxylic acid-functionalized graphene films laminated on polydimethylsiloxane membranes. ACS Nano 2011(5):1003–1011
67. Galindo-Rodriguez SA, Allemann E, Fessi H, Doelker E (2005) Polymeric nanoparticles for oral delivery of drugs and vaccines: a critical evaluation of in vivo studies. Crit Rev Ther Drug 2005(22):419–464
68. Leamon CP, Weigl D, Hendren RW (1999) Folate copolymermediated transfection of cultured cells. Bioconjug Chem 1999(10):947–957
69. Schaffer DV, Lauffenburger DA (1998) Optimization of cell surface binding enhances efficiency and specificity of molecular conjugate gene delivery. J Biol Chem 1998(273): 28004–28009
70. Hobbs SK, Monsky WL, Yuan F, Roberts WG, Griffith L, Torchilin VL, Jain RK (1998) Regulation of transport pathways in tumor vessels: role of tumor type and microenvironment. Proc Natl Acad Sci U S A 1998(95):4607–4612
71. Peer D, Karp JM, Hong S, Farokhzad OC, Margalit R, Langer R (2007) Nanocarriers as an emerging platform for cancer therapy. Nat Nanotechnol 2007(2):751–760
72. Smith B, Uhl K (2009) Drug delivery in the twenty-first century: a new paradigm. Clin Pharmacol Ther 2009(85):451–455
73. Lawson LB, Freytag LC, Clements JD, Clements JD (2007) Use of nanocarriers for transdermal vaccine delivery. Clin Pharmacol Ther 2007(82):641–643
74. Orringer DA, Koo YE, Chen T, Kopelman R, Sagher O, Philbert MA (2009) Small solutions for big problems: the application of nanoparticles to brain tumor diagnosis and therapy. Clin Pharmacol Ther 2009(85):531–534
75. Madsen S-K, Mooney DJ (2000) Delivering DNA with polymer matrices: applications in tissue engineering and gene therapy. Pharm Sci Technol Today 2000(3):381–384
76. Kala D, Gupta S, Nagraik R, Verma V, Thakur A, Kaushal A (2020) Diagnosis of scrub typhus: recent advancements and challenges. 3 Biotech 10:396

Bioprinting for Therapeutics 12

Ritesh Verma, Neetu Dhanda, Raksha Rathore, Deepa Suhag, Fayu Wan, Atul Thakur, and Preeti Thakur

Abstract

The pharmaceutical sector is fueled by novel, high-value products that result from research and development. We can all agree that the pharmaceutical sector is the heart of the medical sciences, everything from nourishment to necessary medications, paracetamol to corona vaccine, are all related to the pharmaceutical industry. A relatively new but rapidly developing technique, bioprinting is mostly employed in tissue engineering and regenerative medicine. The advantages of modifying preset tissue forms and architectures with computer assistance and the creation of extracellular matrix (ECM)-like habitats for cells are combined in 3D bioprinting techniques. The main use of bioprinting is to regenerate or restore wounded and damaged tissues by creating implantable organs and tissues. The medical and cosmetic sectors stand to gain greatly from the technique. The ability of bioprinting has not yet been thoroughly examined in fields like patient-specific cell placement and bio-robotics. But the question arises, with the right application

R. Verma (✉)
Amity Centre for Nanotechnology, Amity University Haryana, Gurgaon, Haryana, India

Amity School of Applied Sciences, Amity University Haryana, Gurugram, Haryana, India
e-mail: rverma1@ggn.amity.edu

N. Dhanda · R. Rathore · D. Suhag · A. Thakur
Amity Centre for Nanotechnology, Amity University Haryana, Gurgaon, Haryana, India
e-mail: dsuhag@ggn.amity.edu; athakur1@ggn.amity.edu

F. Wan
School of Electronics and Information Engineering, Nanjing University of Information Science & Technology, Nanjing, China

P. Thakur
Amity School of Applied Sciences, Amity University Haryana, Gurugram, Haryana, India
e-mail: pthakur@ggn.amity.edu

D. Suhag et al. (eds.), *Integrated Nanomaterials and their Applications*,
https://doi.org/10.1007/978-981-99-6105-4_12

of bioinks and other methodologies, can bioprinting change biomedicine and biomedical engineering, or is it only useful in research settings. This comprehensive study covers current approaches and potential future developments in bioprinting, with a focus on the potential application of 3D bioprinting in other medical disciplines and regenerative medicine. In this chapter, potential uses for experimental systems and techniques that are currently being developed for drug delivery, cancer research, and the bioprinting of various types of tissues are examined. This assessment also explores potential future opportunities for the efficient redesign of technical and medical procedures through the use of bioprinting.

Keywords

Bioprinting · Regenerative medicine · Tissue engineering · Implantable organs

12.1 Introduction

With the advent of bioprinting technologies, tissue engineering has merged science fiction and reality in the modern era of automation and technology [1]. Bioprinting is an adaption of additive (3D) printing that uses layer-by-layer deposition of biological materials and biologically active substances, such as cells, growth hormones, and other substances. This technique allows for the creation of 3D models of tissues and organs that closely resemble their human counterparts in terms of structure [2]. By setting up preset places and structures in computer-aided design software (CAD), the architecture is methodically finished. How effective the bioprinting method is can be determined by the value of the structures produced. The three primary factors to take into account while creating a product are cellular interactions, bioactive chemicals, and the bioprinting apparatus [3]. The primary biomaterial used in 3D printing, called bioink, is formed of cells that are meant to be deposited. These cells are often contained in a carrier matrix that attempts to replicate the physical and biochemical environment of native tissues in order to improve cell adhesion, proliferation, and differentiation [4].

In the last 10 years, the bioprinting sector has made tremendous achievements, and 3D bioprinting is currently one of the most intriguing and promising technologies with the potential to have a significant impact on a range of medical applications. If scientists could use bioprinting technology to create living, de novo organs like the heart, liver, kidneys, lungs, and skin, the need for organ transplants would be decreased. Taking cells directly from the patient would also prevent immune system assault and organ rejection. Another creative industrial use of 3D bioprinting is found in the pharmaceutical industry. The market for 3D bioprinting is anticipated to reach USD 1647 million [4] by 2024 as a result of the development of 3D bioprinters and biomaterials as well as their use in the pharmaceutical and cosmetics sectors [5, 6]. We briefly addressed the most cutting-edge bioprinting techniques in this chapter, along with other essential elements important to the application of 3D bioprinting to construct 3D tissues and organs. With a focus on the potential for creating in vitro models as tools for drug discovery in the pharmaceutical industry, we also discuss some of the significant challenges and fascinating opportunities that 3D bioprinting technologies present for developing realistic tissue and organs across a variety of market segments.

12.2 History

The concept of 3D-printed organs and bioprinting was first explored in 1996. When Dr. Gabor Forgacs discovered that biological components adhere together and have liquid-like qualities during experiments with Charles Hull (who later founded 3D Systems), the general public started to exhibit interest in bioprinted items that use ingredients extracted directly from patient tissues. Early studies employed synthetic scaffolds, but after investigation into a new technique called Dynamic Optical Projection Stereolithography (DOPsL), it was discovered that biological tissues, like blood cells, could be printed in a matter of seconds.

12.2.1 Which Organs Have Been Previously Bioprinted?

There have not been any additional reports of 3D-printed organs since the first human blood vessel was made. Minor components such as arteries, bones, cartilage, tissues, and veins may have been printed for implantation, but since regenerative medicine professionals prefer to have a thorough understanding of the technology before moving forward, they are currently waiting for new case studies before thinking about printing more complex organs.

12.3 3D Bioprinting

Using a suitable imaging tool, such as computed tomography (CT) or magnetic resonance imaging (MRI), to ascertain the anatomical makeup of the target tissue is the first stage in the bioprinting process. The picture is then transformed using specialist software into a CAD drawing of cross-sectional layers so that the printing machine may add the layers one at a time. The bioprinting device then uses a specific printing method, such as inkjet 3D bioprinting, micro-extrusion 3D bioprinting, laser-assisted 3D bioprinting, or stereolithography, along with a combination of printing materials, such as scaffold and bioink, and other additive factors to produce the tissue (Fig. 12.1). Precision, stability, and tissue viability vary via various methods. In order to replicate the in vivo conditions required to maintain tissue viability during the maturation process, the generated tissue is then further processed in a bioreactor.

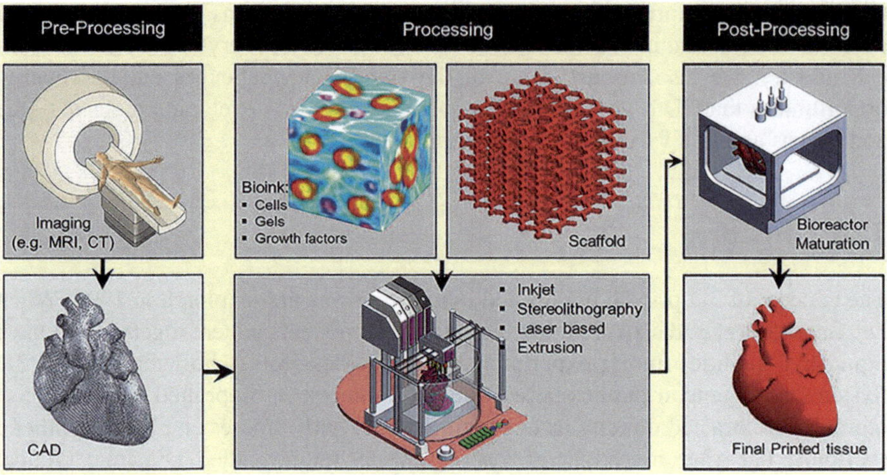

Fig. 12.1 Schematic representation of the bioprinting process [7]

12.3.1 Techniques for 3D Bioprinting

The basis for 3D bioprinting technology was laid by conventional 2D printing on paper and then by 3D printing of non-biological materials. Therefore, it is not surprising that the engineering component has more advanced technology than bioink material technology. However, because bioink was first developed for printing on non-biological materials, each printing method still has a number of limitations regarding material compatibility when using other building materials instead of bioink [8–10].

12.3.1.1 Micro-Extrusion 3D Bioprinting

It is a regularly used pressure-assisted technique for printing non-biological materials. Typically, a computer-controlled robotic arm uses a pneumatic or mechanical (piston or screw-driven) approach to provide pressure to a nozzle in order to disperse the selected bioinks for the bioprinting process. Most frequently, a glass or plastic cartridge is used to hold the bioinks. The bioink is ejected via the nozzle in the form of a thin filament and deposited on the substrate based on a CAD design that regulates the location and course of nozzle movement to produce the tissue in the required 3D shape. This technology, which developed from conventional 3D printing, has been demonstrated to be the most efficient approach to manufacture large-scale constructs because of its structural soundness, making it more suitable for scaling up for organ manufacturing [11]. Rapid printing also allows for the use of a wide variety of bioinks, including both scaffold-based and scaffold-free bioinks. The resolution of this technique is just 100 m though [12]. Due to the relatively high extrusion pressure via the nozzle, the bioink components are subjected to experience extreme shear stress, which might induce cellular viability loss and tissue structure deformation [13].

12.3.1.2 Inkjet 3D Bioprinting

Because of its structural soundness, this technology, which evolved from traditional 3D printing, has been shown to be the most effective method for producing large-scale structures, making it more ideal for scaling up for organ manufacturing [11]. Additionally, a range of bioinks, including scaffold-based and scaffold-free bioinks, can be used with rapid printing. However, this method only has a 100 m resolution [12]. The bioink components are subjected to significant shear stress as a result of the relatively high extrusion pressure through the nozzle, which could result in cellular viability loss and tissue structural distortion. Eventually, the vapor bubble swiftly bursts and expels a bioink droplet as a result of the pressure-induced expansion it experiences [14]. These temperature changes have little effect on the survival of cells since the processes only last a few microseconds. It has been determined if the bioprinted cells can keep their genetic, phenotypic, and functional characteristics [15]. In contrast, a voltage pulse in a piezoelectric inkjet bioprinter causes the piezoelectric actuator to change form. Then, the volume of the fluid chamber containing the bioink abruptly changes, releasing a droplet [16]. The viability of the cells could be maintained even after the surgery without any issues [17]. Similar principles govern how electrostatic actuators work. When an electrical pulse is supplied between the pressure plate and an electrode, the pressure plate flexes. About 80–95% yield in the cells created utilizing the thermal, piezoelectric, and electrostatic inkjet bioprinting techniques demonstrated a robust cell viability. After the voltage pulse is stopped, the pressure plate reverts to its original form, expelling a bioink droplet.

Inkjet bioprinting techniques show promise since they enable high-resolution printing and have fine control over the ejection of droplets and pico-liter sized ink droplets.

12.3.1.3 Laser-Assisted 3D Bioprinting (LAB)

Laser-assisted 3D bioprinting (LAB) is another non-contact printing technique. The laser-assisted bioprinter's ribbon has an absorbent coating, such as gold, on it. When a laser pulse is directed and travels through the transparent ribbon, a hydrogel droplet is induced and eventually transmitted to the receiving substrate. The final structure is constructed layer by layer by repeating this process utilizing laser pulses to form a jet. This technology, which enables cells to be deposited at a density of up to 108 cells/mL with a single-cell resolution and at a high pace, enables high-throughput cell and biomaterial patterning [18–20]. Additionally, cell transfer and selection are made possible by the real-time cell monitoring offered by laser-assisted printing. However, the extreme heat generated by the laser beam might damage the printed tissue's cells and reduce their vitality.

12.3.1.4 Stereolithography-Based Bioprinting (SLB)

Photopolymerizable liquid polymers are utilized in stereolithography-based bioprinting (SLB). A predefined pattern of laser or UV light is focused on the polymers, causing the polymers to cross-link and harden. Each polymerization cycle generates a little layer of the structure, and this polymerization cycle is then

Table 12.1 Advantages and disadvantages of major 3D bioprinting techniques

Bioprinting techniques	Advantages	Disadvantages	Refs.
Micro-extrusion 3D bioprinting	The ability to print tissues with extremely high cell densities and scaffold-free bioink, as well as the ability to print high-viscosity bioinks by altering the driving pressure, all contribute to the creation of organs and tissues with good structural integrity	High shear stress from the pressure-driven dispensing has a negative impact on the cells' survival and resolution, making it impossible to build a microcapillary network	[21, 22]
Inkjet 3D bioprinting	Non-contact based, lowering the possibility of contamination Integration of multiple printing heads for diverse tissue architectures. Enables fabrication of a vasculature like structure and high-speed printing	Needs bioink with low viscosity (0.1 Pa s^{-1}), non-uniform droplet size. This makes it more challenging to install very viscous hydrogel ECM components	[23, 24]
Laser-assisted 3D bioprinting	Bioprinting of micro-patterned peptides, DNA, and cells with single-cell resolution is made possible by the printed structures' great precision and resolution. The capacity to print tissues with extremely high cell densities and without restrictions due to viscosity	The heat produced by laser energy may have an impact on cell survival	[25]
Stereolithography-based bioprinting (SLB)	Good resolution with no clogging during printing	Takes a strong radiation to cross-link molecules, which is a lengthy process	[26, 27]
Acoustic bioprinting	Convenient, cost effective, and biocompatible method for patterning	–	[28]

repeated to create the 3D structure layer by layer. The main advantages of this technique over alternative methods are its high resolution and lack of extremely high shear stress. Cells can be harmed by the cross-linking process, which exposes them to strong UV light (Table 12.1).

12.3.1.5 Acoustic Bioprinting

The acoustic bioprinting process shields the biomaterials from destructive stress, such as heat, high voltage, high pressure, and any kind of shear stress. Droplets are released through a nozzle using a gentle acoustic field [29]. Mild acoustic waves, however, cannot expel viscous or highly cell-concentrated bioink droplets. Few research have been conducted on this methodology.

12.4 Bioinks

The most challenging obstacle still standing in the way of the advancement of 3D bioprinting technology is the production of printing biomaterials, sometimes referred to as "bioink," which is a crucial component of this technology. The ideal properties for bioink must meet both the physical and biological material criteria in order to enable cellular activity comparable to that observed in real creatures, such as cellular proliferation, differentiation, migration, and maturation. The physical properties include viscosity, structural toughness, printing ability, degradation, and functionality. Examples of biological properties include bioactivity, cytocompatibility, and biocompatibility [30]. The most crucial aspect of the bioprinting procedure, the bioink viscosity, needs to be constantly tuned in order to regulate the bioink flow, cell encapsulation efficiency, and tissue structure stability. To meet the requirements of various printing processes, there were several bioink formulas available. Hydrogels are intriguing possibilities for the creation of bioinks due to their biocompatibility, low cytotoxicity, hydrophilicity, and ability to build networks of polymers that allow them to adopt ECM with a similar structure [31]. Self-organizing systems exist in organs and tissues. During the embryonic development process, cells engage in biological self-assembly and self-organization without the use of external guidance or guiding structures [32]. In contrast, the complexity of the cell microenvironment is greatly decreased or absent in vitro, which causes distinct cell fusion and delayed aggregation. In bioprinting, biocompatible scaffolds are frequently used to give structural support for cells to attach, grow inside, differentiate into, and eventually produce the tissue on (Fig. 12.2).

Studies revealed that a scaffold's integrin configuration has a significant impact on stem cell development [34]. Replicating a certain tissue type's natural environment encourages stem cells to differentiate into that lineage. The use of appropriate biomaterials to support the biological components is essential for the creation of 3D heterogeneous tissue architectures. The most popular biomaterials for live-cell printing are hydrogels, because of their biocompatibility and capacity to take on a structure resembling that of the ECM [35]. The ultimate goal of bioprinting-based tissue engineering is to somewhat replicate the tissue/organs that grow during embryonic development. However, this unique strategy is still too far from the in vivo equivalents' level of complexity, which is produced by many specialized cells and dynamic extracellular matrix (ECM) composition [36]. In order to print tissue constructions, bioink is a solution of one biomaterial or a combination of many biomaterials (for example, in the form of a hydrogel). Bioink encapsulates the appropriate cell types throughout the printing process. Natural or artificial biomaterials, or a combination of the two, are used to create bioinks (Table 12.2).

In order to create tissue that closely matches it is in vivo counterpart in terms of both form and function, it is crucial to maximize the biological, mechanical, and

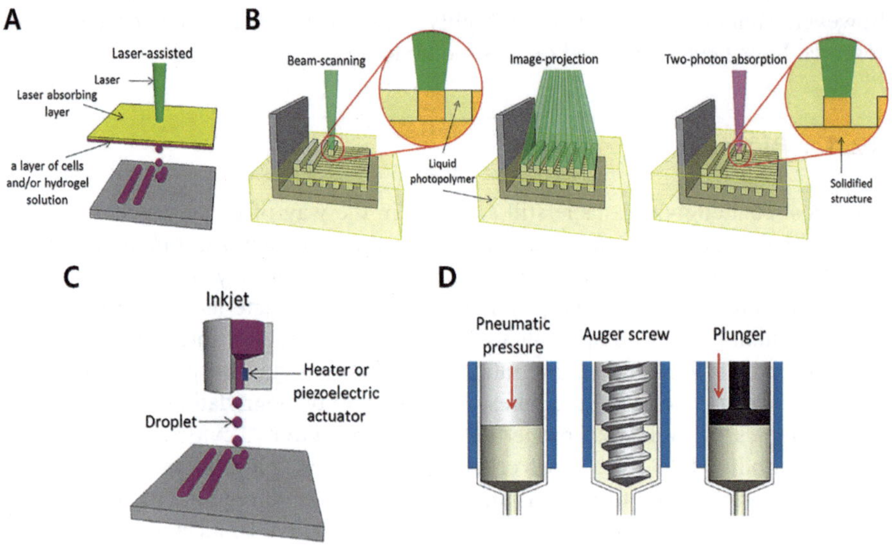

Fig. 12.2 Organ and tissue 3D bioprinting. Schematic image of printing methods. (**a**) LIFT-based (**b**) SLA-based (**c**) Inkjet-based (**d**) Extrusion-based printing system [33]

Table 12.2 Bioink material building blocks [37]

Bioink material	Overview	Advantages	Disadvantages
Chitosan	Polysaccharide derived from the shellfish's exoskeleton (e.g., shrimp). Fungal fermentation can produce chitosan that is not produced from animals	Extremely biocompatible and shows bacterial resistance	Slow pace of gelation
Fibrin/ fibrinogen	High biological relevance insoluble protein produced during blood coagulation	Quick gelation	Only partially printable
Gelatin	Is a protein compound created when some collagen is partially hydrolyzed	Extremely biocompatible high solubility in water, thermally reversible gelation	Lack of stiffness; and inflexible
Graphene	Carbon-based material that can be viewed as a one atom thick sheet of graphite	Flexible and electrically-conductive	Small biological significance
Alginate	Brown algae-derived natural biopolymer	Mild cross-linking (Ca^{2+}) conditions, quick gelation extremely biocompatible	Dynamics of slow deterioration; inadequate cell adhesion

rheological properties of bioinks. For diverse uses and cell types, several bioinks can be required. There are a number of basic qualities that should be considered while choosing a bioink, including the following:

12.4.1 Viscosity

The bioink matrix should fit in all bioprinting stages, both as a fluid during cell encapsulation and as a solid following dispensing, according to the bioprinting procedure [38].

12.4.2 Gelation and Stabilization Processes

The viability and print resolution may be affected by how the bioink solidifies after extrusion. The bioink should be cell-safe and still have tissue-matching qualities after printing, and the process should be rapid. Ionic, thermal, stereocomplex, photocrosslinking, enzymatic, and click chemistry are just a few of the gelation techniques that may be used, and the characteristics and composition of the bioink material can also affect these techniques [39, 40].

12.4.3 Biocompatibility

Hydrophilic and cell-adhesive materials encourage cell survival and growth. Furthermore, the use of natural or synthetic bioinks has a significant influence on how biological interactions proceed. There is batch-to-batch variability even though natural-based bioinks can endure challenging manufacturing conditions (such high temperatures and organic solvents). On the other hand, batch-to-batch unpredictability is solved by synthetic polymers, which have a tremendous potential for large-scale manufacturing but lack the natural cell attachment sites [41].

12.4.4 Mechanical Characteristics

When the mechanical properties of their surrounding environment, such as matrix elasticity are altered, cells can adapt their behavior [42]. The ability to control cell activity inside the printed tissue construct, such as their form and rate of proliferation, is therefore possible by modifying the mechanical characteristics of bioinks. This is important for the growth of a functional tissue. Another important mechanical property is shear-thinning, a non-Newtonian property that implies a reduction in viscosity as the shear rate increases and leads to rearrangement of the polymer chain. For instance, Chen et al. [43] combined a shear-thinning hybrid bioink which was created by combining hard gellan gum, flexible sodium alginate, bioactive thixotropic magnesium phosphate-based gel, and thixotropic TMP-BG. The mechanical, rheological, and bioactive characteristics of the bioink were improved for printability and cell survival.

A number of organic and synthetic biomaterials have been employed as bioinks. Bioink has recently been created using materials that are sensitive, dynamic, and supramolecular. Morgan et al. provide a comprehensive summary of the most recent discoveries in bioinks, including their mechanical properties and dynamic nature [44]. The potential for modifying synthetic polymers to increase bioactivity is enormous. However, when they decay, they may emit toxic byproducts and lose mechanical properties. Furthermore, self-assembling peptides are prospective biomaterials for constructing 3D scaffolds with structural and mechanical properties similar to extracellular matrices. Cofino et al. [45], for example, optimized methyl-cellulose and RAD16-I-based biomaterials to build 3D predefined structures to generate bioink with controllable viscosity. The resulting objects have exceptional shape fidelity and stability. Standardized bioink formulations are desperately needed to enable usage in a variety of bioprinting applications.

12.5 Applications of 3D Bioprinting (Fig. 12.3)

12.5.1 Organ Transplantation Using 3D Bioprinting

Building on the huge success of printing industrial prototypes for prostheses and surgical equipment, 3D bioprinting makes excellent progress in generating thick live cellular structures as a transitional stage toward organ-level complexity. Despite the problems associated with the linked biology and engineering, bioprinting offers

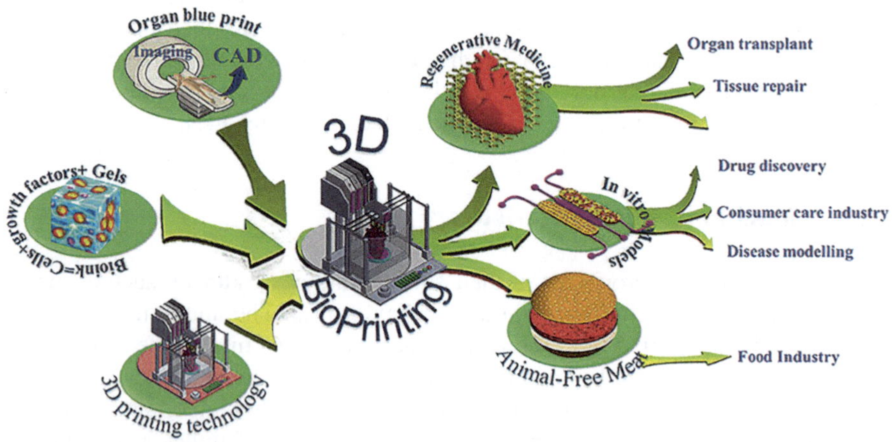

Fig. 12.3 Applications of 3D bioprinting [7]

huge potential in whole-organ printing with an excellent hierarchical organization of cells and creating tissue blocks in a 3D microenvironment. Cells from either patient or adult stem cells are employed to make a bioink capable of producing live tissues. These components are held together by a dissolvable gel or scaffold that may support cells and mold them into the correct form to achieve the intended function. Current advanced imaging technology, like as CT, permits the development of exact CAD models for 3D printing to ensure a flawless fit into the targeted tissue [46, 47]. There have been reports in recent years of the development of several types of thick tissues in a range of forms, with the eventual objective of printing full organs or body parts for organ donation.

Long donor search durations and immunological rejection of the transplanted organ can be avoided by extracting stem cells from transplant recipients and printing them into replacement organs. Recent advances in 3D tissue bioprinting have resulted in organ-level structures such as bone, cornea, cartilage, hearts, and skin [48]. Zhou et al. [49] produced a patient-specific ear-shaped cartilage using expanded microtia chondrocytes and a biodegradable scaffold. Five individuals with microtia had their cartilage 3D-printed, and the results were satisfactory in terms of appearance. Researchers have demonstrated thick, vascularized cardiac patches as well as a cellularized human heart with natural architecture. Pluripotent stem cells were generated from omental tissue samples of patients and subsequently differentiated into cardiomyocytes and endothelial cells. The bioink was made by mixing the two cell types separately with hydrogels for the heart tissue and blood vessels. The functionality of functional vascularized patches was demonstrated to work in line with the patient's anatomy (Fig. 12.4). In order to replace damaged (such as burnt) skin and heal skin ulcers and wounds, much study was undertaken on the skin among the many human tissues. Baltazar et al. [50] described an implantable multilayered vascularized 3D-printed skin transplant. The dermis of the skin was created using a bioink composed of human foreskin dermal fibroblasts, human endothelial cells, and human placental pericytes suspended in rat-tail type I collagen. After that, the epidermis was generated by printing with a second bioink that included human foreskin keratinocytes. In this arrangement, while endothelial cells and pericytes self-assembled into linked microvascular networks, keratinocytes formed a multilayered skin barrier that seemed to speed keratinocyte development.

12.5.2 Organ Models Printed in 3D for Drug Discovery

The translational medical research community is actively attempting to move its focus away from animal models and toward more complex human qualities and circumstances. Unfortunately, while traditional in vitro models are robust and well suited for high-throughput research due to their simplicity, they have little biological relevance to the intricate biological tissues of the human body, resulting in a significant technological difference between lab models and industry/clinic adoptable models. Bioprinting using high-resolution vascularized tissue opens the door to the production of biomimetic habitats and structures that facilitate in vivo

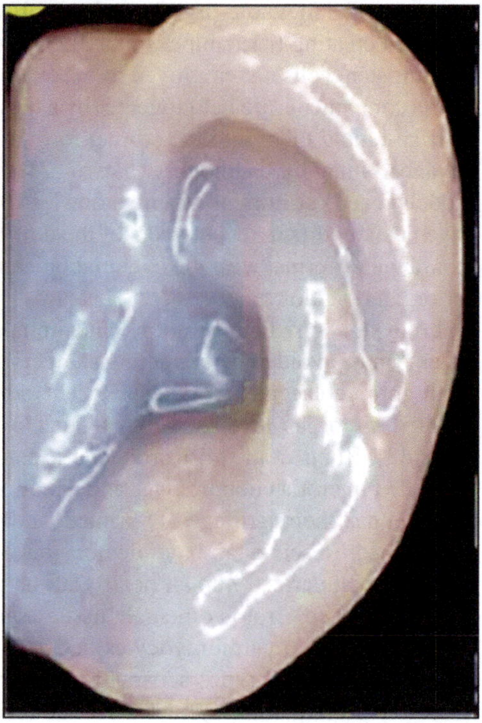

Fig. 12.4 3D printed human organs: making future transplant wait list shorter [48]

interactions between cells and their environs. Bioprinted tissue would be a useful tool for developing physiologically correct in vitro human organ models for drug toxicity studies and disease modeling that precisely reproduce the physiologically complicated human. Because it requires a significant number of distinct cell types to make a medically relevant heterotypic tissue, organotypic bioprinting is often a costly technology for large-scale and high-throughput experiments. Furthermore, a hypoxic environment may develop in the artificial tissue due to inadequate delivery of cell nourishment into the tissue's core in the absence of high-resolution vascularization, which ensures long-term survival. These disadvantages can be solved by creating small in vitro tissue models, sometimes known as "organs-on-a-chip," by combining bioprinting and microfluidic technology. Different organotypic tissues, for example, can be printed concurrently on a segmented microfluidic chip and linked via a vascular network (perfusion channels) to create a "human-on-a-chip" with numerous organs. It is becoming commonly accepted that 3D tissue models are better than 2D counterparts and more biologically relevant. These tissue models are also immune from rigorous ethical issues, making them an appealing choice for many associated industries. They have not yet been adequately verified for toxicity prediction. Systemic validation and standardization are required to guarantee that these powerful models have the potential benefit for high-throughput drug development. Several companies and start-ups have created 3D tissue in vitro models for toxicity assessment and disease modeling in recent years. Organovo Inc., for example, created a customizable bioprinting method to manufacture tissues in various shapes, such as microscale tissues on multi-well tissue culture plates. For example, human primary hepatocytes, hepatic stellate cells, and endothelial cells were used to bioprint liver-like tissue architectures. The liver damage that leads to fibrosis was one of the effects of amethotrexate and thioacetamide exposure that was monitored using a tissue model [51]. To learn more about Kupffer cells' impact on the injury/fibrogenic response to cytokine and pharmacological stimuli, they were added to a separate study [52].

The rapid advancement of bioprinting technology and the widespread use of 3D bioprinter modalities have generated unprecedented interest in using this technology to produce in vitro models for pharmacological research. To develop a neuronal micro-physiological system, Bowser and Moore [53] employed 3D bioprinting technology based on spheroid and magnetic spheres. Spinal cord spheroids made of magnetic nanoparticles are arranged in a three-dimensional hydrogel framework using magnetically assisted bioprinting. The structures had long-range projections that mimicked the architecture observed in vivo as well as small-scale cell-to-cell interactions. Zhuang et al. [54] used an integrated ultraviolet (UV) curing system with extrusion-based bioprinting to allow layer-by-layer UV curing of bioprinted photo-curable GelMA-based hydrogels. By using this technique, cell-laden structures with a high aspect ratio and stability may be made without the need of reinforcing elements like poly(−caprolactone) (PCL) polymer. The research by Heinrich et al. also shown how to construct miniature brains consisting of glioblastoma and macrophages to test therapeutic approaches that focus on the interaction between these two cell types. A hybrid 3D cell-printing system that uses both inkjet- and extrusion-based dispensing modules was developed in order to manufacture a 3D facsimile of human skin inside a transwell system. Polycaprolactone was utilized to produce a collagen-based construct by extrusion-based printing, and keratinocytes were equally distributed across the engineered dermis using an inkjet-based dispensing module. In addition, to mimic the composition and functioning of the original intestinal tissue, human primary intestinal epithelial cells, and myofibroblasts were employed to bioprint 3D intestinal tissue. The tissue model showed significant morphological and physiological characteristics, including a polarized epithelium with close connections and the expression of CYP450 enzymes.

Although many tissues and organ models have been created, the complexity needed to produce replacement tissues and organ models that are medically correct has not yet been obtained or sufficiently characterized. A range of cell types participate in homeostasis and tissue development in vivo in biological systems with linked tissues and organs. Due to the inherent complexity of these systems, simulating the structure and physiology of interrelated human tissues and animal models to enable tracking the physiological processes is difficult. To accomplish differentiation into the needed phenotypes, it is still unknown how much biomimicry of human physiology is necessary or whether all cellular subpopulations must be utilized. Recent advancements in 3D bioprinting technology may provide definite answers to these crucial questions. For example, multimaterial deposition systems may be employed to create intricate heterogeneous cellular architectures. This makes it possible to incorporate vascular and neuronal networks into the in vitro models' structure, accurately simulating the complexity of numerous tissue and organ systems. However, to achieve this ambitious aim, further in-depth study is needed. In addition, combining bioprinting methods with other technologies such as imaging, bioreactor technology, organs-on-a-chip (OOC), artificial intelligence (AI), and semiconductors would enhance tissue engineering's capabilities and speed up the field's progress toward the production of organs and tissues for a variety of applications.

12.5.3 Tumor Modeling

A tumor microenvironment must be produced via 3D bioprinting so that tumor cells may interact with adjacent stromal cells including endothelium and fibroblasts as well as an extracellular matrix in order to appropriately study tumor growth, monitoring, and therapy response. The conventional techniques for mimicking tumors or cancer are monolayer cell cultures or animal models. However, the complicated connections observed in live tissue are absent from monolayer cultures. On the other hand, animal models could respond and react differently from human tissue. Analysis of intracellular molecular interactions, intercellular connections, enzyme kinetics, changes in protein expression, metastatic progression, etc. may be done using a bioprinted tumor model for specific tissues. Zhou et al. examined the dissemination of breast cancer in bone tissue by fabricating biomimetic bone structures using a table-top commercial stereolithographic printer. In order to establish a 3D structure of bone tissue and examine the relationships, breast cancer cells (BrCas) were planted in vitro in this work [55]. Another study used a bioprinted mini-brain to display how glioblastoma multiforme developed in terms of interactions between glioblastoma cells and glioblastoma-associated macrophages [56]. The results of this study indicated that drugs may be developed to stop the communication between these cells in order to inhibit tumor development.

One of the most difficult elements of cancer modeling is simulating the angiogenesis of tumor cells in the absence of vascular networks. Angiogenesis is a pivotal step in the development of cancer because it is essential for the transportation of nutrients and oxygen to the cells [57]. Vasculature can be created via bioprinting techniques, as was previously said, if the right bioinks and growth factors are used. Using patient-derived GBM cells and human endothelial cells in a collagen matrix, Lee et al. may have printed a physiological glioma-vascular niche model to better understand cell–cell interactions and the effects of microenvironmental factors on glioblastoma multiforme [58]. Gelatin and collagen were used to make fluidic vascular channels. In a recent research, Suarez-Martinez et al. examined cancer cell migration, proliferation, and function during microvascular network growth using a bioprinted tumor microenvironment model of mouse breast cancer cells [59]. A bioprinted hydrogel-based vascular flow device was created by Hynes et al. [60] in another recent work as a unique approach to comprehending the progression of cancer. Bone marrow suppression, gastrointestinal distress, liver toxicity, urinary system toxicity, renal toxicity, cardiotoxicity, and neurotoxicity are among the adverse effects of widely recommended cancer therapies. Heterotypic drug interactions were performed on macrophages inside the channels and bioprinted breast cancer cells covered in peptide-conjugated alginate fibers to evaluate the therapeutic effectiveness and toxicity. HER2-positive breast cancer cells embedded in adipose-derived mesenchymal stem cells (ADMSCs) were employed in a new bioprinted model to assess doxorubicin and look into treatment resistance.

On the basis of a biomimetic ECM consisting of adipose cells, endothelial cells, and mammary fibroblasts produced from MSCs, the chemotherapeutic effects of tamoxifen on breast cancer were evaluated. To gauge the effectiveness of the medication, the adenosine triphosphate (ATP) luciferase test was utilized. Temozolomide (TMZ), an anti-cancer drug, and the angiogenic inhibitor sunitinib were both studied in a tumor microenvironment.

The printed tissue was multicellular tumor spheroids of glioblastoma cells on a blood artery layer composed of fibroblasts and endothelial cells. Sun et al. employed HepG2 cells to construct a model of hepatocellular carcinoma in a different recent work to carry out pharmacological pharmacodynamics research [61]. As bioprinting technology develop, it may soon be possible to construct more intricate biological systems and examine how diseases interact with surrounding tissue as well as at the cellular level.

12.6 In Situ Bioprinting

One of the intriguing applications of 3D bioprinting is the direct patterning of de novo tissue onto the appropriate location of the body, such as chronic skin wounds or bone abnormalities. Medical imaging may be used to adjust the architecture of printed tissue to match the wound or defect, allowing heterotypic cellular structures, hydrogels, and soluble components to be correctly deposited inside the faults. This procedure, also referred to as in situ bioprinting or intraoperative bioprinting (IOB), would shorten the distance between the host and implant interfaces and provide clearly defined structures within areas of irregular topographies during the healing process, which can successfully recruit desired cells from surrounding tissues where the patient's body acts as a natural bioreactor [62]. There have been fewer attempts than with the other apps mentioned above. In a recent proof-of-concept work, Albana et al. [63] precisely distributed autologous/allogeneic dermal fibroblasts and epidermal keratinocytes into a damaged region in mice, resembling the layered skin structure. Excisional wounds that were bioprinted using layers of autologous dermal fibroblasts and epidermal keratinocytes in a hydrogel carrier showed faster re-epithelialization, reduced contraction, and rapid wound closure. These results showed that in situ skin bioprinting is viable and revealed potential applications for it in the regeneration of various body parts. Successful in situ bioprinting techniques that enable cell extraction from a small biopsy might dramatically accelerate cell therapy-based healing. Zhao and Xu [64] developed a micro-bioprinting technology that was linked to an endoscope to enable bioprinting within the human body. In order to print tissues with extreme accuracy, they also employed printed circuit micro-electro-mechanical system methods. Using gelatin-alginate hydrogels, human gastric epithelium, and smooth muscle cells as bioinks, two-layer tissue scaffolds were printed in a stomach model to reproduce the anatomical structure of the stomach. Researchers patterned endothelial cells into a mouse calvaria bone defect that was filled with mesenchymal stem cells that contained collagen and vascular endothelial growth factor using laser-assisted bioprinting (LAB). This method allowed for organized microvascular networks to enter bone defects and has a promising vascularization rate for in situ pre-vascularization that aids bone regeneration.

In situ bioprinting, as opposed to in vitro printing, is a contact-based technique that necessitates close consideration of bioink properties, bioprinter setup, and cleaning. For instance, in extrusion-based bioprinting, the printing tip could interfere with the region around the fault and result in collateral damage. The bioinks used for in situ bioprinting often need to be biocompatible with fast cross-linkability in order to allow for shorter operating times and to retain the integrity of bioprinted constructs. Numerous biomaterials show great potential for these applications, including collage, fibrinogen, gelatin methacrylamide (GelMA), hyaluronic acid methacrylate (HAMA), and poly (ethylene glycol). Since it takes living tissue more than 10 days to go through angiogenesis, vascularization is a highly challenging process, particularly for in situ bioprinting. Utilizing oxygen-generating biomaterials or oxygen-filled microparticles that may be bioprinted inside of the bioink might give a temporary oxygen supply prior to angiogenesis. Another technique creates sacrificial porosity structures inside the bioprinted tissue using mesh filaments.

12.7 Microfluidics and Organ-on-a-Chip Converge with Bioprinting

Recent bioprinting research has taken use of well-proven microfluidic technology to develop bioprinting devices that allow precision dispensing of low-viscosity bioink in a well-defined template under very controlled circumstances [65, 66]. Microfluidic dispensing technique is used in several commercial bioprinters. As an example, Aspect Biosystems developed the RX1TM bioprinter, which has precise motion and pressure control as well as high-speed microscale resolution. The RX1TM bioprinter was used to create 3D neural tissues using hiPSC-derived neural aggregates, according to Abelseth et al. [67]. The exact manipulation of the microstructures and microarchitecture of tissue constructs would be made feasible by complex 3D-shaped scaffolds, enabling the development of various tissues and organs as in vitro drug discovery models. As a bottom-up strategy, organ-on-a-chip and organoids are miniature in vitro models of human organs that are produced by spatially immobilizing various types of living cells to produce heterogeneous functional structures within a prefabricated chip and scaffold. Multiorgans-on-a-chip, which are more complex heterogeneous tissue structures, might also be made using

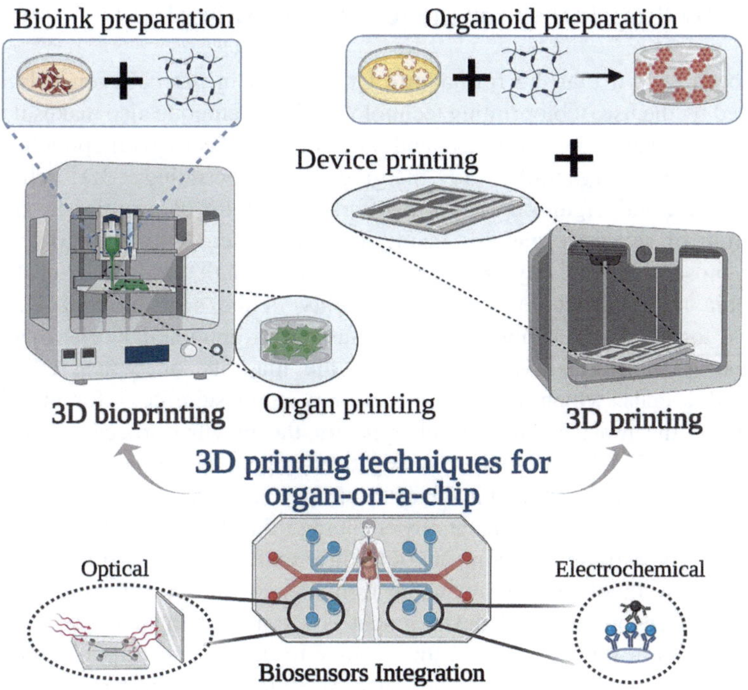

Fig. 12.5 Steps for 3D-bioprinting organ-on-a-chip [68]

this technique. Tissue engineering might significantly benefit from the use of 3D printing technology in this specific discipline. Other tiny organ models, such as those of the heart, kidney, liver, and vasculature, have recently been realized (i.e., printed). Human cell-derived organoids are efficient instruments for disease modeling, drug screening, and individualized treatment. By employing these organoids as the building blocks for 3D bioprinting, it would be able to scale up the deposition of these tissue structures. Maloney et al. described an immersion printing technique to bioprint tissue organoids in 96-well plates (Fig. 12.5) [68, 69].

To maintain a spherical shape, a hydrogel made of hyaluronic acid and collagen is bioprinted into a viscous gelatin solution. This prevents the bioink from interacting with the well walls. Reid et al. [70] created tumoroid arrays using 3D bioprinting to study the carcinogenesis and microenvironmental redirection of breast cancer cells. It has been shown that the formation of tumoroids in 3D collagen gels is much improved by the use of bioprinting technology. This technique also makes it possible to precisely create tumoroid arrays and co-print cancer cells with epithelial cells to create chimeric organoids. By merging 3D bioprinting, 3D cell culture, microfluidics, and organ-on-a-chip, it offers a great lot of promise to enable the integration of several organoids inside a single system with tiny footprints and greater biosensing capacity.

12.8 3D Bioprinting for Meat without Animals

It has been shown that the formation of tumoroids in 3D collagen gels is much improved by the use of bioprinting technology. This technique also makes it possible to precisely create tumoroid arrays and co-print cancer cells with epithelial cells to create chimeric organoids. By merging 3D bioprinting, 3D cell culture, microfluidics, and organ-on-a-chip, it offers a great lot of promise to enable the integration of several organoids inside a single system with tiny footprints and greater biosensing capacity. However, because to high production costs, public neophobia may restrict its economic viability in the near future [71]. Skeletal muscles, together with adipocytes, fibroblasts, and endothelial cells that provide the meat its nutritional value, make up the majority of typical edible meat. Myosatellite cells, which serve as the primary adult stem cells for muscle, have emerged as the most promising cell type for the in vitro process of developing muscle tissues [72]. Myosatellites are separated from a biopsy taken from a suitable animal, and they are cultivated in the best culture conditions with a consistent supply of nutrients and growth hormones to encourage the development of multinuclear myotubes. Myotube maturation, ongoing differentiation, and further development by fusing new myoblasts result in the formation of muscle fibers (Fig. 12.6). A scaffold that encourages cell development is a crucial part of tissue engineering. Similar to this, a pliable scaffold that is easy to detach from the completed meat product, enables contraction, and optimizes medium diffusion is required for myoblast development [73]. Alternatively, the scaffolding material needs to be composed of natural ingredients and be eatable. Finding low-cost, large-quantity food-grade culture medium is a big challenge for IVM. Animal-derived sera have long been a standard component of cell culture medium. While using this system, concerns over rules and morality are raised. Alternately, plant-based growth medium might take the

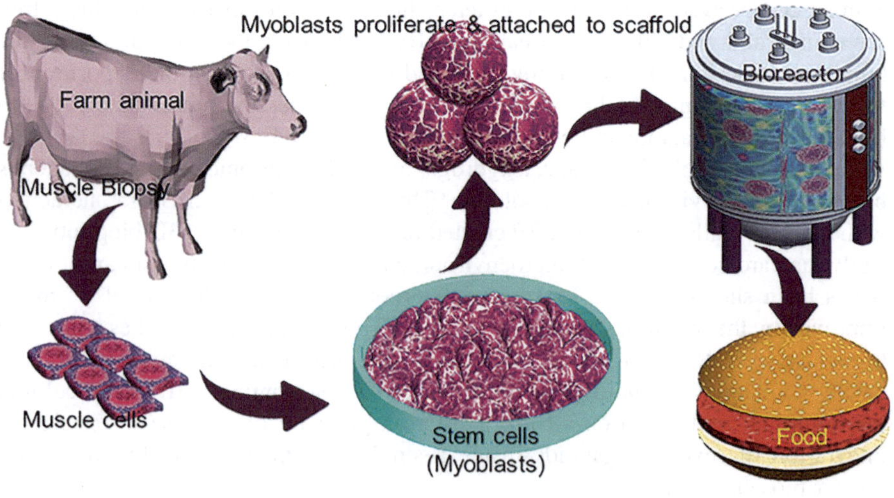

Fig. 12.6 Schematic of the in vitro meat production process [7]

place of the divisive animal-based growth media. For in vitro meat to be accepted by consumers, its nutritional value must be on par with or higher than that of regular meat. It should be emphasized that in vitro meat can be supplemented with necessary elements like vitamins and minerals [74].

The primary technological challenge facing the IVM sector is scaling up the product for commercialization. Since lab-grown beef is currently fairly pricey, its marketability is diminished. The advancement of bioreactor technology, however, has caused prices to decline in recent years, which is optimistic for commercialization. The primary obstacles keeping the IVM technology from scaling up are the high cost of the microcarriers and culture medium as well as the absence of an adequate large-scale bioreactor for mass manufacturing. Finally, before the general public embraces IVM, regulatory standards and laws that reassure customers and reduce mistrust among start-ups functioning in the field are required.

12.9 Limitations and Challenges of 3D Bioprinting Technology

The ultimate aim of 3D bioprinting is to develop a method that can generate intricate, functioning organs in 3D, which may subsequently be utilized as a source for tissue grafts, organ transplants, and drug testing models as an alternative for using animals. This technology is still in its early stages, but it is progressing swiftly because of the extensive research being done in areas like printing engineering, tissue engineering, and cell sciences. The scaling up of bioprinting structures to functional tissues is still hampered by a number of significant barriers, despite the significant developments and countless discoveries. The ultimate aim of 3D bioprinting is to develop a method that can generate intricate, functioning organs in 3D, which may subsequently be utilized as a source for tissue grafts, organ transplants, and drug testing models as an alternative for using animals. This technology is still in its early stages, but it is progressing swiftly because of the extensive research being done in areas like printing engineering, tissue engineering, and cell sciences. The scaling up of bioprinting structures to functional tissues is still hampered by a number of significant barriers, despite the significant developments and countless discoveries [75, 76].

The current limitation of 3D printing resolution, which is 20 m, whereas the blood capillary can be as small as 3 m, makes it difficult to fabricate blood capillaries. To generate vascularized human tissue, several potential techniques are being used. For instance, adding angiogenic growth factors to bioinks to encourage the formation of blood vessels after printing. Benmeridja et al. suggested using micro-vascularized units as building blocks in conjunction with 3D bioprinting [48]. In this study, human umbilical vein endothelial cells (HUVEC) and adipose-derived stem cells (ADS) were cocultured to produce dense, viable adipose tissue spheroids with a capillary-like network. Using a microfluidic device, vasculogenesis was promoted in a new way, although the hydrogels used to do so do not support cell–cell interactions and have an effect on the stability of phenotypic traits. It is still not viable to create functioning vasculature in time to sustain the bioprinted tissue because of the vascular network's complexity and small size.

Biomaterials are essential in 3D bioprinting because they support the structural and functional properties of the printed tissue and preserve structural integrity and biocompatibility during tissue printing and maturation. The printed materials now available on the market, however, cannot completely replicate the original ECM compositions that support the cellular structure. It is crucial to develop novel printable biomaterials that can be handled mechanically by cells and can be produced with living cells. Finding enough cells is a challenging undertaking since tissue printing requires a large number of cells. Since bioprinting would affect stem cell development at various stages of the process, a stem cell source would be the most promising choice. Another limitation at the moment is the 3D bioprinting industry's low throughput and high cost. All current approaches need manual cell seeding and bioink loading, which is essential for high-throughput manufacture of 3D objects. 3D organoids will probably be used in future large-scale drug discovery technologies. To create organoids, however, microscopic tissue culture plates are still employed.

12.10 Conclusion and Outlook for the Future

Enormous of developments in tissue engineering, it is now possible to regenerate de novo tissue or organs in vitro for the first time in the history of medicine. Despite many challenges, the successful demonstration of printed tissue architectures over the past 10 years points to a very exciting and promising technique that deserves additional study in a number of medical and industrial application areas. As tissue engineering is multidisciplinary, systematic research on bioink optimization, bioreactor engineering, and cell culture environment by engineers, scientists, and doctors is crucial to enabling high-throughput production that is connected to efficient screening tests. It is vital to develop printing systems that can swiftly print hybrid materials (bioinks) with sustained biocompatibility and repeatability since the structure of live tissue and organs is so intricate. Combining bioprinting technologies with other enabling techniques such as 3D cell culture, bioreactor technology, microfluidics, and organ-on-a-chip may be able to achieve this. With the help of this technology, now the vast gap between the lab and the factory can someday be reduced, allowing for the fulfillment of clinical and industrial demands as well as the advancement of regenerative medicine and enhanced drug development [77]. Advanced bio-fabrication technologies underpin this technology. 3D bioprinting might signal a paradigm shift for the twenty-first century in a number of biological sectors. To accelerate the development of this technology and realize this ambition, the scientific and technical communities must effectively collaborate to exchange information.

References

1. Singh S, Choudhury D, Yu F, Mironov V, Naing MW (2020) In situ bioprinting—bioprinting from benchside to bedside? Acta Biomater 101:14–25
2. Vyas D, Udyawar D (2019) A review on current state of art of bioprinting. In: Kumar L, Pandey P, Wimpenny D (eds) 3D printing and additive manufacturing technologies. Springer, Singapore, pp 195–201. https://doi.org/10.1007/978-981-13-0305-0_17
3. Yilmaz B, Tahmasebifar A, Baran ET (2019) Bioprinting technologies in tissue engineering. In: Advances in biochemical engineering/biotechnology. Springer, pp 279–319
4. Cho DW, Kim BS, Jang J, Gao G, Han W, Singh N (2019) 3D bioprinting techniques. In: 3D bioprinting. Springer, Cham, pp 25–29. https://doi.org/10.1007/978-3-030-32222-9_4
5. Kala D, Sharma TK, Gupta S, Verma V, Thakur A, Kaushal A, Trukhanov AV, Trukhanov SV (2021) Graphene oxide nanoparticles modified paper electrode as a biosensing platform for detection of the htrA gene of O. tsutsugamushi. Sensors 21:4366
6. Kačarević ŽP, Rider P, Alkildani S, Retnasingh S, Smeets R, Jung O, Ivanišević Z, Barbeck M (2018) An introduction to 3D bioprinting: possibilities, challenges and future aspects. Materials 11(11):2199
7. Ramadan Q, Zourob M (2021) 3D bioprinting at the frontier of regenerative medicine, pharmaceutical, and food industries. Frontiers 2:607648
8. Singh A, Wan F, Yadav K, Salvi A, Thakur P, Thakur A (2023) Synergistic effect of ZnO nanoparticles with Cu2+ doping on antibacterial and photocatalytic activity. Inorg Chem Commun 157:11425
9. Thakur A, Verma R, Wan F, Ravelo B, Edelman I, Ovchinnikov S, Thakur P (2023) Investigation of structural, elastic and magnetic properties of Cu2+ ions substituted cobalt nanoferrite. J Magn Magn Mater 581:170980
10. Dababneh A, Ozbolat IT (2014) Bioprinting technology: A current state-of-the-art review. Pubmed 136(6):061016
11. Leberfinger AN, Dinda S, Wu YC, Koduru SV, Ozbolat V, Ravnic DJ, Ozbolat IT (2019) Bioprinting functional tissues. Acta Biomater 95:32–49
12. Malda J, Visser J, Melchels FP, Groll J, Hennink WE, Dhert WJ, Hutmacher DW (2013) 25th anniversary article: engineering hydrogels for biofabrication. Adv Mater 25(36):5011–5028
13. Ozbolat IT, Yu Y (2013) Bioprinting toward organ fabrication: challenges and future trends. IEEE Trans Biomed Eng 60(3):691–699
14. Yilmaz B, Tahmasebifar A, Baran ET (2019) Bioprinting technologies in tissue engineering. In: Silva A, Moreira J, Lobo J, Almeida H (eds) Current applications of pharmaceutical biotechnology. Springer, Cham, pp 279–319
15. Xu T, Zhao W, Zhu J, Albanna MZ, Yoo JJ, Atala A (2013) Complex heterogeneous tissue constructs containing multiple cell types prepared by inkjet printing technology. Biomaterials 34(1):130–139
16. Singh M, Haverinen HM, Dhagat P, Jabbour GE (2010) Inkjet printing-process and its applications. Adv Mater 22(6):673–685
17. Li X, Chen J, Liu B, Wang X, Ren D, Xu T (2018) 3D printing and biofabrication. Springer eBooks, pp 283–301
18. Verma R, Thakur P, Sun AC, Thakur A (2023) Investigation of structural, microstructural and electrical characteristics of hydrothermally synthesized Li0.5-0.5xCoxFe2.5-0.5xO4 ferrite nanoparticles. Phys B: Cond Mater 661:414926
19. Kala D, Sharma TK, Gupta S, Nagraik R, Verma V, Thakur A, Kaushal A (2020) AuNPs/CNF-modified DNA biosensor for early and quick detection of O. tsutsugamushi in patients suffering from scrub typhus. 3 Biotech 10:446
20. Guillotin B, Souquet A, Catros S, Surdo S, Pippenger BE, Bellance S, Bareille R, Rémy M, Bordenave L, Amédée J, Guillemot F (2010) Laser assisted bioprinting of engineered tissue with high cell density and microscale organization. Biomaterials 31(28):7250–7256

21. Shim J, Lee J, Kim JM, Cho D (2012) Bioprinting of a mechanically enhanced three-dimensional dual cell-laden construct for osteochondral tissue engineering using a multi-head tissue/organ building system. J Micromech Microeng 22(8):085014
22. Bhise NS, Manoharan V, Massa S, Tamayol A, Ghaderi M, Miscuglio M, Lang Q, Zhang Y, Shin SR, Calzone G, Annabi N, Shupe T, Bishop CE, Atala A, Dokmeci MR, Khademhosseini A (2016) A liver-on-a-chip platform with bioprinted hepatic spheroids. Biofabrication 8(1): 014101
23. Boland T, Xu T, Damon B, Cui X (2006) Application of inkjet printing to tissue engineering. Biotechnol J 1(9):910–917
24. Derby B (2008) Bioprinting: inkjet printing proteins and hybrid cell-containing materials and structures. J Mater Chem 18(47):5717
25. Schiele NR, Corr DT, Huang Y, Raof NA, Xie Y, Chrisey DB (2010) Laser-based direct-write techniques for cell printing. Biofabrication 2(3):032001
26. Chan VWS, Zorlutuna P, Jeong JM, Kong H, Bashir R (2010) Three-dimensional photopatterning of hydrogels using stereolithography for long-term cell encapsulation. Lab Chip 10(16):2062
27. Elomaa L, Pan C, Shanjani Y, Malkovskiy AV, Seppälä J, Yang Y (2015) Three-dimensional fabrication of cell-laden biodegradable poly (ethylene glycol-co-depsipeptide) hydrogels by visible light stereolithography. J Mater Chem B 3(42):8348–8358
28. Jamee R, Araf Y, Naser IB, Promon SK (2021) The promising rise of bioprinting in revolutionalizing medical science: advances and possibilities. Regenerat Therapy 18:133–145
29. Demirci U, Montesano G (2007) Single cell epitaxy by acoustic picolitre droplets. Lab Chip 7(9):1139
30. Chang RPH, Nam J, Sun W (2008) Effects of dispensing pressure and nozzle diameter on cell survival from solid freeform fabrication–based direct cell writing. Tissue Eng 14(1):41–48
31. Shakeel A, Singh A, Das S, Suhag D, Sharma AK, Rajput SK, Mukherjee M (2017) Synthesis and morphological insight of new biocompatible smart hydrogels. J Polym Res 24:1–10
32. Mironov V, Visconti RP, Kasyanov V, Forgacs G, Drake CL, Markwald RR (2009) Organ printing: tissue spheroids as building blocks. Biomaterials 30(12):2164–2174
33. Kim JM, Kong JS, Han W, Kim BY, Cho D (2020) 3D cell printing of tissue/organ-mimicking constructs for therapeutic and drug testing applications. Int J Mol Sci 21(20):7757
34. Prowse AB, Chong F, Gray PH, Munro TP (2011) Stem cell integrins: implications for ex-vivo culture and cellular therapies. Stem Cell Res 6(1):1–12
35. Murphy SD, Skardal A, Atala A (2013) Evaluation of hydrogels for bio-printing applications. J Biomed Mater Res A 101A(1):272–284
36. Kaushik S, Gandhi S, Chauhan M, Ma S, Das S, Ghosh D, Chandrasekharan A, Alam B, Parmar AS, Sharma A, Santhoshkumar T, Suhag D (2020) Water-templated, polysaccharide-rich bioartificial 3D microarchitectures as extra-cellular matrix bioautomatons. ACS Appl Mater Interfaces 12(18):20912–20921
37. Sears NA, Seshadri DR, Dhavalikar P, Cosgriff-Hernandez E (2016) A review of three-dimensional printing in tissue engineering. Tissue Eng Part B Rev 22(4):298–310
38. Highley CB, Rodell CB, Burdick JA (2015) Direct 3D printing of shear-thinning hydrogels into self-healing hydrogels. Adv Mater 27(34):5075–5079
39. Chahar D, Thakur P, Sun AC, Thakur A (2023) Investigation of structural, electrical and magnetic properties of nickel substituted Co-Zn nanoferrites. J Mater Sci Mater Electron 34:901
40. Kang HW, Lee JS, Ko KI, Kengla C, Yoo JJ, Atala A (2016) A 3D bioprinting system to produce human-tissue constructs with structural integrity. Nat Biotechnol 2016(34):313–322
41. Carrow JK, Kerativitayanan P, Jaiswal M, Lokhande G, Gaharwar AK (2015) Polymers for bioprinting. Elsevier eBooks, pp 229–248
42. Engler AJ, Sen S, Sweeney HL, Discher DE (2006) Matrix elasticity directs stem cell lineage specification. Cell 126(4):677–689
43. Chen YY, Xiong X, Liu X, Cui R, Wang C, Zhao G, Zhi W, Lu M, Duan K, Weng J, Qu S, Ge J (2020) 3D bioprinting of shear-thinning hybrid bioinks with excellent bioactivity derived from

gellan/alginate and thixotropic magnesium phosphate-based gels. J Mater Chem B 8(25): 5500–5514
44. Morgan FJ, Moroni L, Baker MJ (2020) Dynamic bioinks to advance bioprinting. Adv Healthc Mater 9(15):1901798
45. Cofino C, Pérez-Amodio S, Semino CE, Engel E, Mateos-Timoneda MA (2019) Development of a self-assembled peptide/methylcellulose-based bioink for 3D bioprinting. Macromol Mater Eng 304(11):1900353
46. Verma R, Thakur P, Chauhan A, Jasrotia R, Thakur A (2023) A review on MXene and its composite for electromagnetic interference (EMI) shielding applications. Carbon 208:170–190
47. Vignesh U, Mehrotra D, Dichen A, V., & Howlader, D. (2017) Three dimensional reconstruction of late post traumatic orbital wall defects by customized implants using CAD-CAM, 3D stereolithographic models: A case report. J Oral Biol Craniofac Res 7(3):212–218
48. Benmeridja L, De Moor L, De Maere E, Vanlauwe F, Ryx M, Tytgat L, Vercruysse C, Dubruel P, Van Vlierberghe S, Blondeel P, Declercq H (2020) High-throughput fabrication of vascularized adipose microtissues for 3D bioprinting. J Tissue Eng Regen Med 14(6): 840–854
49. Zhou G, Jiang H, Yin Z, Liu Y, Zhang Q, Zhang C, Pan B, Zhou J, Zhou X, Sun H, Li D, He A, Zhang Z, Zhang W, Liu W, Cao Y (2018) In vitro regeneration of patient-specific ear-shaped cartilage and its first clinical application for auricular reconstruction. EBioMedicine 28:287–302
50. Baltazar T, Mulligan DC, Catarino CM, Xie CB, Kirkiles-Smith NC, Lee VS, Hotta SYK, Dai G, Xu X, Ferreira FC, Saltzman WM, Pober JS, Karande P (2020) Three dimensional bioprinting of a vascularized and Perfusable skin graft using human keratinocytes, fibroblasts, pericytes, and endothelial cells. Tissue Eng A 26(5–6):227–238
51. Norona LM, Nguyen DG, Gerber DE, Presnell SC, LeCluyse EL (2016) Editor's highlight: modeling compound-induced fibrogenesis in vitro using three-dimensional bioprinted human liver tissues. Toxicol Sci 154(2):354–367
52. Norona LM, Nguyen DG, Gerber DE, Presnell SC, Mosedale M, Watkins PB (2019) Bioprinted liver provides early insight into the role of Kupffer cells in TGF-β1 and methotrexate-induced fibrogenesis. PLoS One 14(1):e0208958
53. Bowser DA, Moore M (2019) Biofabrication of neural microphysiological systems using magnetic spheroid bioprinting. Biofabrication 12(1):015002
54. Zhuang P, Ng WL, An J, Chua CK, Tan LP (2019) Layer-by-layer ultraviolet assisted extrusion-based (UAE) bioprinting of hydrogel constructs with high aspect ratio for soft tissue engineering applications. PLoS One 14(6):e0216776
55. Zhou X, Zhu W, Nowicki M, Miao S, Cui H, Holmes B et al (2016) 3D bioprinting a cell-laden bone matrix for breast cancer metastasis study. ACS Appl Mater Interfaces 8(44):30017–30026
56. Heinrich MA, Bansal R, Lammers T, Zhang YS, Schiffelers RM, Prakash J (2019) 3D-Bioprinted mini-brain: a glioblastoma model to study cellular interactions and therapeutics. Adv Mater 31(14):e1806590
57. Yadav P, Bhaduri A, Thakur A (2023) Manganese oxide nanoparticles: an insight into structure, synthesis and applications. ChemBioEng 10(4):510–528
58. Lee VS, Dai G, Zou H, Yoo S (2015) Generation of 3-D glioblastoma-vascular niche using 3-D bioprinting
59. Suarez-Martinez AD, Sole-Gras M, Dykes SS, Wakefield ZR, Bauer KT, Lampejo A, Siemann DW, Huang Y, Murfee WL (2020) A novel tumor microenvironment model that combines bioprinting and tissue culture to investigate cancer cell and microvascular interactions. FASEB J 34(S1):1
60. Hynes WF, Pepona M, Robertson C, Alvarado J, Dubbin K, Triplett M et al (2020) Examining metastatic behavior within 3D bioprinted vasculature for the validation of a 3D computational flow model. Sci Adv 6(35):eabb3308
61. Sun L, Yang H, Wang Y, Zhang X, Jin B, Xie F et al (2020) Application of a 3D bioprinted hepatocellular carcinoma cell model in antitumor drug research. Front Oncol 10:878

62. Lee VS, Dai G (2017) Printing of three-dimensional tissue analogs for regenerative medicine. Ann Biomed Eng 45(1):115–131
63. Albanna M, Binder KW, Murphy SD, Kim J, Qasem S, Zhao W, Tan J, El-Amin IB, Dice D, Marco J, Green JL, Xu T, Skardal A, Holmes JF, Jackson JE, Atala A, Yoo JJ (2019) In situ bioprinting of autologous skin cells accelerates wound healing of extensive excisional full-thickness wounds. Sci Rep 9(1):1856
64. Zhao W, Xu T (2020) Preliminary engineering for in situ in vivo bioprinting: a novel micro bioprinting platform for in situ in vivo bioprinting at a gastric wound site. Biofabrication 12(4): 045020
65. Petrov D, Edelman I, Thakur A, Thakur P, Sukhachev A, Ovchinnikov S (2023) Correlation between magnetic and electric properties in the series of CoxZn1-xFe2O4 nanoparticles. JETP Lett 117:765–768
66. Costantini M, Testa S, Mozetic P, Barbetta A, Fuoco C, Fornetti E, Tamiro F, Bernardini S, Jaroszewicz J, Swieszkowski W, Trombetta M, Castagnoli L, Seliktar D, Garstecki P, Cesareni G, Cannata S, Rainer A, Gargioli C (2017) Microfluidic-enhanced 3D bioprinting of aligned myoblast-laden hydrogels leads to functionally organized myofibers in vitro and in vivo. Biomaterials 131:98–110
67. Abelseth E, Abelseth L, De La Vega LEF, Beyer S, Wadsworth SC, Willerth SM (2019) 3D printing of neural tissues derived from human induced pluripotent stem cells using a fibrin-based bioink. ACS Biomater Sci Eng 5(1):234–243
68. Carvalho V, Gonçalves IC, Lage TCC, Rodrigues RO, Minas G, Teixeira SR, Moita AS, Hori T, Kaji H, Lima R (2021) 3D printing techniques and their applications to organ-on-a-Chip platforms: A systematic review. Sensors 21(9):3304
69. Maloney EK, Clark CT, Sivakumar H, Yoo K, Aleman J, Rajan SAP, Forsythe S, Mazzocchi A, Laxton AW, Tatter SB, Strowd RE, Votanopoulos KI, Skardal A (2020) Immersion bioprinting of tumor organoids in multi-well plates for increasing chemotherapy screening throughput. Micromachines 11(2):208
70. Reid JL, Palmer X, Mollica PA, Northam N, Sachs PC, Bruno R (2019) A 3D bioprinter platform for mechanistic analysis of tumoroids and chimeric mammary organoids. Sci Rep 9(1): 7466
71. Bhat ZF, Kumar S, Fayaz H (2015) In vitro meat production: challenges and benefits over conventional meat production. J Integr Agric 14(2):241–248
72. Kadim IT, Mahgoub O, Baqir S, Faye B, Purchas RW (2015) Cultured meat from muscle stem cells: a review of challenges and prospects. J Integr Agric 14(2):222–233
73. Engler AJ, Griffin MA, Sen S, Bönnemann CG, Sweeney HL, Discher DE (2004) Myotubes differentiate optimally on substrates with tissue-like stiffness. J Cell Biol 166(6):877–887
74. Young JF, Therkildsen M, Ekstrand B, Che BN, Larsen MV, Oksbjerg N, Stagsted J (2013) Novel aspects of health promoting compounds in meat. Meat Sci 95(4):904–911
75. Agarwal S, Saha S, Balla VK, Pal A, Barui A, Bodhak S (2020) Current developments in 3D bioprinting for tissue and organ regeneration–a review. Front Mech Eng 6:90
76. Taneja S, Thakur P, Kumar D, Slimani Y, Thakur A (2023) Comparison of RE doping (Nd and Sm) on structural and magnetic properties of nickel-zinc-bismuth nanoferrites. J Rare Earths
77. Thakur P, Thakur A (2022) Nanomaterials, their types and properties. In: Thakur A, Thakur P, Khurana SP (eds) Synthesis and applications of nanoparticles. Springer, Singapore

Nanomaterials for Precision Medicine

13

Abhigyan Satyam and Deepa Suhag

Abstract

Precision medicine is characterized as the personalization of illness management with an emphasis on the genetic, environmental, and lifestyle characteristics of specific individuals, including customized medical choices and interventions. This approach involves a number of steps, including the collection and analysis of a large amount of data, selection of the best and most specific medication dose for each patient, and development of focused and reliable analytical tools for the tracking of clinical, genetic, and environmental variables. Even at the molecular level, very accurate diagnosis and therapy are now possible because of recent advancements in medical research. As a result, a brand-new subspecialty of "medicine" called "precision" has emerged, with unique clinical ramifications and challenges that can be effectively addressed by recently discovered nanomaterials. The advancement is demonstrated by the omics-based biomarker discovery and the pharmaco-omics-based medication development programs. In the realm of oncology, precision medicine has proven most beneficial in the formulation of diverse treatment protocols. The COVID-19 pandemic has demonstrated the best application of precision medicine through the use of numerous biomarkers, including IL-6 and c-reactive protein, in the assessment of illness severity, the development of various therapeutics, and the evaluation of vaccination efficacy. Additionally, the care of infectious diseases, chronic

A. Satyam (✉)
Division of Rheumatology and Clinical Immunology, Beth Israel Deaconess Medical Center, Harvard Medical School, Boston, MA, USA
e-mail: asatyam@bidmc.harvard.edu

D. Suhag
Amity Centre for Nanotechnology, Amity University Haryana, Gurgaon, Haryana, India
e-mail: dsuhag@ggn.amity.edu

D. Suhag et al. (eds.), *Integrated Nanomaterials and their Applications*,
https://doi.org/10.1007/978-981-99-6105-4_13

illnesses like asthma, connective tissue diseases, cardiovascular diseases, diabetes, and obesity is finding a role for precision medicine. There are still several obstacles standing in the way of precision medicine's future, including the associated costs, ethical considerations, the security of Big Data, the fusion of many platforms to combine data, and the lack of qualified personnel to manage the data and algorithms.

Keywords

Precision medicine · Accurate diagnosis · Vaccination efficacy · Modern medicine

13.1 Introduction

Precision medicine aims to reduce individual variability caused by factors like genetic make-up, environment, and lifestyle, offering personalized medical care to everyone. [1, 2].

In reality, it has been proven that each person has a predisposition to certain diseases, which corresponds to a different reaction to pharmacological treatments and a different set of undesirable effects. For this reason, interdisciplinary methods are necessary to achieve significant strides in the field of precision medicine. A few of the fields involved include massive data management techniques, new equipment (such as new high-performance techniques, instrument components, and materials), nano-sensors, biosensors, smart materials, novel drugs, and delivery systems [3, 4]. The majority of the original efforts were focused on assembling a sizeable data set regarding the clinical history of a particular patient, including an exhaustive list of all biochemical and instrumental exams performed on them throughout their life as well as those linked with their parents. This will enable the gathering of relevant information and links for a certain illness. The present challenge is the development of new technologies that could allow the revolutionary concept of precision medicine to be used regularly. The success of this technique depends on the capacity to obtain data from a single patient in real time with a powerful diagnostic value for an early diagnosis of a specific ailment. The point-of-care approach (POC) effectively addresses the demand for new diagnostic and analytical tools that may be employed anywhere by untrained people [5, 6].

The need for on-site new generation prompt diagnostics has increased because current diagnostic procedures are frequently pricey, time-consuming, and performed in various healthcare facilities, making patients wait an unacceptably long time between the test and the results and restricting early pathology detection and prompt pharmacological treatment and follow-up. The most effective analyzers for commercial point-of-care applications right now include those designed to detect hyperglycemia, pregnancy, hormones for fertility monitoring, viruses (like HIV, OraQuick In-Home HIV Test), malaria, and typhoid [7, 8]. However, in recent years, a lot of pharmaceutical and healthcare businesses have focused on creating diagnostic gadgets in response to consumer demand. In the realm of biosensors for personalized therapy, the development of wearable (bio)analytical equipment offers a tremendous opportunity. The advantages of this approach include non-invasive, continuous monitoring of physiological parameters and biomarkers that may be collected from biofluids like sweat and saliva. Additionally, the creation of highly predictive diagnostic tools for specific disorders is made possible by the capacity to perform a variety of tests on a particular biomarkers panel while concurrently measuring a number of significant analytes. In reality, a clinical diagnosis is based on a set of evidence that, when combined, creates a precise clinical framework. A single analytical instrument has been used in many multiplexing approaches to identify a panel of biomarkers.

With the development of intelligent polymeric materials that are more biocompatible and adjustable and whose features may be tuned to have properties that are particularly specific to individuals, precision medicine may proceed more swiftly [9, 10]. Because of the hydrophilic functional groups on their polymeric backbone, hydrogels have the capacity to absorb excessive quantities of solvent and expand in size, but cross-linked networks retain their stability in the presence of water.

13.2 Precision Medicine/Personalized Medicine

With the least amount of iatrogenic damage and cost load on society, it aims to achieve optimal therapeutic outcomes. The ultimate goal is to provide rational pharmacotherapy, which entails giving the right drug to the right patient at the right time at the right amount while avoiding causing any harm to the patient [11]. The potential for precision medicine has significantly grown as a result of recent advances in genetics, molecular biology, and biochemistry. The COVID-19 pandemic has highlighted the importance of making swift, focused decisions even more, and the development of therapies and a corona virus vaccine has given precision medical practices a strong foundation. The following section titles are only a few of the many uses of precision medicine in both diagnostic and therapeutic fields (Fig. 13.1):

13.2.1 Creation of Novel Biomarkers Based on Omics Signatures

Omics technology has made it easier to find omics-based biomarkers with greater accuracy. The notion of a biomarker may be translated in the context of omics data by examining the associations between the traits of the omics profile or signature and certain medical conditions, and then building computer models utilizing the selected traits. The omics signature can be employed as a potential biomarker for screening, diagnosis, prognosis, drug development, therapeutic targeting, and treatment response [13]. Some molecular markers discovered using omics have successfully moved from the lab to the bedside. The top of the list is the MammaPrint, a 70-gene expression signature diagnostic tool to estimate breast cancer prognosis and guide

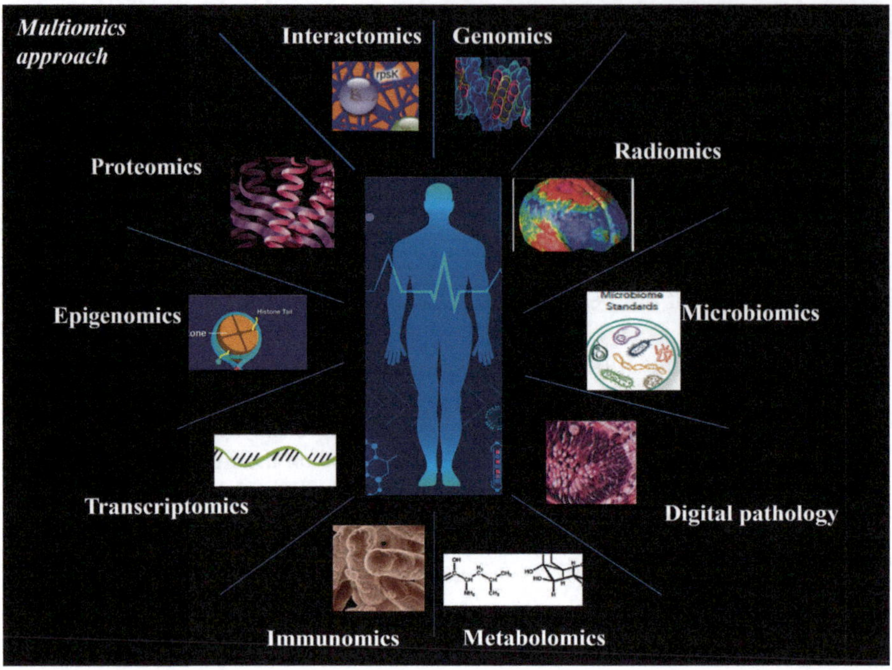

Fig. 13.1 Multi-omics approach with precision medicine [12]

targeted therapy. SEPT9 promoter methylation is a recently discovered epigenetic diagnostic marker for colorectal cancer [14]. The OVA1 protein is used in another effective omics signature-based biomarker that forecasts the likelihood of malignancy in women who arrive with an ovarian adnexal tumor prior to the scheduled operation [15]. Utilizing very sensitive omics methods, the phrase "liquid biopsy" describes the analysis of a tiny subset of biological components in non-invasive samples like blood and urine. Any of the omics may be studied with the same specificity as the corresponding organ tissue, including circulating tumor cells and cell-free DNA. Omics-based precision medicine procedures include cancer treatment at various stages and prenatal screening for inborn or inherited diseases. In contrast, the Clustered Regularly Interspaced Short Palindromic Repeats (CRISPR)-associated protein (CRISPR-Cas) system, a gene editing technology with a promising future in precision medicine, would perform miracles utilizing the data from multi-omics. It would contribute to editing both genes and epigenetic modifications, allowing for the "on and off" switching of genes pharmacological therapies developed using pharmaco-omics.

The emphasis has shifted from general, one-size-fits-all medicines to more specialized, tailored treatments thanks to pharmaco-omics. Pharmacogenetic guided dosage is particularly important in situations where there is a documented drug–gene interaction and in drugs with a constrained therapeutic window. Pharmaco-omics methods can be applied to accelerate clinical trials and identify novel therapeutic targets. Omics data can be used to identify the patient sub-population where the drug would be most effective. This would speed up the approval process for new pharmaceuticals, lower costs and failure rates, more effective treatments, and fewer drug withdrawals after approval. The combination drug BiDilR (hydralazine + isosorbide dinitrate) was used to treat heart failure, but it was ineffective since it only worked for a limited number of patients. The drug underwent more testing in this population, and in 2005 it was only permitted for use by persons with African ancestry [16]. Companion diagnostics, or the diagnostic tests required to choose patients or monitor them throughout therapy, are frequently designed concurrently with the medicine in order to ensure proper drug usage and to speed up the approval process. Crizotinib, a drug used to treat non-small-cell lung cancer, received FDA approval 4.9 months after its New Drug Application (NDA) was submitted in August 2011 [17]. Despite only having 347 participants in the study, it was developed with companion diagnostics connected to Anaplastic Lymphoma Kinase (ALK) testing, and 2013 saw its regular approval.

13.3 Oncology Precision Medicine

The ability of traditional chemotherapy and radiation to effectively treat cancer depends on how quickly cancer cells divide. Hair follicles, blood cells, and digestive tract cells commonly experience a bystander impact as a result, which can have negative effects and be dangerous. Targeted treatment, which is often referred to as precision oncology, employs target identification at extracellular, cell membrane, intracellular, and intranuclear levels to lessen the usual side effects of chemotherapy while maintaining the effect on the cancer cell. Extracellular targets include ligands that bind to receptors on the cell surface and control cellular processes. Vascular endothelial growth factor-A (VEGF-A) is essential for the formation of solid tumors, which necessitates vascularization. Bevacizumab, a monoclonal antibody, binds to soluble VEGF-A and prevents it from communicating with VEGF receptors and neovascularization pathways. Whether used alone or in conjunction with chemotherapy, this drug has been effective in treating a number of malignancies, including colorectal, lung, renal, glioblastoma multiforme, and ovarian tumors [18, 19]. The receptors on the cell surface contribute to the cell's survival by activating signaling pathways farther down. Targeting cancer cells with excessive or abnormal receptor expression with receptor-directed monoclonal antibodies reduces the toxicity of normal cells. The epidermal growth factor-receptor tyrosine kinase (EGF-RTK), commonly known as the ErbB receptor family, is made up of four members: the EGFR, ErbB2, ErbB3, and ErbB4. It is one of the most fully studied receptors for precision medicine.

One of them, overexpression of ErbB2 (Her2/neu), by homodimerizing with other HER receptors and activating the signaling pathways downstream, promotes carcinogenesis. Breast and stomach cancers express her-2/neu excessively in 20–25% of cases. Furthermore, Her-2/neu overexpression has been discovered in a wide range of solid tumors, including malignancies of the esophagus, lung, bladder, uterus, cervix, and head and neck [20, 21]. This overexpression has been the target of monoclonal antibodies like trastuzumab and pertuzumab, which have shown to improve survival in advanced malignancies as well as in the adjuvant context. A monoclonal antibody can also be used to target the Her-2 receptor together with a chemotherapeutic medication that is linked to it. The term "antibody-drug conjugation" (ADC) refers to this method. The antibody and a chemical that targets microtubules are combined as an ADC known as trastuzumab emtansine. When the ADC binds to the Her-2/neu receptor, trastuzumab is released, and the internalization of the microtubule agent causes cell cytotoxicity, there is a double impact. Several strategies, including monoclonal antibodies, ADCs, and intracellular tyrosine kinase inhibitors, are being used to target Her-2/neu. The cornerstone of EGFR-directed treatment for colorectal cancer is the concept of mutant receptor or wild-type receptor. In the case of mutant EGFR receptors (rat sarcoma mutations), the effect of inhibiting surface receptors is balanced by constitutive activation of the downstream signaling cascade. Anti-EGFR drugs (cetuximab, panitumumab) are only given to those with EGFR wild-type status in clinical settings. The intracellular targets for precision treatment include 24 TKIs, as discussed earlier. By phosphorylating the tyrosine residues on the protein substrates, the RTK starts the downstream signaling cascades. Mutations, translocations, or RTK amplifications promote malignant cell invasion, metastasis, and aberrant signaling. TKIs that aim to suppress RTK have been used to treat a range of cancers depending on the type and function of RTK. The second technique for precision treatment that makes use of the host immune system (ICIs) is the use of immune checkpoint inhibitors. Thanks to the pioneering work of James Allison and Tasuko Honjo, immunotherapy has been acknowledged as the fourth modality of treatment, joining surgery, chemotherapy, and radiation therapy. All because of genetic and epigenetic changes, neoantigens are expressed on the cell surface of cancer cells. In response, the body produces T-cells as a defense mechanism. To stop autoimmune disease and an inflammatory state, the body has defense mechanisms called immunological checkpoints which regulates the T-cell response. The cancer cell successfully evades the host immune system by expressing the ligands of these immunological checkpoints. In other cases, ICIs directs the body's T-cell defense toward the tumor. Even though there are currently no known biomarkers for enhancing the ICI in clinical usage, the expression of the programmed death ligand (PD-L1), a high tumor mutational load, and a deficiency in mismatch repair genes are usually utilized to begin ICI therapy. Since its indications are not tumor-specific, ICI has been approved for treatment in a wide range of solid tumors [22, 23].

Over time, breakthroughs in molecular knowledge of the pathophysiology of cancer led to the discovery of several targeted medications that are now used in the clinical practice of precision medicine in oncology.

13.4 Chronic Disease Precision Medicine

Chronic diseases attract a variety of pharmacologic therapies due to their increased frequency and frequent consequences, making them an ideal substrate on which to implement a precision medicine approach. The multi-omics repertoire is constantly expanding, covering everything from illness risk prediction to tailored therapies. Patients stand to gain from early, accurate diagnosis, treatment, and preventive measures as we better grasp the clinical consequences of genes and biomarkers.

13.5 Asthma and Personalized Medication

The CD4 T helper cell subsets comprise the Th1 and Th2 populations are well established. Th2 cell activity (mucus secretion and IgE synthesis) mediates the generation of IL-4 (IgE production), IL-5 (eosinophil activation), and IL-13, which are all connected to asthma. This defines a "type 2 (T2) high" endotype together with the recent discovery that type 2 innate lymphoid cells are potent IL-5 and IL-13 producers. The endotype designated as "type-2 (T2) low" is characterized by the secretion of IL-1 and IL-17. Each endotype is thus defined by a distinct biological mechanism linking clinical features to a molecular route [24, 25]. It is said that these two dissimilar groups have asthma syndrome. Precision medicine has made it feasible to utilize monoclonal antibodies as a customized kind of treatment for refractory type 2 high asthmatics. Benralizumab, mepolizumab, and reslizumab all inhibit the IL-5 receptor, whereas dupilumab, an IL-4RA antagonist, inhibits the IL-4 and IL-13 signals. Omalizumab binds to the IgE in the blood. Chemokine receptor (CXCR2) antagonists and dual CXCR1/CXCR2 antagonists are potential therapies for type 2 moderate or "neutrophilic" asthma.

13.6 Medical Precision and Diabetes

Monogenic diabetes mellitus, with its several subtypes, offers a chance for precision medicine, which bases treatment choices on identified gene abnormalities. A moderate, non-progressive form of diabetes with high fasting glucose that responds to diet and exercise is known as maturity-onset diabetes of the young type 2 (MODY-2). Hepatic nuclear factor 1 (HNF-1) and HNF-4 genes are responsible for diabetes that develops over time and responds to low dosage sulphonylureas in MODY-3 and MODY-1, respectively [26]. Precision medicine and systemic autoimmune connective tissue diseases A broad variety of symptoms, many of which overlap, a varied course, and progressive remissions and relapses are the hallmarks of AICTDs, or systemic autoimmune connective tissue diseases. There has recently been a drive to categorize them into subgroups based on the quantifiable chemical fingerprints that genetic technology can measure, such as "Inflammatory," "Lymphoid," "Interferon," and "Undefined Normal like." Additionally, it would enable the targeting of a certain subset of patients for therapeutic interventions including on-the-shelf or experimental therapies. By analyzing gene expression, methylation, and genetic data, it is possible to create a molecular taxonomy of AICTDs, which might represent a new paradigm in precision medicine-based therapy [27].

13.7 Precision Medicine in Emergency Care

Decisions in emergency care settings must be made rapidly and with limited margin for mistake, in contrast to those with chronic conditions. Overriding the need for study is the requirement for action. Both the seasoned doctor and the inexperienced clinician are put to the test in this demanding environment. In such a high-stakes scenario, precision medicine technologies linked with clinical medicine would be invaluable. A biomarker score could be used to prioritize a patient with dyspnea (e.g., Manchester Triage Score). The following diagnostic biomarker panel would be used to distinguish causes based on a "urgent" score in conjunction with clinical, epidemiologic, and sociodemographic parameters:

1. Decisions in emergency care settings must be made rapidly and with limited margin for mistake, in contrast to those with chronic conditions.
2. D-Dimer: Removing the need for unnecessary imaging with a result of less than 500 g/dL would rule out venous thromboembolism as the cause.
3. Elevated levels of the brain natriuretic peptide could signal heart failure and need the use of diuretics and a 2D ECHO.
4. A universal indicator of inflammation, C-Reactive protein (CRP) levels can forecast how an infection or inflammation will respond to therapy.
5. Procalcitonin: Elevated concentrations would trigger microbiological cultures and direct antibiotic therapy. Values are unaffected by concurrent steroid use, unlike CRP.
6. In the setting of community-acquired pneumonia and other respiratory tract infections, pro-adrenomedullin is a more recent biomarker with prognostic relevance that is viewed as superior to that of CRP and procalcitonin (PCT).

Additionally, other necessary care settings can gain from this. Vasculopathic alterations linked to the APOE 2 genotype that affect the severity and clinical trajectory of intracerebral hemorrhage have been identified by GWAS. This information could help with treatment planning and talks about point-of-care objectives. Three SNPs that have been linked to recanalization in stroke patients treated with r-tPA have been discovered through genomic investigations in these individuals and could be used to inform therapy choices. Decisions about coronary imaging and transplant alternatives are aided by the identification of miRNAs that are differentially altered in individuals with ischemic and non-ischemic cardiomyopathy by transcriptomics in heart failure [28]. The ApoE 4 homozygous genotype, toll-like receptor-7 loss-of-function mutations, and gene changes on chromosome 3 and 9 are also linked to severe COVID-19. As discussed, CRP, IL-6, and procalcitonin play a part in predicting poor outcomes in COVID-19. The presence of biomarkers such as

MCP-1, CRP, procalcitonin, suPAR, and NGAL facilitates distinguishing sepsis from SIRS, forecasting severe sepsis and septic shock, and predicting death in sepsis.

Several other encouraging signs led to the concept of emergency precision medicine, which would, at its core, include a functional electronic health record system, receiving inputs concerning date of prior visits, baseline patient data, research updates, and primary care screening data. When paired with acute Omics (rapid turnaround testing and targeted testing), this would aid in improving complicated decision making in an urgent situation. The patient gains from obtaining better medical care since the doctor can make judgments and execute treatments more effectively with the use of information from a variety of sources, including scores, clinical symptoms, and biomarker levels. Precision medicine is used in emergency conditions using a procedure known as systematic molecular phenotyping approach.

13.8 Infectious Illness with Precision Medicine

A device called the Medical Tricorder has a second handheld scanner that could transmit data about a person's vital signs to the tricorder itself. All vital organ functions could be examined, and any potentially dangerous pathogens may be found. Also, its data bank held information on non-human races, made possible to cure other life-forms. It was a success for Dr. McCoy's ship in the science fiction television series Star Trek. Metagenomics—the study of genomes arising from environmental samples—is already being employed to reach this far-fetched future aim. During metagenomic next-generation sequencing (mNGS), all of the nucleic acids in a sample are run, which may contain diverse populations of microbes. Reference genomes are used to establish which bacteria are present and in what amounts. Sequencing reads are frequently compared to the human genome before digital subtraction to eliminate sequences by removing sequences originating from the host origin. The remaining readings are then mapped to sequence databases in order to detect infections. When culture, PCR, serological tests, and oligonucleotide microarrays failed to detect a novel arena virus in a cluster of three transplant recipient deaths in 2008, this approach was used to identify a new arena virus. Metagenomic sequencing is likely to make it easier to detect and track infections in both acute and chronic illnesses, and may even substitute bacterial isolate culture. The discovery of resistant microorganisms is another objective of precision medicine.

13.9 Genetic Disease Screening for Newborns: The Emergency Medical Genome

Rapid genome sequencing (STATseq), a groundbreaking use of medical genomics, provides a technique for combining genomic research with symptom and sign data to pinpoint genetic disorders. It permits a 2-day genome assessment (24-h whole-genome sequencing +18-h bioinformatics analysis) of severely ill newborns suspected of having genetic disorders. This technology allows for a rapid diagnosis and application of treatment, if any is available, hence avoiding unnecessary intensive care in circumstances when no treatment is available and instead offering genetic counseling [29]. This tool is also essential for research and the production of management advice in a quickly growing sector. Other specialty have followed suit, pursuing direct-to-consumer testing for a variety of diseases.

13.10 Wearable Biosensors

By enabling the assessment of clinically relevant data while continuously monitoring a patient's health, wearable and implantable technology will play a significant part in the advancement of precision medicine. Wearable biosensors offer a large potential to give continuous, physiological information through dynamic, non-invasive monitoring of biomarkers (such as metabolites, hormones, and microorganisms) in biofluids such as tears, saliva, urine, sweat, and interstitial fluids. Recent applications, in particular, have concentrated on the utilization of flexible and novel materials, sophisticated microfluidic technologies, and electrochemical and optical biosensors. Numerous wearable biosensor instances are mentioned in the literature, and there are more publications about this application field than ever before. One of the earliest instances was given in 2012 by researchers, who developed a graphene-based biosensor for wireless bacteria detection on tooth enamel. Another study published which demonstrated how an instrumented mouthguard could non-invasively monitor salivary uric acid levels. On a mouthguard platform, anatomically tiny instrumentation electronics including a potentiostat, microcontroller, and Bluetooth low energy transmitter were integrated with an enzyme-modified screen-printed electrode system. The mouthguard biosensor system provides exceptional sensitivity, stability, and selectivity for uric acid detection in human saliva and spans the healthy and hyperuricemia concentration ranges.

Recognizing the potential of this research field, the Google[x] team in Mountain View, California, developed a "smart" contact lens that might 1 day provide non-invasive blood glucose monitoring for diabetics and correct presbyopia patients' eyesight. The device comprises of a lens made of standard lens hydrogel material with a tiny wireless chip, a scaled-down glucose sensor, and a tiny battery buried between two layers of the lens perimeter, away from the iris and pupil. Tear fluid can enter the sensor via a pinhole in the lens, resulting in blood glucose readings that can be communicated to a smartphone and, perhaps, directly to a doctor [30]. Sweat is a biofluid that has piqued the curiosity of many researchers. Given the dispersion of sweat glands across the body, this matrix is conveniently accessible for chemical sensing applications. The first biosensors for tracking biomarkers in sweat were skin-applied patches that could detect a variety of target analytes. Koh et al., for example, developed a technique for successfully and rapidly collecting sweat from the skin's surface pores [31]. The device feeds this sweat to multiple channels and reservoirs for multiparametric sensing of markers of interest, with possibilities for wireless connections to other devices for image capture and analysis. Because of the combination of biocompatible adhesives, soft device mechanics, including flexible and elastic properties, and waterproof interfaces, the devices may be mounted in a variety of locations on the body without causing chemical or physical discomfort. These devices use colorimetric detection and wireless data transfer to quantify total sweat loss, pH, lactate, chloride, and glucose concentrations. Gao et al. developed a mechanically flexible and fully integrated sensor array for multiplexed in situ perspiration analysis that simultaneously and selectively monitors skin temperature, electrolytes (such as sodium and potassium ions), and sweat metabolites (such as glucose and lactate) (to calibrate the sensors' response) [32]. This innovation combined silicon integrated circuits condensed on a flexible circuit board with plastic-based skin-interface sensors for complex signal processing. Using wearable technology, a thorough sweat profile of human subjects engaged in prolonged indoor and outdoor physical activity was measured, as well as a real-time physiological evaluation of the individuals. Another way for evaluating biomarkers in sweat is temporary tattoos. The sensor is the first of its kind, combining interstitial glucose reverse iontophoretic extraction with an enzyme-based amperometric biosensor. Humans are utilized as test subjects for the iontophoretic biosensing tattoo platform, and blood sugar changes induced by meal ingestion are detected.

13.11 Cancer Precision Medicine

13.11.1 Next-Generation Sequencing

The phrase "next-generation sequencing," or NGS, encompasses genomic, proteomic, epigenetic, and targeted sequencing. The ability to affordably obtain the full cancer genome is really beneficial. Correlations between genomic sequencing (RNA sequence and transcriptional level, genetic mutation, DNA methylation, protein function and level), clinical investigation (biological behavior, drug sensitivity/resistance, end point), and clinical investigation will allow for the identification of potential biomarkers for cancer precision therapy. Biomarker validation and patient categorization are critical components of precision medicine for generating clinical treatment strategies. The "Omics Era" has already begun with NGS, due to small labs capable of examining hundreds of molecules at the same time. The Cancer Genome Atlas and the Cancer Genome Project have been constructed using NGS. They aim to catalog the genetic variations that cause cancer and to characterize the topography of the cancer genome. This includes point mutations, copy number variation, and translocations, all analyzed through genome sequencing and analytics (Fig. 13.2) [33].

Sequencing can also be used to find possible driver genes and chromosomal structural rearrangements. These genetic changes may have an impact on a number of biological functions, including cell signaling, metabolism, and gene expression. Diagnostics, prognostics, prevention, and treatability are all areas of precision medicine where NGS has clinical value [33]. NGS is crucial for identifying new biomarkers that can be used to predict drug effects and divide individuals into clear-

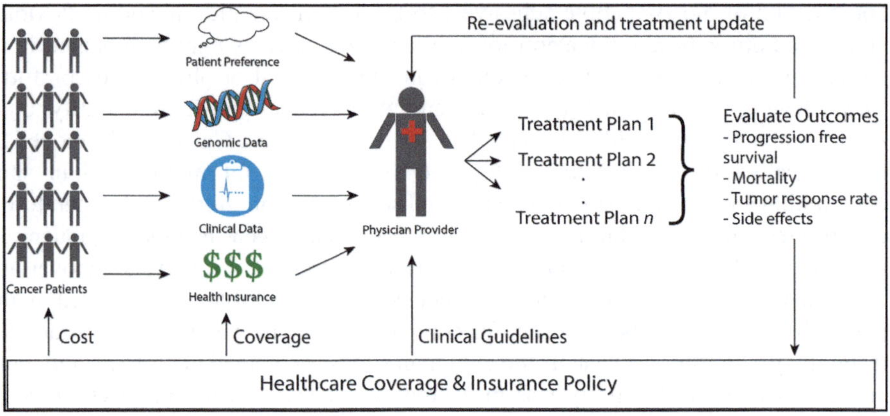

Fig. 13.2 Cancer precision medicine [33]

cut risk categories. Therapeutic predictors can be found by comparing NGS results to a patient's pharmacological effectiveness. These biomarkers offer a very useful reference for doctors to use when making clinical judgments regarding recommended therapy following the validation of preclinical models. More and more clinics are using NGS information to make more accurate cancer diagnosis and treatment decisions. As a result, tumor patient classification based on genetic traits is increasingly becoming more common in clinical practice.

13.11.2 Cancer Defined Molecularly

The clinicopathologic traits and tumor anatomical origins are used in the traditional classification of cancer, from which therapeutic approaches can be devised. However, unique genetic variations lead to varied clinical outcomes, and treatment responses might differ even among individuals with comparable histologic features and tumor stages. As a result, drug responses and patient prognoses are difficult to predict [33]. Recent studies have shown that FGFR3 genetic aberrations such as amplification, mutation, or FGFR3-TACC3 fusion can operate as an oncogenic driver in a range of cancers such as urothelial carcinoma, multiple myeloma, cervical cancer, and HNSC. Because of these significant discoveries, medical experts are considering reclassifying cancer in order to treat patients accurately and uniquely depending on their molecular features. By improving the molecur-based classification of cancer, the intersection of genetics, bioinformatics, and targeted treatments has been quickly extending the reach of precision medicine. This classification of cancer uses molecular biology rather than conventional anatomical and histological criteria, which is more therapeutically useful.

A new perspective on oncology has been provided by the global gene expression, genomic, and epigenomic investigations made possible by the rapidly developing high-throughput technology [29]. The well-known heterogeneity in tumor genetics and pathology demands radically different clinical prognosis and treatment approaches. As a result, a new cancer classification is emerging, which is based on cutting-edge genetic sequencing technologies. Tumor treatment may now be accurately targeted to an oncogenic mutation (driver gene) using various genetic regulation approaches such as microRNA replacement therapy. Other oncogenic mutations, on the other hand, may exist and may or may not react to a certain therapeutic intervention or gene-regulation. Nonetheless, a single driver gene may have a significant impact on the tumor's prognosis. As a result, in certain cases cancer can be suppressed by altering the expression of a single driver gene. This cancerous tumor behavior, which some people have nicknamed "oncogene addiction," provides the scientific basis for targeted gene therapy.

Cancers develop as a result of acquiring somatic mutations in their genomes that change how important cancer genes—known as the driving genes—perform. But it is possible that the tumors formed from other organs have the same driving mutation and react to the same treatment. In this case, the same driver gene might have a role in the emergence of colorectal cancer (or EGFR mutation), brain and neck cancer, non-small-cell lung cancer, digestive track cancer, and cancer of the stomach. This driver gene can be therapeutically downregulated by delivering a chosen microRNA (like microRNA-100) through nanocarriers. To successfully find the biomarkers, this approach may make use of multi-omics, which include the genome, proteome, transcriptome, epigenome, and microbiome. As a result, the interaction between oncology and genetics will be appropriately matched.

13.11.3 Precision Therapy for Cancer

Lack of preclinical models that may accurately predict therapeutic clinical action in cancer patients, primarily as a result of tumor heterogeneity and genetic variety, is a major issue in modern cancer therapy. The classification of cancer genetic changes is not used to inform traditional cancer therapies like chemoradiotherapy. Despite the fact that cancer cell lines originate from patient tumors, in vitro modification and genetic changes will produce cell cultures with uniform phenotypes. The anatomical sites of the tumors, the classification of malignant tumors (TNM classification, where "T" stands for the primary tumor site, "N" stands for the involvement of regional lymph nodes, and "M" stands for the presence or absence of distant metastatic spread), the histological grade, risk profiles, and clinical characteristics are frequently used to categorize patients receiving traditional cancer therapy, as shown in Fig. 13.3.

The treatment approach typically adheres to a broad regimen without accounting for tumor heterogeneity and genetic characteristics. This type of traditional treatment, sometimes known as "one-size-fits-all" drug, may be hazardous to some people [34]. $N = 1$ medicine, sometimes referred to as precision medicine, is centered on a person's genetic profile. A particular genetic mutation, or "driver gene," can be identified at the individual level as the root cause of a number of cancers, including breast, lung, ovarian, colon, and other cancers, by genome

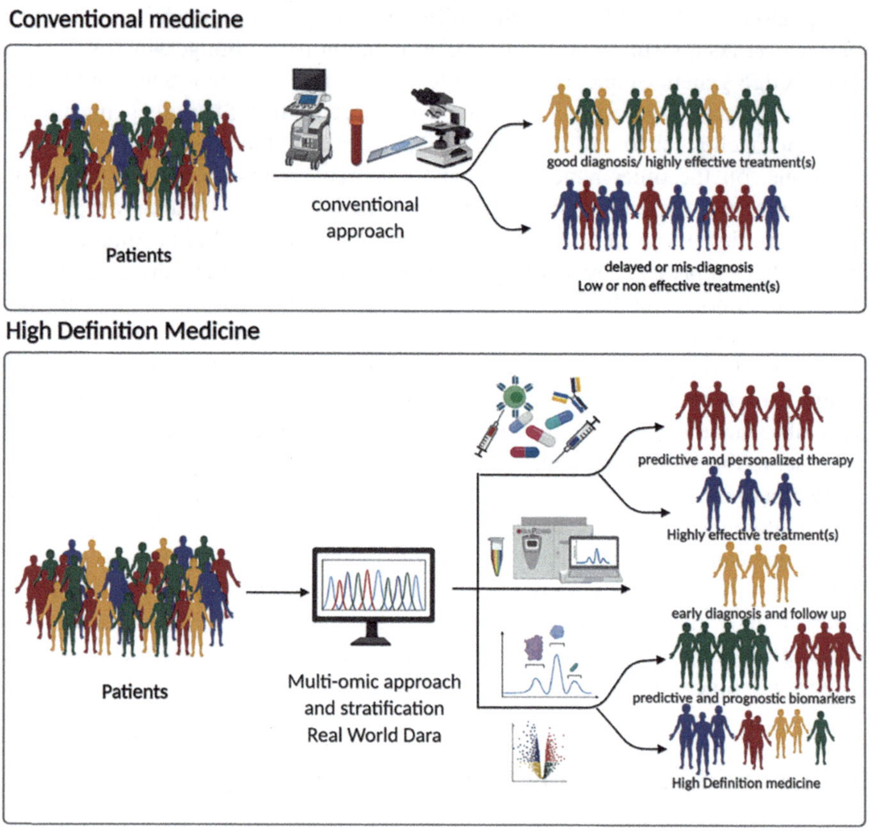

Fig. 13.3 Outlining the distinctions between precision medicine and conventional medicine [12]

sequencing. This method advances tumor identification by gene-mapping analysis, which categorize patients according to particular driver genes. In this approach, by grouping people into subpopulations that react differentially to a certain treatment, the medication can be therapeutically applied more precisely. The so-called tailored treatment can provide patients a more accurate prognosis of the diseases they may acquire.

13.11.4 Nanomaterials for Personalized Cancer Therapy

One of the numerous problems in precision medicine that cannot be readily overcome by current medical approaches is targeted gene delivery. In the past 10 years, a number of therapeutic approaches based on chemical and physical characteristics at the nanoscale, such as magnetic hyperthermia and photothermal treatments, have showed promise for addressing some of the fundamental needs in medical diagnostics and therapy. These include medication and gene delivery, medical imaging, cell and gene targeting, and cell targeting. The discovery that specifically tailored nanocarriers and their surface modifications can considerably advance diagnostics and therapeutics has ushered in a new age in medical history, despite the fact that major issues with these cutting-edge materials have not yet been entirely overcome [20]. These particulate carrier systems were created with special characteristics. These enable intelligent initiation the release of drugs, efficient gene transport, sensitive biosensing, super-resolution nano-imaging, signal nano-transduction, and the analysis of biological nanoanalytes [35]. The utilization of customized nanostructures created for specific biological and clinical purposes in cancer diagnosis and therapy is one of the key areas of interest in the so-called nanomedicine. Because of their much smaller scales (10 nm), large surface areas, novel structures, and properties specific to the nanoscale, as well as their extensive functionality, manipulability, and capacity for biomimicry, nanoparticles stand out from more traditional materials like bulk, single crystals, and thin films [1, 36]. Due to this, it offers a lot of flexibility in terms of structural architecture, tunable properties, microenvironmental interactivity, functions at the nano-bio interface, and even the potential to build nanorobots for specialized cellular tasks, pointing to a promising future for both basic science research and clinical applications. Cancer precision treatment, which has made good use of the unique nano characteristics, has prompted the creation of a variety of nanoparticles. In hindsight, the bulk of the first generation (1G) nanomaterials are single, unfunctionalized, and as-produced basic nanoparticles like SiO_2, Fe_3O_4, ZnO, and quantum dots with constrained characteristics and applications. To address the complex medical issues, it is necessary to functionalize nanoparticles with a variety of moieties, including anticancer drugs, biological molecules, fluorescent dyes, tumor-specific ligands, and genetic entities. These components need to be appropriately built at the nanoscale in order for them to function either separately or together for the best medical theranostics. Thus, the term "multifunctional nano-system" has become the accepted definition of second generation (2G) nanoparticles. Some molecular systems may be constructed as a tube or ring, for instance. These structures are controlled by molecular interactions, and they may be "turned on-off" by structural reversal, pH, light, and electrical fields. A "microbivore" is a clever nanomachine that might act as a synthetic phagocyte, or white blood cell. Third generation (3G) nanomaterials could be used to describe the following autonomous nanodevices.

13.12 New Methods for Clinical Trials in Precision Medicine

Oncology trials have traditionally been medication-focused, with the aim of finding commonalities between patients (such as their tumor type or, more recently, a shared genetic abnormality) and enlisting them in a study with a specific treatment regimen. Tumor-specific clinical investigations required a relatively broad population to be included in gene-agnostic trials due to the substantial variability in genetic subgroups, the microenvironment, baseline features, comorbidities, and other factors. Through reactions on the sidewalls of graphene or at the termini of nanotubes, covalent functionalization has grown in popularity as a method for modifying carbon nanomaterials.

13.13 Octopus, Platform, Basket, Umbrella, and Master Protocols

Recently, basket designs have focused on a widespread genetic anomaly [37]. The 75% objective response rate shown across tumor types with larotrectinib, which targets NTRK fusions, best represents the potential of the basket gene-directed, histology-agnostic paradigm, even if other single-gene targets have demonstrated to be noticeably less effective. In umbrella studies, a single histology is employed, but different medications are used according on the genetic differences in patient subgroups [38]. Other trial types include octopus trials (also called "complete phase I trials"), which have multiple arms testing various drug combinations, and master protocols, which cover trials with multiple histologic arms (previously known as "broad phase II trials") or multiple platform, basket, or umbre designs. Platform trials search tumors with various histologies for genomic or other biomarkers using a single analytical approach, such as NGS. Randomization has advanced along with the development of Bayesian adaptation, which allows for dynamic modifications of randomization based on sparse patient populations and immediate results.

The ultimate goal of precision medicine is an individual, patient-centered (rather than drug-centered) research based on the best available biomarkers. In "N-of-1" studies, each patient's response to therapy is assessed independently based on their molecular, immunological, and other biologic characteristics. In these trials, specially formulated pharmaceutical combinations are employed for certain individuals. To ascertain effectiveness, "N-of-1" studies must assess the "method" of matching patients to drugs rather than treatments, which differ from patient to patient.

13.13.1 Observable Facts

The growth of data mining and real-world registries is a result of improved computer data "processing" capabilities. Pembrolizumab for solid tumors with a mismatch repair gene deficit and palbociclib for male breast cancer were both approved by the FDA based, at least in part, on the same data. Real-world data has a striking potential to speed up the approval process for medications if it can properly forecast the results of upcoming trials.

13.14 Individual Vaccines (Vaccinomics)

The accumulation of somatic mutations in cancer can result in the production of cancer-specific neo-epitopes. Since autologous T-cells typically identify these neo-epitopes as foreign molecules, they make great targets for cancer vaccines.

Even while every malignancy has unique mutations of its own, only a few number of neoantigens are present in all malignancies. Theoretically, rapid genetic change mapping, rational selection of vaccine targets like neo-epitopes, and on-demand production of immunizations tailored to a patient's individual tumor will soon be made possible by technology breakthrough. It could also be feasible to employ commercially available immunizations for tumors with related epitopes as an alternative.

Clinical studies are now being carried out to evaluate different tailored vaccines. For instance, utilizing computational prediction of neo-epitopes, researchers developed customized RNA mutanome vaccines for patients with malignant melanoma. Two of the five patients treated demonstrated objective responses to the vaccination alone, while the third patient had a CR to therapy with the immunization combined with PD1 blocking [39]. A study of vaccine-induced polyfunctional CD^{4+} and CD^{8+} T-cells targeting different neoantigens in individuals with melanoma found that four out of six vaccinated patients showed no signs of recurrence 25 months after immunization [40]. The first therapeutic cancer vaccine licensed by the FDA, sipuleucel-T, is generated from autologous peripheral blood mononuclear cells and a recombinant fusion protein of prostatic acid phosphatase and granulocyte-macrophage colony-stimulating factor ex vivo. Sipuleucel-T is utilized to treat metastatic castration-resistant prostate cancer, according to the results of a randomized, double-blind, placebo-controlled phase III study. Sipuleucel-T-treated patients lived longer than placebo-treated patients (25.8 vs. 21.7 months, respectively; $p = 0.03$).

13.15 Recommendations for the Best Use of Precision Medicine

Tumors are tremendously varied and complicated, and standard clinical research and practice models typically do not result in the most effective therapy, as genomic studies have demonstrated. Studies in precision medicine highlight the significant difficulties in developing trials for this novel paradigm. First, in these studies, patients are matched to medications at a rate that ranges from 5% to 49%, with the majority of matches occurring between 15% and 20%. The failure to match patients is caused in part by the following factors: (1) enrolling patients with advanced disease who deteriorate or die before the study is finished; (2) using small gene panels that result in few actionable alterations; (3) taking a long time to receive and interpret genomic results; and (4) having trouble obtaining targeted therapy drugs and/or low drug availability. The following are a few answers provided by studies with greater matching rates, including I-PREDICT 12 (matching rate, 49%): Utilizing a large NGS panel with more than 200 genes, just-in-time electronic molecular tumor boards that are created immediately in response to a doctor's request, and the use of biomarkers to match patients to chemotherapy, hormone therapy, and immunotherapy (in addition to gene targeted agents) are just a few of the strategies used [41–43]. The great majority of these trials have shown improved clinical outcomes when treatments are matched to drugs as opposed to when they are not specific to the complicated molecular biology of malignancies, it is important to note that the use of customized medication combinations that target a larger percentage of the aberrations present in a specific malignancy is linked with better results than more selective matching.

The following are other significant obstacles to precision medicine implementation: (a) Based on histology and/or genetic co-alterations, there may be variations in how patients respond to matching therapy. Selected genetic indicators are predictive in particular tumor histologies, in contrast to molecular abnormalities (such as NTRK fusions, MSI-H) that indicate tumor agnostic response to treatment. (b) Genomic landscapes' variability, complexity, and ongoing evolution. The significant diversity between the primary tumor and metastatic locations makes it possible that molecular profiling of tumor tissue collected from a single lesion is not always predictive of the systemic illness. Due to the pressure from targeted treatments, the tumor's molecular profile is also constantly changing, with newly arising resistant clones and novel molecular alterations hastening the progression of the illness [44]. (c) The need for thorough patient screening to find specific or uncommon genetic abnormalities (such NTRK fusions). (d) Inadequate technology and resources to fully understand the causes of cancer in particular patients; incomplete biologic/molecular profiles to guide therapy selection; (e) Major holdups in the start of clinical studies; (f) variances in drug absorption and side effects among different racial and ethnic groupings; (g) there is an absence of agreement among assays from various diagnostic companies/laboratories; (h) and, most importantly, there is a shortage of access to medical care. Clinical trials recruit about 3–5% of cancer patients, although enrolment is limited by onerous qualifying standards and limited medication availability [45, 46].

13.16 Future Innovations in India

On January 3, 2020, the Genome India Project was launched in India. In partnership with 20 academic institutions, it seeks to map the population's genomes using genome-wide sequencing. The Indian Cancer Genome Atlas and the Indian Genome Variation Consortium are examining the differences between the genomes of the Indian population and the world genomes. The Unique Methods of Management and Treatment of Inherited Disorders project and the September 2019 launch of NIDAN (National Inherited Disease Administration Kendra Network) are recent initiatives that provide as examples of advancements in this discipline. These significant breakthroughs bode well for the future of precision medicine in this nation.

13.16.1 Difficulties with Precision Medicine

Like any new endeavor, precision medicine must overcome several obstacles. In addition to being expensive, the massive amount of data that must be gathered and analyzed also requires a lot of labor and technological expertise. A significant problem is data security and anonymization. Therefore, ethical issues are crucial. Another distraction comes from data noise. At the earliest, data analysis turnaround time is 26 h, which is still a long time for judgments in acute care settings. Calculating the veracity of such data is equally challenging. Capacity building must proceed at breakneck speed, particularly in light of the accessibility of high-caliber artificial intelligence (AI), machine learning (ML), and laboratory equipment. The pursuit of precision-based medicines may undermine population health. Despite recent developments and the expanding influence of precision medicine, there are still significant obstacles to be addressed before it can be widely used in clinical practice.

The following are the difficulties that precision medicine must overcome:

1. There are countless economic, legal, social, and technical problems that demand creative answers.
2. Providing solid evidence in favor of precision medicine. More investigation is currently required to support the use of precision medicine, which would result in noticeably better outcomes.
3. Considerable funding is required to enhance data collection, storage, sharing, and EHR integration (electronic health records).
4. Although it has not yet been accomplished, it will be essential in many healthcare systems to incorporate genomic information into clinical care and research.
5. Lack of physician trust in their capacity to make therapeutic judgments when genetic or genomic information is present makes education a significant barrier. Education initiatives across the various medical specialties are required.
6. Efforts must be made to ensure participant trust and engagement. Currently, patient worry, the apprehension of pointless and expensive testing, potential procedures resulting from a genetic result, and privacy concerns are barriers to full patient engagement.
7. Improving ethnic diversity is necessary to effectively translate genetic findings.

13.17 Conclusion

It only seems a matter of time before precision medicine and omics technologies take the lead in the selection of diagnostic and therapeutic choices, given the magnitude of the field and its rate of growth. As a result, modern physicians need to be flexible, revived, empathetic, patient-centered, aware about omics, able to apply big data, fully utilize AI and ML, evaluate complicated material, and present it to the patient in a coherent and persuasive "capsule." We must build intelligent big data platforms, collect uniformly formatted data, educate healthcare professionals in genomics, supply infrastructure, and include patients and the general public if precision medicine is to survive and succeed. According to Sir Charles Darwin, "The species that survives is not the one that is the strongest or the most clever, it is the one most adaptable to change." On the therapy front, a number of gene therapies are getting closer to approval despite a number of setbacks. Pharmacogenetics, which uses a person's DNA to prescribe the safest, most effective prescription for them, is being pioneered by a few medical centers, which are also building the infrastructure required to make it operate in a clinical context. However, some scientists contend that we first need further proof of the benefits. Patient information is the fuel for precision medicine. People can impact their own health care and the course of research by using the genetic information and health records of patients and healthy volunteers. Gaining participants' trust is essential for the success of large-scale initiatives like the Precision Medicine Initiative of the US National Institutes of Health. Large-scale initiatives like the Precision Medicine Initiative of the US National Institutes of Health must win participants' trust if they are to be successful. But when the promised anonymity turns out to be nothing more than a fig leaf, it becomes difficult.

References

1. Calabretta MM, Zangheri M, Lopreside A, Marchegiani E, Montali L, Simoni P, Roda A (2020) Precision medicine, bioanalytics and nanomaterials: toward a new generation of personalized portable diagnostics. Analyst 145(8):2841–2853
2. Thakur P, Thakur A (2022) Introduction to nanotechnology. In: Thakur A, Thakur P, Khurana SP (eds) Synthesis and applications of nanoparticles. Springer, Singapore. https://doi.org/10.1007/978-981-16-6819-7_1
3. Naithani N, Atal AT, Tilak T, Vasudevan B, Misra P, Sinha S (2021) Precision medicine: uses and challenges. Med J Armed Forces India 77(3):258–265
4. Ghasemi M, Nabipour I, Omrani A, Alipour Z, Assadi M (2016) Precision medicine and molecular imaging: new targeted approaches toward cancer therapeutic and diagnosis. Am J Nucl Med Mol Imaging 6(6):310–327. PMID: 28078184; PMCID: PMC5218860
5. Ahmed MU, Saaem I, Wu PC, Brown AS (2014) Personalized diagnostics and biosensors: a review of the biology and technology needed for personalized medicine. Crit Rev Biotechnol 34(2):180–196
6. Rebelo R, Barbosa AM, Correlo VM, Reis RL (2021) An outlook on implantable biosensors for personalized medicine. Engineering 7(12):1696–1699
7. Rodrigues D, Barbosa AM, Rebelo R, Lee SJ, Reis RL, Correlo VM (2020) Skin-integrated wearable systems and implantable biosensors: a comprehensive review. Biosensors 10(7):79
8. Johnson KP, Wei WI, Weeraratne D, Frisse ME, Misulis KE, Rhee KY, Zhao J, Snowdon JL (2021) Precision medicine, AI, and the future of personalized health care. Clin Transl Sci 14(1): 86–93
9. Shakeel A, Singh A, Das S, Suhag D, Sharma AK, Rajput SK, Mukherjee M (2017) Synthesis and morphological insight of new biocompatible smart hydrogels. J Polym Res 24:1–10
10. Suhag D, Bhatia R, Das S, Shakeel A, Ghosh A, Singh A, Mukherjee M (2015) Physically cross-linked pH-responsive hydrogels with tunable formulations for controlled drug delivery. RSC Adv 5(66):53963–53972
11. Hodson R (2016) Precision medicine. Nature 537(7619):S49
12. De Maria Marchiano R, Di Sante G, Piro G, Carbone C, Tortora G, Boldrini L, Pietragalla A, Daniele G, Tredicine M, Cesario A, Valentini V, Gallo D, Babini G, D'Oria M, Scambia G (2021) Translational research in the era of precision medicine: where we are and where we will go. J Pers Med 11(3):216
13. Ghosh D, Poisson LM (2009) "Omics" data and levels of evidence for biomarker discovery. Genomics 93(1):13–16

14. Beltrán-García J, Osca-Verdegal R, Mena-Mollá S, García-Giménez JL (2019) Epigenetic IVD tests for personalized precision medicine in cancer. Front Genet 10:621
15. Zhang Z, Chan DW (2010) The road from discovery to clinical diagnostics: lessons learned from the first FDA-cleared in vitro diagnostic multivariate index assay of proteomic biomarkers. Cancer Epidemiol Biomark Prev 19(12):2995–2999
16. Shaw AT, Yeap BY, Mino-Kenudson M, Digumarthy SR, Costa DP, Heist RS, Solomon B, Stubbs H, Admane S, McDermott MW, Settleman J, Kobayashi S, Mark EJ, Rodig SJ, Chirieac LR, Kwak EL, Lynch TJ, Iafrate AJ (2009) Clinical features and outcome of patients with non–small-cell lung cancer who harbor EML4-ALK. J Clin Oncol 27(26):4247–4253
17. Kazandjian D, Gong Y, Keegan P, Pazdur R, Blumenthal GM (2019) Prognostic value of the lung immune prognostic index for patients treated for metastatic non–small cell lung cancer. JAMA Oncol 5(10):1481
18. Saltz LB, Clarke S, Díaz-Rubio E, Scheithauer W, Figer A, Wong R, Koski S, Lichinitser M, Yang T, Rivera F, Couture F, Sirzén F, Cassidy J (2008) Bevacizumab in combination with oxaliplatin-based chemotherapy as first-line therapy in metastatic colorectal cancer: a randomized phase III study. J Clin Oncol 26(12):2013–2019
19. Escudier B, Bellmunt J, Negrier S, Bajetta E, Melichar B, Bracarda S, Ravaud A, Golding S, Jethwa S, Sneller V (2010) Phase III trial of bevacizumab plus interferon alfa-2a in patients with metastatic renal cell carcinoma (AVOREN): final analysis of overall survival. J Clin Oncol 28(13):2144–2150
20. Grávalos C, Jimeno A (2008) HER2 in gastric cancer: a new prognostic factor and a novel therapeutic target. Ann Oncol 19(9):1523–1529
21. Schuell B, Gruenberger T, Scheithauer W, Zielinski CC, Wrba F (2006) HER2/neu protein expression in colorectal cancer. BMC Cancer 6(1):123
22. Huang LL, Shi Y (2020) Tyrosine kinase inhibitors for solid tumors in the past 20 years (2001–2020). J Hematol Oncol 13(1):143
23. Meyers DE, Banerji S (2020) Biomarkers of immune checkpoint inhibitor efficacy in cancer. Curr Oncol 27(12):106–114
24. Ahmad A, Uddin S, Steinhoff M (2020) CAR-T cell therapies: an overview of clinical studies supporting their approved use against acute lymphoblastic leukemia and large B-cell lymphomas. Int J Mol Sci 21(11):3906
25. Kuruvilla M, Lee FC, Lee GB (2019) Understanding asthma phenotypes, endotypes, and mechanisms of disease. Clin Rev Allergy Immunol 56(2):219–233
26. Gloyn AL, Drucker DJ (2018) Precision medicine in the management of type 2 diabetes. Lancet Diabetes Endocrinol 6(11):891–900
27. Barturen G, Babaei S, Català-Moll F, Martínez-Bueno M, Makowska Z, Martorell-Marugán J, Carmona-Sáez P, Toro-Domínguez D, Carnero-Montoro E, Teruel M, Kerick M, Acosta-Herrera M, Lann LL, Jamin C, Rodríguez-Ubreva J, Garcia-Gomez A, Kageyama J, Buttgereit A, Hayat S et al (2021) Integrative analysis reveals a molecular stratification of systemic autoimmune diseases. Arthr Rheumatol 73(6):1073–1085
28. Schuetz P, Aujesky D, Müller CP, Müller B (2015) Biomarker-guided personalised emergency medicine for all – hope for another hype? Schweizerische Medizinische Wochenschrift
29. Palacios G, Druce J, Du L, Tran T, Birch C, Briese T, Conlan S, Quan P, Hui J, Marshall J, Simons JF, Egholm M, Paddock CD, Shieh W, Goldsmith CS, Zaki SR, Catton M, Lipkin WI (2008) A new arenavirus in a cluster of fatal transplant-associated diseases. N Engl J Med 358(10):991–998
30. Saunders C, Miller NR, Soden SE, Dinwiddie DL, Noll A, Alnadi NA, Andraws N, Patterson M, Krivohlavek LA, Fellis J, Humphray S, Saffrey P, Kingsbury Z, Weir J, Betley J, Grocock RJ, Margulies EH, Farrow EG, Artman M et al (2012) Rapid whole-genome sequencing for genetic disease diagnosis in neonatal intensive care units. Sci Transl Med 4(154):154ra135
31. Koh A, Kang D, Xue Y, Lee S, Pielak RM, Kim J, Hwang T, Min S, Banks A, Bastien P, Manco M, Wang L, Ammann KR, Jang KL, Won P, Han S, Ghaffari R, Paik U, Slepian MJ et al

(2016) A soft, wearable microfluidic device for the capture, storage, and colorimetric sensing of sweat. Sci Transl Med 8(366):366ra165
32. Gao W, Emaminejad S, Nyein H et al (2016) Fully integrated wearable sensor arrays for multiplexed in situ perspiration analysis. Nature 529:509–514
33. Morash M, Mitchell H, Beltran H, Elemento O, Pathak J (2018) The role of next-generation sequencing in precision medicine: a review of outcomes in oncology. J Pers Med 8(3):30
34. Garraway LA, Lander ES (2013) Lessons from the cancer genome. Cell 153(1):17–37
35. Sakamoto J, Van De Ven AL, Godin B, Blanco E, Serda RE, Grattoni A, Ziemys A, Bouamrani A, Hu Y, Ranganathan SI, De Rosa E, Martinez JO, Smid C, Buchanan R, Lee SJ, Srinivasan S, Landry MJ, Meyn A, Tasciotti E et al (2010) Enabling individualized therapy through nanotechnology. Pharmacol Res 62(2):57–89
36. Kala D, Gupta S, Kaushal A (2022) Nanotechnology in healthcare. In: Thakur A, Thakur P, Khurana SP (eds) Synthesis and applications of nanoparticles. Springer, Singapore
37. Culver HR, Clegg J, Peppas NA (2017) Analyte-responsive hydrogels: intelligent materials for biosensing and drug delivery. Acc Chem Res 50(2):170–178
38. Drilon A, Laetsch TW, Kummar S, DuBois SG, Lassen U, Demetri GD, Nathenson MJ, Doebele RC, Farago AF, Pappo AS, Turpin B, Dowlati A, Brose MS, Mascarenhas L, Federman N, Berlin J, El-Deiry WS, Baik CS, Deeken JF et al (2018) Efficacy of Larotrectinib inTRKFusion–positive cancers in adults and children. N Engl J Med 378(8):731–739
39. Schwaederle M, Parker BA, Schwab R, Daniels GA, Piccioni D, Kesari S, Helsten T, Bazhenova L, Romero J, Fanta PT, Lippman SM, Kurzrock R (2016) Precision oncology: the UC San Diego Moores cancer center PREDICT experience. Mol Cancer Ther 15(4):743–752
40. Sahin, U., Derhovanessian, E., Miller, M., Kloke, B., Simon, P., Löwer, M., Bukur, V., Tadmor, A. D., Luxemburger, U., Schrörs, B., Omokoko, T., Vormehr, M., Albrecht, C., Paruzynski, A., Kuhn, A., Buck, J., Heesch, S., Schreeb, K. H., Müller, F., . . . Türeci, Ö. (2017). Personalized RNA mutanome vaccines mobilize poly-specific therapeutic immunity against cancer. Nature, 547(7662), 222–226
41. Ott PA, Hu Z, Keskin DB, Shukla SA, Sun J, Bozym DJ, Zhang W, Luoma AM, Giobbie-Hurder A, Peter L, Chen CH, Olive O, Carter TA, Li S, Lieb DJ, Eisenhaure T, Gjini E, Stevens JR, Lane WS et al (2017) An immunogenic personal neoantigen vaccine for patients with melanoma. Nature 547(7662):217–221
42. Sicklick JK, Kato S, Okamura R, Schwaederle M, Hahn SK, Williams C, De P, Krie A, Piccioni D, Miller VA, Ross JS, Benson AF, Webster J, Stephens PJ, Lee JSH, Fanta PT, Lippman SM, Leyland-Jones B, Kurzrock R (2019) Molecular profiling of cancer patients enables personalized combination therapy: the I-PREDICT study. Nat Med 25(5):744–750
43. Schwaederle M, Zhao M, Lee JSH, Eggermont AM, Schilsky RL, Mendelsohn J, Lazar V, Kurzrock R (2015) Impact of precision medicine in diverse cancers: a meta-analysis of phase II clinical trials. J Clin Oncol 33(32):3817–3825
44. Kobayashi S, Boggon TJ, Dayaram T, Jänne PA, Kocher O, Meyerson M, Johnson BE, Eck MJ, Tenen DG, Halmos B (2005) EGFR Mutation and resistance of non–small-cell lung cancer to gefitinib. N Engl J Med 352(8):786–792
45. Napolitano A, Vincenzi B (2019) Secondary KIT mutations: the GIST of drug resistance and sensitivity. Br J Cancer 120(6):577–578
46. Murthy V, Krumholz HM, Gross CP (2004) Participation in cancer clinical trials. JAMA 291(22):2720

Nanotechnology for AI in Healthcare

14

Swati Kaushik and Deepa Suhag

Abstract

The adoption of AI technologies, which are widely used in modern business and daily life, is becoming incremental in the healthcare industry. In patient care and administrative processes, artificial intelligence offers a variety of uses that help healthcare practitioners improve on existing solutions and find issues' answers more rapidly. Even though hospitals and other healthcare organizations may use a range of techniques as a consequence, the bulk of AI and healthcare technologies are crucial to the industry. And even though several papers on the application of artificial intelligence in healthcare claim that it can execute specific processes, including disease diagnosis, just as well as or better than humans, there is still some time before AI in healthcare replaces humans for a variety of medical duties. But a lot of people are still uncertain. How could artificial intelligence improve medical technology? What role will AI play in healthcare going forward? What part does AI presently play? Will it eventually take the place of people in medical care and life-saving procedures? Not whether AI will be useful in healthcare, but rather how to assure its adoption in everyday clinical practice, is the most crucial topic. Clinicians may eventually gravitate toward professions requiring the highest level of cognitive function and unique human abilities. Only those healthcare providers that choose not to engage with AI in healthcare may not gain completely from it.

S. Kaushik (✉)
Arjuna Natural Pvt. Ltd., Kochi, Kerala, India

D. Suhag
Amity Centre for Nanotechnology, Amity University Haryana, Gurgaon, Haryana, India
e-mail: dsuhag@ggn.amity.edu

D. Suhag et al. (eds.), *Integrated Nanomaterials and their Applications*,
https://doi.org/10.1007/978-981-99-6105-4_14

Keywords

Artificial intelligence · Advanced healthcare · Patient care · Disease diagnosis

14.1 Introduction

Artificial intelligence and nanotechnology have been in the development for many years, not just as a field of technology in and of themselves, but also in the various industries and spaces where they can now be seen. Complex systems that can operate on the nano-meter scale are typically found in technology that is unsuitable for the many facets of AI [1]. However, there are a few emerging areas where AI and nanotechnology are combining. In addition to combining these two technologies, linked work will boost study in a single discipline, potentially leading to a slew of new communication technologies and insights [2]. The use of machine learning (ML) algorithms and other cognitive technologies in healthcare is referred to as artificial intelligence (AI). AI can be defined simply as the ability of computers and other devices to replicate human cognition and to learn, think, and make decisions or take actions. Artificial intelligence and nanotechnology are fields that can aid in the pursuit of precise medicine tailoring the best treatment for each patient, particularly cancer patients [3]. Recent progress in these fields enables better nanomaterial design and patient data acquisition for cancer medicine. Diagnostic nanomaterials were used to create a patient-specific disease outline, which was then used to improve treatment outcomes by combining a variety of healing nanotechnologies. Despite this, the high interpatient and intratumor heterogeneity makes rational design of therapeutic and diagnostic platforms to analyze the output extremely difficult. Using classification algorithms and pattern analysis for improved therapeutic and diagnostic accuracy, integrating AI approaches will bridge the gap. Therefore, AI in healthcare refers to the application of machines to analyze and take action on medical data, frequently with the aim of forecasting a specific outcome. AI may assist doctors and other healthcare workers in providing more accurate diagnostic and treatment suggestions by utilizing patient data and other information. By analyzing vast volumes of data to provide better preventative care ideas for patients, AI may also aid in making healthcare more proactive and predictive. Because of the crucial part it plays in a healthy, productive society, healthcare is one of the most important businesses in the larger big data ecosystem. Patients might live or die depending on how AI is used in medicine. AI can help healthcare workers with everyday tasks, including doctors, nurses, and other healthcare providers. AI in healthcare can improve overall patient outcomes, preventative care, and quality of life in addition to offering more precise treatment and diagnostic modalities. By examining data from the public sector, the healthcare industry, and other sources, artificial intelligence (AI) can help anticipate and observe the rise of hazardous illnesses. AI has the potential to be tremendously useful in the fight against pandemics and epidemics, as well as in the improvement of global public health.

Artificial intelligence must have access to a lot of data in order to be helpful in the healthcare industry. Another difficulty is the possibility of bias if the population-representativeness of the data used to train the algorithms is low. It could be challenging to compare results or merge data from many sources due to the absence of standards among various artificial intelligence systems.

14.2 AI and Nanotechnology

Medical practice is gradually evolving as a result of artificial intelligence (AI). AI applications are advancing into fields that were previously believed to be the sole purview of human expertise as a result of recent developments in digitized data collecting, machine learning, and computing infrastructure. While some strategies are still only ideas, others have already been developed and put into practice. Nanotechnology is the atomic-scale manipulation of matter [4]. Nanomaterials have fascinating characteristics and are frequently used in technical applications [5, 6]. Electronics and nanotechnology are commonly mentioned together [7–9]. Amazing new and innovative concepts in healthcare will result from the integration of AI with nanotechnology. Gene therapy, diagnostics, medicine delivery, and a wide range of other applications will all undergo revolutionary change thanks to nanotechnology [10, 11]. Making an accurate risk predictor out of the electronic health record EHRs are the gold mine of patient data, but developers and providers have long struggled to safely, precisely, and quickly analyze and extract this wealth of knowledge. It has become incredibly challenging to comprehend precisely how to engage in crucial risk clinical decision support, stratification, and predictive analytics due to integrity and data quality challenges, including a jumble of data formats, unstructured and structured inputs, and inadequate records.

14.3 How Artificial Intelligence Works?

Rapid data digitization improvements have taken place in the pharmaceutical industry during the past few years. A tremendous amount of data has been documented as a result, which needs to be properly processed and examined before being used in experimental study. Massive volumes of data can be handled by AI with improved automation. It has developed into a potential tool for reverse engineering and ongoing medication delivery system optimization. AI applications have considerably enhanced the effectiveness of treatment. A technology-based system called artificial intelligence (AI) consists of a number of sophisticated tools and networks that mimic human intelligence. By using software and systems that can learn from the input data, reason about it, and interpret it, AI may make decisions independently for a specific goal. AI employs a wide range of methodology, including the basic machine learning paradigm (ML) [12].

ML use algorithms to find patterns in a collection of input data. A subset of machine learning (ML), known as deep learning, uses artificial neural networks

(ANNs) to simulate how the human brain transmits electrical impulses. Various tools have been created by scientists based on the network that serves as the fundamental structure of AI technology. The IBM Watson supercomputer is one of the tools based on the AI system; it has access to patient medical records and can compare them to a huge database to find anomalies or diseases. IBM, for example, can identify breast cancer in under 60s.

Applications of Artificial Intelligence Can Help Nanotechnology. Although the concept of nanotechnology is straightforward, its application may be difficult. Atomic level operation is initially far more challenging than other levels of operation. Different laws of physics are in play [13, 14] and even little mistakes can have disastrous consequences. Environment-related error rates are introduced and have an impact on the output's accuracy. It is essential to use the right instruments in a delicate environment like the human body to achieve the best results. Evidently, we are talking about machine learning algorithms. When precision is lost, you must approximate and use statistics. Artificial intelligence models have a significant role to play in the analysis of noisy data. In this situation, an algorithm has to give the best estimate. AI integration in healthcare is being handled by medical data scientists.

14.4 Microscopy and AI

The signals that are generated are of low quality, which is the main issue with AFM (atomic force microscopy). These microscopes research erratic and complex atomic interactions. The analysis of tip-to-sample contacts and the resulting signal is the goal of machine learning techniques. The data scientist approach aims to address these problems. The method is known as functional recognition scanning probe microscopy. It is a great illustration of how computer vision is used in the medical field [15]. In order to streamline data, artificial neural networks (ANNs) examine the behavior of scanned material (Fig. 14.2). Simple interactions between the tip and the sample are ideal for AFM. It is never simple in medical imaging, but ANNs try to make the data more understandable and take different factors into consideration. There are many uses for AI, and in a few years, it will be difficult to picture a world without deep learning and AI in healthcare. There will undoubtedly be a large demand for data scientists in the deep learning in healthcare field.

14.5 Chemical Modeling and AI

Many problems that at first glance seem simple at the macroscale turn out to be complicated at the nanoscale. Atom and molecule modeling at the nanoscale is challenging. It is a difficult undertaking to thoroughly represent a chemical system, even for artificial intelligence in healthcare. Error rates can be reduced using data science techniques to a tolerable level. Nanoelements' structural properties have been identified using machine learning algorithms. Because of the crucial applications for these components, all calculations must be accurate. When traditional measurements are insufficient, scientists frequently use machine learning.

14.6 Nanofibers

A good diagnosis can aid in the management of any medical issue. An accurate and timely diagnosis is crucial for effective treatment of any condition [16]. According to the World Health Organization, an in vitro diagnostic tool that is sensitive, specific, user-friendly, repeatable, robust, portable, and economically viable should be able to make a diagnosis quickly. We desperately require rapid point-of-care diagnostic technologies for medical diagnostics. Studies suggest that nanofibers offer more binding sites, increasing sensitivity and decreasing the detection threshold [17]. Due to their three-dimensional structure, electrospun nanofibers have a larger surface area, higher enzyme loading, and lower mass transfer resistance [18]. Composite nanofibers have also been intensively studied because the synergistic advantages shown when metal oxide nanoparticles or nanofibers are used in combination with polymer nanofibers. Over the past 10 years, research on nanofibers has become increasingly important for applications in tissue engineering and drug delivery. Nanofibers are very beneficial from a therapeutic standpoint due to their high drug loading capacity and straightforward production process [19]. A second approach for delivering two active chemicals in the same formulation is provided by combination systems, such as embedding nanoparticles within nanofiber architectures. Depending on the required release period for therapy, the formulation characteristics also make it simple to change the pace at which the active components release. Antibiotics, cancer treatments, analgesics, hemostatic agents, and different proteins are delivered via nanofiber systems for tissue engineering. In addition, other uses, such as coating medical devices, offer fresh perspectives on the therapeutic application of nanofibers. The most popular method for creating nanofibers is electrospinning, which offers a foundation for scaling up processes from laboratory to commercial applications. The restrictions on systemic routes are the primary limit for nanofibers (Fig. 14.1).

Nanofibers are primarily made for the local delivery of active substances. Nanofibers are ideal three-dimensional structures that are excellent for tissue regeneration, regardless of the therapeutic goal. For cell regeneration, particularly in applications for wound healing, they offer matrix structure.

14.6.1 Medical Diagnosis

Researchers used electrospinning to generate polyvinyl alcohol (PVA)/polyethyleneimine (PEI) nanofibers on the surface of gold electrodes for enzyme-based glucose monitoring. The glucose oxidase enzyme was immobilized on the nanofibers. Gold nanoparticles were put to the nanofibers to boost conductivity. The combination of metal nanoparticles with electrospun nanofibers is reported to

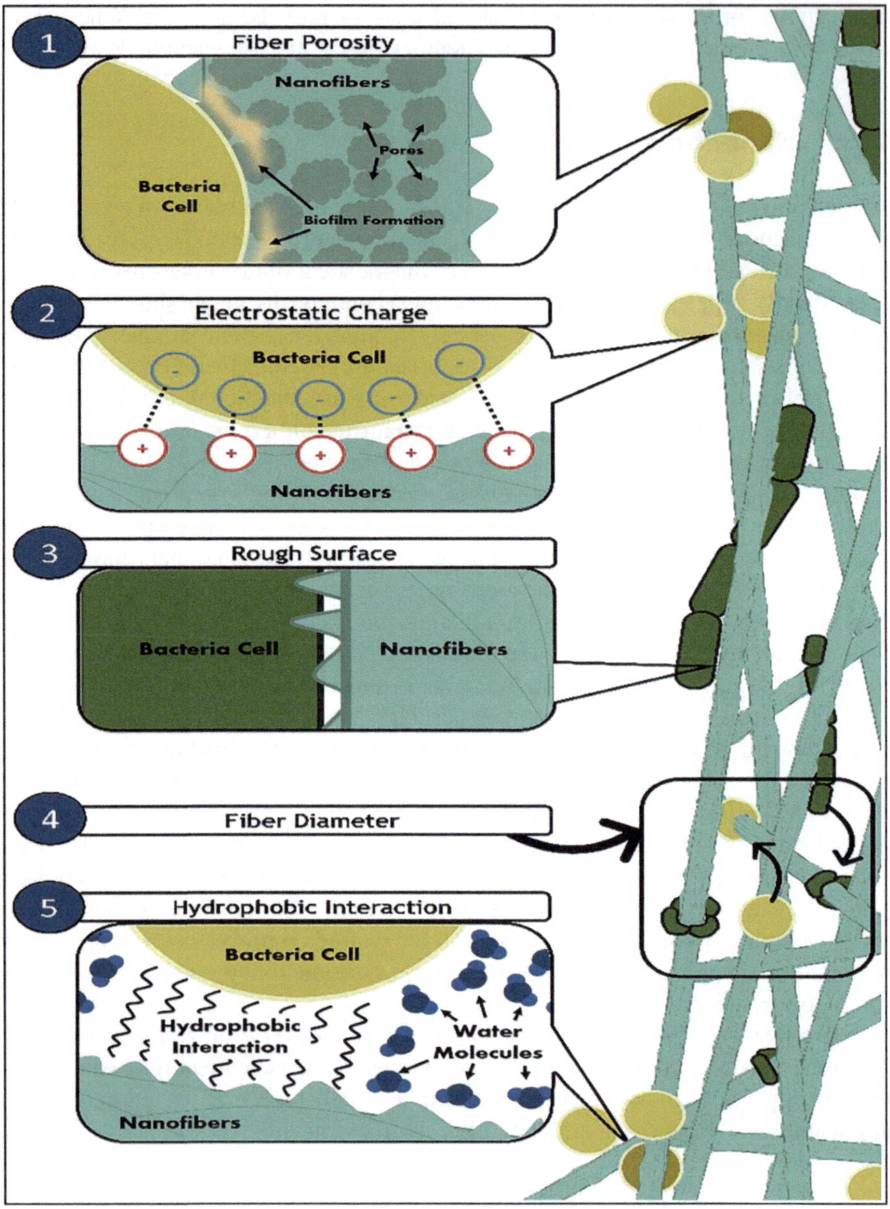

Fig. 14.1 Nanofibers action toward bacteria, nanofibers are the newest popular polymers that are widely employed for research and commercial applications [20]

increase mechanical strength, heat resistance, optical, and electrical characteristics. The nanosensor has a very low detection limit for glucose, great selectivity for glucose, and good operating and storage stability [21]. Blood glucose levels must be assessed in order to identify and treat diabetes mellitus. Glucose is measured using enzyme-based sensors, which provide accurate, rapid, and precise glucose detection. Enzymatic activity, however, may be impacted by environmental conditions including temperature, pH, humidity, etc. [22]. Frequently employed for this purpose, electrochemical sensors based on the enzyme glucose oxidase have limitations such low stability and repeatability and expensive cost. One of the researchers used the adsorption of cupric oxide nanoparticles on carbon nanofibers to create a nonenzymatic sensor for glucose measurement. Because they are simpler to manufacture on a large scale, more amenable to surface functionalization, and more wettable than carbon nanotubes, carbon nanofibers are used to make biosensors. The sensor demonstrated excellent sensitivity and a low detection limit. The glucose analysis performed by the sensor was accurate and exact. It also successfully detected glucose in human serum, proving its versatility [23]. Cobaltosic oxide (Co_3O_4)-polyvinyl pyrrolidone (PVP) nanofibers were created by Ding et al. utilizing electrospinning and calcination for nonenzymatic glucose monitoring. Co_3O_4 fibers exhibit very good catalytic activity. The biosensor demonstrated good sensitivity, selectivity, reproducibility, and a low limit of detection in addition to a quick reaction time of <7 s. The sensor proved successful in determining the concentration of glucose in human serum samples with similar accuracy to that of the commercial glucometer [24]. Continuous glucose monitoring may be useful to assess changes in blood glucose levels. Implantable nanosensors are used for this purpose due to their small size and possibility for less intrusive insertion. The human epidermal growth factor (hEGF) receptor is expressed in around 33% of breast cancers, making it a valuable biomarker. Zinc oxide nanofibers have a lot of promise as biological sensors because of their high surface-area-to-volume ratio, environmental friendliness, biocompatibility, excellent electron transport, and chemical stability. Mesoporous zinc oxide nanofibers were produced by Ali et al. utilizing the electrospinning process. Zinc oxide nanofibers were electrophoretically generated on ITO glass following treatment with oxygen plasma and covalent coupling with an antibody against the hEGF receptor. The biosensor showed high selectivity, reproducibility, and femtomolar sensitivity. Compared to the enzyme-linked immunosorbent test (ELISA), the sensor's sensitivity for identifying breast cancer biomarkers was three orders of magnitude greater [25].

Malaria is a serious problem, particularly in developing countries. Malaria is usually diagnosed by microscopy, however, it is unable to detect low parasite blood levels. The pricey ELISA, flow cytometry, mass spectrophotometry, and polymerase chain reaction are among more recently developed technologies. Brince Paul et al. developed a low-cost, ultrasensitive nanosensor to detect malaria. The researchers produced copper-doped zinc oxide nanofibers by electrospinning and functionalized them with mercaptopropylphosphonic acid (MPA). Functionalization with MPA was done to help with antibody immobilization. Monoclonal antibodies against protein-2 rich in histidine were immobilized. Protein-2 with a lot of histidine is released into the blood by Plasmodium falciparum. Copper doping helped to improve conductivity and pre-concentrate the target analyte on the surface of nanofibers because the electric field created at the copper oxide/zinc oxide interface increased conductivity. Zinc oxide has a high isoelectric point (9.5), which facilitates the attraction and binding of the target analyte molecules through electrostatic interactions. The developed nano-biosensor exhibited an attogram/ml limit of detection and was highly selective for histidine-rich protein-2 [26].

14.6.2 Therapy

For the delivery of medicinal medicines, nanofibers have a wide range of advantages. Their enormous surface area aids in improving the solubility and bioavailability of drugs. They can also be used to create customized delivery systems that deliver drugs precisely where they are needed [27]. The medicine can be loaded onto electrospun nanofibers using a variety of techniques, including blending, coaxial electrospinning, emulsion electrospinning, chemical immobilization, and physical adsorption [28, 29]. These approaches are contrasted in Table 14.1.

One capability that makes nanofibers ideal for therapy is their capacity to prevent drug burst release. Also, biocompatibility, biodegradable makeup, which does away with the requirement for surgical removal once the medication load has been released is important. Mechanical toughness and the kind of polymer, nanofiber size, core-shell morphology, mechanism of drug loading, and nanofiber shape all affect how drugs are released from nanofibers. Applications of nanofibers in therapy include oral/oromucosal applications, transdermal and topical applications, intravenous delivery, ophthalmic applications, wound healing, targeted drug delivery system, coating on implants and stents, gene delivery, veterinary drug delivery among others [30–32] (Table 14.2).

14.7 Improved Drug Delivery with AI

Because it increases the therapeutic efficacy of pharmaceuticals, it is essential to optimize the drug delivery mechanism. By overcoming the inherent instability of bioactive chemicals, which compromises the medicine's pharmacokinetic qualities, an improved drug delivery strategy improves the physicochemical properties of the drug. Artificial intelligence (AI) has recently been used by scientists to design, characterize, and optimize medicine delivery. To obtain the greatest therapeutic effect, drug delivery uses a variety of strategies, including formulations, manufacturing processes, storage of bioactive chemicals, and transportation to the target areas [39]. Recently, developments based on nanotechnology have been closely linked to improvements in drug delivery [40]. The development of computer methodologies has sped up the entire process, even though scientists have used a variety of scientific disciplines to optimize drug delivery systems, including pharmaceutical sciences, biosciences, and engineering. AI is used by researchers to design, describe, and produce medication delivery nanosystems [41]. In the past 10 years, AI has been used to analyze and understand genetic and biological data. In

Table 14.1 Comparison of several electrospinning drug loading techniques

	Blending	Coaxial-electrospinning	Emulsion electrospinning	Chemical immobilization	Physical absorption
Definition	Drug dissolution in a polymer solution, and then electrospinning	When a concentric spinneret with two distinct diameters is utilized, core-shell nanofibers can be produced	Emulsification of an oily solution comprising polymer and an aqueous solution containing active	Use of primary amino, hydroxyl, and carboxylic groups in the conjugation of drugs and polymers	Hydrogen bonds, electrostatic interactions, hydrophobic bonds, and van der Waals forces are used to bind the drug to the matrix
Advantages	Simpler than alternative methods	Provides the drug trapped in the core with protection and a sustained release of the substance contained in the core Since the contact between the aqueous and organic phases is kept to a minimum, labile actives' bioactivity is maintained	There is little to no drug contact with the organic solvent and no need for a common solvent	Covalent binding aids in preventing burst release of active, making it suitable for slow and extended delivery of genes and growth factors	Most straightforward procedure, enables instant discharge of active
Disadvantages	It is difficult to distribute drugs evenly across the nanofiber matrix		Increased risk of labile materials being damaged as a result of shearing forces between the aqueous and oily phases	Compounds immobilized may be partially inactivated	

Table 14.2 Examples of nanofibers in wound healing

System	Polymer	Drug	Result	References
Nanofibrous wound dressing	Using PVP	Free ciprofloxacin monohydrochloride monohydrate	Superior wound healing quicker healing than transparent films	[33]
Chitosan/PVA nanofibers decorated with silver nanoparticles	Chitosan/PVA	Sulfanilamide	Silver nanoparticles and sulphanilamides combined antibacterial action results in a faster rate of wound healing in rats	[34–36]
Nanofibers	Chitosan ethylenediaminetetraacetic acid/PVA	Lysozyme	Smooth texture, rapid release of lysozyme, and expedited wound healing In rats as compared to gauze treated wounds	[37]
Nanofibers	Gelatin	Cefazolin	High entrapment efficiency, great oxygen permeability, and sustained drug release	[38]

addition to accelerating the drug development process, it has made it possible to identify functional small compounds quickly and anticipate their behavior.

A significant chemical space with more than 10 compounds can be used to develop a drug. Researchers use AI to swiftly assess therapeutic targets, identify lead compounds with therapeutic promise against a disease, and enhance medication delivery. This essay's main focus is on how AI may improve the delivery of medical care. Integrated circuits, a power supply, sensors, and a secure data backup that is kept up to date utilizing AI-based technologies make up the Nanorobots [42]. Medications are delivered via nanorobots. They are built to avoid accidents, find their intended recipient, connect to them, deliver chemical medications, and then leave the body. Advanced micro- and nanorobots may navigate the targeted regions (like pH) based on physiological conditions. This method reduces systemic negative effects while increasing pharmaceutical effectiveness. Nanorobots that can be implanted provide pharmaceuticals in a regulated and dose-dependent way. Researchers employed artificial neural networks (ANNs), logistic regression, and network-based modeling to investigate pharmaceutical combinations and improve the overall dosage regimen. Combination drug therapy, which requires many drugs to be taken at regular times, is typically more successful than using only one. For instance, a cancer combination therapy may involve six or seven drugs. Researchers have also developed a quadratic phenotypic optimization platform to identify the optimal drug combination for the treatment of bortezomib-resistant multiple myeloma. The creation of a strong drug delivery combination may result from analysis of data on drug interactions and their potential for synergistic or antagonistic effects. The Master Regulator Inference Algorithm, which uses "Mater regulator genes," was successfully used by scientists to predict 56% synergism (Fig. 14.2).

Network-based Laplacian is another platform for controlling synergistic medication combinations used for efficient drug delivery [43]. These AI-driven algorithms

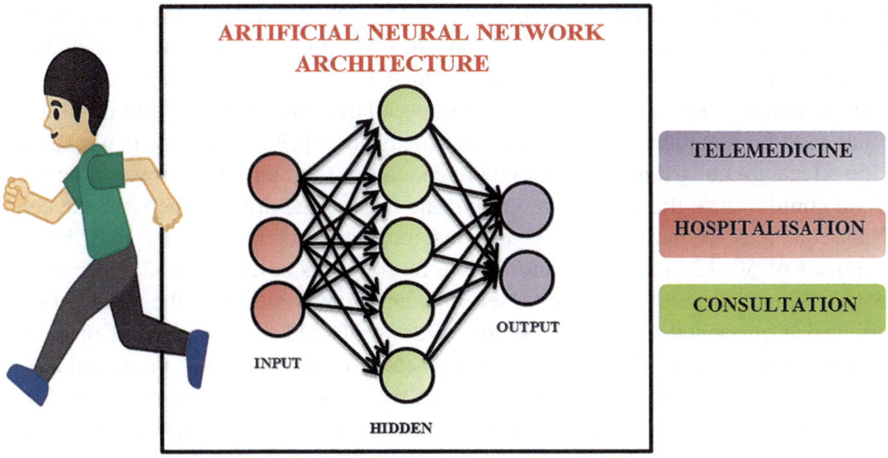

Fig. 14.2 Artificial neural network architecture

could theoretically forecast up to twenty-eight synergistic combinations for anticancer treatments. Combination synergy estimation was devised to predict likely synergistic antimalarial combinations [44].

14.8 The Role of Nanobots in the Future of Medicine

A nanobot is a tiny machine with a diameter of 1–100 nm that is made to carry out specific, occasionally exact tasks. The size of a ribosome is about 30 nm, that of an antibody is about 10 nm, and that of a hemoglobin molecule is about 5 nm [45].

The Greek word for a little person is where the word "nano" comes from. The idea of nanotechnology was initially presented by Nobel Prize–winning physicist Richard Feynman in 1959 in a talk titled "There's plenty of room at the bottom" [46]. However, since Feynman's initial theories, nanotechnology has made considerable advancements. More than 12 years ago, the first nanobot experiment was completed. It dealt with a nanorobot designed to correct genetic flaws in DNA through gene replacement. Although the idea may seem futuristic, it is nonetheless a valid premise [47].

The nanobot's job is to scan nucleotides, look for broken pieces, get rid of those pieces, and then fix the DNA. In the brand-new and fast evolving field of gene therapy, nanobots might eventually have a big impact [48]. A therapy for germs using nanobots is another futuristic technology concept. Imagine a nanobot with the ability to enter bacteria and kill it with heat while causing no harm to surrounding cells. What does it seem to be, in your opinion? Our antibacterial nanobot is a retractable heating sphere that can be injected into a patient and activates when microbes are present. However, occasionally the production of nanobots is motivated by the bacteria themselves. Some nanobots are designed to kill pathogens. For instance, many bacteria possess chemical rotors that they may employ to drive flagella or cilia, which result in movements like a corkscrew or bead. This propulsion strategy enabled rotating synthetic microrobots using flexible filaments, helical microstructures, or tumblers rotating on an axis similar to a bacterial flagellum. Additionally, certain nanoliposome-containing bacteria that are magnetotactic are used to deliver medications. Doxorubicin, popularly referred to as "red devil," is a clear and vividly colored injectable cancer medication [49]. Doxorubicin has serious side effects that might include heart damage, nausea, and hair loss. It would be great if we could reduce these negative effects and deliver the drugs directly to the cells. Nanobots may be advantageous. For instance, a robot nanocarrier may have a diameter of 50–120 nm. The nanobot is a mesoporous silicon sphere that contains doxorubicin molecules. When exposed to near-infrared radiation, the nanocarrier is effectively propelled thanks to a small covering of gold that covers half of it. It is capable of moving at 950 body lengths per second, which is quite rapid. Additionally, the nanocarrier has exceptional cell and tissue penetration as well as the ability to switch between on and off regulated release depending on the environment. Drug side effects are reduced because of the nanobot's capacity to infiltrate the tumor cell by cell. The ability of micro- or nano-surgical tools to access or remove cellular tissues from remote regions of the body would be extremely helpful to surgeons. A 700-mm star-shaped microgripper is a small medical device made of ferromagnetic material and can be accurately controlled with a magnet from up to 10 cm distant. When the robot arrives at the proper location and is propelled by mechanical, electrical, or chemical cues, the microgripper shuts, trapping a cell within. Microgrippers might be used to gather samples and take tissue biopsies without requiring invasive surgical procedures.

In the distant future, nanobots are anticipated to move throughout our bodies. Before they are utilized regularly for disease diagnosis and treatment, they still need to be developed. The staff at Nanobot Medical Animation Studio is continually eager to take on new projects linked to nanotechnology and biotechnology [50].

14.8.1 Nanorobot Structures and Components

There are two basic methods used to build nanorobots. They are the "Self-assembly" and the "Positional assembly." In the positional assembly, billions of molecules are put together and the nanobots can automatically combine them into their natural configuration. In the self-assembly, the robotic arm of the nanomachine that is used to pick and assemble the molecules is controlled manually by an operator [51]. Additionally, nanobots are equipped with swarm intelligence for decentralized activity, a strategy that was modeled after the cooperative behaviors of social creatures like ants, bees, and other insects. But these particular parts and substructures are necessary for the nanobots to perform flawlessly. Components of Nanorobots include power supply, fuel buffer tank, sensors, motors, manipulators, onboard computers, pumps, pressure tanks, and structural support.

14.8.1.1 Nanorobot Substructures

1. **Payload:** The region where the nanobot stores and distributes a tiny dose of medication or a substance.
2. **A little camera:** A tiny camera might be incorporated into the nanorobot. When guiding the nanorobot manually through the body, the operator can.
3. **Electrodes:** By harnessing the electrolytes in the blood, the electrode attached on the nanorobot may create a battery.
4. **Lasers:** These lasers have the ability to destroy dangerous substances such as cancer cells, blood clots, and arterial plaque.
5. **Generating ultrasonic signals:** Use when kidney stones are the target of the nanorobots' attack.
6. The nanorobot will require a form of propulsion to enter the body because they move against the blood flow [52].

14.8.1.2 The First-Ever Self-Propelled Nanobots to Deliver Nanoparticles

In a world first, University of California, San Diego researchers have demonstrated that man-made, microscopic devices may enter a living thing and deliver their therapeutic payload without causing any negative side effects. These tiny devices have been successfully implanted inside the body of a living mouse using micromotor-powered nanobots that are pushed by gas bubbles created by a reaction with the contents of the stomach in which they were put [53]. The teeny robots used in the study were tubular, 5 mm in diameter, 20 mm long, and zinc-coated. When the nanobots were consumed by the mouse and reached the stomach, the hydrogen bubbles they produced when the zinc interacted with the hydrochloric acid in the digestive secretions propelled them like little rockets. The nanobots moved outside into the stomach lining at a speed of up to 60 mm/s, where they were lodged, disintegrated, and transported a nanoparticle compound straight into the gut tissue. The nanobots that were able to penetrate the stomach walls of the mouse's stomach remained firmly adhered to the lining for a full 12 h after ingestion, according to the researchers, proving their effectiveness and durability. Furthermore, once the mouse was finally killed and its stomach was dissected and examined, the presence of the nanobots showed no signs of elevated toxin levels or tissue damage. The researchers said that this was in line with their expectations, especially given that zinc acts as a vitamin with many purposes.

The self-propelled, nanoparticle-delivering micromachines developed by the University of California are the first ones ever made. Although nanobots had previously been employed on biological tissue, such as in the elimination of the Hepatitis C virus, and other nanobots have been created to be pushed by outside forces within a live thing, the University of California micromachines are the first of their kind. This aspect, along with the fact that this is the start of a tried-and-true approach to administer targeted drug administration, leads the research team to conclude that their success thus far justifies additional study. For everyone else, this is a fascinating piece of technology that, in the not-too-distant future, may contribute to the medical care of people. This research is still in its early stages, therefore many more consistently positive tests must be conducted before organizations like the US Food and Drug Administration will ever consider approving its usage in humans. But these initial efforts are crucial for what might someday become a regular, focused, and secure replacement for conventional high-dose drugs.

14.8.1.3 Applications of Nanorobotics in Cancer Therapy and Detection

Nanorobots created with improved detecting capabilities will be able to speed up a cancer diagnosis and hence improve the prognosis of the disease because cancer survival rates increase with early detection [54]. It is possible to create nanobots with integrated chemical sensors to find tumor cells in the body. Integrated communication technology is now used in proposed designs to create two-way signaling. This means that in addition to relaying data they have gathered and responding to acoustic signals, nanobots will also receive programming instructions from outside sound waves. Through carefully placed nanobots in the body that can log data from active nanobots moving through the bloodstream, a straightforward reporting interface might be created. In vivo adjustments to the instructions could be made to offer active targeting for healing or monitoring [55]. Nanorobots equipped with chemical sensors can also be used for medical treatment. Therapy can be administered at both the primary and metastatic stages of cancer by using specific programming to detect various amounts of cancer biomarkers including e-cadherins and beta-catenin. The advantage of nanobots is that they may produce focused medical care. Because healthy cells are destroyed during current cancer treatments, there are serious side effects. Designing nanorobots with chemotactic sensors on their surfaces that correspond to particular antigens on cancer cells can result in targeted treatment.

14.8.1.4 Applications of Nanorobotics in Biohazard Defense

Among other things, nanorobots will be helpful for protecting against biohazards and improving the response to epidemic disease. Nanobots using protein-based biosensors will be able to transmit real-time information in areas with limited public infrastructure and without access to laboratory analysis. This is especially important for biomedical surveillance during humanitarian operations to remote or unstable countries or areas hit hard by epidemic illness. Additionally, nanorobotics may reduce pollution and carry out quarantine screening with success. In the event of an influenza pandemic, elevated levels of the alpha-NAGA enzyme in the blood might be used as a biomarker for the influenza infection. The larger concentration would activate the nanorobot prognostic method, which transmits electromagnetic back propagating signals to portable devices like a cell phone. By redistributing the information and revealing the location of the sick individual, this would hasten the quarantine procedure for contamination.

14.8.1.5 The New Era of Nanomedicine Is Dominated by Nanobots

Traditional water-soluble medications can make treatment challenging because they don't absorb well in the sick areas. However, applications of nanomedicine, such as diagnostic nanomachines, enable direct access to sick regions by enabling the monitoring of the interior chemistry of human organs [55]. Additionally, because technologies like nanobots can be fitted with wireless transmitters, doctors have the option to alter the treatment plan if a patient's condition deteriorates. The neurological system of a patient could possibly be implanted with nanobots in medicine to track heartbeat and brainwave activity. Nanobots can totally replace pacemakers by treating the heart's cells directly, according to scientists. Artificial antibodies, synthetic white blood cells (WBCs) and red blood cells (RBCs), and antiviral nanobots are just a few of the possibilities for medical nanobot research [43, 56]. The main benefit that nanobots offer is that they are incredibly robust. Due to their small size, which prevents mechanical damage, they should theoretically last for years without experiencing any problems. Nanobots and nanomedicine have numerous benefits [57]. As a result, numerous top businesses are funding research and development in this field. Leading businesses' strategic cooperation has accelerated the market for nanomedicine's expansion. In medicine, there is very little room for error.

To be more precise, everything must be perfect. Professional physicians have a wealth of knowledge and training in a range of procedures and other treatments that they may provide to patients. But some medical procedures need such a high degree of precision that not even the greatest doctors can do them. Fortunately, our modern world already has nanorobots and nanomachines. It ensures that patients will get the treatment they need. They will thus have the chance to live better and longer lives.

14.9 Future Prospects for AI and Nanotechnology in Healthcare

There are a few methods in which data science may advance healthcare. In theory, the integration of AI and nanotechnology into healthcare presents an intriguing prospect; however, numerous potential applications remain to be substantiated through empirical evidence. The field is frequently referred to as nanomedicine. As we have already mentioned, nanobots are still a thing of the future, especially for practical applications like medicine. Even while the outlook is positive, there might still be problems when these technologies are put to action. Research on nanotechnology must show that there won't be any negative consequences on the patient's health. Only a handful of the harmful compounds and metals used in nanotechnology include cadmium, arsenic, and lead [58–60].

14.9.1 Smart Drugs

Although the notion of smart pills is not new, a "pill cam" was approved by the FDA in 2001—researchers are developing creative new uses for it. For instance, MIT researchers created a wirelessly controllable ingestible sensor pill. The medication would function as a "closed-loop monitoring and therapy" solution, altering the dosage of a certain medication in accordance with information obtained from the body (e.g., gastrointestinal system). The most recent FDA-approved smart pill that keeps track of when medication was taken is an illustration of this technology in action. The product enables users to track their own medication history using a smartphone or to let doctors and caregivers access that information online. It is approved for adults with schizophrenia and bipolar disorder. Nanotechnology-based therapeutic strategies have recently been used to fight against the coronavirus pandemic [61].

14.9.2 Personalized Medicine

Doctors study the human body in medical colleges using their textbooks. Each sufferer is unique in actuality. Everyone can have a customized experience thanks to nanotechnology. Doctors now have a new set of instruments, whether it be for identifying a tumor, administering medication, or transplanting organs. These instruments will be equipped with hands, eyes, and a brain thanks to machine learning! Even after the procedure, tailored care will be provided through artificial intelligence in healthcare. The health of the patient is monitored, and the information can be used to enhance the procedure for future clients.

14.9.3 Limitations with AI-Based Technology

One of the challenges addressed by AI-based systems is the variety and unpredictability of data sources. Pharmaceutical companies have access to data sets with millions of molecules, yet standard ML algorithms are unable to assess these sorts of data. Although the computer model built on the quantitative structure-activity relationship (QSAR) can predict a wide variety of molecules and removes some restrictions, it also has several serious flaws. There are certain downsides, including small training sets, erroneous experimental data in training sets, and a lack of experimental validation.

14.10 Conclusion

Healthcare advancements that would not otherwise be impossible could be made possible by nanotechnology. In several medical task domains, for example remote surgery, drug delivery, microscopes, and many more, AI has improved clinical diagnosis and decision-making performance. How swiftly AI applications co-evolve with a healthcare system under severe financial strain while accepting swift advancements in molecular and genomic research will decide the influence of this performance on the landscape of medical practice, including illness diagnosis and treatment. The medical education system will need to prepare clinicians for their new duties as information integrators, translators, and patient advocates. Clinicians will need to learn how to play these new roles.

To accomplish so, they will be provided with the means. Because it is yet unclear who will ultimately be in charge of, certify, or profit from the use of AI, it is crucial to strike a balance between legal protections and market forces that will maximize patient benefits. It may be stated that nanotechnology would not become as important without data science. Nanotechnology works at the atomic level. As physics' principles vary, data that has to be examined is frequently noisy and imprecise. In particular, atomic force microscopy produces pictures that are worthless. The data is subsequently filtered by data scientists using computer vision methods. Nanobots sound much more fascinating. They are powered by nanosensors and artificial intelligence. These tiny devices can treat cancer and provide drugs without endangering the patient. Surgeons cannot carry out procedures at the atomic level, despite being skilled and accurate. In the future, genetic illnesses may be prevented via the application of data science and gene modification.

References

1. Yu K, Beam AL, Kohane IS (2018) Artificial intelligence in healthcare. Nat Biomed Eng 2(10): 719–731
2. Chang E, Harford JB, Eaton MP, Boisseau P, Meyer M, Hayeshi R, Swai H, Lee DH (2015) Nanomedicine: past, present and future—a global perspective. Biochem Biophys Res Commun 468(3):511–517

3. Srinivasu PN, Sandhya N, Jhaveri RH, Raut R (2022) From Blackbox to explainable AI in healthcare: existing tools and case studies. Mob Inf Syst 2022:1–20
4. Thakur P, Chahar D, Taneja S, Bhalla N, Thakur A (2020) A review on MnZn ferrites: synthesis, characterization and applications. Ceram Int 46:15740–15763
5. Kumari S, Dhanda N, Thakur A, Gupta V, Singh S, Kumar R, Hameed S, Thakur P (2023) Nano Ca–Mg–Zn ferrites as tuneable photocatalyst for UV light-induced degradation of rhodamine B dye and antimicrobial behavior for water purification. Ceram Int 49(8):12469–12480
6. Kharbanda S, Dhanda N, Sun A-CA, Thakur A, Thakur P (2023) Multiferroic Perovskite bismuth ferrite nanostructures: a review on synthesis and applications. J Magn Magn Mater 572:170569
7. Verma R, Thakur P, Chauhan A, Jasrotia R, Thakur A (2023) A review on MXene and its' composites for electromagnetic interference (EMI) shielding applications. Carbon 208:170–190
8. Saini A, Thakur A, Thakur A (2017) Miniaturization and bandwidth enhancement of a microstrip patch antenna using magneto-dielectric materials for proximity Fuze application. J Electron Mater 46(3):1902–1907
9. Bharti MK, Chalia S, Thakur P, Hermosa GC, Sun A-CA, Thakur A (2021) Low-loss characteristics and sustained magneto-dielectric behaviour of cobalt ferrite nanoparticles over 1–6 GHz frequency range. Ceram Int 47(15):22164–22171
10. Thakur A, Thakur P, Hsu J-H (2011) Smart magnetodielectric nano-materials for the very high frequency applications. J Alloys Compounds 509:5315–5319
11. Pathania A, Thakur P, Trukhanov AV, Trukhanov SV, Panina LV, Luders U, Thakur A (2019) Development of tungsten doped Ni-Zn nano-ferrites with fast response and recovery time for hydrogen gas sensing application. Results Phys 15:102531
12. Tamang S, Humbert-Droz M, Gianfrancesco MA, Izadi Z, Schmajuk G, Robinson P (2023) Practical considerations for developing clinical natural language processing systems for population health management and measurement. JMIR Med Inform 11:e37805
13. Thakur P, Taneja S, Sindhu D, Lüders U, Sharma A, Ravelo B, Thakur A (2020) Manganese zinc ferrites: a short review on synthesis and characterization. J Supercond Nov Magn 33:1569–1584
14. Taneja S, Thakur P, Kumar R, Hemalatha S, Slimani Y, Raveloe B, Thakur A (2022) Nanostructured rare earth Nd^{3+} doped nickel–zinc–bismuth spinel ferrites: structural, electrical and dielectric studies. Ceram Int 48:27731–27738
15. Krull A, Hirsch P, Rother C, Schiffrin A, Krull C (2020) Artificial-intelligence-driven scanning probe microscopy. Commun Phys 3(1):54
16. Rošic R, Kocbek P, Pelipenko J, Kristl J, Baumgartner S (2013) Nanofibers and their biomedical use. Acta Pharma 63(3):295–304
17. Mondal K, Ali MA, Agrawal VV, Malhotra BD, Sharma A (2014) Highly sensitive biofunctionalized mesoporous electrospun TiO_2 nanofiber based interface for biosensing. ACS Appl Mater Interfaces 6:2516–2527
18. Su X, Ren J, Meng X, Ren X, Tang F (2013) A novel platform for enhanced biosensing based on the synergy effects of electrospun polymer nanofibers and graphene oxides. Analyst 138: 1459–1466
19. Haidar M, Eroglu H (2017) Nanofibers: new insights for drug delivery and tissue engineering. Curr Top Med Chem 17(13):1564–1579
20. Hamdan N, Yamin A, Hamid SA, Khodir WKWA, Guarino V (2021) Functionalized antimicrobial nanofibers: design criteria and recent advances. J Funct Biomater 12(4):59. https://doi.org/10.3390/jfb12040059
21. Sapountzi E, Braiek M, Vocanson F, Chateaux J, Jaffrezic-Renault N, Lagarde F (2017) Gold nanoparticles assembly on electrospun poly(vinyl alcohol)/poly(ethyleneimine)/glucose oxidase nanofibers for ultrasensitive electrochemical glucose biosensing. Sens Actuat B Chem 238: 392–401
22. Senthamizhan A, Balusamy B, Uyar T (2016) Glucose sensors based on electrospun nanofibers: a review. Anal Bioanal Chem 408:1285–1306

23. Zhang J, Zhu X, Dong H, Zhang X, Wang W, Chen Z (2013) In situ growth cupric oxide nanoparticles on carbon nanofibers for sensitive nonenzymatic sensing of glucose. Electrochim Acta 105:433–438
24. Ding Y, Wang Y, Su L, Bellagamba M, Zhang H, Lei Y (2010) Electrospun Co_3O_4 nanofibers for sensitive and selective glucose detection. Biosens Bioelectron 26:542–548
25. Ali MA, Mondal K, Singh C, Malhotra BD, Sharma A (2015) Anti-epidermal growth factor receptor conjugated mesoporous zinc oxide nanofibers for breast cancer diagnostics. Nanoscale 7:7234–7245
26. Paul KB, Kumar S, Tripathy S, Vanjari SRK, Singh SG, Singh SG (2016) A highly sensitive self assembled monolayer modified copper doped zinc oxide nanofiber interface for detection of plasmodium falciparum histidine-rich protein-2: targeted towards rapid, early diagnosis of malaria. Biosens Bioelectron 80:39–46
27. Goonoo N, Bhaw-Luximon A, Jhurry D (2014) Drug loading and release from electrospun biodegradable nanofibers. J Biomed Nanotechnol 10(9):2173–2199
28. Hu X, Liu S, Zhou G, Huang Y, Xie Z, Jing X (2014) Electrospinning of polymeric nanofibers for drug delivery applications. J Control Release 185:12–21
29. Zhang Z, Hu J, Ma PX (2012) Nanofiber-based delivery of bioactive agents and stem cells to bone sites. Adv Drug Deliv Rev 64:1129–1141
30. Dwivedi C, Pandey H, Pandey AC, Ramteke PW (2016) Nanofibre based smart pharmaceutical scaffolds for wound repair and regenerations. Curr Pharm Des 22(11):1460–1471
31. Athanassiou A, Heredia-Guerrero JA, Perotto G, Valentini P, Pompa PP, Spanò R, Goldoni L, Bertorelli R (2017) Transparent ciprofloxacin-povidone antibiotic films and nanofiber mats as potential skin and wound care dressings. Eur J Pharm Sci 104:133–144
32. Ganesh M, Aziz A, Ubaidulla U, Hemalatha P, Saravanakumar A, Ravikumar R, Peng MM, Choi EH, Jang HM (2016) Sulfanilamide and silver nanoparticles-loaded polyvinyl alcohol-chitosan composite electrospun nanofibers: synthesis and evaluation on synergism in wound healing. J Ind Eng Chem 39:127–135
33. Charernsriwilaiwat N, Rojanarata T, Ngawhirunpat T, Opanasopit P (2014) Electrospun chitosan/polyvinyl alcohol nanofibre mats for wound healing. Int Wound J 11(2):215–222
34. Rath G, Hussain T, Chauhan G, Garg T, Goyal AK (2015) Fabrication and characterization of cefazolin-loaded nanofibrous mats for the recovery of post-surgical wound. Artif Cells Nanomed Biotechnol 44:1783–1792
35. Shahdeo D, Kesarwani V, Suhag D, Ahmed J, Alshehri SM, Gandhi S (2021) Self-assembled chitosan polymer intercalating peptide functionalized gold nanoparticles as nanoprobe for efficient imaging of urokinase plasminogen activator receptors in cancer diagnostics. Carbohydr Polym 266:118138
36. Shahdeo D, Kesarwani V, Suhag D, Ahmed J, Alshehri SM, Gandhi S (2022) Corrigendum to self-assembled chitosan polymer intercalating peptide functionalized gold nanoparticles as nanoprobe for efficient imaging of urokinase plasminogen activator receptor in cancer diagnostics. Carbohydr Polym 284:119239
37. Paul D, Sanap G, Shenoy S, Kalyane D, Kalia K, Tekade RK (2021) Artificial intelligence in drug discovery and development. Drug Discov Today 26(1):80–93
38. Ramesh A, Kambhampati C, Monson JRT, Drew PJ (2004) Artificial intelligence in medicine. Ann R Coll Surg Engl 86(5):334–338
39. Colombo S (2020) Applications of artificial intelligence in drug delivery and pharmaceutical development. In: Artificial Intelligence in Healthcare. Elsevier, Amsterdam, pp 85–116
40. Thakur A, Thakur P, Khurana SP (eds) (2022) Synthesis and applications of nanoparticles. Springer, Singapore. https://doi.org/10.1007/978-981-16-6819-7
41. Hassanzadeh P (2019) The significance of artificial intelligence in drug delivery system design. Adv Drug Deliv Rev 151–152:169–190
42. Fleming N (2018) How artificial intelligence is changing drug discovery. Nature 557:S55–S57
43. Liu S, Zhao Y, Hao W, Zhang X, Ming D (2020) Micro- and nanotechnology for neural electrode-tissue interfaces. Biosens Bioelectron 170:112645

44. Anjum S, Ishaque S, Fatima H, Farooq W, Hano C, Abbasi BH, Anjum I (2021) Emerging applications of nanotechnology in healthcare systems: grand challenges and perspectives. Pharmaceuticals 14(8):707
45. Zhang Y, Li M, Gao X, Chen Y, Liu T (2019) Nanotechnology in cancer diagnosis: progress, challenges and opportunities. J Hematol Oncol 12(1):137
46. Thakur P, Thakur A (2022) Nanomaterials, their types and properties. In: Thakur A, Thakur P, Khurana SP (eds) Synthesis and applications of nanoparticles. Springer, Singapore, pp 19–44. https://doi.org/10.1007/978-981-16-6819-7_2
47. Kala D, Gupta S, Kaushal A (2022) Nanotechnology in healthcare. In: Thakur A, Thakur P, Khurana SP (eds) Synthesis and applications of nanoparticles. Springer, Singapore
48. Saadeh YS, Vyas D (2014) Nanorobotic applications in medicine: current proposals and designs. Am J Robot Surg 1(1):4–11
49. Termanini RD (2020) The miraculous anatomy of the digital immunity ecosystem. In: Storing digital binary data in cellular DNA
50. El-Readi MZ, Althubiti M (2019) Cancer nanomedicine: a new era of successful targeted therapy. J Nanomater 2019:1–13
51. Cavalcanti A, Shirinzadeh B, Freitas RA, Hogg T (2008) Nanorobot architecture for medical target identification. Nanotechnology 19(1):015103
52. Gao W, Dong R, Thamphiwatana S, Li J, Gao W, Zhang L, Wang J (2015) Artificial micromotors in the Mouse's stomach: a step toward in vivo use of synthetic motors. ACS Nano 9(1):117–123
53. Innovation (2019) Nanobot micromotors deliver medical payload in living creature for the first time. Innovation Toronto https://www.innovationtoronto.com/2015/01/nanobot-micromotors-deliver-medical-payload-in-living-creature-for-the-first-time/
54. Saxena S et al (2015) Design, architecture and application of nanorobotics in oncology. Indian J Cancer 52:236–241
55. Yang X (2007) Nano- and microparticle-based imaging of cardiovascular interventions: overview. Radiology 243(2):340–347
56. Saha M (2009) Nanomedicine: promising tiny machine for the healthcare in future – a review. Oman Med J 24:242–247
57. Robotloginadm & Robotloginadm (2020) What is Nanobots & its biggest contribution in the medical industry. Robotsnet
58. Thakur P, Verma Y, Thakur A (2022) Toxicity of nanomaterials: an overview. In: Thakur A, Thakur P, Khurana SP (eds) Synthesis and applications of nanoparticles. Springer, Singapore, pp 535–544. https://doi.org/10.1007/978-981-16-6819-7_25
59. Thakur A, Punia P, Dhar R, Aggarwal RK, Thakur P (2022) Separation of cadmium and chromium heavy metals from industrial wastewater by using Ni-Zn nanoferrites. Adv Nano Res 12:457–465
60. Thakur SS, Pathania P, Thakur P, Thakur A, Hsu J-H (2015) Improved structural, electrical and magnetic properties of Mn–Zn–Cd nanoferrites. Ceram Int 41:5072–5078
61. Statnano (2020) Could nanotechnology help win the fight against the coronavirus? StatNano

15 Nanomaterials in Bioimaging and Diagnostics

Adeeba Shakeel and Sonali Rawat

Abstract

The downsizing of materials has given rise to various effects, particularly the high surface area-to-volume ratio combined with other unique tunable properties that made nanomaterials appropriate tools for several biomedical applications. By improving existing clinically relevant technologies, nanotechnology has significantly advanced imaging, early detection, diagnosis, and prognosis of diseases. A pressing need for early detection and diagnosis of diseases endlessly drives the advancements of imaging modalities and contrast agents. Current challenges include fast and detailed imaging of tissue microstructures and lesion characterizations that could be achieved via developing nontoxic contrast agents with longer circulation time. With the nanoparticles' unique biophysical properties, it is possible to enhance biomedical imaging and manipulate nanoparticles for molecular specificity to make a tissue-specific diagnosis. Importantly, subtle variation in the size or composition of these nanoparticles results in abundant changes in their optical, magnetic, or electrical properties that allow the unique possibility of tuning them.

Keywords

Nanomaterials · Biomedical applications · Contrast agents · Tissue microstructures

A. Shakeel (✉)
Biotechnology Research Centre, Technology Innovation Institute, Abu Dhabi, United Arab Emirates

S. Rawat
Stem Cells Facility, DBT-CoESCR, All India Institute of Medical Sciences, New Delhi, India

D. Suhag et al. (eds.), *Integrated Nanomaterials and their Applications*,
https://doi.org/10.1007/978-981-99-6105-4_15

15.1 Introduction

Nanoparticles possess unique optical and physic-chemical properties that may potentiate applications in biomedicine, particularly in diagnostics, therapy, and imaging [1]. Advances in biomolecular diagnostics strategies have greatly focused on single molecule detection and characterization of DNA, RNA, or proteins through improved nanoparticle-based platforms. Nanoparticles improve analytical capability compared to traditional techniques with high resolution and medium-high throughput. Several applications have been identified, spanning from monitoring a disease or treatment to identifying problem areas in the body (Fig. 15.1).

Nanomaterials extensively explored for health care diagnostics are mostly either purely organic or inorganic materials or a combination of both, i.e., hybrid nanomaterials. Organic nanomaterials, including carbon nanotubes, nanocrystals,

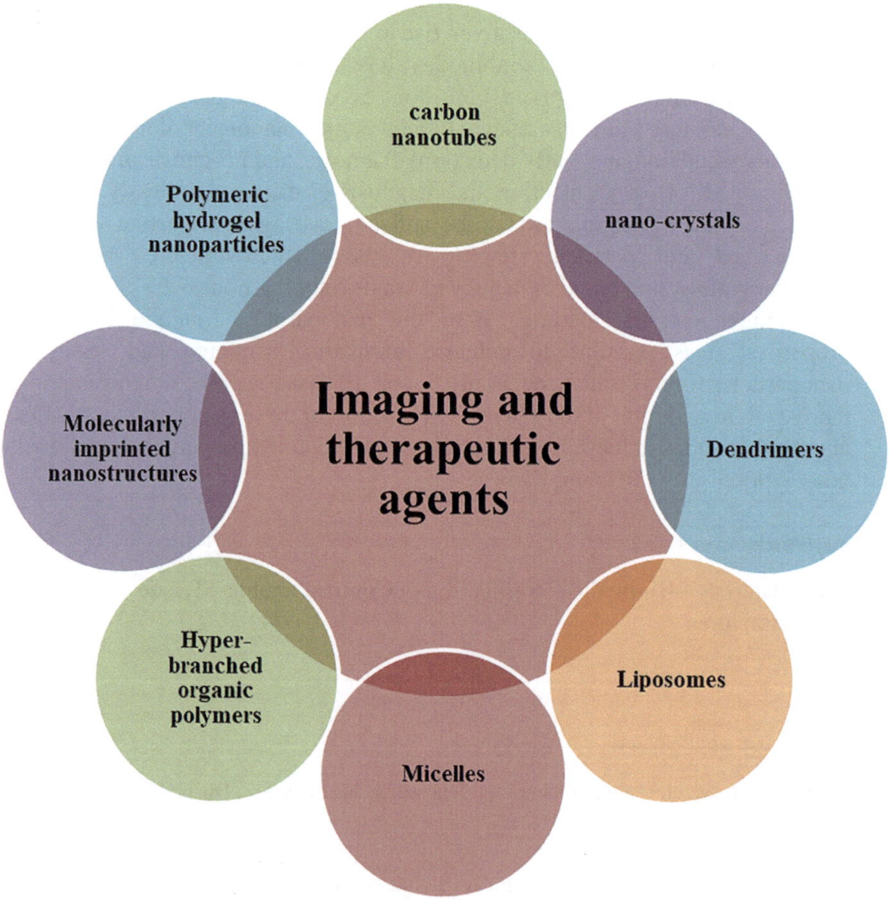

Fig. 15.1 Nanomaterials used for imaging and therapeutic purposes

dendrimers, liposomes, micelles, hyper-branched organic polymers, molecularly imprinted nanostructures, and polymeric hydrogel nanoparticles, have been largely employed as imaging and therapeutic agents [2, 3]. Whereas, inorganic nanomaterials, like metallic nanoparticles, quantum dots, superparamagnetic iron oxide nanoparticles, and metal oxides, have also gathered great attention for health care diagnostics, particularly in biosensing. These materials have been extensively used for various molecular imaging techniques, including magnetic resonance imaging (MRI) [4], positron emission tomography, single-photon emission tomography, optical imaging, and ultrasound imaging [5–7].

The unique characteristics of nanomaterials also allow a wide range of diagnostic approaches, such as surface modification to obtain more available binding sites for immobilized binding receptor molecules and signal enhancement using nanoparticle-labeled disease biomarkers to improve sensitivity and specificity of the bio-detection assays [8]. Nanoparticles are the key players of next-generation multimodal technologies as they provide abundant surface area for functionalization and surface ligands to facilitate site-specific localization and stimuli-responsive behavior. Nanomaterials have been expanding the current situation of molecular diagnostics, point-of-care diagnosis, disease care with therapeutics, and personalized medicine. Integrating biomarker discovery into the nanodevices and allowing the combination and modifications of nanomaterials have significantly improved the clinical and research-based applications for cancer, cardiovascular, infectious, and neurological diseases in recent years. Moreover, Moreover, nanomaterials allow mass production with enhanced functionality, significantly lower costs, and environment-friendly manufacturing processes to improve healthcare [9]. The further engineered nanomaterials, in combination with the current developments, will spectacularly impact healthcare diagnostics and result in synergetic medical solutions [10]. This chapter focusses on the recent applications of nanoparticles in enhancing the bioimaging and diagnostics in clinical settings (Fig. 15.2).

15.2 Nanoparticles for Magnetic Resonance Imaging (MRI)

MRI is a routinely used non-invasive imaging technology that works on the principle of nuclear magnetic resonance (NMR) [12]. MRI employs contrast agents to acquire images by highlighting areas of interest based on contrast uptake. The MRI signals are generated by the protons of the water molecules, while the image contrast is produced by the differences in the intensity of this signal among different tissues depending on the concentration, relaxation times (T_1 and T_2), and mobility of the water molecules within each tissue [13]. Gd(III)-based T_1 contrast agents (GBCAs) have been extensively used for a long time; however, increasing reports on events of nephrotoxicity of these have forced scientists to look for alternatives [14]. Recent developments in the use of nanoparticles have designated them as better contrast agents for MRI and radiotherapy [15]. The large surface-area-to-volume ratio of nanoparticles enhances their interaction rates giving better biological effectiveness of the radiation dose. Additionally, the single-domain magnetization of some

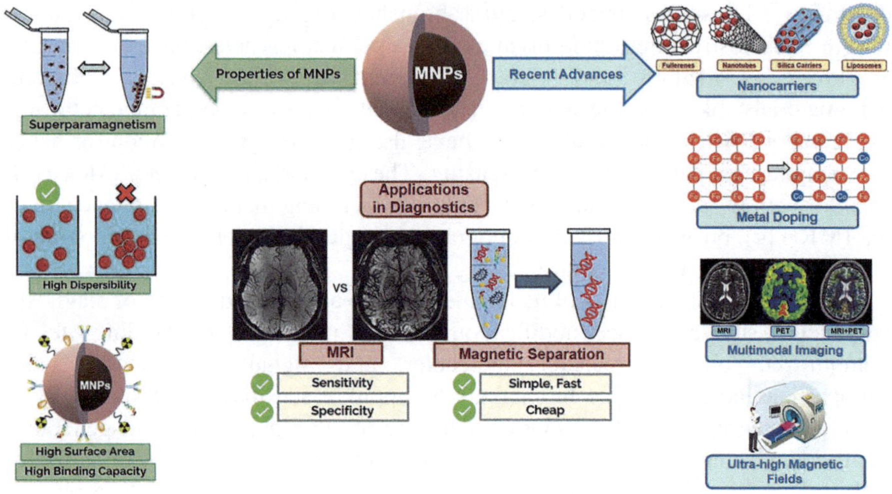

Fig. 15.2 Magnetic nanoparticles in diagnostics [11]

nanoparticles enhances spin interactions with ^{1}H in surrounding water molecules, enhancing MRI contrast [16]. There are several varieties of nanoparticles in various stages of preclinical development, however, regulatory agencies have approved only a few for clinical use, thus demonstrating the challenge of clinical translation [17].

Iron oxide nanoparticles (IONPs)-based contrast agents are being widely used for T_2-weighted magnetic resonance imaging (MRI) in clinical diagnosis due to their high biocompatibility and excellent magnetic properties [1]. Control of the size of the IOPNPs in the 1–100 nm range results in controlled magnetic properties ranging from paramagnetic to ferrimagnetic, thereby significantly improving their clinical performance and application as a T_1 or T_2 contrast agents. Moreover, the ease of functionalization of IONPs surfaces with different types of ligands such as antibodies, peptides, sugars, etc. have made carrying out more specific molecular MRI possible. Also, IONPs functionalized with few epithelial growth factor receptor antibodies have been projected for the diagnosis of various types of cancer, including breast, stomach, colon, kidney, liver or brain cancer [18, 19]. IONPs can effectively accumulate in tumors due to the Enhanced Permeability and Retention effect [20]. Efremova et al. developed IONPs for targeted breast cancer diagnosis in a heterotopic model, wherein IONPs accumulated passively inside the tumor 24 h post intravenous injection using T_2-weighted MR images [21]. Similar studies have been conducted using orthotopic models of breast cancer, pancreatic cancer, and glioblastoma multiforme (GBM). Kucharzyk et al. developed IONPS composite made with three different bioengineered spider silks functionalized with the adhesion peptides carrying a positive or negative charge [22]. When subjected to the alternating magnetic field, these particles generated 2.5 times higher heat causing an almost three times higher percentage of apoptosis in cancer cells than the control particles. Quiet recently Norouzi et al. synthesized biocompatible magnetic iron oxide nanoparticles (IONPs) stabilized with trimethoxysilylpropyl-ethylenediamine triacetic acid (EDT) were developed as a carrier of Doxorubicin for glioblastoma multiforme chemotherapy [23]. The combination not only helped to overcome the blood–brain barrier and multidrug resistance of glioma cells but also provided site-specific magnetic targeting.

In addition to cancer diagnosis, different types of IONPs are being employed for other clinical applications, such as the detection of brain inflammation or the early diagnosis of thrombosis [24, 25]. Furthermore, it was seen that zwitterion-coated superparamagnetic IONPs (3 nm core/1 nm shell) show considerable contrast enhancement compared to the clinical dose of GBCAs in T_1-weighted MRI imaging and angiography in murine models as effective renal clearance [26]. Suzuki et al. invented a fucoidan-coated USPIONs to visualize by MRI arterial thrombi in the early stage of the disease [27]. Highlighting the efficacy of IONPs as drug delivery system, Hussein-Al-Ali et al. developed IONPs having a chitosan shell that encapsulated chlorambucil as an anti-cancer drug for the targeted delivery to leukemia cancer cells [28]. In a similar work, Ebadi et al. formulated polymer-coated IONPs for the targeted delivery of a liver anti-cancer drug, sorafenib, showing positive results [29]. In combination with MRI, IONPs have also been used in many other in vivo applications, such as imaging of activated microglia during brain inflammation [30], tracking of stem cells [31], image-guided treatment of anemia using bacteria loaded with IONPs [32], or to carry out vascular imaging [33].

Despite all these advances, very little clinical translation has been achieved. The main reasons behind this relative failure are likely related to the lack of comprehensive toxicity assays and preclinical imaging studies using appropriate animal models (Fig. 15.3).

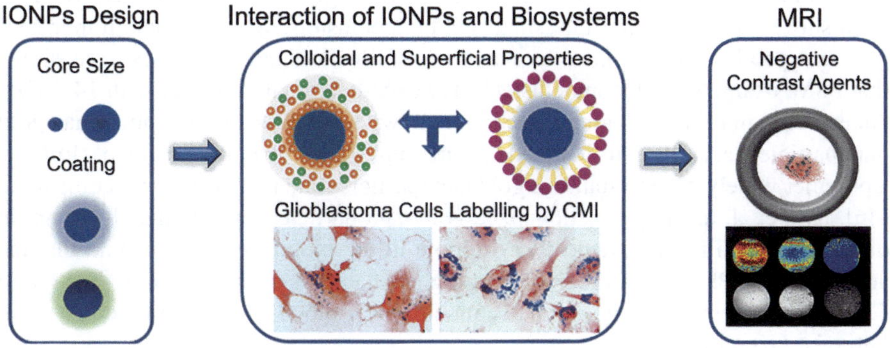

Fig. 15.3 Influence of magnetic nanoparticle size and coating on cellular uptake for in vitro MRI [34]

15.3 Nanoparticles CT Imaging

In clinical practice, CT is one of the most commonly used non-invasive imaging modalities because of its high spatial resolution, wide availability, and rapid acquisition of images [35]. It uses the interaction of X-ray with the body or contrast agent. X-ray attenuation is related to a material's atomic number, so higher Z elements such as Ba or I-based CT contrast agents are used clinically. Besides suffering from poor contrast generation at high voltages, these contrast agents also suffer from rapid clearance and toxicity issues.

Using nano-sized iodinated contrast agents such as micellar, polymeric, and liposomal contrast agents in CT can now overcome current limitations. Gold nanomaterials such as spheres, rods, shells, and cages have also been studied because of their excellent biocompatibility and facile synthesis [36, 37]. Since the gold nanoparticles have K-edge energy level of 80.7 keV, they exhibit better CT contrast effect than iodine-based contrast agents with quite low energy levels [38, 39]. Depending on the particle size and crystal structure, there are two types of gold nanoparticles, clusters, and colloidal particles and their size can be tuned to around 1–2 nm level [40]. Their ultrasmall size is the key to their clinical translations to avoid unnecessary retention in the body and subsequent toxicity. Advancing surface chemistry and nanotechnology have enabled gold nanoparticles to be developed as radiosensitizers [41]. PEG or zwitterions stabilized AuNPs display remarkable circulation in vivo and show high tumor accumulation thanks to their enhanced permeability and retention (EPR). To further enhance gold nanoparticles' accumulation and tumor targeting specificity, active targeting ligands for multiple tumor biomarkers such as anti-Her2 antibodies (Herceptin) [42], folic acid [43], and RG [44] have been conjugated to gold nanoparticles to construct active tumor targeting CT contrast agents. Due to long retention time of gold nanoparticles, efforts have been shifted to developing ultrasmall gold nanoclusters, which display tumor contrast and good renal clearance. In recent work, Basilion and colleagues have shown that PSMA-targeted gold nanoclusters have excellent renal clearance with 14% ID/g Au excreted in urine 24 h after injection, and two times less liver retention than gold nanoparticles, according to ICP and CT imaging [45]. Polyethylene glycol (PEG) is a polymer widely used to stabilize gold nanoparticles and render them biocompatible [46]. Kim et al. recently functionalized gold nanoparticles (approximately 30 nm in diameter) with PEG to extend their lifetime in the bloodstream and found that the attenuation of PEG–GNPs is 5.7 times higher than that of the current iodine-based CT contrast agent, Ultravist and had a much longer blood circulation time (4 h) than Ultravist (10 min) [47].

Contrast agents such as radiolabeled gold nanoparticles that facilitate multimodal imaging capabilities like PET/CT or single-photon emission computed tomography (SPECT)/CT have garnered more interest owing to enhanced contrast generated from orthogonal imaging methods. Gold nanoparticles coated with dendrimer (labeled with ^{99m}Tc) shows improved imaging capability due to high sensitivity SPECT imaging that enhances the information from CT imaging (54). Recently, Zhu et al. constructed functional technetium-99m (^{99m}Tc) labeled polyethyleneimine

(PEI)-entrapped gold nanoparticles (Au PENs) with pH-responsive charge conversion property for enhanced single-photon emission computed tomography (SPECT)/ computed tomography (CT) dual mode imaging of cancer cells [48]. Although several studies using AuNPs for cancer imaging or therapy did not exhibit significant toxicity, the long-term effect of AuNPs in the body is still uncertain, and few studies have investigated a prolonged time for toxicity and renal clearance. Moreover, the cost associated with the gold nanoparticles also restricts their efficient translation.

Micelles and lipoproteins are another micelle-related class of nanoparticles that have notably been adapted as CT contrast agents. et al. doxorubicin loaded An amphiphilic copolymer poly(−caprolactone)-ss-poly(2-(dimethylamino) ethyl methacrylate), PCLSS-PDMAEMA [49]. The drug-loaded gold nanoparticle micelles possessed better in vitro CT imaging properties than the conventional contrast agent, Omnipaque, serving as a promising "theragnostic platform" for efficient cancer thermotherapy and diagnosis. Similarly, Torchilin described a method to prepare polymeric iodine-containing PEG-based micelles giving a long-circulating blood pool imaging agent for CT.

Falling into the class of metallic nanoparticles, Bismuth (Bi) is another suitable candidate for an X-ray contrast agent due to its high atomic number, high density, and good X-ray attenuation (absorption $k_{edge} = 90.5$ keV). Moreover, the photocatalysis and photothermal properties of the Bi also enable it to be used as a multimodal contrast agent. Tungsten, tantalum, ytterbium, gadolinium, silver, iron, platinum, and copper are a few other nanoparticles that are being investigated for their potential in CT and SPECT. Such metallic nanoparticles provide targeted and dual/multipurpose contrast enhancement properties and therapeutic effects. Research has shown that they have strong potential to be able to take place at the market; however, more research is needed to shed light to the clinical potential of metallic nanoparticles keeping in mind the factor of long-term toxicity.

15.4 Nanoparticles for Ultrasound Imaging and Therapies

Compared to other imaging methods, ultrasound imaging also known as ultrasonography (US, is frequently used in clinical imaging due to its inherent merits, such as its wide adaptability, availability, non-radioactivity, non-invasive nature, and real-time monitoring. Microbubbles are the most widely used US contrast agents, as they appear bright on US images because of their high compression and strong scattering [50]. However, conventional microbubbles are not only too large to exit the vasculature but also suffer from low stability. Nanoparticles have appeared as useful tools for wide application in the US and can be applied for extravascular ultrasonic imaging or therapeutic purposes by incorporating multimodal agents. US imaging can direct the US therapeutic process and monitor the efficiency of therapy for cancer and many other serious diseases. US possesses important physicochemical properties for other therapeutic applications, such as sonodynamic therapy (SDT) [51], high-intensity focused US (HIFU) [52], US-triggered drug release [53] and

US-enhanced gene transfection [54] Therefore, nanoparticle-based US imaging and therapeutic modalities have been extensively used in the clinical settings.

Gas stabilizing nanoparticles (GSNs) are a distinct class of ultrasound responsive materials that gained a lot of popularity specially for US imaging and therapy of cancer [12]. They are engineered to form gas pockets either on their surface or within hydrophobic cavities, which can function as heterogeneous nucleation sites to produce micron-sized cavitating bubbles under ultrasound irradiation [55]. Their fluorocarbon-free nature and size as small as <50 nm ensures excellent signal stability in ambient storage for months and after intravenous injection for several days [56]. Additionally, GSNs, when administered either solely or in combination with a sonosensitizer have shown promising outcomes in sonodynamic therapy [57]. Researchers also have a proven record in localized drug delivery applications using external ultrasound pulses however, their role in HIFU is still limited due to the requirement of HIFU activation pulses, which complicates the ultrasound imaging system and brings safety issues [58].

Another class of nanomaterials that garnered much interest for US were Mesoporous Silica Nanoparticles (MSNs), made of silica with a porous structure that allows the introduction of therapeutic agents in their network of cavities. Earlier efforts to encapsulate perfluoropentane (PFP), within a hollow silica shell considerably increased the US image contrast and imaging lifetime and exhibited targeting ability after modification with targeting small molecules in vitro (breast tissue from mastectomy) [59]. Moreover, Chen et al. designed exosome-like silica nanoparticles that differed significantly from other silica structures used for US imaging in vitro [60] while Janus porous silicon NPs designed by Tamarov et al. generated microbubbles under US radiation and prolonged the blood circulation time [61]. Multiwalled carbon nanotubes that have been functionalized for high compatibility were found to provide enhanced and long-lasting imaging without toxicity, demonstrating their potential as US contrast imaging agents with a wide range of frequencies. Additionally, ex vivo and in vivo studies conducted using calcium carbonate ($CaCO_3$) hybrid nanoparticles loaded with doxorubicin (DOX-$CaCO_3$-MNPs) exhibited simultaneous tumor-specific US imaging and effective antitumor therapeutic function [62].

Apart from these, few organic NPs have also shown promising results in US based imaging and therapeutics. Li et al. outlined the applications of Perfluorooctyl Bromide nanoparticles NPs for enhanced US contrast imaging and drug delivery [63]. They showed that PFOB NPs coated with lipids were useful for drug delivery while PFOB NPs coated with polymers were more fit for US contrast imaging. Moreover, PFP/PLGA-PEG-folic acid NPs have been designed for targeted US imaging by phase shift induced by low-intensity focused US irradiation.

15.5 Nanoparticles for Multimodal Imaging

Over the years it has been realized that no single imaging technique can offer the optimum combination of properties (e.g., resolution, sensitivity, cost, availability), thus efforts to combine modality or multimodal imaging are rapidly growing and so is the requirement for advanced and specific multimodal imaging agents. The large surface areas of nanoparticles make them ideal for these applications since different modality reporters and targeting vectors can be incorporated to alter localization properties. Multimodal imaging with an optical contrast agent coupled to MRI or PET/SPECT has been heavily explored to improve imaging resolution and diagnosis sensitivity. Radioisotope/MR imaging agents are typically designed by beginning with a magnetic nanoparticle core and then either attaching radioactive isotopes to the surface or incorporating them into the core. Voulgari et al. created a poly (methacrylic acid)-graft poly (ethylene glycol) copolymer-derived iron oxide nanoparticle system for cisplatin delivery to tumors that outperformed free cisplatin in vivo, especially in the presence of an external magnetic field [64].

In a recent report, photo uncaging nanoparticles (PUNPs) with CdSe/ZnS QDs and SPIONs enabled the combination of NIR fluorescence imaging with MRI and were successfully used for in vitro and in vivo imaging. UNPs and QDs exhibit excellent fluorescence signal after injection, whereas SPIONs exhibit excellent MRI contrast. In another case, combining Gd-DTPA with Ag2Se QDs enabled T_1-weighted MRI and fluorescence imaging with high tissue penetration and excellent temporal and spatial resolution. There are also many progresses in the combination of optical imaging with radioisotopes, The Cornell Dots being the prime example. Developed in 2014 and still under clinical trials, these agents combine targeting peptides, radioisotopes, and an optical dye (cyanine 5 dye) into a silica nanoparticle. A PET/NIR dual-modality imaging system was developed by doping Cu to gold nanoclusters to form a self-illuminating imaging agent that allowed successful imaging of tumor without a need for external excitation. Interestingly, it has been seen that a combination of optical with MRI contrast agents overcomes the low sensitivity of MRI and allows the collection of real-time information during in vitro assessment, allowing for rapid screening and identification of key characteristics for iterative design. Very recently, a precision delivery strategy using the magnetic targeting and ultrasound-triggered release were designed. This multifunctional nano-system with responsiveness to the magnetic field and ultrasound by synergizing different functions of its components was used for targeted and controlled delivery of thrombolytic drugs.

Another interesting study carried out by Chen et al. described nano-agents for trimodal imaging of ischemic myocardium in rats [65]. The team prepared a targeted nanoprobe (IMTP-Fe_3O_4-PFH NPs) with enhanced ultrasound (US), photoacoustic (PA), and magnetic resonance (MR) performance for direct and non-invasive visual imaging of ischemic myocardium in a rat model. Post-injection of NPs through the tail vein of model rats, in vivo imaging results, showed a significantly enhanced US/PA/MR signal. Despite such advances, the increased nanoparticle complexity significantly limits potential clinical translation and must be improved in quality and efficiency to allow successful translation to clinical imaging and diagnostics.

15.6 Nanotechnology-Assisted Cell Tracking

Long-term, continuous in vivo cell tracking can provide detailed knowledge about cellular functions and medicinal treatments. Numerous biomedical imaging methods have been investigated for these applications, from the bench to the bedside, including positron emission tomography (PET), single-photon emission computed tomography (SPECT), magnetic resonance imaging (MRI), magnetic particle imaging (MPI), photoacoustic (PA) imaging, and fluorescence imaging. As a result, translational research must develop adaptable contrast agents as long-term cell trackers to observe the target for at least a few weeks. Direct and indirect labeling are now the two main categories of cell labeling techniques used. Each tactic has benefits and drawbacks of its own. Generally speaking, the direct labeling method has the benefits of simple preparation, high labeling effectiveness, and wide availability of exogenous contrast agents, whereas the indirect labeling method, which involves genetic change, can enable persistent cell tagging. Among them, bioluminescence is the most common and well-suited indirect labeling technology, a natural light source based on the oxidation of its luciferin substrate by the enzyme luciferase.

Nonetheless, bioluminescence imaging technology still faces a lot of obstacles and restrictions. For instance, unstable bioluminescence experiences signal degradation, and the imaging process needs a highly sensitive CCD lens. Their slow progress in translational research is further hampered by their weak signals, which require repeated luciferin injections, long detection times, expensive costs, and the possibility that transgenic markers will affect cells, genes, or antibodies.

Direct cell labeling with inorganic NPs designed with several tracers has drawn more interest in the last decade for enhancing the monitoring and imaging of tissue and/or organ regeneration. The multi-labeling approach, also known as the employment of a combination of two or more chemical elements to be tracked using various imaging techniques, does in fact provide a way to go around a particular technique's limitations in terms of imaging detection. Examples include inorganic gold NPs with red-emitting firefly luciferase (RfLuc)-based bioluminescence (BL) tags and inorganic NPs with gold and gadolinium elements that have been created for the non-invasive labeling and tracking of human mesenchymal stem cells (hMSCs) in a mouse model of pulmonary disease. To assess the in vivo migration, homing, functions, and survival of transplanted hMSCs and gain a better understanding of pulmonary fibrosis therapy, multimodal imaging with CT, MRI, and BL was successfully utilized [66, 67].

A wide range of research interests has been sparked by the investigation of exogenous contrast agents, such as nanoparticle (NP)-based cell trackers, for biological and/or preclinical investigation. Clinical studies evaluated numerous nanoparticles, including superparamagnetic iron oxide NPs (SPIONs), carbon NPs, silver NPs, and CdS/ZnS quantum dots [68]. Phase 4 clinical studies for central venous catheters based on silver nanoparticles (NPs) and SPIONs for MRI-guiding pancreatic cancer and catheter-related infections have been completed. This suggests the preclinical investigation and development of NP-based trackers or medicines is crucial.

For a variety of reasons, the use of artificial nanoparticles in biological research has grown significantly during the past 5 years. Unique physical, chemical, and optical properties have been found at the nanoscale scale. In order to accurately regulate the size and shape of nanoparticles in order to modify their absorption and emission properties, new synthetic techniques have been devised. The high surface-to-volume ratio of nanoparticles has been used in conjunction with surface modification, or "biofunctionalization," to enable multivalent ligand binding to target biomolecules.[113] The molecular targeting of nanoparticle beacons is now enabling a wide range of scientific applications.

Bioluminescence imaging, which enables in vivo cell tracking together with precise anatomical knowledge of specific organs, is not as spatially precise as magnetic resonance imaging (MRI), which has a spatial resolution of 50 μm. In vivo cell tracking has shown encouraging results using iron oxide nanoparticles (IONs), gadolinium (Gd)-based nanoparticles, manganese (Mn)-based nanoparticles, F-based nanoparticles, and SPIONs as MRI contrast agents. High spatial resolution, desirable penetration depth, strong biocompatibility, efficacy, and non-ionizing radiation are all characteristics of these contrast agents.

Mesenchymal stem cells (MSCs) generated from mouse bone marrow were recently tagged by Ashraf et al. using SPIONs enclosed in polyelectrolyte multilayers. The encapsulated SPIONs worked well as a labeling agent for stem cells at a low feeding concentration (100 g Fe mL^{-1}) since they had a two-fold higher absorption efficiency of Fe than those labeled with naked SPIONs [69]. Rivas et al. modified polyacrylic acid (PAA)-coated ultrasmall iron oxide nanoparticles (USPIO-PAA-GlcN) with glucosamine to improve cellular absorption and biocompatibility, whereas PAA-coated SPIONs (SPIONs-PAA) and PAA-coated USPIONs (USPIONs-PAA) were utilized for comparison. Due to their severe aggregation behavior in cell culture media, the SPIONs-PAA were only detected in some cells with non-uniform distribution after incubation with MSCs [70].

In a recent study, Daldrup-Link et al. used matrix-associated stem cell implants (MASIs) tagged with ferumoxytol to predict cartilage repair results with MRI in a minipig model. Compared to unlabeled MASIs, the ferumoxytol-labeled MASIs had significantly shorter T_2 times (22.2 ms 3.2 vs. 27.9 ms 1.8; $P = 0.001$). After transplanting it into cartilage defects, this made it easier to image the MASI. Additionally, due to the quick loss of iron after cell death, ferumoxytol-labeled apoptotic MASIs had a longer T_2 relaxation duration than labeled live MASIs 2 weeks after implantation (26.6 ms 4.9 vs. 20.8 ms 5.3; $P = 0.001$) [71].

Metallic nanoparticles have tremendous potential as X-ray contrast imaging agents because of their strong X-ray absorption and low toxicity profiles that have been reported in animals for brief periods. Due to their possible biocompatibility, low short-term toxicity, high absorption coefficient, and higher physical density compared to iodine (gold 79(Z), 5.16 cm^2/g, 19.32 g/cm^3; iodine 53(Z), 1.94 cm^2/g,

4.9 g/cm^3), gold nanoparticles have attracted much attention. As a result, there is a high demand for the safe synthesis of these nanoparticles to allay worries about potential toxicity brought on by reducing agents and reaction conditions [72].

Paclitaxel (PTX)-loaded human serum albumin (HSA) nanoparticles are sold under the brand name "Abraxane," and it has been clinically proven to be effective in treating cancer [73]. To increase the biocompatibility of some therapeutic medications, HSA is typically used to encapsulate them [74]. For cancer imaging and therapy, Wang et al. (2016) [71] created a targeted agent using PTX and AIEgens. Theranostic nanoplatforms had the following four components: AIEgens coupled to HSA for imaging, HSA modified by cyclic arginine-glycine-aspartic acid (cRGD) for specific recognition, HSA functionalized polypyrrole (PPy) for thermotherapy, and PTX acting as a chemotherapeutic agent and mediating protein synthesis are some examples of HSA-modified HSA. First, by putting AIEgens into a hydrophobic pocket of HSA, which gave AIEgens a high fluorescence, HSA-AIEgens were created. Interestingly, the final product displayed significant fluorescence intensity even after conjugation with quencher-PPy thanks to the exceptional properties of AIEgens. The imaging capabilities of the AIR genes-based nanoplatforms were then examined to determine the effectiveness of tumor treatment in mice. The mice's entire bodies had AIEgens fluorescence after receiving an intravenous injection for almost 1 h; this fluorescence was connected to the high levels of NPs in the blood. The nano-agent then demonstrated greater accumulation in tumors and improved fluorescence vs. time, demonstrating its capacity to target tumors precisely. Additionally, 48 h after injection, the nano-agent's fluorescence in mice was still robust, demonstrating AIEgens' high bioimaging capability and exceptional in vivo retention.

In vivo fluorescence imaging increasingly uses quantum dots as fluorophores. Fluorescence imaging provides a number of benefits over other imaging modalities since it is non-invasive, has good sensitivity, and uses widely accessible and reasonably priced equipment [75]. Its depth of tissue penetration is constrained by its optical nature. Numerous in vivo investigations have confirmed the effectiveness of QDs. In the lungs of mice, CdSe/Zns QDs coated with PEG and a lung-targeting peptide have been shown to accumulate significantly [76].

15.7 Conclusion and Future Prospects

For biomedical imaging, the design and development of thoroughly validated nanoparticle-based contrast agents necessitate a variety of skills in fields like radiology, engineering, and chemistry. To realize products, this multidisciplinary research strategy must be continued. To be selective to receptors, nanoparticle-based contrast agents will need knowledge of target identification, colloidal chemistry, biochemistry, analysis, and stability. The process of formulating a biomedical imaging agent may be similar to that used to create novel therapeutic medications, but there will be additional difficulties in addressing colloid stability and the stability of conjugated ligands. As they enter the clinic, nanoparticle contrast agents will face difficulties in repeatable large-scale manufacture with a good value-to-cost ratio. Publications, for instance, frequently highlight a contrast agent's fivefold increases in tumor selectivity without mentioning its ten-fold increase in liver concentration. When possible, rudimentary toxicity testing using blood samples should be incorporated into in vivo research because they are typically relatively straightforward.

The potential for nanoparticle application in biological imaging is apparent. In order to combat non-specific organ absorption and RES, synthesis methodologies should continue to generate nontoxic, biocompatible, and biodegradable nanoparticles. The dosages of nanoparticles could be reduced to enhance imaging performance by raising the contrast-to-noise ratio, directing the agent to tissues of interest, or enhancing instrument sensitivity to increase the possibility of translation into the clinic.

References

1. Thakur P, Thakur A (2022) Introduction to nanotechnology. In: Thakur A, Thakur P, Khurana SP (eds) Synthesis and applications of nanoparticles. Springer, Singapore, pp 1–17. https://doi.org/10.1007/978-981-16-6819-7_1
2. Thakur P, Thakur A (2022) Nanomaterials, their types and properties. In: Thakur A, Thakur P, Khurana SP (eds) Synthesis and applications of nanoparticles. Springer, Singapore, pp 19–44. https://doi.org/10.1007/978-981-16-6819-7_2
3. Kala D, Gupta S, Kaushal A (2022) Nanotechnology in healthcare. In: Thakur A, Thakur P, Khurana SP (eds) Synthesis and applications of nanoparticles. Springer, Singapore, pp 405–416. https://doi.org/10.1007/978-981-16-6819-7_19
4. Avasthi A, Caro C, Pozo-Torres E, Leal M, García-Martín ML (2020) Magnetic nanoparticles as MRI contrast agents. Top Curr Chem 378(3):40
5. Miedema IHC, Zwezerijnen G, Huisman MC, Doeleman E, Mathijssen RH, Kiessling F, Hu Q, Van Dongen GA, Rijcken CJ, Vugts DJ, Van Oordt CWMDH (2022) PET-CT imaging of polymeric nanoparticle tumor accumulation in patients. Adv Mater 34(21):2201043
6. Forte E, Fiorenza D, Torino E, Di Polidoro AC, Cavaliere C, Netti PA, Salvatore M, Aiello M (2019) Radiolabeled PET/MRI nanoparticles for tumor imaging. J Clin Med 9(1):89
7. Wang Y, Sun Z, Chen Z, Wu Y, Gu Y, Lin S, Wang Y (2019) In vivo photoacoustic/single-photon emission computed tomography imaging for dynamic monitoring of aggregation-enhanced photothermal Nanoagents. Anal Chem 91(3):2128–2134
8. Ahmad F, Salem-Bekhit MM, Khan F, Alshehri S, Khan A, Ghoneim MM, Wu H, Taha EI, El-Bagory IM (2022) Unique properties of surface-functionalized nanoparticles for bio-application: functionalization mechanisms and importance in application. Nano 12(8):1333
9. Mashaghi S, Jadidi T, Koenderink G, Mashaghi A (2013) Lipid nanotechnology. Int J Mol Sci 14(2):4242–4282
10. Sangam S, Garg P, Sanyal T, Pahari S, Khurana SMP, Mukherjee M (2022) Graphene quantum dots and their hybrid hydrogels: a multifaceted platform for Theranostic applications. In: Thakur A, Thakur P, Khurana SP (eds) Synthesis and applications of nanoparticles. Springer, Singapore, pp 417–453. https://doi.org/10.1007/978-981-16-6819-7_20
11. Farinha P, Coelho J, Reis CP, Gaspar MM (2021) A comprehensive updated review on magnetic nanoparticles in diagnostics. Nano 11(12):3432
12. Blanco E, Shen H, Ferrari M (2015) Principles of nanoparticle design for overcoming biological barriers to drug delivery. Nat Biotechnol 33(9):941–951
13. Enriquez-Navas PM, García-Martín ML (2012) Application of inorganic nanoparticles for diagnosis based on MRI. In: Frontiers of nanoscience. Elsevier BV, pp 233–245

14. Kanal E, Tweedle MF (2015) Residual or retained gadolinium: practical implications for radiologists and our patients. Radiology 275(3):630–634
15. Schuemann J, Bagley AF, Berbeco R, Bromma K, Butterworth KT, Byrne HL, Chithrani DB, Cho SH, Cook JC, Favaudon V, Gholami YH, Gargioni E, Hainfeld JF, Hespeels F, Heuskin A, Ibeh U, Kuncic Z, Kunjachan S, Lacombe S et al (2020) Roadmap for metal nanoparticles in radiation therapy: current status, translational challenges, and future directions. Phys Med Biol 65(21):21RM02
16. Wang YJ (2011) Superparamagnetic iron oxide based MRI contrast agents: current status of clinical application. Quant Imaging Med Surg 1(1):35–40
17. Hua SZ, De Matos MIS, Metselaar JM, Storm G (2018) Current trends and challenges in the clinical translation of nanoparticulate nanomedicines: pathways for translational development and commercialization. Front Pharmacol 9:790
18. Naumenko V, Garanina AS, Nikitin AA, Vodopyanov S, Vorobyeva NS, Tsareva Y, Kunin MA, Ilyasov A, Semkina AS, Chekhonin VP, Abakumov MA, Majouga AG (2018) Biodistribution and tumors MRI contrast enhancement of magnetic nanocubes, nanoclusters, and nanorods in multiple mice models. Contrast Media Mol Imaging 2018:1–12
19. Wu L, Wang C, Li Y (2022) Iron oxide nanoparticle targeting mechanism and its application in tumor magnetic resonance imaging and therapy. Nanomedicine 17(21):1567–1583
20. Caro C, Pozo D (2015) Polysaccharide colloids as smart vehicles in cancer therapy. Curr Pharm Des 21(33):4822–4836
21. Efremova MV, Naumenko V, Spasova M, Garanina AS, Abakumov MA, Blokhina AA, Mel'nikov PA, Prelovskaya A, Heidelmann M, Li Z, Ma Z, Shchetinin I, Golovin YI, Kireev II, Savchenko AG, Chekhonin VP, Klyachko NL, Farle M, Majouga AG, Wiedwald U (2018) Magnetite-gold nanohybrids as ideal all-in-one platforms for theranostics. Sci Rep 8(1):11295
22. Kucharczyk K, Kaczmarek K, Józefczak A, Slachcinski M, Mackiewicz A, Dams-Kozlowska H (2021) Hyperthermia treatment of cancer cells by the application of targeted silk/iron oxide composite spheres. Mater Sci Eng C 120:111654
23. Norouzi M, Yathindranath V, Thliveris JA, Kopec BM, Siahaan TJ, Miller DR (2020) Doxorubicin-loaded iron oxide nanoparticles for glioblastoma therapy: a combinational approach for enhanced delivery of nanoparticles. Sci Rep 10(1):11292
24. Zhu F, Hu Y, Yu L, Zhou X, Wu J, Tang Y, Qin D, Fan Q, Wu A (2021) Nanoparticles: a hope for the treatment of inflammation in CNS. Front Pharmacol 12:683935
25. Russell P, Hagemeyer CE, Esser L, Voelcker NH (2022) Theranostic nanoparticles for the management of thrombosis. Theranostics 12(6):2773–2800
26. Wei H, Bruns OT, Kaul MG, Hansen E, Mariya B, Wiśniowska A, Chen O, Chen Y, Li N, Okada S, Cordero JM, Heine M, Farrar CT, Montana DM, Adam G, Ittrich H, Jasanoff A, Nielsen P, Bawendi MG (2017) Exceedingly small iron oxide nanoparticles as positive MRI contrast agents. Proc Natl Acad Sci U S A 114(9):2325–2330
27. Suzuki M, Bachelet-Violette L, Rouzet F, Beilvert A, Autret G, Maire M, Ménager C, Louedec L, Choqueux C, Saboural P, Haddad O, Chauvierre C, Chaubet F, Michel J, Serfaty JM, Letourneur D (2015) Ultrasmall superparamagnetic iron oxide nanoparticles coated with fucoidan for molecular MRI of intraluminal thrombus. Nanomedicine 10(1):73–87
28. Hussein-Al-Ali SH, Hussein MZ, Saifullah B, Arulselvan P (2021) Chlorambucil-iron oxide nanoparticles as a drug delivery system for leukemia cancer cells. Int J Nanomedicine 16:6205–6216
29. Ebadi M, Saifullah B, Buskara K, Hussein MZ, Fakurazi S, Pastorin G (2020) Release of a liver anticancer drug, sorafenib from its PVA/LDH- and PEG/LDH-coated iron oxide nanoparticles for drug delivery applications. Sci Rep 10(1):21521
30. Tang TK, Valenzuela A, Petit F, Chow S, Leung K, Gorin FA, Louie AY, Dhenain M (2018) In vivo MRI of functionalized iron oxide nanoparticles for brain inflammation. Contrast Media Mol Imaging 2018:1–10
31. Gu L, Li X, Jiang J, Guo G, Wu H, Wu M, Zhu H (2018) Stem cell tracking using effective self-assembled peptide-modified superparamagnetic nanoparticles. Nanoscale 10(34):15967–15979

32. Garcés V, Rodríguez-Nogales A, González AM, Gálvez N, Rodríguez-Cabezas ME, García-Martín ML, Gutiérrez L, Rondón D, Olivares M, Gálvez J, Domínguez-Vera JM (2018) Bacteria-carried iron oxide nanoparticles for treatment of anemia. Bioconjug Chem 29(5): 1785–1791
33. Nimi N, Saraswathy A, Nazeer SS, Francis N, Shenoy SJ, Jayasree RS (2018) Multifunctional hybrid nanoconstruct of zerovalent iron and carbon dots for magnetic resonance angiography and optical imaging: an in vivo study. Biomaterials 171:46–56
34. Cortés-Llanos B, Ocampo SM, De La Cueva L, Calvo GF, Belmonte-Beitia J, Pérez L, Salas G, Ayuso-Sacido A (2021) Influence of coating and size of magnetic nanoparticles on cellular uptake for in vitro MRI. Nano 11(11):2888
35. Lee N, Choi SH, Hyeon T (2013) Nano-sized CT contrast agents. Adv Mater 25(19):2641–2660
36. Lusic H, Grinstaff MW (2013) X-ray-computed tomography contrast agents. Chem Rev 113(3): 1641–1666
37. Dreaden EC, Alkilany AM, Huang X, Murphy CJ, El-Sayed MA (2012) The golden age: gold nanoparticles for biomedicine. Chem Soc Rev 41(7):2740–2779
38. Xi D, Dong S, Meng X, Lu Q, Meng L, Ye J (2012) Gold nanoparticles as computerized tomography (CT) contrast agents. RSC Adv 2(33):12515
39. Taghavi H, Bakhshandeh M, Montazerabadi A, Moghadam HE, Shahri SBM, Keshtkar MM (2020) Comparison of gold nanoparticles and iodinated contrast media in radiation dose reduction and contrast enhancement in computed tomography. Iran J Radiol 17(1):e92446
40. Fitzgerald PB, Colborn RE, Edic PM, Lambert JL, Torres AS, Bonitatibus PJ, Yeh BM (2016) CT image contrast of high-Z elements: phantom imaging studies and clinical implications. Radiology 278(3):723–733
41. Song G, Cheng L, Chao YC, Yang K, Liu Z (2017) Emerging nanotechnology and advanced materials for cancer radiation therapy. Adv Mater 29(32):1700996
42. Chen J, Chen J, Bhattacharjee S, Cao Z, Wang H, Swanson SJ, Zong H, Baker JR, Wang S (2020) Functionalized nanoparticles with targeted antibody to enhance imaging of breast cancer in vivo. J Nanobiotechnol 18(1):135
43. Wang H, Zheng L, Peng C, Shen M, Shi X, Zhang G (2013) Folic acid-modified dendrimer-entrapped gold nanoparticles as nanoprobes for targeted CT imaging of human lung adencarcinoma. Biomaterials 34(2):470–480
44. Liang G, Jin X, Zhang S, Xing D (2017) RGD peptide-modified fluorescent gold nanoclusters as highly efficient tumor-targeted radiotherapy sensitizers. Biomaterials 144:95–104
45. Luo D, Wang X, Zeng S, Ramamurthy G, Burda C, Basilion JP (2019) Targeted gold nanocluster-enhanced radiotherapy of prostate cancer. Small 15(34):1900968
46. Scotson C, Muñoz-Hernando M, Duncan S, Ruiz S, Keyes SD, Van Veelen A, Dunlop IE, Roose T (2019) Stabilizing gold nanoparticles for use in X-ray computed tomography imaging of soil systems. R Soc Open Sci 6(10):190769
47. Kim D, Park S, Lee JS, Jeong Y, Jon S (2007) Antibiofouling polymer-coated gold nanoparticles as a contrast agent for in vivo X-ray computed tomography imaging. J Am Chem Soc 129(24):7661–7665
48. Zhu J, Zhao L, Yang J, Chen L, Shi J, Zhao J, Shi X (2019) 99mTc-labeled polyethylenimine-entrapped gold nanoparticles with pH-responsive charge conversion property for enhanced dual mode SPECT/CT imaging of cancer cells. Langmuir 35(41):13405–13412
49. Xiong D, Zhang X, Peng S, Gu H, Zhang L (2018) Smart pH-sensitive micelles based on redox degradable polymers as DOX/GNPs carriers for controlled drug release and CT imaging. Colloids Surf B: Biointerfaces 163:29–40
50. Liu Y, Ai K, Lu L (2012) Nanoparticulate X-ray computed tomography contrast agents: from design validation to in vivo applications. Acc Chem Res 45(10):1817–1827
51. Aslan N, Ceylan B, Koç MM, Findik F (2020) Metallic nanoparticles as X-ray computed tomography (CT) contrast agents: a review. J Mol Struct 1219:128599
52. Qin S, Caskey CF, Ferrara KW (2009) Ultrasound contrast microbubbles in imaging and therapy: physical principles and engineering. Phys Med Biol 54(6):R27–R57

53. Wan Q, Zou C, Hu D, Zhou J, Chen M, Changjun T, Yangzi Q, Yan F, Chuanli C, Zheng H, Zhang B, Liu X, Liang D (2019) Imaging-guided focused ultrasound-induced thermal and sonodynamic effects of nanosonosensitizers for synergistic enhancement of glioblastoma therapy. Biomater Sci 7(7):3007–3015
54. Yildirim A, Shi D, Roy S, Blum NT, Chattaraj R, Cha JN, Goodwin AL (2018) Nanoparticle-mediated acoustic cavitation enables high intensity focused ultrasound ablation without tissue heating. ACS Appl Mater Interfaces 10(43):36786–36795
55. Batchelor DVB, Abou-Saleh RH, Coletta PL, McLaughlan JR, Peyman SA, Evans SD (2020) Nested nanobubbles for ultrasound-triggered drug release. ACS Appl Mater Interfaces 12(26): 29085–29093
56. Mira JJ, Wu L, Sabuncu S, Sapre AA, Civitci F, Ibsen S, Esener SC, Yildirim A (2020) Gas-stabilizing sub-100 nm mesoporous silica nanoparticles for ultrasound THERANOSTICS. ACS Omega 5(38):24762–24772
57. Ho YJ, Wu CH, Jin QF, Lin CY, Chiang PH, Wu N et al (2020) Superhydrophobic drug-loaded mesoporous silica nanoparticles capped with β-cyclodextrin for ultrasound image-guided combined antivascular and chemo-sonodynamic therapy. Biomaterials 232:119723
58. Grundy M, Bau L, Hill CC, Paverd C, Mannaris C, Kwan JJ, Crake C, Coviello C, Coussios CC, Carlisle R (2021) Improved therapeutic antibody delivery to xenograft tumors using cavitation nucleated by gas-entrapping nanoparticles. Nanomedicine 16(1):37–50
59. Haar GT, Coussios C (2007) High intensity focused ultrasound: physical principles and devices. Int J Hyper 23(2):89–104
60. Chen F, Ma M, Wang J, Wang F, Chern SX, Zhao ER et al (2017) Exosome-like silica nanoparticles: a novel ultrasound contrast agent for stem cell imaging. Nanoscale 9(1): 402–411.74
61. Tamarov K, Sviridov A, Xu W, Malo MK, Andreev V, Timoshenko VY, Lehto V (2017) Nano air seeds trapped in mesoporous Janus nanoparticles facilitate cavitation and enhance ultrasound imaging. ACS Appl Mater Interfaces 9(40):35234–35243
62. Ma J, Yu M, Zhang ZG, Cai W, Zhang Z, Zhu H, Cheng Q, Tian Z, Pang D (2018) Gd-DTPA-coupled Ag_2Se quantum dots for dual-modality magnetic resonance imaging and fluorescence imaging in the second near-infrared window. Nanoscale 10(22):10699–10704
63. Li X, Sui Z, Li X, Xu W, Guo Q, Sun J, Jing F (2018) Perfluorooctylbromide nanoparticles for ultrasound imaging and drug delivery. Int J Nanomedicine 13:3053–3067
64. Voulgari E, Bakandritsos A, Galtsidis S, Zoumpourlis V, Burke BP, Clemente GS, Cawthorne C, Archibald SJ, Tucek J, Zboril R, Kantarelou V, Karydas AG, Avgoustakis K (2016) Synthesis, characterization and in vivo evaluation of a magnetic cisplatin delivery nanosystem based on PMAA-graft-PEG copolymers. J Control Release 243:342–356
65. Chen X, Zhang Y, Zhang H, Zhang L, Liu L, Cao Y, Ran H, Tian J (2021) A non-invasive nanoparticles for multimodal imaging of ischemic myocardium in rats. J Nanobiotechnol 19(1): 82
66. Phillips EH, Penate-Medina O, Zanzonico P, Carvajal RD, Mohan P, Ye Y, Humm JL, Gonen M, Kalaigian H, Schöder H, Strauss HW, Larson SM, Wiesner U, Bradbury MS (2014) Clinical translation of an ultrasmall inorganic optical-PET imaging nanoparticle probe. Sci Transl Med 6(260):260ra149
67. Lewis CM, Graves SE, Hernandez R, Valdovinos HF, Barnhart TE, Cai W, Meyerand ME, Nickles RJ, Suzuki M (2015) 52Mn production for PET/MRI tracking of human stem cells expressing divalent metal transporter 1 (DMT_1). Theranostics 5(3):227–239
68. Li L, Jiang WG, Luo K, Song H, Lan F, Wu Y, Gu Z (2013) Superparamagnetic iron oxide nanoparticles as MRI contrast agents for non-invasive stem cell labeling and tracking. Theranostics 3(8):595–615
69. Ashraf S, Taylor AC, Sharkey J, Barrow M, Mayans O, Wilm B, Poptani H, Rosseinsky MJ, Adams DJ, Lévy R (2019) In vivo fate of free and encapsulated iron oxide nanoparticles after injection of labelled stem cells. Nanoscale Adv 1(1):367–377

70. Mertens ME, Frese J, Bölükbas DA, Hrdlicka L, Golombek S, Koch S, Mela P, Jockenhövel S, Kiessling F, Lammers T (2014) FMN-coated fluorescent USPIO for cell labeling and non-invasive MR imaging in tissue engineering. Theranostics 4(10):1002–1013
71. Wang K, Fan X, Zhao L, Zhang X, Zhang X, Li Z, Yuan Q, Zhang Q, Huang Z, Xie W, Zhang Y, Wei Y (2016) Aggregation induced emission Fluorogens based Nanotheranostics for targeted and imaging-guided chemo-photothermal combination therapy. Small 12(47): 6568–6575
72. Sangam S, Gupta A, Shakeel A, Bhattacharya R, Sharma A, Suhag D, Chakrabarti SK, Garg S, Chattopadhyay S, Basu B, Kumar V, Rajput SK, Dutta MK, Mukherjee M (2018) Sustainable synthesis of single crystalline sulphur-doped graphene quantum dots for bioimaging and beyond. Green Chem 20(18):4245–4259
73. Ma P, Mumper RJ (2013) Paclitaxel Nano-Delivery Systems: A Comprehensive Review. J Nanomed Nanotechnol 4(2):1000164. https://doi.org/10.4172/2157-7439.1000164. PMID: 24163786; PMCID: PMC3806207
74. Desai N (2016) Nanoparticle albumin-bound paclitaxel (Abraxane®). Springer eBooks, pp 101–119
75. Jaiswal JK, Goldman ER, Mattoussi H, Simon SM (2004) Use of quantum dots for live cell imaging. Nat Methods 1(1):73–78
76. Shahdeo D, Kesarwani V, Suhag D, Ahmed J, Alshehri SM, Gandhi S (2021) Self-assembled chitosan polymer intercalating peptide functionalized gold nanoparticles as nanoprobe for efficient imaging of urokinase plasminogen activator receptor in cancer diagnostics. Carbohydr Polym 266:118138

Bioinspired Nanomaterials

16

Fayu Wan, Saarthak Kharbanda, Preeti Thakur, and Atul Thakur

Abstract

Nanotechnology has advanced, and several methods of creating nanomaterials have been created. Among them, bioinspired or biomimetic nanoplatforms have been employed for a variety of biomedical purposes. Bioinspiration is the creation of new materials, devices, and structures that are inspired by biological evolution and refinement over millions of years. The goal is to improve biological system modeling and simulation in order to gain a better understanding of nature's critical structural features, such as a wing, for use in future bioinspired designs. Bioinspiration differs from biomimicry in that the latter seeks to precisely replicate biological material designs. Bioinspired research is a return to science's classical roots: it is a field that focuses on observing the remarkable functions that distinguish living organisms and attempting to abstract and imitate those functions. By combining science and nature, it is possible to produce a wide variety of new materials with amazing qualities by taking the best traits from natural species. In addition to 2D nanomaterials remarkable qualities, bioinspired 2D nanomaterials and technologies have produced important advancements.

F. Wan
School of Electronics and Information Engineering, Nanjing University of Information Science & Technology, Nanjing, China

S. Kharbanda (✉) · P. Thakur
Amity Centre for Nanotechnology, Amity University Haryana, Gurgaon, Haryana, India
e-mail: pthakur@ggn.amity.edu

A. Thakur
School of Electronics and Information Engineering, Nanjing University of Information Science & Technology, Nanjing, China

Amity Centre for Nanotechnology, Amity University Haryana, Gurgaon, Haryana, India
e-mail: athakur1@ggn.amity.edu

D. Suhag et al. (eds.), *Integrated Nanomaterials and their Applications*,
https://doi.org/10.1007/978-981-99-6105-4_16

Bioinspired nanomaterials and nanostructures have received significant attention from the domains of biology and medicine. This chapter gives a broad review of the creation of bioinspired nanomaterials and nanostructures, as well as the applications of these materials in organ chips, DNA nanorobots, immune cell adhesion, smart nanochannels, and bioinspired mineralization. Finally, the possibility of biologically inspired nanomaterials and nanostructures is examined.

Keywords

Biomimicry · Bioinspiration · 2D-nanomaterials · Nanostructure

16.1 Introduction

The creation of new technologies using concepts distilled from the study of biological systems is known as biomimetics. A flow of ideas from the biological sciences into engineering gives rise to "biomimetic technologies," which take the advantage of natural selection's millions of years of design work in living systems [1]. A crucial aspect of biomimicry or bioinspiration is taking inspiration from nature's designed principles. This involves developing incredibly complex engineering models at various length scales and using the wealth of knowledge to solve some of humanity's most serious issues [2]. When you take into account the variety of functions that natural materials and systems have evolved to carry out, such as structural support, signal transduction, actuation, sensing, catalysis, trafficking, gating, light-harvesting, charge transfer, molecular recognition, self-assembly, self-organization, self-replication, or combinations of two or more of these functions, you can see that nature's "solution manual" is quite comprehensive. As a class of simple-to-use biomaterials, bioinspired nanomaterials have proven to be effective in addressing a variety of biomedical problems by serving as adaptable instruments for biosensing, bioimaging, bio-catalysis, antibacterial treatment, and biotherapy [3]. The surface of the functional nanomaterial can be subjected to a variety of chemical modifications or biological interactions, enabling it to recognize a particular spot in complicated surroundings both in vitro and in vivo. Unlike many biological systems, which demand a very unique lock-and-key technique for molecular interaction, surface engineering offers a new way to make specificity of bio-functional nanomaterials [4]. The attachment of biological molecules, such as proteins, antibodies, and nucleic acids, to the surface of nanomaterials is a crucial prerequisite for ensuring the biosafety of such materials. An advanced nanomaterial's key characteristics are its low toxicity, excellent biocompatibility, and distinctive stimuli responses including light, electricity, and magnetism, which broaden its range of bio-applications [5].

Recently, bioinspired nanoparticles with improved surface bio-physicochemical characteristics have been produced using nanotechnology and biomimetic techniques for drug delivery and vaccine development. These biomimetic nanoparticles have a number of advantages over conventional nanomaterials, including their diversity, adaptability, and reproducibility as well as their usefulness, complexity, and biocompatibility [6]. Biologically inspired nanoparticles can perform well inefficient nano-therapies or nano-vaccines for infectious diseases because of their inherent activity.

This chapter reviews the creation of bioinspired nanomaterials and divides them into two subclasses: bio-templates and bio-mimics. By taking cues from nature's extraordinary finesse, scientists have begun developing nanomaterials with a range of biological uses. The primary factor in the extraordinary functionalities of these bioinspired nanomaterials is the fact that the human biological system is made up of nanoscale self-assemblies of biological components. Over the past few decades, tremendous progress has been made in the creation of biomimicry and bioinspired materials, including nanomaterials for tissue engineering and orthopedic implants, smart robots, and a new class of materials that mimic the capacity of living things to adapt and self-regulate [7].

16.2 Bioinspired Nanovesicles

Lipid-based nanovesicles have been extensively exploited as significant drug delivery carriers in the field of nanomedicine due to their superior biocompatibility and sample preparation techniques. Two lipids make up the exterior layer of nanovesicles, while an aqueous cavity fills the inside. The first described lipid nanovesicle was a liposome, which has been utilized to carry medications and genes [8]. An innovative class of drug delivery vehicles is comprised of bioinspired nanovesicles, which also include biomimetics and cell-derived nanovesicles [9]. Exosomes are fused to form these vesicles after whole cells are extruded, followed by the creation of nanoparticles coated with cell-derived membranes. As a result, there is little chance of an immunological reaction and a high loading capacity is guaranteed. They can also be employed for gene delivery, immunotherapy, tumor targeting, and medication delivery because of their great biocompatibility.

16.2.1 Extracellular Vesicles

Extracellular vesicles, a variety of substances produced by cells, are crucial for cell-to-cell communication. The smallest extracellular vesicles are exosomes, which have a size range of 50–150 nm. Because they may be transferred from one place to another, they are utilized as medicine delivery agents. In some circumstances, they are also used as diagnostic markers. It is occasionally challenging to employ it as the only drug delivery method, nevertheless, because of difficulties that many researchers confront, such as poor loading capacity and getting a smaller number of exosomes under normal circumstances. Therefore, bioinspired exosomes come into play as an alternative to naturally generated exosomes, and they have already been cited as an effective treatment for a number of conditions [10].

16.2.2 Exosomes

Exosomes produced by many cell types, including immune cells and mesenchymal stem cells (MSCs) under particular circumstances, have been found to have various therapeutic effects [11]. The major histocompatibility complex on the heads of exosomes generated from B cells triggers T-cell reactions. This implies that exosomes may be employed as an immunomodulatory agent, as shown by exosomes originating from dendritic cells (DC) that were combined with tumor antigens, inducing immune responses and reducing the survival of established tumors. Exosomes are employed for immunotherapy and immunomodulation, as well as loading cargo and displaying antigens on their surfaces. For example, melanoma (skin cancer)-derived exosomes from mouse models loaded with immunomodulatory CpG DNA exhibited antigens on their surface that were more successful in eliminating tumors than either exosomes or DNA alone. Bovine milk is employed in the cost-effective and large-scale manufacturing of exosomes via bioinspired synthesis. When xenograft mice with A549 lung cancer were treated with aferin A (WFA) three times per week, anticancer activity enhanced. When compared to non-targeted exosomes (50%), exosomes treated with the ligand folic acid (FA) showed a 74% anticancer impact. Other organizations were also effective in loading medications into milk-derived exosomes for oral delivery, such as anthocyanidins and paclitaxel [12].

16.3 Wound Healing Dressing Mat with a Bioinspired Design

Wound therapy is complicated since some conditions, such as diabetes and cardiovascular disease, aggravate the condition. Site dressing is vital in the healing process because it mimics ECM, adheres to the site, and eventually migrates to the wound, aiding in the healing and skin regeneration process. Polycaprolactone (PCL) has been widely employed in wound healing due to its strong biocompatibility, biodegradability, and simplicity of availability. It may be combined with other polymers to increase its mechanical qualities and tissue regeneration capabilities [13]. Electrospinning was utilized in this study to construct biodegradable and ecologically friendly polyhydroxy butyrate/poly-3-caprolactone (PHB/PCL) mats that imitate the extracellular matrix (ECM) and give structural and biochemical evidence for tissue regeneration [14]. The aforementioned mats were changed by melanin-TiO_2 nanostructures, which were inspired by the natural component melanin, which is frequently employed as an antimicrobial agent. The antibacterial activity of these coated mats was substantial against both gram-positive and gram-negative bacteria strains. They demonstrated high water retention, hydrophilicity, and in vitro activity, all of which indicated contact and attachment [15].

16.4 Carbon Dots/Biodots Inspired by Nature

Quantum dots have been widely employed in the biomedical area as a potent carrier due to their optical qualities dependent on their size. One of the green chemistry approaches for producing fluorescent carbon quantum dots (C-dots) on a large scale using waste material and natural resources as carbon starting materials (such as orange, ginger, and sugarcane juice) is the hydrothermal approach. The preparation methods are challenging, time-consuming, and have low reproducibility. Watermelon peel, milk, lignin, sugarcane juice, coffee grounds, chicken eggs, food scraps, bananas, hair, ginger, onion scraps, honey, bread, candle soot, chitosan, and gelatine have also been used as carbon sources [16–19]. Because citrus limetta juice has a lot of sugar as a carbohydrate, which is the beginning point for C-dot manufacturing, Asiya F. Shaikh employed a hydrothermal method to quickly create C-dots that are extremely luminous. In vitro activity, investigations have shown that they have anti-adhesion and anti-biofilm formation abilities against Candida albicans growing on polystyrene surfaces. In summary, this innovative method provides a fresh method for producing C-dots from natural carbon sources (Fig. 16.1) [20].

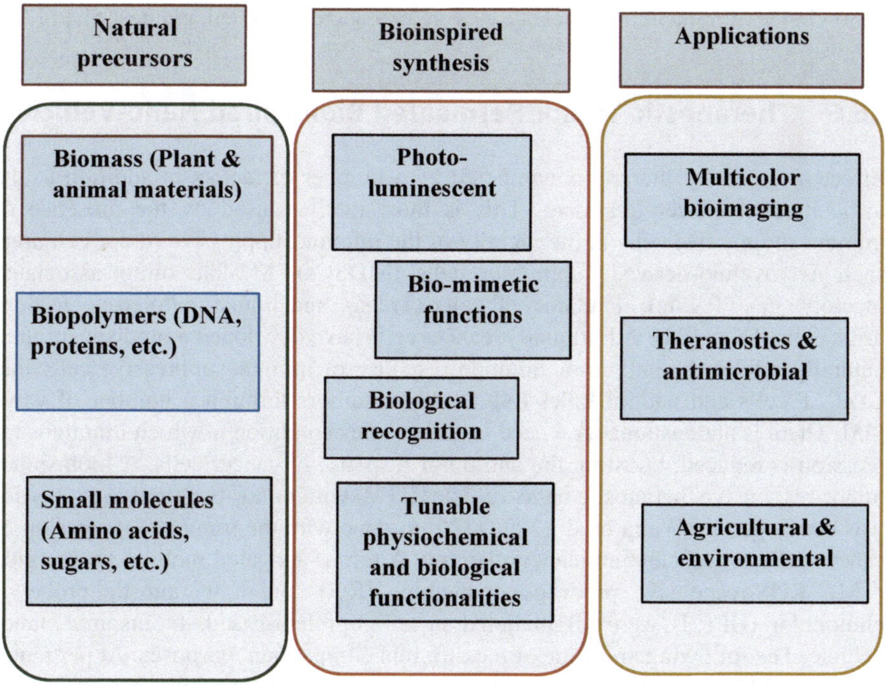

Fig. 16.1 Bioinspired carbon dots/biodots natural precursors, bioinspired synthesis and applications

16.5 Nanoparticles of Viral Origin

Viruses are increasingly being utilized in the creation of bioinspired/biomimetic nanoplatforms due to their special features. Viral nanomaterials (VLPs) can be made from virus nanoparticles (VNPs) and virus-like particles [21]. The latter is utilized for the delivery of pharmaceuticals and bioimaging agents, as well as the creation of inorganic NPs [22]. The capsid, a protein coat that covers a virus, is a smart material because of its monodispersity, symmetry, and polyvalency. The helical virus is a prefabricated scaffold with a distinctive design and a high surface area-to-volume ratio that makes it possible for it to produce different kinds of nanostructures. The perfect starting point for the production of new nanomaterials that mix inorganic or organic moieties precisely and controllably is planted virus capsids. Furthermore, icosahedral three-dimensional designs with structural symmetry are formed when spherical plant virus capsid proteins are put together. They may be applied for a variety of biological purposes with just a few straightforward adjustments. This leads to the utilization of TMV, a helical virus that causes tobacco mosaic virus disease in tobacco plants, to create bioinspired nanomaterials. Due to its elongated hollow tube-like shape (4 nm in diameter), it can be readily converted into nanorods. Three-nanometer-diameter cobalt and nickel nanowires are produced in the middle hollow area. Using a unique recipe, Tsukamoto and associates produced bimetallic Co-Pt and Fe-Pt alloy nanowires in the hollow channel of the TMV. Co-Pt and Fe-Pt nanowire development and nucleation were extensively studied and described [23].

16.6 Theranostic Tumor Permeated Bioinspired Nano-Vehicle

Effective oncology therapy is hampered by a number of factors in addition to the difficulties of cancer treatment. This is most likely caused by the presence of immunosuppressive cells in the vicinity of the immunosuppressive tumor cell area, such as myeloid-derived suppressor cells (MDSCs), M2-like tumor-associated macrophages (TAMs), regulatory T cells (Tregs), and immature/tolerogenic dendritic cells (DCs) [24]. Additionally, cancer cells have developed a predisposition to naturally inhibit the antitumor immune response of immunosuppressive cells like $CD8^+$ T cells and natural killer (NK) cells in tumors through a number of ways [25]. There is unquestionably a need to address this condition in which immunosuppression is reduced, boosting the antitumor response of cancer cells. A bioinspired tumor-responsive theranostic nano-vehicle (BTV) with tumor-penetrating properties was developed by Wang et al. (2021) [26] to cope with the immunosuppression of cancer cells for efficient anticancer therapy. A tumor-activated melittin pro-peptide (TM), ROS-responsive prodrug gemcitabine (RG), and a theranostic probe of photochlor (HPPH) were all included in a lipoprotein-based, bioinspired nano-vehicle. The following are some of the different compounds' purposes: At particular locations, TM has been enzymatically transformed to active melittin, increasing the action of pharmaceuticals. This boosts tumor penetration and accumulation capability. RG (prodrug) was converted into active gemcitabine, an immunomodulator.

When HPPH was exposed to radiation, it produced singlet oxygen, which boosted the anticancer activity of the resulting nano-vehicle. This allowed HPPH to serve as a ranostic probe in BTN for in vivo systemic tumor monitoring. Surprisingly, this combination therapy markedly decreased tumor immune suppression and increased the infiltration of cytotoxic lymphocytes, which is essential for reducing tumor immune suppression and dramatically inhibiting tumor development. In a word, this innovative design opens the door for the delivery of a nanoplatform with exceptional immune suppression-relieving qualities that may be applied in efficient anticancer therapy.

16.7 Bioinspired Polymeric Nanoparticles

16.7.1 Nanoparticles of Poly (Lactic-Co-Glycolic Acid)

Organ-specific medication targeting is still a challenge in the domains of drug delivery and nanomedicine. Scientists developed programmable bioinspired nanoparticles (P-BiNPs) that can target bone localization and deliver a payload to homotypic cancer cells in animal models. They started by encouraging cancer cells to over express integrin on their outermost surface. These programmed cancer cell membranes were then applied to the polymeric NPs, where they were absorbed by prostate cancer cells, enhancing the therapeutic potential of the medicine, the clarity of the image, and eventually minimizing adverse effects. These NPs were coated with biologically inspired nanomaterials, which prolonged their circulation duration, prevented immune system recognition, and enhanced biocompatibility [27].

16.7.2 Alginate Nanoparticles

The discipline of nanotechnology has extensively generated and studied natural and synthetic nanoparticles (NPs) for medication delivery [28]. Alginate, chitosan, gelatin, collagen, poly (D, L-lactide), poly (D, L-glycolide), poly (lactic acid), and poly (lactide-co-glycolide) are all included [29]. A form of polymeric drug delivery vehicle called alginate nanoparticles improves the bioavailability and, consequently, the effectiveness of many medications. Alginate, a naturally occurring sugar polysaccharide, has mucoadhesive properties and can stick to the plasma membrane as a result of the polymer's cationic makeup [30]. Various chemical strategies have been developed throughout the years to produce appropriate alginate NPs. However, these traditional procedures lack adequate control, call for NP dispersion and size, and ultimately call for the use of corrosive organic solvents that can be hazardous to the in vivo environment. Bubble bursting is a natural occurrence in the marine environment that produces nano- and micro-sized particles. The main cause of wave breaking is bubble film breakdown and jetting. By adopting the microbubble-bursting technique, alginate NPs with a size range of 80–200 nm were created. The study team created microbubbles of various sizes in the most precise way conceivable using a T junction microfluidic device. The size achieved was proportional to the process's alginate solution's viscosity [31].

16.8 Metallic Nanoparticles Inspired by Nature

16.8.1 Bioinspired Silver Nanoparticles (AgNPs)

Using various fractions of Elaeagnus umbellate extract (EU), Ali et al. (2020) [32] reduced silver nitrate to silver NPs in order to create bioinspired silver nanoparticles (AgNPs). The developed NPs were morphologically controlled, and their utility was evaluated according to their size and form. Additionally, the size, shape, and bactericidal activity of the NPs were investigated using scanning electron microscopy (SEM) and atomic force microscopy (AFM). They observed that the NPs were monodisperse, non-toxic, and roughly 40 nm in size. Also, they were effective against both gram-positive and gram-negative strains of Escherichia coli (E. coli) and Staphylococcus aureus (S. aureus). Electrostatic interaction took place between the NPs and the bacterial cell wall. The findings demonstrated that the accumulation of NPs on the cell surface of E. coli was quicker and more pronounced than that of S. aureus, which was likely caused by the two bacterial strains' differing cell wall compositions. When NPs entered bacteria, they reacted with the sulfhydryl groups, denatured the proteins, and subsequently changed the activity of the enzymes.

16.8.2 Biologically Inspired Gold Nanoparticle

Graphene, one of the most sophisticated nanomaterials, has proven to have important scientific relevance for potential future applications of nanotechnology [33]. One of the best biocompatible nanoplatforms, it is used in antibacterial, antiviral, cancer-targeting, medication delivery, and photo thermal treatment [34]. Researchers made gold nanoparticles (AuNPs) with decreased graphene oxide coatings. They used syzygium cumin seed extract (GO) to lower graphene oxide and chloroauric acid. Biophysical techniques including UV-Vis spectroscopy (UV-Vis), dynamic light scattering (DLS), and Fourier transform infrared spectroscopy (FTIR) were used to carefully analyze its physicochemical properties. The results showed that the deposition of graphene oxide on AuNPs during their creation was successful. The human colorectal cancer cell line (HCT 116) and lung cancer cell line (A549), as well as the gram-positive bacterial strains S. aureus and Bacillus subtilis, were each exposed to the antibacterial and anticancer activities. According to cytotoxicity and antibacterial toxicological tests, the synthesized nanocomposite showed significant anticancer activity against the A549 cell line and gram-negative bacterial strain E. coli compared to the other strains [35].

16.8.3 Iron Oxide Nanoparticles Influenced by Biology

Superparamagnetic iron oxide NPs (SPIONs) have found extensive application as a result of their unique properties. Due to the harsh chemical reagents used in their production, these NPs are less magnetic, more costly, and dangerous, which limits their use in the biomedical field. Russell J. Wilson created SPION-loaded silica nano-capsules using a bimodal catalytic peptide surfactant stabilized nano-emulsion template. The stability of the nano-emulsions was due to the peptide's surface property and electrostatic repulsion. SPIONs were preloaded into the oil phase of the nano-emulsions. Catalytic peptides support both the bio-solidification process and the creation of iron oxide-containing silica shell nano-capsules. The simple encapsulation of iron oxide into silica nano-capsules serves as a simple indicator of the potential of these developed nanoplatforms to distribute medicines.

16.9 Green Metallic Nanoparticle Synthesis with Bioinspired Design

There is a growing possibility for bioinspired synthesis of metallic nanoparticles due to the efficiency and cheap cost of biosynthesis as well as the simplicity with which several proteins, lipids, carbohydrates, and antibodies may be added to change them [37]. The harmful side-effects and high energy needs of the abrasive compounds and stabilizers utilized in the manufacture and manufacturing of these metallic NPs provide explain this. The NPs must be spread equally, and their size must be controlled for industrial processing. There have been several efforts to utilize food by-products like orange peels and banana peels [38]. In addition to their function as reducing agents in synthesis; they may readily be recovered into a number of valuable chemicals through these processes. Using an aqueous extract of the fruit's rind, which is used to treat diarrhea, cholera, and inflammation, a group of researchers produced silver (Ag), gold (Au), and platinum (Pt) [39]. They performed this in order to examine the metals' antibacterial properties both attached and detached from different antibiotic classes. The results showed a correlation between antibiotics and NPs, with AgNPs outperforming AuNPs and PtNPs in terms of their ability to combat gram-negative bacterial strains. Likewise, all kinds of NPs showed synergistic interactions with different antibiotic classes. Bacillus spp., which had previously been found to be resistant to streptomycin, was also made susceptible by the combination of AuNPs and antibiotics. Collectively, metallic NPs increased the susceptibility of bacterial strains to antibiotics [40].

16.10 Bioinspired Nanomaterials as Bio-templates

Scientists have paid a lot of attention to recent studies on protein cages, which are synthetic, highly symmetrical, multifunctional protein structures with three distinct interfaces: the inside, the outside, and the interface between subunits [41] (Fig. 16.2).

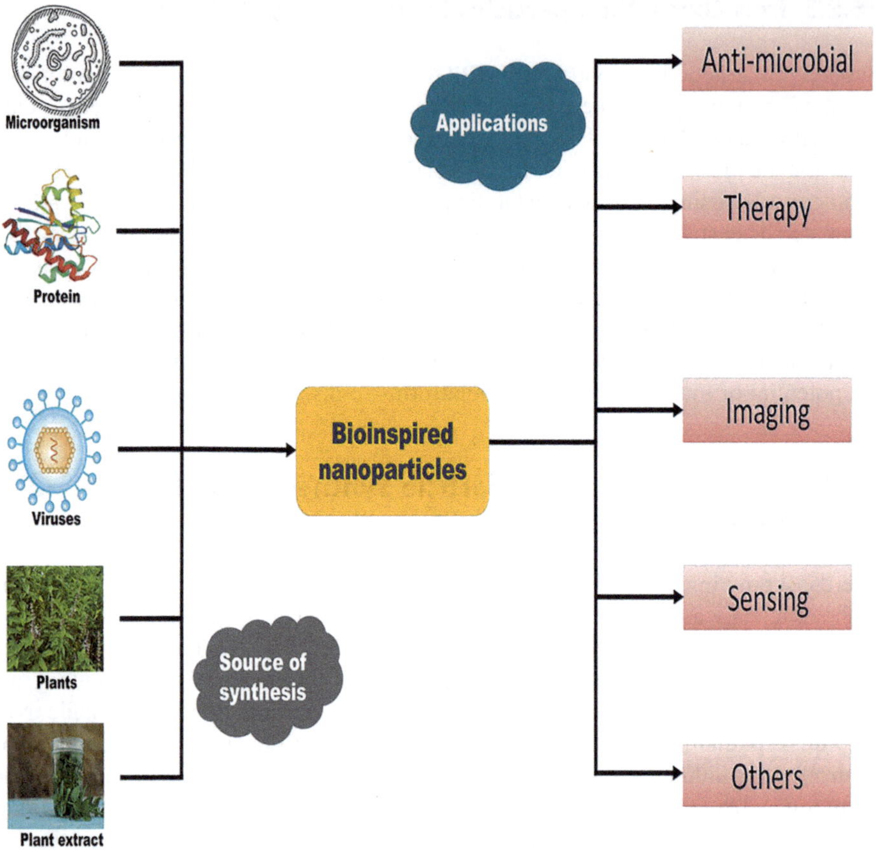

Fig. 16.2 Applications of bioinspired nanoparticles [36]

Chemical and genetic modifications of these subunits can result in the development of a single cage with various functionalities and a range of applications in industries spanning from medical to electronics. Ferritins, a class of proteins whose main job is to store and sequester iron (Fe), were the first protein cages to be used in the production of inorganic nanoparticles [42]. Many studies looked at how these protein complexes self-assemble. In one of the most recent studies, Huard et al. developed a revolutionary engineering approach called reverse metal-templated interface redesign (rMeTIR) [43]. It alters a naturally occurring protein-protein interface such that it only reacts to a metal ion in a particular way. Through the binding of divalent copper, they were able to control the self-assembly of the ferritin protein cage. Here, copper serves as a template for ferritin assembly, similar to how RNA sequences serve as structural templates for viral capsid construction. It is possible to demonstrate the critical role of conserved hydrogen-bonding interactions in providing geometrical specificity for cage assembly by making protein cages in this way. It is also possible to gain a fundamental understanding of the structure and

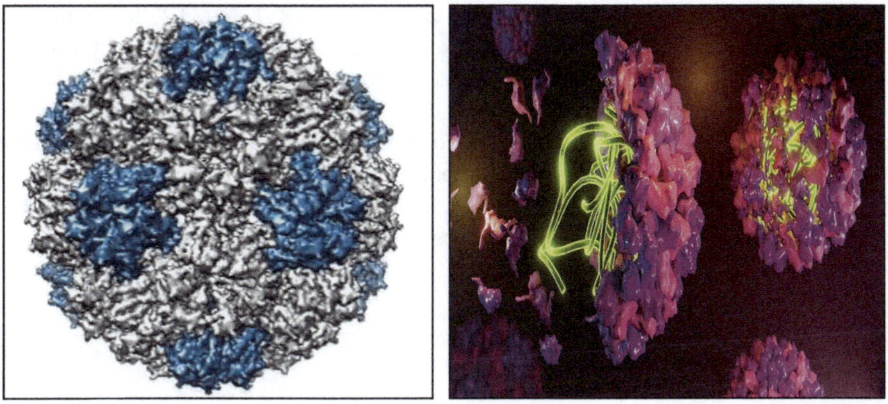

Fig. 16.3 Protein cages are small, monodisperse nano-compartments with a clearly defined structure [44]

stability of the isolated ferritin monomer as well as the uniform chemical modification of the cage interior under physiological conditions (Fig. 16.3).

16.11 Amino Acid/Protein Cage

The storage and transport of nucleic acids between diverse chemical environments, biomineralization, sequestration, and the production of anode materials for lithium-ion batteries are some of the principal applications of protein cages [45, 46]. Following Mann and colleagues [47], additional investigations used the biomineralization technique to create metallic and inorganic nanoparticles inside the apoferritin framework. Some of these techniques include sequential ion sequestration, just like in the production of ZnSe nanoparticles. The nanoparticles only form in this circumstance if Zn^{2+} is delivered before Se_2 due to the protein's directing electrostatic force that concentrates cations in the cage's interior.

16.12 Protein Cage for Nanomedicine

Protein-based nanomedicine platforms, which are created from naturally self-assembled protein subunits of the same protein or a mix of proteins, have attracted the greatest interest recently in the design of drug delivery systems because of their biocompatibility, biodegradability, and low toxicity [47]. Drugs can be kept inside a protein cage and then given just to the cells that will benefit from them. Because protein cages are the structural shells of viruses after the nucleic acids have been removed, potential synthetic cargo can be stored there.

The ferritin/apoferritin protein cage is a naturally occurring cage that may be utilized to deliver drugs since it is stable under physiological settings. Apoferritin is created when the iron atoms in ferritin are taken out [48, 49]. At pH 2, the subunits of ferritin/24 apoferritin can segregate before coming back together to form a full shell as the pH of the fluid is progressively adjusted to neutral [50]. This is a crucial characteristic of these proteins. Low pH disassembles the ferritin/apoferritin cage, allowing the medication to be assembled into the cage (pH 2). When the pH is adjusted to basic (pH 8.5), the cage is put back together. To increase therapeutic efficacy, stability, and biocompatibility, a dissociation-reassembly technique has been seen as a viable method for encapsulating tiny medicines, biomarkers, and even nanoparticles [51].

16.13 Bio Batteries

In rechargeable lithium batteries, biomaterials have been employed as electrode materials [52]. These materials' nanostructure enhances electrochemical activity, which enhances battery performance. For instance, nanosized amorphous $FePO_4$ particles, which are desirable cathode materials for rechargeable Li batteries due to their high specific capacity and safety, were successfully transformed from the M13 virus and employed as a template to create the particles [53]. An extensively researched virus for nanotechnology is M13. It consists of a main coat protein (p8) encasing a single-stranded circular DNA. Minor coat proteins P3, P6, P7, and P9 cap the viral end. Major and minor coat proteins serve as inorganic material nucleation sites. For instance, the p3 protein of the M13 virus was discovered to bind conducting single-walled carbon nanotubes (SWNTs), while the p8 protein of the virus was altered to bind $FePO_4$ cathode material. The virus-nanoparticle hybrid material exhibits better cyclability and retains a high capacity (about 130 $mAhg^{-1}$ at 10 °C) at high current rates due to the high specific capacity of $FePO_4$ and the high conductivity of the SWNTs. The hybrid cathode materials as a result are electrochemically stable. Similar to this, to produce hybrid virus-Au-Co_3O_4 cathode materials, the p8 protein may be changed to incorporate a Co nucleation site as well as an Au binding motif. Nearly 890 $mAhg^{-1}$ is the estimated specific capacity of Co_3O_4. Regardless of the nanostructure of Co_3O_4, the reaction product, Li_2O, which is inert in bulk, becomes electrochemically active. Au nanoparticles promote electrochemical activity by improving electrical conductivity and catalytic action. Despite the apparent advantages of employing biomaterials in the manufacture of bio batteries, the high prices connected with them impede commercialization [54]. The development of optimum architectures for increased performance of large-scale commercial devices remains a basic issue.

16.14 Nanomaterials Inspired by Biological Systems

As a result of being motivated by engineering principles found in nature, scientists have begun to develop materials that explicitly clone biological processes to regulate the design of devices for medical, environmental, and energy purposes with extraordinary precision. For example, scientists have imitated the water-repellent properties of some naturally occurring surfaces, such as the lotus plant's leaves and water striders' feet, to create structured hydrophobic surfaces [55]. Biologists, physicists, chemists, mathematicians, and computational scientists have united to better understand the physics behind these processes. Here, we look at the application of nanoparticles in the rapidly evolving fields of tissue engineering and biomineralization. The creation of these materials was motivated by the natural capacity of biological proteins and peptides to self-assemble.

16.15 Bioinspired Nanomaterials for Tissue Engineering and Biomineralization

Nanomaterials are a great option as scaffolds for tissue regeneration and engineering since the tissues and organs of the human body are composed of nanostructured proteins. The development of peptide nanotubes (NTs), which resemble protein-like tubular structures and have applications in tissue engineering, was inspired by the spontaneous self-assembly of proteins and peptides to generate solid nanofibrils or hollow nanotubes (NTs) [56, 57]. Dipeptides from the diphenylalanine motif of the Alzheimer's -amyloid peptide dissolve in 1,1,1,3,3,3-hexafluoro-2-propanol and are diluted to a final concentration of less than or equal to 2 mg mL. This produces micrometer-long multi-walled NTs with diameters ranging from 80 to 300 nm [58]. Amazing mechanical, thermal, and chemical strength characterizes these NTs. With deterioration starting only around 200 °C, dry NTs have a high Young's modulus of about 19–27 GPa and an average point stiffness of 160 N m1 [59]. This is an order of stiffness higher than biological microtubules, which provide cells with their rigid cytoskeleton and have Young's modulus of about 1 GPa. An order of magnitude stiffer than this material is biological microtubules, which provide cells with their rigid cytoskeleton and have Young's modulus of about 1 GPa. According to certain studies, the stacking of aromatic side chains to give stability and mechanical strength, micro- and nano-electromechanical systems, different functional devices, and tissue engineering may all make use of these peptide NTs. Another advantage of using biological NTs for biological sensing is that they are biocompatible. For instance, the placement of peptide NTs on graphite electrodes increased the electrochemical activity toward the redox process of potassium hexacyanoferrate. Due to the aforementioned property, as well as their low cost and strong solubility, the peptide NTs are economically feasible. Similar to this, thiol-modified peptide NTs mounted on gold electrode surfaces and coated with enzymes showed improved sensitivity and repeatability for the detection of glucose and ethanol, with short detection times, great stability, enormous current densities, and superior sensitivity [60, 61]. Long peptide surfactants with a hydrophobic tail made up of four or more successively hydrophobic amino acids and a hydrophilic head made up of two to three charged amino acids are known as linear peptides. These peptides include AIN D, V6D, G8DD, and KV6 as examples. At dosages of 4–6 mM, these peptides form a network of cationic or anionic open-ended NTs of 30–50 nm diameters in water. These tubes feature peptide bilayer walls that are 4–5 nm thick. They are held together by hydrogen bonding and stabilized by their hydrophobic characteristics. By adopting sheet topologies and forming homogeneous tubes in water with 30 nm thick walls, biotinylated peptides have been demonstrated by Matsumura et al. [62] to function as protein scaffolds. Nano-fibers used in tissue engineering as scaffolds include those formed of collagen, gelatin, and chitosan as well as synthetic biodegradable polymers including polylactic acid, poly (d,l-lactic co-glycolic acid), polyvinyl alcohol, and polycaprolactone. a few of these items have also received FDA approval.

A three-dimensional scaffold for tissue engineering is made using carbon-based nanomaterials due to their unique mechanical, electrical, and optical characteristics. These materials can be used to activate artificial scaffolds because of their high electrical conductivity and structural similarity to the extracellular matrix (ECM) found in living things. For instance, graphene, graphene oxide (GO), and reduced graphene oxide (rGO) form ECM-like microenvironments that encourage cell adhesion, proliferation, and differentiation [62, 63].

According to Kalbacova et al. [64] human mesenchymal stem cells (hMSCs) showed a spindle-like shape when grown on graphene, which improved their ability to develop into osteoblast lineage and showed higher adhesion. Since the surface oxygen content of graphene varies, a mixture of hydrogen bonds, electrostatic interactions, and interactions are used to attach it to proteins. Its highly wrinkled nanoscale structure also significantly increases osteogenic differentiation. Hydrogels, which are more often used for tissue engineering because of their outstanding biocompatibility and biodegradability, have also been reinforced with graphene [64]. The superior electrical properties of graphene and CNTs have an impact on the behavior of electro-active brain cells. Neuroendocrine PC12 cells showed neurite outgrowth and expansion on rGO [65]. Similar to this, the introduction of multi-walled and single-walled NTs (MWNTs and SWNTs, respectively) significantly boosted network activity by boosting the frequency of spontaneous action potentials and continuous postsynaptic currents in hippocampus neurons [66, 67]. Better protein adsorption is also made possible by the huge surface area of the NTs.

16.16 Conclusion and Potential Outcomes

The highly evolved biological world serves as a very efficient testing ground for the creation of novel materials because of its richness and intriguing reactions to its environment. Science and nature meet in a mystical doorway that motivates us to develop novel, multifunctional nanomaterials that transcend the limits of the current material systems. The field of nanomedicine offers potential applications for innovative bioinspired nanoplatforms. They are being utilized for a number of medicinal purposes, including cancer treatment, antimicrobial therapy, immunotherapy, biosensing, and diagnostics. By applying NPs to the membranes of cancer cells, membrane coating works to activate the body's natural defense system to create cancer immune therapies that are comparable to nano-vaccines. Additionally, aggressive tumor-specific targeting, increased blood flow, and reticuloendothelial system escape may be advantageous to them. A number of factors need to be carefully considered before the successful commercial development of bioinspired materials for practical applications, including the high cost of manufacturing large-scale devices using biomaterials and the short lifespan of biological substances like enzymes that may cause these materials to degrade in devices. Despite the fact that bio mimetic structures are fascinating and offer new opportunities for designing novel materials with distinctive properties for a variety of applications, future research in this field should focus on the development of cleaner, more sophisticated, yet cost-effective methods for designing them. To this purpose, design principles should be considered in addition to the structure of biological structures. Working together, biologists, chemists, physicists, material scientists, and computer scientists can successfully progress in this field of study. Last but not least, a fascinating research field will be the development of novel methods to duplicate the controlled self-assembly, adaptability, and self-healing characteristics of biological materials in synthetic materials inspired by biological materials. The ongoing interest in the subject of bioinspired nanomaterials will motivate the creation of novel materials as natural prototypes with unique properties for sustainable applications, driving multidisciplinary research and education.

References

1. Zhang Y, Mei J, Li H, Liao T, Bell J, Sun Z (2020) Bioinspired 2D nanomaterials for sustainable applications. Adv Mater 32(18):1902806
2. Kumari S, Dhanda N, Thakur A, Singh S, Thakur P (2023) Investigation of calcium substitution on magnetic and dielectric properties of Mg–Zn nano ferrites. Mater Chem Phys 297:127394. ISSN 0254-0584. https://doi.org/10.1016/j.matchemphys.2023.127394
3. Lai Y, Fenglin L, Zhifeng Z, Madiha S, Zhiai X, Haijun Y (2021) Bio-inspired amyloid polypeptides: from self-assembly to nanostructure design and biotechnological applications. Appl Mater Today 22(2021):100966
4. Kankala RK, Han Y, Na J, Lee C, Sun Z, Wang S, Kimura T, Kim K, Yamauchi Y, Chen A, Wu KC (2020) Nanoarchitectured structure and surface biofunctionality of mesoporous silica nanoparticles. Adv Mater 32(23):1907035
5. Kumari S, Dhanda N, Thakur A, Gupta V, Singh S, Kumar R, Hameed S, Thakur P (2023) Nano Ca–Mg–Zn ferrites as tuneable photocatalyst for UV light-induced degradation of rhodamine B dye and antimicrobial behavior for water purification. Ceram Int 49(8):12469–12480., ISSN 0272-8842. https://doi.org/10.1016/j.ceramint.2022.12.107
6. Thakur A, Thakur P, Hsu JH (2012) Magnetic behaviour of $Ni_{0.4}Zn_{0.6}Co_{0.1}Fe_{1.9}O_4$ spinel nano-ferrite. J Appl Phys 111:07A305. https://doi.org/10.1063/1.3670606
7. Nel AE, Mädler L, Velegol D, Xia T, Hoek EM, Somasundaran P, Klaessig F, Castranova V, Thompson M (2009) Understanding biophysicochemical interactions at the nano–bio interface. Nat Mater 8(7):543–557
8. Grimaldi N, Andrade FH, Segovia N, Ferrer-Tasies L, Sala S, Veciana J, Ventosa N (2016) Lipid-based nanovesicles for nanomedicine. Chem Soc Rev 45(23):6520–6545
9. Zhao Q, Hai B, Zhang X, Xu J, Koehler B, Liu F (2020) Biomimetic nanovesicles made from iPS cell-derived mesenchymal stem cells for targeted therapy of triple-negative breast cancer. Nanomedicine 24:102146

10. Pathania A, Bhardwaj S, Thakur SS, Mattei JL, Quefelec P, Panina LV, Thakur P, Thakur A (2018) Investigation of structural, optical, magnetic and electrical properties of tungsten doped Ni–Zn nano-ferrites. Phys B Condens Matter 531:45–50. https://doi.org/10.1016/j.physb.2017.12.008
11. Sun Y, Shi H, Yin S, Ji C, Zhang X, Zhang B, Wu P, Shi Y, Mao F, Yan Y, Xu W, Qian H (2018) Human mesenchymal stem cell derived exosomes alleviate type 2 diabetes mellitus by reversing peripheral insulin resistance and relieving β-cell destruction. ACS Nano 12(8): 7613–7628
12. Aqil F, Kausar H, Singh IP, Jeyabalan J, Kyakulaga AH, Munagala R, Gupta RC (2016b) Exosomal formulation enhances therapeutic response of celastrol against lung cancer. Exp Mol Pathol 101(1):12–21
13. Sawadkar P, Mohanakrishnan J, Rajasekar P, Rahmani B, Kohli N, Bozec L, García-Gareta E (2020) A synergistic relationship between polycaprolactone and natural polymers enhances the physical properties and biological activity of scaffolds. ACS Appl Mater Interfaces 12(12): 13587–13597
14. Kaushik S, Gandhi S, Chauhan M, Ma S, Das S, Ghosh D, Chandrasekharan A, Alam B, Parmar AS, Sharma A, Santhoshkumar T, Suhag D (2020) Water-templated, polysaccharide-rich bioartificial 3D microarchitectures as extra-cellular matrix bioautomatons. ACS Appl Mater Interfaces 12(18):20912–20921
15. Avossa J, Pota G, Vitiello G, Macagnano A, Zanfardino A, Di Napoli M, Pezzella A, Lucchini T, Varcamonti M, Luciani G (2021) Multifunctional mats by antimicrobial nanoparticles decoration for bioinspired smart wound dressing solutions. Mater Sci Eng C 123:111954
16. Choi Y, Zheng XX, Tan YN (2020) Bioinspired carbon dots (biodots): emerging fluorophores with tailored multiple functionalities for biomedical, agricultural and environmental applications. Mol Syst Design Eng 5(1):67–90
17. Zhou J, Zheng H, Han H, Zou M, Li C (2012) Facile synthesis of fluorescent carbon dots using watermelon peel as a carbon source. Mater Lett 66(1):222–224
18. Wang M, Wan Y, Zhang K, Fu Q, Wang L, Zeng J, Xia Z, Gao D (2019) Green synthesis of carbon dots using the flowers of Osmanthus fragrans (Thunb.) Lour. as precursors: application in Fe^{3+} and ascorbic acid determination and cell imaging. Anal Bioanal Chem 411(12): 2715–2727
19. Thakur P, Chahar D, Taneja S, Bhalla N, Thakur A (2020) A review on MnZn ferrites: synthesis, characterization and applications. Ceram Int 46:15740–15763. https://doi.org/10.1016/j.ceramint.2020.03.287
20. Shaikh AF, Tamboli MS, Patil R, Bhan A, Ambekar JD, Kale BB (2019) Bioinspired carbon quantum dots: an antibiofilm agents. J Nanosci Nanotechnol 19(4):2339–2345
21. Srivastava N, Saxena SK (2022) Microbicidal nanoparticles. In: Thakur A, Thakur P, Khurana SP (eds) Synthesis and applications of nanoparticles. Springer, Singapore
22. Liepold LO, Anderson SA, Willits DA, Oltrogge LM, Frank JA, Douglas T, Young MJ (2007) Viral capsids as MRI contrast agents. Magn Reson Med 58(5):871–879
23. Tsukamoto R, Muraoka M, Seki M, Tabata H, Yamashita I (2007) Synthesis of CoPt and $FePt_3$ nanowires using the central channel of tobacco mosaic virus as a biotemplate. Chem Mater 19(10):2389–2391
24. Mathur P, Thakur A, Singh M (2008) Low temperature synthesis of $Mn_{0.4}Zn_{0.6}In_{0.5}Fe_{1.5}O_4$ nanoferrite for high-frequency applications. J Phys Chem Solids 69:187–192. https://doi.org/10.1016/j.jpcs.2007.08.014
25. Thakur P, Sharma R, Kumar M, Katyal SC, Barman PB, Sharma V, Sharma P (2019) Structural, morphological, magnetic and optical study of co-precipitated Nd^{3+} doped Mn-Zn ferrite nanoparticles. J Magn Magn Mater 479:317–325. https://doi.org/10.1016/j.jmmm.2019.02.048
26. Wang H, Li J, Wang Z, Wang Y, Xu X, Gong X, Wang J, Zhang Z, Li Y (2021) Tumor-permeated bioinspired theranostic nanovehicle remodels tumor immunosuppression for cancer therapy. Biomaterials 269:120609

27. Hu CMJ, Zhang L, Aryal S, Cheung C, Fang RH, Zhang L (2011) Erythrocyte membrane-camouflaged polymeric nanoparticles as a biomimetic delivery platform. Proc Natl Acad Sci U S A 108(27):10980–10985
28. Krishnia L, Thakur P, Thakur A (2022) Synthesis of nanoparticles by physical route. In: Synthesis and applications of nanoparticles. Springer, Singapore, pp 45–59
29. Soppimath K, Aminabhavi TM, Kulkarni AR, Rudzinski WE (2001) Biodegradable polymeric nanoparticles as drug delivery devices. J Control Release 70(1–2):1–20
30. Malafaya PB, Silva GA, Reis RL (2007) Natural–origin polymers as carriers and scaffolds for biomolecules and cell delivery in tissue engineering applications. Adv Drug Deliv Rev 59(4–5): 207–233
31. Thakur P, Gahlawat N, Punia P et al (2022) Cobalt Nanoferrites: a review on synthesis, characterization, and applications. J Supercond Nov Magn 35:2639–2669. https://doi.org/10.1007/s10948-022-06334-1
32. Ali S, Perveen S, Ali M, Jiao T, Sharma AS, Hassan H, Devaraj S, Li H, Chen Q (2020) Bioinspired morphology-controlled silver nanoparticles for antimicrobial application. Mater Sci Eng C 108:110421
33. Mao H, Laurent S, Chen W, Akhavan O, Imani M, Ashkarran AA, Mahmoudi M (2013) Graphene: promises, facts, opportunities, and challenges in nanomedicine. Chem Rev 113(5): 3407–3424
34. Thakur P, Thakur A (2022) Nanomaterials, their types and properties. In: Thakur A, Thakur P, Khurana SP (eds) Synthesis and applications of nanoparticles. Springer, Singapore. https://doi.org/10.1007/978-981-16-6819-7_2
35. Kadiyala NK, Mandal BK, Ranjan S, DasGupta N (2018) Bioinspired gold nanoparticles decorated reduced graphene oxide nanocomposite using Syzygium cumini seed extract: evaluation of its biological applications. Mater Sci Eng C 93:191–205
36. Wu G, Hui X, Hu L, Bai Y, Rahaman A, Yang X, Chen C (2022) Recent advancement of bioinspired nanomaterials and their applications: a review. Front Bioeng Biotechnol 10:952523
37. Thakur A, Chahar D, Thakur P (2022) Synthesis of nanomaterials by biological route. In: Thakur A, Thakur P, Khurana SP (eds) Synthesis and applications of nanoparticles. Springer, Singapore. https://doi.org/10.1007/978-981-16-6819-7_5
38. Wilson RB, Hui Y, Whittaker AK, Zhao C (2021) Facile bioinspired synthesis of iron oxide encapsulating silica nanocapsules. J Colloid Interface Sci 601:78–84
39. Giri AK, Jena B, Biswal B, Pradhan AK, Arakha M, Acharya S, Acharya L (2022) Green synthesis and characterization of silver nanoparticles using Eugenia roxburghii DC. extract and activity against biofilm-producing bacteria. Sci Rep 12(1):8383
40. Pedraza-Chaverri J, Cárdenas-Rodríguez N, Orozco-Ibarra M, Pérez-Rojas JM (2008) Medicinal properties of mangosteen (Garcinia mangostana). Food Chem Toxicol 46(10):3227–3239
41. Nishanthi R, Malathi S, Paul SJ, Palani P (2019) Green synthesis and characterization of bioinspired silver, gold and platinum nanoparticles and evaluation of their synergistic antibacterial activity after combining with different classes of antibiotics. Mater Sci Eng C 96:693–707
42. Chen A, Bao Y, Ge X, Shin Y, Du D, Lin Y (2012) Magnetic particle-based immunoassay of phosphorylated p53 using protein cage templated lead phosphate and carbon nanospheres for signal amplification. RSC Adv 2(29):11029
43. Uchida M, Klem MT, Allen MG, Suci PA, Flenniken ML, Gillitzer E, Varpness Z, Liepold LO, Young MJ, Douglas T (2007) Biological containers: protein cages as multifunctional nanoplatforms. Adv Mater 19(8):1025–1042
44. Liu Q, Shaukat A, Kyllönen D, Kostiainen MA (2021) Polyelectrolyte encapsulation and confinement within protein cage-inspired Nanocompartments. Pharmaceutics 2021(13):1551
45. Kharbanda S, Dhanda N, An- Sun, Thakur A, Thakur P (2023) Multiferroic perovskite bismuth ferrite nanostructures: a review on synthesis and applications. J Magn Magn Mater 572:170569. ISSN 0304-8853. https://doi.org/10.1016/j.jmmm.2023.170569

46. Nam KT, Kim D, Yoo PJ, Chiang C, Meethong N, Hammond PT, Chiang Y, Belcher AM (2006) Virus-enabled synthesis and assembly of nanowires for lithium ion battery electrodes. Science 312(5775):885–888
47. Wong KP, Mann S (1996) Biomimetic synthesis of cadmium sulfide-ferritin nanocomposites. Adv Mater 8(11):928–932
48. Suci PA, Kang S, Young MJ, Douglas T (2009) A streptavidin−protein cage Janus particle for polarized targeting and modular functionalization. J Am Chem Soc 131(26):9164–9165
49. Dhanda N, Thakur P, Kumar R, Fatima T, Hameed S, Slimani Y, Sun A-CA, Thakur A (2023) Green-synthesis of Ni-Co nanoferrites using aloe vera extract: structural, optical, magnetic, and antimicrobial studies. Appl Organomet Chem 37:e7110. https://doi.org/10.1002/aoc.7110
50. Dhanda N, Thakur P, An-Cheng S, Thakur A (2023) Structural, optical and magnetic properties along with antifungal activity of Ag-doped Ni-Co nanoferrites synthesized by eco-friendly route. J Magn Magn Mater 572:170598.,ISSN 0304-8853. https://doi.org/10.1016/j.jmmm.2023.170598
51. Aime S, Frullano L, Geninatti CS (2002) Compartmentalization of a gadolinium complex in the apoferritin cavity: a route to obtain high relaxivity contrast agents for magnetic resonance imaging. Angew Chem Int Ed Engl 41(6):1017–1019. https://doi.org/10.1002/1521-3773(20020315)41:6<1017::aid-anie1017>3.0.co;2-p. PMID: 12491298
52. Dhanda N, Kumari S, Kumar R, Kumar D, An-Cheng S, Thakur P, Thakur A (2023) Influence of Ni over magnetically benign Co ferrite system and study of its structural, optical, and magnetic behaviour. Inorganic Chem Commun 151:110569, . ISSN 1387-7003. https://doi.org/10.1016/j.inoche.2023.110569
53. Lee DY, Lee Y, Oh D, Chen TY, Ceder G, Belcher AM (2010) Biologically activated Noble metal alloys at the nanoscale: for lithium ion battery anodes. Nano Lett 10(7):2433–2440
54. Lee DY, Yi H, Kim W, Kang K, Yun DH, Strano MS, Ceder G, Belcher AM (2009) Fabricating genetically engineered high-power lithium-ion batteries using multiple virus genes. Science 324(5930):1051–1055
55. Zhou H, Fan T, Zhang D (2011) Biotemplated materials for sustainable energy and environment: current status and challenges. ChemSusChem 4(10):1344–1387
56. Bhattacharya P, Gohil S, Mazher J, Ghosh S, Ayyub P (2008) Universal, geometry-driven hydrophobic behaviour of bare metal nanowire clusters. Nanotechnology 19(7):075709
57. Dhanda N, Thakur P, Thakur A (2023) Green synthesis of cobalt ferrite: a study of structural and optical properties. Mater Today Proc 73(Part 2):237–240, . ISSN 2214-7853. https://doi.org/10.1016/j.matpr.2022.07.202
58. Hartgerink JD, Granja JR, Milligan ARA, Ghadiri MR (1996) Self-assembling peptide nanotubes. J Am Chem Soc 118(1):43–50
59. Adler-Abramovich L, Mandler D, Sedman VL, Allen S, Tendler SJB, Gazit E (2006) Thermal and chemical stability of diphenylalanine peptide nanotubes: implications for nanotechnological applications. Langmuir 22(3):1313–1320
60. Niu L, Chen X, Allen S, Tendler SJB (2007) Using the bending beam model to estimate the elasticity of diphenylalanine nanotubes. Langmuir 23(14):7443–7446
61. Yemini M, Mandler D, Gazit E, Rishpon J (2005) Peptide nanotube-modified electrodes for enzyme−biosensor applications. Anal Chem 77(16):5155–5159
62. Matsumura S, Uemura S, Mihara H (2005) Construction of biotinylated peptide nanotubes for arranging proteins. Mol BioSyst 1(2):146
63. Ku SH, Lee M, Park CB (2013) Carbon-based nanomaterials for tissue engineering. Adv Healthc Mater 2(2):244–260
64. Kalbacova MH, Broz A, Kong J, Kalbac M (2010) Graphene substrates promote adherence of human osteoblasts and mesenchymal stromal cells. Carbon 48(15):4323–4329
65. Lim HS, Huang N, Lim SS, Harrison I, Chia CH (2011) Fabrication and characterization of graphene hydrogel via hydrothermal approach as a scaffold for preliminary study of cell growth. Int J Nanomed 6:1817

66. Agarwal S, Zhou X, Ye F, He Q, Chen G, Soo J, Boey FYC, Zhang H, Chen P (2010) Interfacing live cells with nanocarbon substrates. Langmuir 26(4):2244–2247
67. Hu H, Ni Y, Mandal SK, Montana V, Zhao B, Haddon RC, Parpura V (2005) Polyethyleneimine functionalized single-walled carbon nanotubes as a substrate for neuronal growth. J Phys Chem B 109(10):4285–4289

Nanomaterials for Environmental Applications 17

Yassine Slimani and Essia Hannachi

Abstract

The rapid growth in industrialization and population has resulted in the generation and permanence of a wide variety of inorganic and organic pollutants, posing potential risks to the environment and human health. The degradation (either removal or reduction) and sensing of these pollutants from various environmental media (e.g., air, water, and soil) are nowadays important problems worldwide. To address this issue, several nanotechnologies and numerous nanostructured materials and composites have been established for efficient environmental remediation applications. This is ascribed to their large surface area and improved characteristics and efficiency (like high reactive feature). The current chapter reports nanotechnologies such as filtration, photocatalysis, chemical reactions, absorption, adsorption, nanosensing, etc. It also offers an overview of the use of nanostructured materials and composites in the remediation of the environment. Novel nanostructured materials, involving inorganic-based nanomaterials (NMs) (e.g., zero-valent metal NMs, metal oxide-based NMs, silica-based NMs), carbon-based NMs, polymeric-based NMs, as well as their nanocomposites were extensively reported and discussed. These materials are efficiently employed for remediating various environmental pollutants such as halogenated herbicides, volatile organic compounds, chlorinated organic compounds, aromatic and aliphatic hydrocarbons, pharmaceuticals, dyes, heavy metals,

The original version of this chapter has been revised. The correction to the chapter can be found at https://doi.org/10.1007/978-981-99-6105-4_23

Y. Slimani (✉)
Department of Biophysics, Institute for Research and Medical Consultations (IRMC), Imam Abdulrahman Bin Faisal University, Dammam, Saudi Arabia
e-mail: yaslimani@iau.edu.sa

E. Hannachi
Department of Nuclear Medicine Research, Institute for Research and Medical Consultations (IRMC), Imam Abdulrahman Bin Faisal University, Dammam, Saudi Arabia

D. Suhag et al. (eds.), *Integrated Nanomaterials and their Applications*,
https://doi.org/10.1007/978-981-99-6105-4_17

biological substances (e.g., antibiotics, parasites, bacteria, and viruses), as well as gases (NO_x, CO, SO_2, etc.). Their efficiency, advantages, and limitations were compared and discussed. Moreover, some potential directions for future studies were proposed.

Keywords

Nanomaterials · Nanocomposites · Polymers · Carbonaceous-based materials · Nanotechnology · Pollutants · Environmental remediation

17.1 Introduction

Environmental contamination is without a doubt one of the most serious issues confronting civilization nowadays. Novel techniques for the remediation of pollutants in the environment (soil, water, and air) are continually being developed. Organic substances, wastewater, industrial waste, poisonous gases, spills of oil, fertilizers, herbicides, pesticides, heavy metals, and particulate matter are only a few of the numerous pollutants to be concerned about [1]. Because diverse substances could be used in remediating the environment, a broad range of methodologies are able to be used to accomplish this aim. Because of the complexities of the mixture of diverse substances, minimal reactivity, and significant volatility, capturing and degrading environmental contaminants might be difficult. Recently, scientists have concentrated on the utilization of nanostructured materials to establish novel environmental remediation technologies [1].

Because of the peculiar physical features of nanomaterials, nanotechnology has received a great deal of interest in recent years. Because of their larger surface-to-volume ratio, nanostructured materials have more potent reactivity and hence superior efficacy in comparison to their bulk equivalents. Furthermore, nanostructured materials have the ability to harness distinct surface chemical properties in comparison to standard techniques, allowing them to be coated or functionalized with functional groups that may target particular molecules of concern (contaminants) for effective remediation. Furthermore, purposeful tweaking of the physical features of nanoparticles (e.g., chemical composition, porosity, size, and shape) might bestow further favorable traits that could improve the material's efficacy for pollutant remediation. The complex surface modification chemistry of nanomaterials, together with their customizable physical properties, provides substantial benefits compared to current approaches dealing with environmental contaminants. Techniques designed as a mixture of multiple distinct materials (composites/hybrids), acquiring certain desirable features from each of their constituents, are possibly more stable, selective, and efficient over techniques founded on a single nano-platform. Once we compare the utilization of nanomaterials alone, gluing nanomaterials to a scaffold may be an alternate strategy to boost the material's stability. The functionalization of a compound with specific substances accountable for addressing pollutant molecules of concern may boost the material's efficacy and selectivity [2–4].

The material used for remediating pollution should not necessarily be another pollutant itself after its use. So, for this application area, biodegradable materials are of particular interest [5]. In addition to increasing the trust and recognition by consumers of a specific technology, the use of biodegradable materials may also provide a more green and safe alternate route to remedy environmental contaminants, since there is no creation of product wastes to be discarded after use. In addition, there is a particularly strong appeal for new technologies that can leverage the targeted capture of contaminants to overcome low-efficiency results generated by direct targeting. Therefore, in order to obtain materials that can cope with a range of challenges related to the removal of contamination, several studies have been conducted, which focus on the use of nanotechnology principles and combine them with chemical/physical or mechanical modifications of material surfaces [4–6]. Some of the most significant challenges that should be considered while developing novel nanomaterials for the remediation of the environment include target selective capture, cost efficiency, easy preparation, green synthesis, nontoxicity, biodegradable feature, recycling, and possible recovery after utilization. Despite the promising benefits of nanostructured materials stated above, some nanomaterials are fundamentally nonstable in normal conditions and hence require specific procedures for the preparation of nanomaterials at the nanoscale. To avoid agglomerations and to improve monodispersity and stability it is necessary to carry out additional operations. Another factor that could limit their application is the feasible poisonousness of metal-based NPs included in remediation, together with their byproduct and recovery expenditures originating at the remediated location. Therefore, it is important to obtain a thorough knowledge of the nanomaterial features, their manufacturing processes, and performance improvement to identify excellent nanomaterials that are able to cope with environmental challenges.

This chapter aims to deliver an overall overview of the latest developments made in the design of nanostructured materials and/or composites employed for the treatment of the environment from different kinds of contaminants. Filtration, photocatalysis, chemical redox processes, adsorption, and absorption, as described in Fig. 17.1, are among a series of feasible routes that could remove contaminants [1, 3, 5]. For the methods mentioned in Fig. 17.1, some kinds of materials that could be utilized are also reported. For the case of distinct kinds of products that might be used for environmental remediation, it is to the best of our knowledge that no classification has been performed. Accordingly, the current chapter will focus on the different nanotechnologies as well as the diverse kinds of materials involved in environmental remediation applications. Particularly, it will focus on three major classes of nanomaterials reported in the literature: inorganic, polymer-based, and carbon-based nanomaterials and nanocomposites.

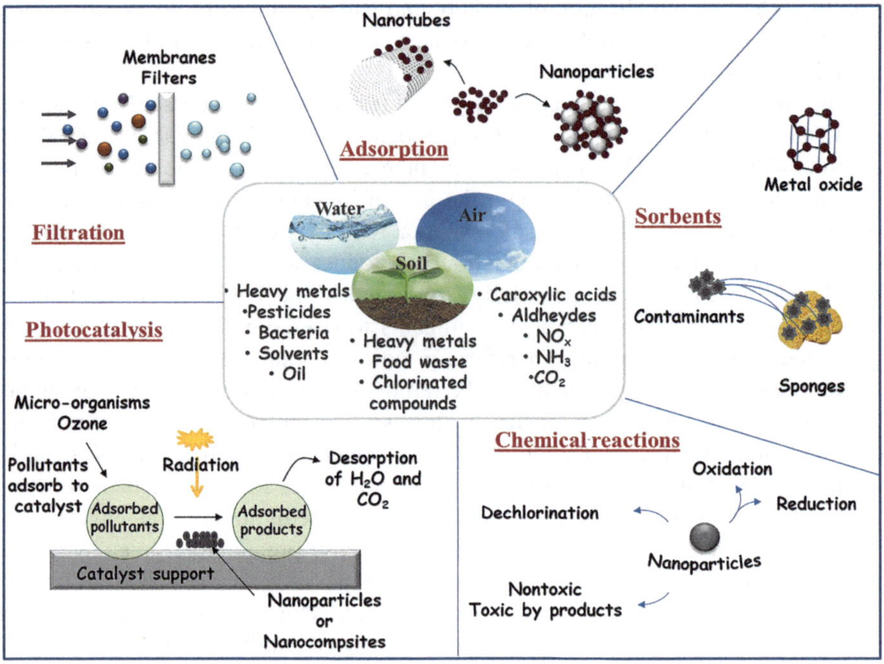

Fig. 17.1 Technologies used for environmental remediation

17.2 Nanotechnologies Used for Environmental Remediation

17.2.1 Photocatalysis

Advanced reduction processes (ARPs) and advanced oxidation processes (AOPs) are very potential environmental remediation alternatives, particularly for mediums that enable the transmission of light, such as water and atmosphere [7, 8]. The generation of highly reactive oxygen species (ROS), primarily superoxide radical anions ($O_2^{\bullet -}$) and hydroxyl radicals ($HO^{\bullet}$), characterizes the above processes. In an aqueous medium, $HO^{\bullet}$ is the most powerful oxidant, $E_0\left(HO^{\bullet}, \frac{H^+}{H_2O}\right) = 2.73\ \text{V vs.NHE}$ [9]. Positive vacancies on the photocatalyst surface (usually designated as h^+) are likewise highly oxidant, $E_0 = 2.53$ V vs. SHE [10]. The occurrence of ROS assures an elevated reactivity (quick reactions) and poor selectivity (no differentiation among reducible and oxidizable molecules). Currently, AOPs are being widely utilized, more particularly for the remediation of our environment, and therefore our emphasis will be on them in upcoming sections. In AOPs, heterogeneous photocatalysis has been extensively utilized, with complete mineralization of both inorganic and organic contaminants being achieved in a number of cases [7, 8]. AOPs utilize various kinds of near-ultraviolet/visible

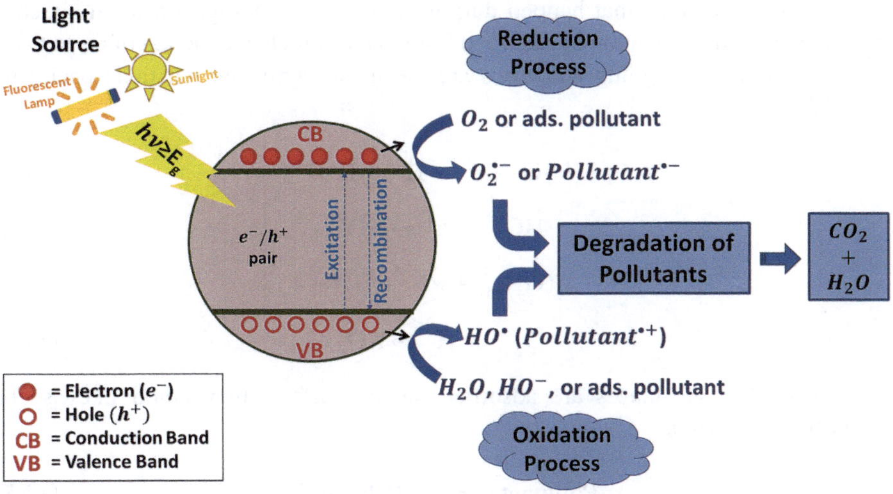

Fig. 17.2 Photocatalysis processes of a semiconducting photocatalyst in water containing pollutants

lamps (like LED) and, preferably, sunlight. The effectiveness of the photocatalytic processes is enhanced if there are sufficient overlaps among the photocatalyst absorption and the radiation spectrum of sunlight. Indeed, this is among the major defies in this area, and it has been addressed in a variety of ways, like altering the crystalline structure of the photocatalysts by doping and/or introducing defects.

The utilization of semiconducting materials, preferably in the form of nanoparticles or films placed on substrates, is relied upon for heterogeneous photocatalytic processes. Nanoparticles tend to be agglomerated when they are utilized, due to the decrease of some surfaces and thus their reactivity. The reactive surface of the hemi-sphere in contact with the surface will not exist once nanoparticles are made as films, considering that they display a sphere-like shape, due to a considerable deficit of the active surface in the product. The bandgap energy (E_g) separating the conduction band (CB) and the valence band (VB) should be identical to or smaller than the incident photon energy (hν), which means $E_g \leq h\nu$. As demonstrated in Fig. 17.2, the photons are adsorbed and the electrons (e^-) propagate from VB to CB, creating vacancies (holes, h^+) in VB and forming pairs of electron-hole ($e^- - h^+$).

Pairs of $e^- - h^+$ are extremely nonstable and typically annihilate in a few nanoseconds, going back to the starting state before the absorption of photons. A relatively small fraction of $e^- - h^+$ separates, which is governed by the experiment settings and, in particular, the material being studied. The magnitude of E_g is the essential parameter for the success of photocatalysts. To prevent undue energy consumption of irradiation lamps, a preference should be given to (nano)materials having E_g magnitudes within the energy spectrum of photons supplied by sunlight.

The main reactions that happen during the $e^- - h^+$ segregation in an aqueous media involve the oxidation of adsorbed H_2O to $HO^\bullet$ and the reduction of O_2 to $O_2^{\bullet -}$ (Fig. 17.2). The foregoing equations are the most significant reactions occurring [11, 12]:

$$H_2O + h^+ \rightarrow H^+ + HO^\bullet \tag{17.1}$$

$$HO^- + h^+ \rightarrow HO^\bullet \tag{17.2}$$

$$(O_2)_{ads} + e^- \rightarrow O_2^{\bullet -} \tag{17.3}$$

$$O_2^{\bullet -} + H^+ \rightarrow HO_2^\bullet \tag{17.4}$$

When other substances are adsorbed on the surface, they could likewise be reduced or oxidized:

$$\text{Pollutant} + h^+ \rightarrow \text{Pollutant}^{\bullet +} \tag{17.5}$$

$$\text{Pollutant} + e^- \rightarrow \text{Pollutant}^{\bullet -} \tag{17.6}$$

The reactions of adsorbed substances could enhance the effectiveness of a pollutant's removal, but they could open new pathways which result in products other than those resulting from ROS reactions and could also have different toxicities and stabilities.

During heterogeneous photocatalytic processes, additional ROS can also be produced. For instance, H_2O_2 is created by the dismutation of $HO_2^\bullet$ and $O_2^{\bullet -}$ and can then be reduced to $HO^\bullet$ as shown in the following equations [13, 14]:

$$O_2^{\bullet -} + HO_2^\bullet + H^+ \rightarrow H_2O_2 + O_2 \tag{17.7}$$

$$H_2O_2 + H^+ + e^- \rightarrow HO^\bullet + H_2O \tag{17.8}$$

The findings from experiments on the great differences in reactivity dependences with $\left[O_2^{\bullet -}\right]$ and $[HO^\bullet]$ or other ROS support the view that $HO^\bullet$ is primarily accountable for degrading organic contaminants [11].

$HO^\bullet$ commonly combines with non-saturated functional groups to generate $HO^\bullet$ adducts, but it could alternatively interact with $H^\bullet$ abstraction to form radical entities with H-deficiency, or with an e^- oxidation to form a radical cation and HO^-:

$$\text{Pollutant} + HO^\bullet \rightarrow \text{Pollutant}^{\bullet +} + HO^- \tag{17.9}$$

$$\text{Pollutant} + HO^\bullet \rightarrow \text{Pollutant}(-H)^\bullet + H_2O \tag{17.10}$$

$$\text{Pollutant} + HO^\bullet \rightarrow \text{Pollutant}(-OH)^\bullet \tag{17.11}$$

The intermediates Pollutant$^{\bullet +}$, (Pollutant$(-H)^\bullet$, and (Pollutant $-$ OH)$^\bullet$ are of short life and would endure chemical reactions via distinct reaction routes, resulting in a number of byproducts where the majority of them will hydroxylate and thus being soluble in water, greatly polar, and easily adsorbed on mineral surfaces like clays, potentially increasing their residence time in the environment. When a sufficient period is given for the reactions to occur, it may eventually result in mineralization, which means the conversion toward the most oxidized ionic substances (SO_4^{2-}, PO_4^{3-}, NO_3^-, CO_3^{2-}, and so on). Nevertheless, some kinds of refractory materials, like triazines, have been shown to create a stable intermediate that will not react anymore [15].

For different situations, $HO^•$ reactions occur extremely quickly, with bimolecular rate constants of $k > 10^9\ M^{-1}s^{-1}$ or more, which is near to or at the diffusion-controlled limit in water, and in a non-selective fashion. Throughout the oxidation of aromatics, h^+ and $HO^•$ are shown to differ in regioselectivity and could provoke different materials [16]. Lately, it has been demonstrated that for similar processes, there are competing routes among the oxidation of chemisorbed species by h^+ and the reaction with $HO^•$ [17, 18].

It is especially essential to note that the heterogeneous photocatalytic process is a superficial process, which means that the reactive substances must be diffused from the core of the solution to the surface, become adsorbed (perhaps with some changes in the structure), endure reactions, desorb, and diffuse back inside the solution [11]. The prospect of ROS being formed at the surface and migrating within the solution has been investigated, and it was demonstrated that $HO^•$ is diffused just a few hundred Angstroms down from the surface inside the aqueous medium. ROS migrates just a couple of atomic lengths beyond the surface, according to ESR experiments. As a result, if the molecules are not adsorbed, they will not undergo reactions through heterogeneous photocatalysis. Since the adsorption/desorption equilibrium would be shifted as the reaction progresses, only a small amount of starting adsorption is required.

Titanium dioxide (TiO_2) is one of the widely utilized photocatalytic (nano)-materials for the remediation of the environment (water and atmosphere), both in suspensions and thin films [19]. It has good photochemical stability, less toxic, inexpensive, easily accessible, simple to recover, and so on. Its chief limitation is the limited level of spectrum overlapping with sunlight radiation, which is approximately 5% at ground level, reducing the degradation efficacy of TiO_2 photocatalysts under sunlight [19]. As a result, significant attempts are being made to enhance its photocatalytic efficacy by forming nanocomposites with other products [20–22], producing two-dimensional (thin films and nanosheets) [23] and three-dimensional (nanowires and nanotubes) nanostructures [24] and introducing imperfections into the crystalline system [25]. TiO_2 nanomaterials tend to agglomerate, resulting in a loss of photoactivity and active surface. Hence, sonication may be employed prior to or throughout the procedure to prevent agglomeration [26]. The resultant separation is not definite, since the equilibrium is favorable to reaggregate; however, the agglomeration process's kinetics could be sluggish. Prolonged ultrasound periods may not always aid deagglomeration. A significant issue regarding the utilization of nanoparticles, particularly in suspended form, was the ability to remove them out of the medium before discarding the photocatalysts [27]. Ultracentrifugation and nanofiltration are frequent treatments for this; however, they require large energy costs, which need to be prevented.

Given that the reactions occur among entities that are adsorbed on the nano-photocatalyst's surface, the photocatalytic conversion of a contaminant could be appropriately characterized via an altered Langmuir-Hinshelwood kinetic model, as shown in the following expression [28]:

$$r = k_{\mathrm{LH}} \cdot \theta = k_{\mathrm{LH}} \cdot \frac{K_{LH} \cdot C}{(1 + K_{\mathrm{LH}} \cdot C)} \tag{17.12}$$

Here θ (cm^2) represents the material's surface coverage, k_{LH} (mol/s cm^2) is the kinetic rate constant per unit of surface area, and C is the contaminant concentration after the equilibrium of adsorption is achieved. When the Langmuir model was rigorously obeyed, the Langmuir-Hinshelwood adsorption constant (K_{LH}) will be equivalent to the Langmuir adsorption constant (K_{eq}, see Eq. 17.15). Typically, the amount of contaminants is very small, hence $K_{\mathrm{LH}} \cdot C << 1$, and the process rate subsequently falls to pseudo-first-order kinetics:

$$r = k_{\mathrm{LH}} \cdot K_{\mathrm{LH}} \cdot C = k_{\mathrm{app}} \cdot C \tag{17.13}$$

The mathematical model of first order typically fits the experimental kinetics very well. This could lead to estimating the rate constant and half-life duration of various processes and comparing rates of eliminating diverse contaminants. Generally, the size variations and the modification in the active surface display a significant impact on the rate as well as the degradation concentration. In this area, a large number of studies and techniques are under investigation to improve the photocatalytic performance of nanomaterials and nanocomposites under sunlight.

17.2.2 Adsorption

Adsorption represents a surface phenomenon wherein a material known as "adsorbate" which is submerged in an aqueous solution (it could even be a gas) diffuses in the direction of the adsorbent's surface and interacts with the adsorbent, causing it to be held (Fig. 17.3). This occurrence of interaction/retention is greatly determined by the physicochemical features of both adsorbate and adsorbent. They are of two natures: physical and chemical, which are known as physisorption and chemisorption, respectively. Physisorption corresponds to weak interactions which could be easily broken (for example, by low heating). Chemisorption corresponds to more strong interactions (but less strong than chemical bonds) which are hard to be broken. The adsorbate–adsorbent interactions have a significant impact on the former's electronic structure, which determine their interactions with the surface throughout the process of adsorption and could boost their reactivity in heterogeneously catalytic procedures [29]. As a result, the entire process of adsorption is dependent on the available surface and its porosity, two variables that grow

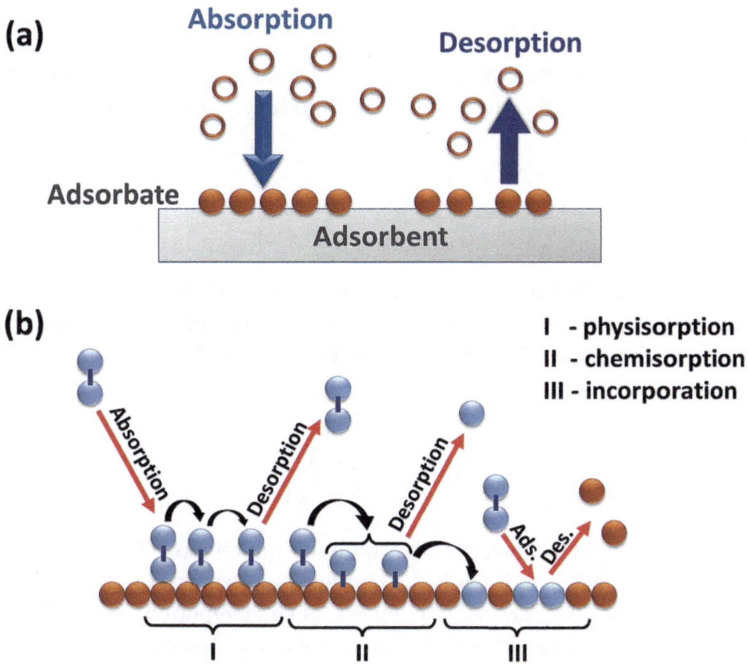

Fig. 17.3 Schematic illustrations of adsorption/desorption processes

dramatically at the nanoscale level. Not every place on the surface is accessible to interact with the adsorbate. Depending upon the chemical features of the adsorbate, the locations of interaction, or binding sites, should meet certain characteristics associated with the density of electrons, acidity, as well as shape. Accordingly, the most popular and most simple method of adsorption is Langmuir's model, which makes the following presumptions: (1) the adsorption–desorption procedure reaches equilibrium, (2) the catalyst's surface is homogeneous, (3) the various linking adsorption points on the surface have the same energy, (4) the adsorbent–adsorbate interactions occur via the same type of functional groups, (5) every single linking point connects with just one adsorbate species, (6) a single layer of adsorbate is created on the surface, and (7) there exists no adjacent interactions among adsorbed species when they are on the surface [15–17]. Even though these hypotheses oversimplify the issue, they result in the widely recognized Langmuir adsorption isotherm. Regardless of its clearness, the term "isotherm" relates to a specific temperature, thus each adsorption investigation should be conducted during temperature-controlled settings that are comparable with the ones that would be used in the environmental remediation processes.

When the equilibrium $A + S \rightleftarrows A_{ads}$ (where A_{ads} is the adsorbed species, S is the adsorbent, and A is the adsorbate) is reached, the rates of adsorption and desorption are equal:

$$r_{ads} = k_{ads} \cdot [A] \cdot [S] = k_{des} \cdot [A_{ads}] \tag{17.14}$$

Here k_{des} and k_{ads} represent, respectively, the rate constants for desorption and adsorption. $[S]$ denotes the number of free binding points per unit of adsorbate surface. Hence:

$$K_{eq} = \frac{k_{ads}}{k_{des}} = \frac{[A_{ads}]}{[A] \cdot [S]} \tag{17.15}$$

The Langmuir isotherm could be stated as mentioned in Eq. (17.16), by considering the overall number of binding locations (occupied and unoccupied) as $[S]_{tot} = [S] + [A_{ads}]$ and identifying the portion of populated binding locations as $\Theta = \frac{[A_{ads}]}{[S]_{tot}}$.

$$\Theta = \frac{K_{eq} \cdot [A]}{1 + K_{eq} \cdot [A]} \tag{17.16}$$

This equation could be linearized by applying the reciprocal. Generally, a linear behavior occurs once $K_{eq} \ll 1$, while the system attains saturation once $K_{eq} \gg 1$ [30].

Considering that the various hypotheses (1)–(7) reported above fail, various models have been developed, which involve surface inhomogeneity, interactions among adsorbate molecules on the surface, the probability of multiple adsorptions, and the possibility of forming multilayers of adsorbate. Therefore, the Freundlich isotherm (Eq. 17.17) accounts for surface heterogeneity because the adsorption heat, ΔH_{ads}, varies with the number of occupied binding points (sites with the most favorable interactions being engaged first), whereas homogeneous surfaces with comparable active sites display an unchanged ΔH_{ads} [29, 30].

$$\frac{x}{m} = K \cdot C^{\frac{1}{n}} \tag{17.17}$$

here, x represents the adsorbate mass, m denotes the adsorbent mass, and K and n are specific constants for each adsorbate and adsorbent at a certain temperature.

The Temkin adsorption isotherm accounts for adsorbate molecules' interactions when they are on the surface, arguing that these interactions decrease the adsorption heat and utilizing this method to adjust the Langmuir model's equilibrium rate constant [29, 30].

The Adsorbent's surface can include functional groups that are ionizable which can be either protonated or deprotonated based on the pH of the environment. The surface is neutral at a pH level that corresponds to the point of zero charge (pH_{PZC}). The surface is charged negatively if $pH > pH_{PZC}$ and charged positively if $pH < pH_{PZC}$. It can be feasible to predict if the interactions among the adsorbate and the binding locations would be favorable or unfavorable by looking at the pH of the environment and the pK_a values of various ionization sites on the molecule. Although it is sometimes overlooked, the research investigation of adsorption process at regulated pH conditions is essential to the efficacy of adsorption as a remediation approach.

Nano-adsorbents, like other nanomaterials employed in environmental remediation, are supposed to be abundant, inexpensive, simple to utilize, highly stable (mechanically and chemically), safe for both human health and the environment, recyclable, and reusable. Most importantly, after they reach the end of their practical lifespan, the produced residues should be simply inertized or valorized to prevent residue accumulation. Nano-adsorbents should also be effective (possess elevated adsorption yields), enable simple on-demand adsorbates' desorption, possess high selectivity wherever feasible, and be easily separated from the environment. It is essential that nanomaterials employed in environmental treatment do not create additional forms of contamination, that is why biodegradable nanostructured materials are being extensively explored [31]. Nano-adsorbents are often hard to separate out except if nanofiltration is employed, which is a costly separation procedure. As a result, magnetic nanoparticles may be integrated with the nano-adsorbents to facilitate their separation using a magnetic field [26, 32].

Like other kinds of nanomaterials, nano-adsorbents could be also synthesized using the well-known top-down and bottom-up approaches [31]. Nevertheless, bottom-up synthesis of nanomaterials is preferable because it provides better control of the procedures as well as the shape, size, and characteristics of the final compounds. The emphasis is nowadays being directed to the biosynthesis of nanostructured materials, with the hope that they would display higher biodegradability or lesser toxicity [33].

Adsorption techniques are effective for removing a wide range of persistent and mobile organic substances from various kinds of liquid and gaseous discharges. The adsorption techniques are often simple and inexpensive to implement. Zeolites, polymers, activated carbons (ACs), and many other materials are presently employed as adsorbents, and many have been created by valorizing various kinds of wastes [34]. Nanomaterials' advancements have resulted in a significant rise in adsorbents' effectiveness and adsorption rate, owing primarily to the significant increase in active surface and porosity, as well as the existence of a much greater number of active sites. Nano-adsorbents are currently employed mostly for the removal/reduction of dyes, oil, and heavy metals from water. They have the potential as well to remediate the so-called toxic forever chemicals like perfluoroalkyl and polyfluoroalkyl substances (PFAS) [35].

Several publications have evaluated the adsorption potential in the remediation of the environment. Even though adsorption has been utilized for this task for a long time, its use in the form of nanomaterials is still in its early stages. There are many basic investigations that have to be conducted to effectively manage the procedure and its related factors. The potential changes in the surface (functionalization) for carrying out specific adsorption procedures, or the successful deposition as films for effective flow chemistry heterogeneously catalytic procedures, appear highly desirable.

17.2.3 Chemical Redox Processes

Chemical redox processes are a class of catalytic procedures that involve an exchange of electrons or ions with the medium to reach the reagents. Generally, a material, either at the nanoscale or not, emits ions inside the volume of the medium or undergoes oxidation to release electrons (e^-). As a result, for both situations, the process could be identified as a homogenous process, occurring inside the volume of the medium rather than on its surface. However, the use of nanomaterials aids since, as described previously in the above parts, the surface-to-volume ratio promotes the procedures occurring in comparison to the employment of aggregates of nanoparticles or macroparticles. The following subsections cover the two primary processes in this category.

17.2.3.1 Fenton and Photo-Fenton Processes

H_2O_2/Fe^{2+} combinations have particularly strong oxidation characteristics in acidic media, which is explained by a sequence of events known as Fenton reactions [36], wherein reactive oxygen species are created, mostly $HO^{\bullet}$ radicals (Eqs. (17.18) and (17.19)). Fe^{3+} regenerates at proper acidic circumstances (pH ~2.7–2.8) and various additional reactive oxygen species are created (Eqs. 17.20–17.23). $HO_2^{\bullet}$ may also regenerate Fe^{2+} as shown in Eq. (17.24). With the existence of a Fe source and H_2O_2, sunlight or lamps simulating sunlight irradiation enable pollutant degradation. Such a process is referred to as the heterogeneous photo-Fenton process, and it has the potential to perform very well in the remediation of the environment. Fe^{3+} complexes could regenerate Fe^{2+} under irradiation ($\lambda \geq 300$ nm), as evident in Eq. (17.26). Lastly, throughout the electro-Fenton process, H_2O_2 is created in situ following Eq. (17.27). and Fe^{2+} is regenerated through Eqs. (17.21) and (17.22).

$$Fe^{2+} + H_2O_2 \rightarrow Fe^{3+} + HO^{\bullet} + HO^- \tag{17.18}$$

$$Fe^{3+} + H_2O_2 \rightarrow Fe^{2+} + HO_2^{\bullet} + H^+ \tag{17.19}$$

$$Fe^{2+} + HO^{\bullet} \rightarrow Fe^{3+} + HO^- \tag{17.20}$$

$$Fe^{3+} + H_2O_2 \rightarrow FeOOH^{2+} + H^+ \tag{17.21}$$

$$FeOOH^{2+} \rightarrow Fe^{2+} + HO_2^{\bullet} \tag{17.22}$$

$$Fe^{2+} + HO_2^{\bullet} \rightarrow Fe^{3+} + HO_2^- \tag{17.23}$$

$$Fe^{3+} + HO_2^{\bullet} \rightarrow Fe^{2+} + H^+ + O_2 \tag{17.24}$$

$$H_2O_2 + HO^{\bullet} \rightarrow HO_2^{\bullet} + H_2O \tag{17.25}$$

$$Fe(OH)^{2+} + h\nu \rightarrow Fe^{2+} + HO^{\bullet} \tag{17.26}$$

$$O_2 + 2H_3O^+ + 2e^- \rightarrow H_2O_2 + 2H_2O \tag{17.27}$$

The various reactive oxygen species produced are extremely reactive and capable of effectively transforming contaminants existing in the media, eventually causing mineralization. Fenton processes offer many benefits such as iron being inexpensive, widely accessible, and non-hazardous, as well as H_2O_2 being simple to manage. Pharmaceuticals, phenols, polychlorinated biphenyls (PCBs), and polycyclic aromatic hydrocarbons (PAHs) have all been demonstrated to be degraded via Fenton processes [37]. One significant disadvantage is that Fe salts precipitate, and the inclusion of Fe^{2+} and H_2O_2 implies high economic costs. Consumption of energy also increases the expenses of electro-Fenton procedures. For that reason, photo-Fenton is widely employed utilizing sunlight, which significantly minimizes related expenses.

17.2.3.2 Metal-Based Processes

Nanoscale metal-based processes provide the advantage of using zero-valence features (Metal^0, M^0), assuming the metal's potential for reduction is adequate. The processes that occur are one-electron reduction or a sequence of sequential one-electron reductions:

$$M^0 + \text{Pollutant} \rightarrow M^+ + \text{Pollutant}^{\bullet -} \tag{17.28}$$

$$M^0 + \text{Pollutant}^{\bullet -} \rightarrow M^+ + \text{Pollutant}^{\Theta} \tag{17.29}$$

$$M^+ + \text{Pollutant}^{\bullet -} \rightarrow M^{2+} + \text{Pollutant}^{\Theta} \tag{17.30}$$

Here "$^{\bullet -}$" denotes a one-electron excess and "$^{\Theta}$" indicates a two electrons excess. The mechanisms described in Eqs. (17.28 and 17.29) often result in the breakdown of C–X bonds, wherein X is an atom that is electronegative with a large electron density charge, like halogen. After this process, the contaminant intermediate takes H^+ or $H^{\bullet}$ from the solvent (typically water) and generates reduced products:

$$H_2O + \text{Pollutant}^{\bullet -} \rightarrow \text{Reduction Products} \tag{17.31}$$

According to this approach, reactions with zero-valence nano-metals may be used to degrade halogenated solvents or halogenated aromatic compounds, which have extremely lengthy environmental lifetimes and are typically resistant to other remediation procedures. Due to Fe's excellent reduction potential, Fe^0 nanomaterials are commonly utilized for this intent [38, 39]. Nano-metals with zero-valence often undergo oxidation on the surface, permitting reactions with other substances that could exist with a larger reduction capability. This might be effective for heavy metals contamination removal. Certain nano-metals with zero-valence have also been demonstrated to exhibit antibacterial or antivirus capabilities, making them appropriate for disinfecting interior environments (such as inside buildings) and remediating microbe or virus-infested water bodies [40].

The usage of bimetallic nanomaterials has been investigated in order to benefit from the existence of metals that have various reduction potentials, which could boost contaminants' degradation [41–43]. This technique additionally contributes to addressing the earlier reported issues of nanomaterials' agglomeration and improving the stability of zero-valence nano-metals. Noble nano-metals have been also utilized to construct bimetallic nanomaterials, with improved rates of pollution reduction; however, this increases significantly the costs because of the cost of the utilized noble metals (Au, Pd, etc.) [41–43].

17.2.4 Nanofiltration

Filtration is a basic physical separation technique that separates objects from fluids (either liquid or liquid) with the help of a medium that serves as a filter having a particular structure that penetrates the fluids and maintains the objects of specific dimensions which depend on the pore size of this filter medium. Membrane filters incorporating material sheets, granular deep bed filters, or surface filters with sieves are some examples of useful filter media [44]. Pressure is applied to drive the process of filtering, either by applying it to the input side or by applying a vacuum to the filtrate side.

As shown in Fig. 17.4, the technologies of membrane filtration are categorized based on membrane selectivity (determined by membrane pore size and the associated size of particles excluded) and the needed applied pressure [46]. They could be classified as follows: microfiltration (MF), ultrafiltration (UF), nanofiltration (NF), and reverse osmosis (RO). Here, MF uses the largest particle/pore sizes, while RO uses the smallest ones. The needed applied pressure varies with the filtering technique and rises as the size of membrane pores decreases as follows: MF > UF > NF > RO.

MF displays an exclusion size ranging between 0.1 and 10 μm and needs a driven pressure of roughly 0.2–5 bar. It is appropriate to eliminate most bacteria and cells, clays, Mycoplasma, *E. coli*, algae, etc. UF exhibits an exclusion size ranging between 0.01 and 0.1 μm and needs a driven pressure of about 1–10 bar. It is appropriate to eliminate some viruses, proteins, natural organic matter, humic acids, etc. MF shows an exclusion size ranging between 0.001 and 0.01 μm and needs a driven pressure of roughly 5–10 bar. It is appropriate to remove multivalent ions, heavy metals, synthetic dyes, carbohydrates, etc. MF presents a smaller exclusion size ranging between 0.0001 and 0.001 μm and needs a much higher driven pressure of about 10 up to 150 bar. It could eliminate practically the different objects other than water due to being built of semi-permeable membranes that only let water pass through.

Pretreatment of the input to the membrane filters is very important because it reduces fouling and enhances the functionality of membranes [47, 48]. For the pretreatment of the input in MF, traditional multimedia filtering processes and coagulation have been largely employed. Furthermore, considering the sequence of graduated size exclusion, each membrane stage could serve as a pretreatment for

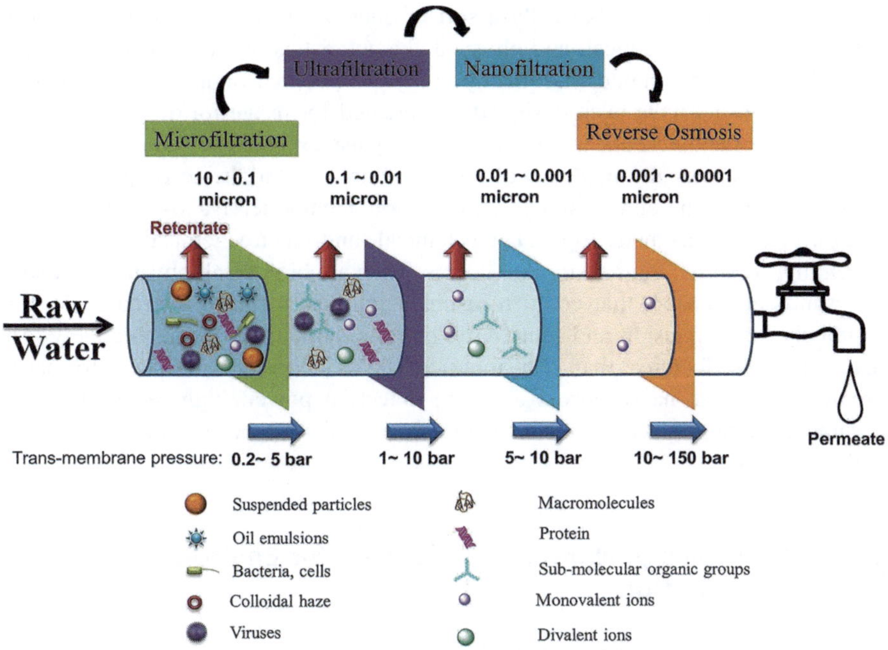

Fig. 17.4 Schematic illustration for water treatment using membrane filtration technologies. (Reproduced with permission from [45])

the succeeding one. Simply, microfiltration could be employed as a pretreatment for ultrafiltration, ultrafiltration as a pretreatment for nanofiltration, and ultrafiltration and nanofiltration as pretreatments for reverse osmosis.

17.2.5 Nano-Sensors

Continuously monitoring air pollution constitutes one of the vital and fundamental requirements in terms of controlling environmental pollution [49]. One of the risks of modern industrialization involves the release and spread of harmful and hazardous gasses (like H_2S, NH_3, NO_2, etc.). Inevitably, industry cautions are sometimes tardy to detect leakage of gasses. Furthermore, numerous issues, arising from the occurrence of heavy metal ions and other contaminants in water, air, and soil, emphasize the requirement to develop some technologies that are able to detect the contaminants earlier than that their amounts achieve dangerous levels in the environment. Once the technologies for pollution detection become more available and inexpensive, decision-making regarding the environment, monitoring of the ecosystem, and process management will be improved. Rapid and precise sensors capable of detecting contaminants at the molecular level will efficiently improve human capabilities to promote long-term human health and environmental sustainability.

Generally, sensors are basically a sort of energy converters, which are able to detect mechanical, chemical, and physical characteristics in their surroundings and show them as output signals (frequently optical or electrical signals). Consequently, numerous sensors have been designed and created for usage for diverse purposes. Interestingly, nano-sensors are one of the widely used sensors nowadays [50–52]. A nano-sensor is basically made up of nanostructured materials, accompanied by an identifying element and a signal transduction technique. Nano-sensors can be categorized into magnetic, optical, mechanical, and electrochemical nano-sensors depending on signal transduction processes. They are biological, physical, or chemical nanoscale sensors that could measure variations at the nanosized scale with extremely high precision and sensitivity, either quantitatively or qualitatively. The main essential features that contribute to the widespread trust in information provided by these nano-sensors are high detection power, high sensitivity, high accuracy, and the capacity to measure numerous kinds of species in the environment at the same time.

17.3 Examples of Nanomaterials Used for Environmental Remediation

17.3.1 Inorganic Nanomaterials

Various metal-based nanostructured materials have been reported for the remediation of a wide range of pollutants; however, most of these investigations have focused on the elimination of chlorinated organic compounds and heavy metals from water. Metal and metal oxide nanostructured materials are very effective nano-adsorbents possessing several advantages like excellent adsorption capacity and rapid kinetics. Some examples of metal-based nanomaterials examined for various environmental remediation applications are outlined in Table 17.1. Over the past years, effective physical and/or chemical synthesis approaches have been widely studied to obtain metal and metal oxide nanostructured materials with controlled shapes, high stability, and monodispersity. Hydrothermal, co-precipitation, sol-gel, etc., are some of the most extensively employed synthesis methods that provide large yields [54, 55].

Silver nanoparticles (Ag-NPs) have been used for disinfecting water due to their good antiviral, antifungal, and antibacterial activities [93–95]. Ag-NPs of very small size (<10 nm) were reported to be extremely toxic to *E. coli* and *P. aeruginosa*. They could additionally hinder viruses from attaching to host cells by attaching to the virus's glycoproteins preferentially. Slightly larger nanoparticles (~10 to 25 nm) showed lesser antibacterial activities [96]. Furthermore, Ag with nano-triangular shapes showed enhanced bactericidal activities in comparison to Ag nanospheres and Ag nanorods, highlighting the relevance of nanoparticle shapes in inducing their desirable effects [97, 98]. Ag-NPs have been combined with a variety of different materials, including polymers and metal oxides, in order to improve the global

Table 17.1 Some examples of metal-based nanomaterials applied for environmental remediation

Products	Applications	References
Silver nanoparticles	Water disinfection (antibacterial activities)	[53–57]
	Degradation of dyes (e.g., Congo red (CR), methylene blue (MB), methyl orange (MO), and rhodamine B (RhB))	
	Nano-sensors (Hg^{2+} and Fe^{3+} metal ions)	
Silver-based nanocomposites (e.g., combination with polymers or other oxide nanomaterials)	Water disinfection (antibacterial activities)	[58–62]
	Adsorption of dyes (like MB, acid red 37)	
	Removal of contaminants (like iodide I^-)	
	Filtration	
TiO_2 NMs	Dye degradation (MB, MO, and auramine O (AO))	[63–67]
	Remediation of eutrophic shallow freshwater systems	
	Antimicrobial and larvicidal activities	
Doped TiO_2 NMs	Dye degradation (MB, RhB, and basic yellow 28 (BY28))	[68–72]
	Degradation of heavy metal ions (like Cr^{6+})	
	Antimicrobial activity (*S. aureus* and *E. coli*)	
	Photocatalytic hydrogen production	
Titanate NMs	Catalytic reduction of gases—nitric oxide	[73–75]
Mixed oxide NMs	Degradation of dyes	[76–79]
Magnetic NMs	Elimination of various heavy metals	[80–85]
	Degradation of dyes	
Bimetallic NMs	Reduction of heavy metal ions (like Cr^{6+})	[86–92]
	Reduction of 4-nitrophenol to 4-aminophenol	
	Detection of Cr^{3+}	
	Degradation of dyes	

effectiveness of the resultant nanocomposites. This point will be further addressed in the upcoming subsections.

Titanium oxide nanomaterials (TiO_2 NMs) are another well-known explored metal-based product for environmental remediation applications. Because of their fascinating features such as energy-converting, gas sensing, electronic, photocatalytic, and semiconducting traits as well as their nontoxicity and lower cost, TiO_2 NMs have been widely investigated for wastewater treatment, surfaces' self-cleaning, and air purification applications [99]. TiO_2 NMs could be activated by an illumination source and hence are considered as good photocatalysts to eliminate organic pollutants from diverse media. TiO_2 NMs can produce highly reactive oxidants such as hydroxyl radicals, which act in disinfecting microorganisms including algae, viruses, bacteria, and fungi. Generally, TiO_2 itself possesses a somehow low photocatalytic activity, hence, it is frequently doped with other transition metal ions to improve its efficiency. As a result, numerous investigations have been conducted on metal ions doped TiO_2 NMs. For instance, Ag-doped TiO_2 nanofibers (NFs) are prepared via a hydrothermal process and then tested as photocatalysts for degrading different dyes (crystal violet (CV), Congo red (CR), malachite green (MG), and Methylene blue (MB)) under direct sunlight irradiation [100] and 2-chlorophenol under UV light illumination [101]. As well their antibacterial activities are tested against *E. coli* and *S. aureus*. In comparison to control TiO_2 NFs, Ag-doped TiO_2 NFs demonstrated both good photodegradation and antibacterial efficiencies. This enhancement had been ascribed to the following possible variables: an appropriate content of silver on the surface that efficiently captures photo-induced holes and electrons, a rapid movement of photo-induced electrons to the adsorbed oxygen available on the surface of the NFs, an increase in the number of surface hydroxyl groups, and an expansion of the response interval to light to the visible spectrum. Recently, Y. Slimani et al. [69] prepared Ce-Sm co-doped TiO_2 NPs with various doping levels via a sol-gel auto-combustion process, which are then tested as photocatalysts for degrading MB dye. In comparison to control TiO_2 NPs, Ce-Sm co-doped TiO_2 NPs demonstrated better photodegradation efficiency. Indeed, excellent performances are obtained for NPs with a doping level of 0.5%, wherein the photodegradation efficiency achieved about 98% within 30 min and a reaction rate constant of 0.0616 min^{-1}. These performances are reported to be much better than many other oxide compositions. It was established that h^+ and $\bullet O^{2-}$ are the utmost important active species in the photocatalytic degradation process. In addition, the reported nanophotocatalysts revealed high stability and can be definitely recovered and reused easily, suggesting their capability for real applications in the future.

Inorganic titanium oxide-based materials, called titanates, have also been studied for removing pollutants. For example, neutral, acidic, and basic titanate nanotubes (NTBs) have been prepared via hydrothermal technique by Chen's group [74], which are tested as catalysts to reduce NO with ammonia. For neutral, acidic, and basic pH environments, manganese oxide was incorporated into different titanate NTBs formulations to produce Mn-doped titanate nanotubes, titanate nanorods, and titanate nanosheets, respectively. The neutral Mn-doped titanate NTBs showed the highest surface area, excellent active species' dispersion, and the greatest active redox performance. Hence, neutral Mn-doped titanate NTBs possessed excellent catalytic reduction efficiency, while those prepared under basic conditions revealed insignificant activity. Different kinds of titanates, $ATiO_3$ where A is a divalent ion like Ba, Ca, Sr, Na, Ni, Co, etc., as well as their composites are widely investigated and numerous of them revealed enhanced environmental remediation performances [75]. $ATiO_3$ NMs combine both piezocatalytic and photocatalytic behaviors, suggesting them as promising multifunctional nanomaterials for practical applications in the environment and energy. To improve their efficiencies, numerous approaches have been performed to alter $ATiO_3$ NMs via doping or substitution, including hetero-cations, combining them with other semiconducting (nano)-materials or carbonaceous materials. Also, the preparation methods, preparation environments, crystalline phase, etc., greatly govern the activity of $ATiO_3$ nano-catalysts. Several studies reported that the cubic $BaTiO_3$ NMs display lesser piezocatalytic and photocatalytic performances under ultrasonication than tetragonal $BaTiO_3$ NMs. In-depth investigations must be performed to figure out the exact facts that lead to the piezocatalytic and photocatalytic activities in $ATiO_3$ NMs. Although numerous research articles investigated the piezoactivity and photoactivity of $ATiO_3$ NMs, the majority of them concentrated on the degradation of organic dyes, while fewer interest was devoted to other kinds of contaminants. Accordingly, the efficacy of $ATiO_3$ NMs for environmental remediation applications should be well verified regarding other inorganic and organic contaminants such as pathogenic species, NO_x, heavy metals, pharmaceutics, herbicides, pesticides, etc. Also, the evaluation of their toxicity in water should be studied.

Other metal oxide nanomaterials also demonstrated good photocatalytic activities for degrading dyes and removing heavy metal ions from polluted environments. For instance, E. Hannachi et al. [102, 103] synthesized a series of NMs made of ZnO NPs doped with rare earth elements like Y, Ce, Yb, etc. via a low-cost simple sol-gel synthesis approach. The findings revealed that an optimal concentration of co-dopants could significantly improve the degradation efficiency as well as accelerate the degradation rate. Numerous other metal oxide nanomaterials revealed enhanced photocatalytic activities [104–106].

Furthermore, mixed oxide nanomaterials have been studied. For instance, mixed TiO_2–SiO_2 oxide materials are synthesized using titanium isopropoxide or titanium butoxide and bamboo as sources of titanium and silica, respectively [76]. These binary materials were tested for their ability to degrade methylene blue (MB) dye photocatalytically. The degradation rate of MB at different treatment durations revealed substantial photoactivity. It has been speculated that these composites could find real use in industrial wastewater treatment systems on a small scale. Overall, although mixed oxide nanomaterials have higher photoactivity in comparison to pure TiO_2 nanomaterials in the majority of cases, their real use is restricted to the mineralization of specific contaminants. In the literature, numerous mixed oxide nanomaterials showed an improved ability to eliminate a broad range of contaminants [77–79].

Moreover, magnetic metal-based and metal oxide-based nanosized adsorbents have particular appeal due to their ease of retention and separation from the water that is treated. Iron and iron oxide-based nanomaterials (NMs) have been widely documented in previous studies for the elimination of various heavy metals (including Hg^{2+}, Cr^{6+}, Pb^{2+}, Cd^{2+}, Co^{2+}, Cu^{2+}, and Ni^{2+}), as well as for the degradation of some dyes [80–85]. Recently, spinel ferrites, hexaferrites, and multiferroic materials received great interest for applications in environmental remediation [107–115]. Nevertheless, there are some limitations to utilizing such kinds of NMs for

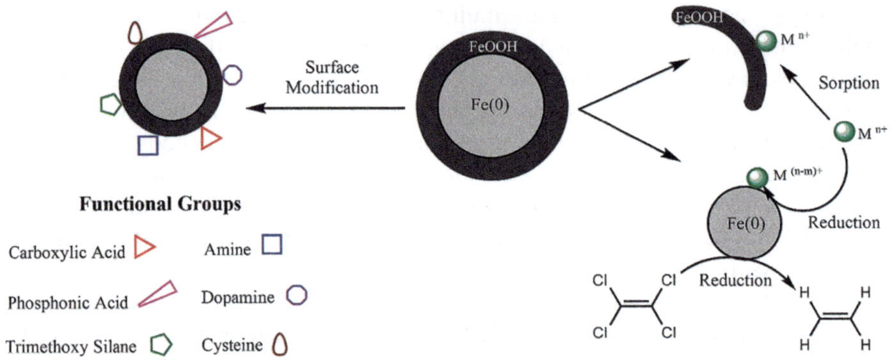

Fig. 17.5 Mechanism of degradation of heavy metals and chlorinated contaminants from aqueous systems utilizing core-shell iron nanoparticles. (Reproduced from [117])

environmental pollutants remediation. Agglomeration is one of the main concerns because it may drastically impact the reactivity of the product, reducing the benefit of employing nanomaterials to improve performance. Another issue to consider when dealing with metal and metal oxide NMs is the potential toxicity of the NMs utilized. Furthermore, the related costs and destiny of the remediation technologies are essential factors to think about when using NMs as remediation products.

Nanoparticles made of iron with a core-shell configuration could be as follows; the core consists of a zero-valent elemental iron (Fe^0) and the shell is comprised of mixed valent (Fe (III) and Fe (II)) oxides [116]. The mechanism whereby these core-shell nanoparticles could be employed to remediate environmental pollutants is depicted in Fig. 17.5 [117]. Electrons donated from the zero-valent Fe^0 core could be used for reducing both heavy metals and chlorinated compounds. Furthermore, the shell part of the nanoparticles may aid in remediating the pollutants, like heavy metals with a larger standard reduction potential (E^0) than the Fe^{2+}/Fe pair.

Some researchers investigated the utilization of sonication of magnetic nanoparticles (MNPs) solution in an attempt to prevent the agglomeration of NPs in order to improve their capacity to remove heavy metals [118]. When subjected to ultrasound treatment, nanoclusters may be dispersed; however, the released nanoparticles are susceptible to re-agglomeration if exposed to extended sonication durations. Hooshyar et al. [118] prepared 12 nm iron spherical NPs and found that sonication may improve the removal efficiency of heavy metals Co^{2+} and Ni^{2+}, with an ideal ultrasound period of around 30 min for the Co^{2+} case and 20 min for the Ni^{2+} case. The greatest removal efficiencies are about 60% and 39% for Co^{2+} and Ni^{2+}, respectively.

Numerous investigations examined the utilization of bimetallic nanoparticles to overcome some of the limitations related to mono-metallic nanoparticles, particularly agglomeration and poor stability. Various surfactants and stabilizers are usually used to boost the stability of NP solutions, but the incorporation of a second metal (or more) into the composition might improve the material's solution stability and eliminate the requirement for surfactants and stabilizers [86, 119]. Enhanced stability may lead to higher capacity and efficiency, as well as a faster rate of pollutants' degradation. The inclusion of extra metals, like Co, Ni, Pt, Ag, or Pd as a way to improve the stability of Fe^0 nanoparticles, has been suggested in recent studies [86–91]. Some noble metals may be coupled with Fe^0 nanoparticles to catalyze hydrogenation and dichlorination reactions with pollutants, which leads to a more effective remediation process. Wang et al. [92] documented an increased rate of dehalogenation of chlorinated organic compounds by incorporating Pd into zero-valent iron nanoparticles, although the costs are being expensive owing to the price of elemental Pd. Additional strategies for improving NPs' stability involve the addition of supporting materials, like the examples provided below in the sub-section polymer-based (nano)materials.

An additional issue related to the utilization of metal-based nanomaterials is the potential toxicity of the chemicals employed to manufacture the nanomaterials as well as the byproducts released through pollutant degradation. In a previous study, Naz et al. [120] showed the effective application of zero-valent metal nanoparticles to remove Cu^{2+} and Ni^{2+} from aqueous mediums. They made up Mn, Fe, and Mn-Fe nanoparticles from using cannabis sativa leaf extract. This strong antioxidant extract includes substances that react with Fe (III) to create zero-valent Fe nanoparticles. Among the various prepared NPs, Mn-Fe NPs revealed increased photocatalytic activity toward MO and CR dyes in comparison to single Mn and Fe NPs. Hence, it is interesting to use natural products in the synthesis of nanomaterials for environmental remediation to alleviate worries about the potential toxicity of chemicals and byproducts when employing chemical production routes. Furthermore, the green synthesis of zero-valent Fe nanoparticles shown by this research group offers the benefits of adding value to natural resources that would otherwise be deemed garbage (like leaf extracts) and offering low-cost nano-catalysts for dye removal from water. In another study, Poguberović and collaborators [121] made bimetallic NPs using cherry, mulberry, and oak leaf extracts for the elimination of As^{3+} and Cr^{6+}. Their study showed rapid kinetics and adsorption rate of the zero-valent Fe NPs, where the highest capacity for Cu^{2+} is about 1047 mg Cu/g with the use of mulberry leaf extracts and the highest capacity for Ni^{2+} is about 777 mg Ni/g with the use of oak leaf extracts. The remarkable disparities in performance may result from the specific raw materials utilized in the synthesis of nano-adsorbents. The best removal capacity of Cu^{2+} was attained at pH $= 7$, whereas the best removal capacity of Ni^{2+} was attained at pH $= 8$. Although the results of this investigation appear encouraging, additional investigations are needed before these materials can be used on a large-scale basis in wastewater treatment. Moreover, pH-sensitive remediation systems could pose some limits in their application to in situ remediation since certain environmental situations might not provide suitable conditions for effective and efficient remediation. Therefore, the researchers were able to create a reasonably harmless and low-cost organic–inorganic combination product for targeting polychlorinated biphenyl pollutants.

The development of mixed magnetic nanomaterials can also lead to enhanced performance for environmental remediation [122–124]. For instance, nano-$NiFe_2O_4$–NiO composites are synthesized via a co-precipitation process and the

Table 17.2 Some studies on the use of silica NMs for environmental remediation of pollutants

Products	Applications	References
Amine-grafted silica materials	Removal of gases (CO_2, H_2S)	[125–132]
Carboxylic acid-functionalized silica	Wastewater treatment (heavy metals, dyes, radioactive ions)	[133–136]
Thiol-functionalized silica	Soil and wastewater treatment (removal of heavy metals)	[137–140]
Amino-functionalized silica	Soil and wastewater treatment (removal of heavy metals and dye degradation)	[127, 141–145]

results showed an optimal photocatalytic degradation of about 95.4% toward MB dye within 2 h for 50%$NiFe_2O_4$-50%NiO composition [122]. NiO–CuO and NiO–Co_3O_4 composite nanophotocatalysts are also prepared and revealed excellent degradation efficiency against cefixime antibiotic [123] and malachite green dye [124], respectively.

Conversely, mesoporous silica materials, because of their flexibility, have also acquired a great interest in numerous kinds of applications, including catalysis and adsorption. These materials provide many advantages for applications in environmental remediation, such as high surface area, controllable size of pores, ease of alteration of the surface, and a large volume of pores. A number of investigations have also documented the application of these kinds of materials in pollutant remediation in the gas phase because of their exceptional performance as adsorbents. Moreover, several papers have presented various surface modifications of mesoporous silica materials. Table 17.2 highlights some of the studies that examine the application of silica NMs in the environmental remediation of various pollutants.

The hydroxyl groups that exist on the surface of silica materials play a key role in gas adsorption, surface modification, and other surface-related phenomena like wetting. Putting functional groups onto the walls of the pores is another recognized method for developing novel catalysts and adsorbents [117]. For instance, A. Ayub et al. [125] used amine-surface-modified silica materials to selectively remove H_2S and CO_2. These materials attained rapid CO_2 uptake by achieving about 82% of their equilibrium CO_2 uptake in 1 min. The materials are able of completely removing H_2S and CO_2 by 100%. Additionally, the materials maintained about 99% of their initial CO_2 uptake during 100-cycle testing in hot air. Anyanwu et al. [126, 127] utilized amine-surface-modified mesoporous silica (SBA-15) grafted at different temperatures for capturing CO_2. The nano-adsorbents grafted at ambient temperature revealed outstanding stability even after ten adsorption–desorption cycling tests, which is comparable to the conventional process that implies grafting at much higher temperatures >75 °C. In another study, Anyanwu et al. [128] employed silica gels grafted with amine for CO_2 capture purposes. Materials with high amine loading and largest pores (1.12 cm^3/g pore volume, 14.9 pore diameter, and 309 m^2/g BET surface area) demonstrated excellent CO_2 adsorption performance, rapid CO_2 uptake rate, and excellent cycle stability. Unexpectedly, the performance of materials exceeded those of the excellent MCM and SBA-type sorbent materials. Bai et al.

[129] synthesized triamine-grafted MCF silica at room temperature. The findings indicated that these materials display excellent CO_2 adsorption capacity, rapid adsorption kinetics, and outstanding cyclic stability. Mesoporous silica SBA-15 materials are grafted by using different sterically hindered amines, which are assessed for removing H_2S [130]. Excellent adsorbance performance is achieved for materials with (*tert*-butylaminopropyl) trimethoxysilane grafting. The significant abundance of amine groups on the surface of silica materials was linked to the observed good efficiency in H_2S and CO_2 removal. Another research group (s) showed [131, 132] the effectiveness of amine-modified alumino-silicates for capturing CO_2 and other carbonyl substances such as ketones and aldehydes. CO_2 capture has been found to be achievable by reversible adsorption of the gaseous molecules into the amino-silica product. Likewise, the creation of hemiaminal or imine is necessary for capturing ketone and aldehyde. The reaction kinetics are very quick, attaining around 90% of the entire capacity of the material in a couple of minutes. As a result, the above-mentioned products offer a feasible replacement for the classical capture of carbon dioxide by aqueous amines and other silica-supported amines since they are simpler to prepare, less costly, and offer higher stability and performance. Nevertheless, this may restrict some of their applications to capture pollutants that do not react with amines. Amine-functionalized porous silica is employed as an aldehyde abatement product to trap aldehydes with low-molecular-weight (like formaldehyde). The obtained findings implied that primary and secondary amines, rather than tertiary amines, are more appropriate to capture aldehydes, which is in accordance with the covalent capturing of the desired pollutant through the creation of imine and hemiaminal intermediates.

An example of mesoporous silica materials and surface properties relevant to adsorption applications is depicted in Fig. 17.6.

In addition to the capture of gases, silica-based compounds are applied for the elimination of organic dyes from wastewater. Numerous research groups [133–136] successfully functionalized mesoporous silica with carboxylic acid groups (–COOH) because the latter may establish hydrogen bonds with a variety of substances, including contaminants, dyes, and metal ions. The best performance for the degradation of methylene blue was attained at a pH of 9. Although these products are good adsorbents under basic circumstances, the pH dependence probably limits their practical applicability. Moreover, other research groups examined the applications of thiol-functionalized silica materials [137–140] and aminopropyl or amino-functionalized silica materials [127, 141–145] in remediating numerous kinds of metal ions like U^{6+}, Hg^{2+}, Pb^{2+}, Cr^{3+}, Al^{2+}, Ni^{2+}, Zn^{2+}, Cu^{2+}, Co^{2+}, and Cd^{2+} as well as for degrading dyes.

17.3.2 Carbon-Based Nanomaterials

Carbon-based nanostructured materials are also utilized for the remediation of pollutants via photocatalysis and adsorption processes. Indeed, carbon-based nanostructured materials possess distinctive chemical, physical, and electronic

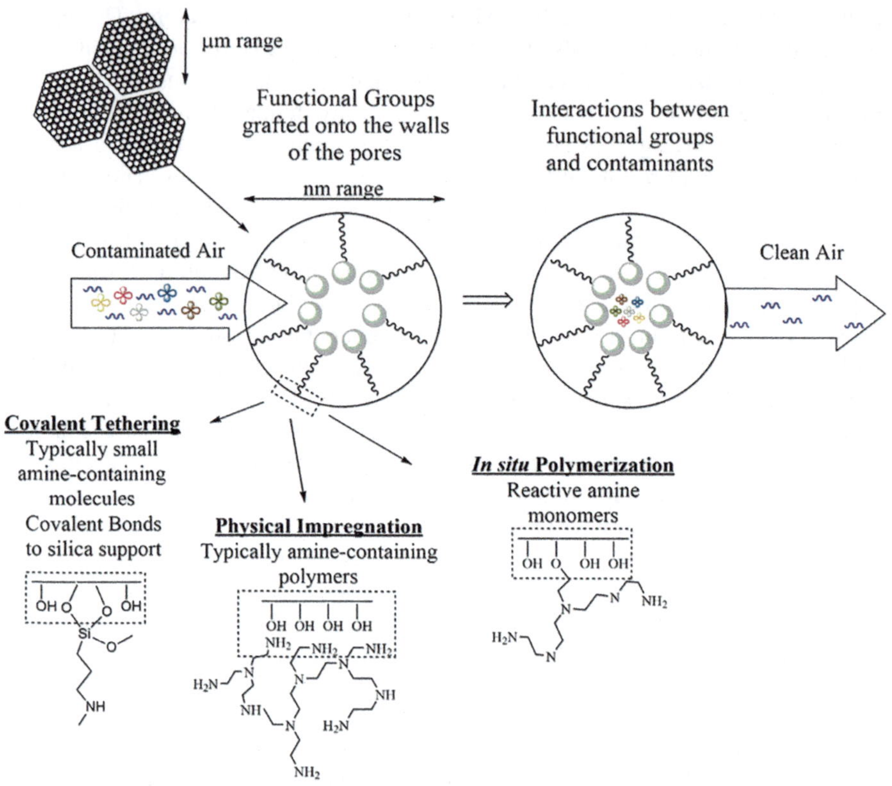

Fig. 17.6 A case of functionalized mesoporous silica materials applied in environmental remediation of pollutants. (Reproduced from [117])

features. Graphene, multi-walled carbon nanotubes (MW-CNTs), single-walled carbon nanotubes (SW-CNTs), and fullerenes C_{540} and C_{60} could all be formed through adjustable hybridization states [146–148]. Some studies reported that the surface modifications, functionalization, or activation of the virgin carbon material are initially needed for the feasibility of graphene and carbon nanotubes (CNTs) in environmental remediation applications. Many investigations have been conducted on MW-CNTs and SW-CNTs. The materials' adsorption capabilities render them especially suitable for removing inorganic and organic contaminants from large volumes of aqueous solutions and from the air.

Numerous papers have been published that detail the application of graphene to create nanocomposite photocatalysts. Pure graphene could be employed as an efficient adsorbent to remove contaminants from aqueous mediums like fluoride. At ambient temperature and pH of 7, graphene's monolayer adsorption capacity for fluoride is about 35.6 mg/g [149]. Although pure graphene could potentially be utilized for the remediation of the environment, a range of strategies depend on the usage of modified graphene to remediate various substances. Surface alterations reduce graphene layers' aggregation and hence improve the effective surface area, rendering the modified graphene potentially better than pure graphene [150]. Graphene oxide (GO) is a kind of modified graphenes that has been reported for the remediation of the environment by the adsorption of a number of water and gaseous pollutants, like pharmaceuticals, heavy metal ions (like Pb^{2+}, Cd^{2+}, As^{3+}, and Cr^{6+}), organic (cationic and anionic) dyes, CO_2 separation, NO_2 detection, etc. [151–156]. On the surface of the carbon of graphene oxides, numerous functional groups that contain oxygen, like hydroxyls, epoxides, and carboxylic acids exist, which play a great role in adsorption. Several researchers demonstrated that nanocomposites comprising of graphene and semiconducting nanoparticles (TiO_2, ZnO, CeO_2, WO_3, etc.) revealed higher photocatalytic activities in comparison to pristine nanoparticles alone (TiO_2, ZnO, CeO_2, WO_3, etc.) [157–162]. These nanocomposites were examined, for example, for the reduction/removal of heavy metal ions (like Cr^{6+}), degradation of dyes (like MG, MB, RhB), antibacterial agents, as well as gas sensors. This could be ascribed to the rise in conductivity. Even magnetic iron oxide-based nanomaterials are combined with graphene oxide to form new potential nanocomposites for environmental remediation applications [163–165]. One should note here that the control of the composition ratio in the nanocomposites between the graphene and the nanoparticles is important to achieve optimal photocatalytic activity. Indeed, exceeding a certain graphene content, a reduction in photocatalytic activity could be noticed. Table 17.3 summarizes some examples of graphene-based nanocomposites used for environmental remediation applications.

Conversely, great efforts have been done to improve the adsorption capabilities of pure CNTs by opening the closed ends. Indeed, unaltered carbon-based nanostructured materials are frequently inert to environmental pollutants. To boost their performance, they are usually altered or coated by other reactive compounds that have the required functional groups or charges. As a result, these hybrid nanomaterials combine numerous properties within a single framework to achieve the required efficiency. SW-CNTs are often organized in a hexagonal pattern where each nanotube is encircled by six others, which results in packages of aligned tubes with a heterogeneous porous framework. Adsorption may occur in four distinct accessible sites for an ordinary open-ended CNT bundle, which are of two sorts: (1) those that have greater energy of adsorption, localized either among two adjacent nanotubes or inside an individual nanotube, and (2) those that possess lower energy of adsorption, localized on the outside surfaces of the external CNTs making up the bundle [179]. Because the exterior sites are directly exposed to the adsorbing substance, adsorption on outside surfaces approaches equilibrium considerably quicker than adsorption on internal sites. MW-CNTs are seldom found in bundles unless particular techniques of synthesis are applied to generate such arrangements. Via nitrogen adsorption–desorption isotherm measurements, Zhong et al. [180] proved that distinct kinds of pores (i.e., inner and aggregated) form a multi-stage process of adsorption. Aggregated pores have been found to be more substantially responsible for the adsorption capabilities of these products than the less available interior pores. Moreover, the oxygen concentration of CNTs might have an impact

Table 17.3 Graphene and CNT-based nanocomposites used for environmental remediation applications

Products	Applications	References
Pristine graphene	Treatment of aqueous solutions (like adsorption of fluoride)	[149]
Graphene oxide	Adsorption of a number of water and gaseous pollutants, like pharmaceuticals, heavy metal ions (like Pb^{2+}, Cd^{2+}, As^{3+}, and Cr^{6+}), organic (cationic and anionic) dyes	[151–156]
	CO_2 separation	
	NO_2 detection	
Semiconductor-graphene oxide nanocomposites (semiconductors like ZnO, TiO_2, WO_3, or CeO_2)	Gas sensing	[157–162]
	Removal of metal ions	
	Dye degradation (like MG, MB, RhB)	
	Water disinfection (antibacterial activity)	
Spinel ferrite-graphene oxide nanocomposites	Removal of heavy metal ions	[163–165]
	Degradation of synthetic dyes	
	Antibacterial applications	
Oxidized CNTs	Adsorption of metal ions	[166–170]
CNT-based nanocomposites:	Dye degradation	[171–178]
• Metal oxide semiconductor-CNTs nanocomposites	Metal ions removal	
• Non-oxide semiconductor-CNTs nanocomposites		
• Magnetic iron oxide-CNTs nanocomposites		
• Bimetallic NMs-CNTs nanocomposites		

on their adsorption ability. CNTs that possess –COOH, –C=O, and –OH groups might favorably influence adsorption capabilities. This could be done based on the particular synthesis techniques and the associated purifying operations. CNTs could be oxidized using a variety of chemicals, including NaOH, KOH, H_2SO_4, NaOCl, H_2O_2, $KMnO_4$, HNO_3, etc., which could significantly enhance the adsorption capacities of metal ions like Cu^{2+}, Ni^{2+}, Cd^{2+}, and Pb^{2+} [166–170].

Recently, several attempts are focused on the combination of CNTs with other kinds of photocatalysts/adsorbents to further enhance their performances, for example, metal oxide semiconductors like TiO_2, ZnO, and WO_3 [171–173], titanates [174, 175], non-oxide semiconductors like CdS [176], magnetic iron oxide-based NMs [177], bimetallic NMs [178], and so on. These nanocomposites demonstrated enhanced adsorption and photocatalytic performances in comparison to individual nanomaterials.

17.3.3 Polymer-Based Nanomaterials

Polymers are largely employed to detect and remove organic contaminants (such as volatile organic compounds, pharmaceuticals, and aromatic hydrocarbons), gases (such as NO_x, SO_2, and CO), pollutant chemicals (such as heavy metals, arsenic, iron, nitrate, and manganese), and a broad range of biologics (such as viruses, parasites, and bacteria). Polymeric hosts (such as surface functionalized ligands, stabilizing agents, emulsifiers, and surfactants) are often employed to enhance the stability and surpass certain limitations of pristine nanoparticles, and they also provide other beneficial characteristics like recyclability, durability, thermal stability, and boosted mechanical strength. Table 17.4 summarizes a few examples of polymer-based nanostructured materials employed for the treatment of environmental pollutants.

Amphiphilic polyurethane nanomaterials have been produced for the removal of polynuclear aromatic hydrocarbons (PAHs) from soils, proving that organic nanomaterials may be tailored to have desirable features [181]. Amphiphilic polyurethane nanomaterials recovered about 80% of the phenanthrene from polluted aquifer sand. An examination of diverse compositions revealed that the affinity of amphiphilic polyurethane nanomaterials for phenanthrene improved as the size of the hydrophobic core rose. In addition, rising the amount of ionic groups on the precursor chains reduced the agglomeration of amphiphilic polyurethane particles in the existence of polyvalent cations [181]. After that, numerous researchers developed diverse potential polyurethane-based nanocomposites for the removal of contaminants from water and soil, as well as for oil separation, by incorporating diverse kinds of oxide NMs like GO, ZnO, and SiO_2 [182–184]. Dendrimers, including, for example, poly(amidoamine) (PAMAM), have been employed in contaminated wastewater treatment [185–189]. These nano-dendritic polymers include functional groups like hydroxamates, carboxylates, and primary amines that may encapsulate a wide variety of solutes in water, such as cations (U^{6+}, Zn^{2+}, Ni^{2+}, Fe^{3+}, Fe^{2+}, Au^+, Ag^+, and Cu^{2+}), as well as they effectively degrade

Table 17.4 Some examples of polymer-based nanomaterials employed for environmental remediation of pollutants

Products	Applications	References
Amphiphilic polyurethane NPs	Removal of polynuclear aromatic hydrocarbons from soils	[181]
Polyurethane-based nanocomposites	Oil adsorption and separation	[182–184]
	Dye degradation	
	Removal of metal ions	
Dendrimers	Wastewater treatment (degradation of dyes and removal of heavy metal ions)	[185–189]
Functionalized biodegradable polymeric nanomaterials	Removal of metal ions, dyes, and oils from water	[190–194]
Polymeric-based nanocomposites	Catalytic degradation of dyes	[195–202]
	Removal of heavy metals	

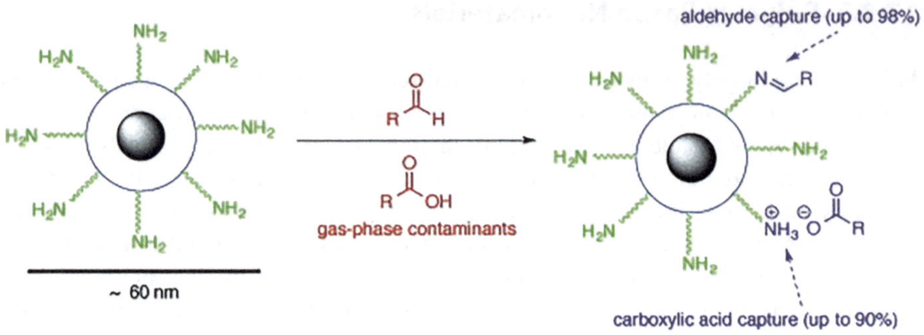

Fig. 17.7 Mechanism of PDLLA-PEG-PEI for aldehydes and carboxylic acid gaseous capturing. (Reproduced from [203])

the dyes. They could be employed as chelating agents and ultrafilters or microfilters that attach with metal ions, to facilitate the purification of water. Functionalized biodegradable and non-toxic polymeric nanomaterials, including cellulose, chitosan, polylactic acid (PLA), gelatin, etc., have been also utilized for specifically capturing volatile organic compounds and removing metal ions, dyes, and oils from water [190–194]. The modified target-specific reactivity of materials enabled the quick and selective capturing or removing of desirable pollutants. The development of ionic or covalent connections among the amine functional groups and the intended aldehydes and/or carboxylic groups is the basis for this technique, which could lead to a more selective and efficient capturing technology. As an example, Fig. 17.7 shows the mechanism of aldehydes and carboxylic acid gaseous captured by poly (D,L-lactic acid)-poly(ethylene glycol)-poly(ethyleneimine) (PDLLA-PEG-PEI) [203].

Polymer-supported nanocomposites (PNCs) are products that use a polymer as a base material and have nano-entities dispersed in or coated on it. These products combine the desired features of the nanomaterials (like high reactivity) and the base polymer (like exquisite mechanical strength). Biodegradability and biocompatibility are the two important factors in the utilization of polymer-supported nanocomposites. Numerous polymers have also been employed for environmental remediation applications by the construction of membranes that include metal and metal oxide nanomaterials [195–198]. Polymers are often utilized as base components in polymeric nanocomposites, while other elements of the nanocomposite, like nanomaterials, take the responsibility of remediating the pollutants. Nevertheless, in some studies, functionalized polymeric nanocomposites have been also identified to be the possible principal agent in the remediation process. Green hybrid polymeric-based nanocomposites, made of ZnO NPs loaded into a hydrogel, have been created by employing acrylamide-grafted on the polymeric chains of gelatin and chitosan [202]. These hydrogel nanocomposites showed a successful degradation of about 91% of Congo red dye in wastewater and under sunlight. Magnetic polymeric-based nanocomposites made of alginate and poly (vinyl alcohol) (PVA) hydrogels and Fe_3O_4 NPs have been prepared by Zhang et al. [204]. The prepared products demonstrated good catalytic performances for the Fenton degradation of organic dyes. Remarkably, the double-cross-linked structure provides the iron oxide NPs much more stability and recyclability in comparison to the conventional Fenton catalytic process. The mass loss of these polymeric-based nanocomposites is about 10% after six cycles. This could prevent the creation of Fe sludge in Fenton-like reactions to the utmost level and offers widespread potential applications for the remediation of wastewater. In another study, biopolymeric hydrogel nanocomposites composed of gelatin-supported Sn-doped Gd_2O_3 have been prepared [199]. The existence of many functional groups on the gelatin's framework, like hydroxyl (–OH) and carboxylic acid (–COOH), facilitates the cross-linking with formaldehyde. These nanocomposites were employed for the catalytic degradation of different dyes (like 2-nitrophenol, 4-nitrophenol, CR, MO, and MB dyes) in aqueous medium $NaBH_4$ reducing agents. For instance, the catalytic degradation of CR dye was successfully performed with a reaction rate of about 0.915 min^{-1}. It has been also stated that these biopolymeric hydrogel nanocomposites can be recycled recovered by a simple discharge of the reduced dye and reutilized for another time. Ibeaho et al. [200] developed multifunctional magnetic hydrogels fabricated by Fe_3O_4 NPs-mediated radical polymerization of poly (ethylene glycol) diacrylate. The obtained products display multifunctional properties integrating photothermal response, superwettability, and magnetism, in addition to the rapid degradation of organic dyes with excelled recyclability. In another study, hydrogel nanocomposites of polyethyleneimine and magnetic NPs have been synthesized [201]. The resultant materials display a rough surface and high porosity, leading to a fast adsorption rate and great adsorption capacity of heavy metals like Pb^{2+}, Cd^{2+}, and Cu^{2+}, which could be further enhanced by raising the pH value of the initial solution.

17.4 Conclusions and Future Perspectives

Nowadays, our ecological equilibrium is being considerably disturbed by the rapid growth in population and in the amount of residues emitted by a large number of industries. The subject of environmental pollution and climate change is one of the dangerous issues of concern worldwide. The pollution of our environment is being a risk to our lives since it could be the main sources of a variety of undesirable diseases. Therefore, an immediate degradation (reduction or removal) of these contaminants is needed. Nevertheless, certain contaminants are extremely difficult to breakdown and require a long period of treatment. The unique function of nanotechnology could be employed for this purpose, and contaminants are able to be changed into safe ones or properly removed, which is attributed to their high surface area, adequate functional groups, and active sites. Additionally, nanotechnology can change strong surface complexes with many types of contaminants and so effectively remove the contaminants from the environment (like an aqueous solution). Accordingly, nanostructured materials and composites could be employed in the environmental remediation processes. For the remediation of environmental pollutants, many types of nanostructured materials and composites are utilized, such as inorganic-based, carbon-based, and polymer-based nanomaterials. Contaminants such as halogenated herbicides, volatile organic compounds, organophosphorus compounds, chlorinated organic chemicals, dyes, heavy metals, pharmaceuticals, etc. could be effectively remediated by using these nanomaterials and nanocomposites. Moreover, processes such as photocatalytic reduction, catalysis, adsorption, (nano)filtration, nanosensing, etc., of pollutants are frequently identified as significant technologies for reducing or removing or capturing pollutants in the environment. In light of this, the current chapter will focus on the important nanotechnologies and nanomaterials employed in environmental remediation applications, as well as their advantages and limitations that must be overcome to achieve a clean and safe environment in the future.

The use of nanostructured materials and composites in the identification and elimination of diseases offers in-line and real-time detection, smaller sizes of samples, reduced costs, quicker treatment period, higher degradation yields, lesser toxicity (even nontoxicity), improved sensitivity and selectivity, and so on. Indeed, nanomaterials of metal and metal oxides have been widely utilized to eliminate heavy metals and organic contaminants by the oxidation or reduction of NMs. The level of degradation could be further improved by functionalizing these NMs with chemical groups capable of selectively capturing specific contaminants in the polluted medium. This technology is efficient and exciting, and it could potentially be employed in the treatment of air and water.

Nanomembranes have been used in the treatment of drinkable water, wastewater reclamation, and reuse, as well as the elimination of pesticides, natural organic matter, pharmaceuticals, dyes, and heavy metals from polluted water. Additional improvements should be done within environmental remediation applications to selectively eliminate the nanomaterials and nanocomposites after use, which should also possess stronger resistance to variations in pH values and concentrations of contaminants existing in polluted water, higher stability over very long durations, and lowest costs. Because of their low basic weight, small size of pores, and large permeability, nanofibrous media are also suitable for a broad variety of applications as filters. Furthermore, nanofibrous membranes exhibit distinctive characteristics including large specific surface area, strong connectivity between pores, and the ability to include active sites or functional groups at the nanoscale. Nanofiber filters could be utilized in a variety of applications, including the extraction of microparticles from wastewater. To extend the lifespan of the membranes, nanofiltration or ultrafiltration could be used. Some research activities are under examination to produce nanostructured materials with different fiber geometries and diameters in order to determine their influence on the efficiency of nano-fibers.

Polymer-supported nanocomposites have been also employed for environmental remediation applications, such as monitoring and detection (sensors) of pollutants, adsorption of contaminants, chemical catalysis degradation, and photocatalytic degradation, which provide a greener environment perspective. Nevertheless, an investigation of the relationship among the surrounding polymers and the encapsulated nanomaterials, as well as their dispersion in contaminated water and air, should be conducted. Furthermore, the manufacture of polymeric-based nanomaterials for large-scale applications is still in progress.

The widespread use of sorbents in the remediation of the environment has revealed their potential to adsorb metals and organic contaminants from polluted air and water. High adsorption selectivity and capacity have been achieved by the use of nanostructured materials like titanate-based NMs, iron-based NMs, and polymeric nano-adsorbents. Sorbent surface modification is being investigated to improve process performances. Improving sorbent reusability and extending their lifetime should be also investigated in order to minimize their costs.

Nano-sensors that are able to detect and identify volatile organic compounds, contaminant chemicals, and toxic gases, as well as bacteria, have been created. Additional improvements in the multifunctional characteristics of nanostructured materials are needed to fulfill requirements for the detection and remediation of contaminants in air and water. Also, fundamental investigations are needed to understand well their capabilities.

Because of their fibrous body and large surface area, SW-CNTs and MW-CNTs have demonstrated outstanding adsorption capabilities to remove numerous chemical and biological pollutants. Tiny NMs made of carbon nanotubes are hard to be separated from aqueous solutions. To achieve this, an ultracentrifugation separation process would be preferably applied, nevertheless, such a procedure requires high energy. Filtration by membranes is an alternate and effective approach for separating carbon nanotubes from an aqueous solution. Nevertheless, these membranes could be easily plugged. Nanocomposites made of carbon nanotubes and magnetic NMs or metal oxide NMs are interesting products for large-scale environmental pollution control. Further attempts will be necessary to design and develop CNT-based nanocomposites for real applications. Nano-dendritic polymers have been created for use in filtering techniques that need low pressure to eliminate or recover contaminants from polluted water, wastewater, and soil, like metal ions, uranium, and perchlorate. The efficiency of composites based on nano-dendritic polymer has not been yet documented and needs to be studied in the future.

References

1. Mohanapriya V, Sakthivel R, Pham NDK, Cheng CK, Le HS, Dong TMH (2022) Nanotechnology-a ray of hope for heavy metals removal. Chemosphere 311:136989
2. Newman L, Ansari AA, Gill SS, Naeem M, Gill R (2023) Phytoremediation: management of environmental contaminants, Volume 7. Springer
3. Chausali N, Saxena J, Prasad R (2023) Nanotechnology as a sustainable approach for combating the environmental effects of climate change. J Agric Food Res 12:100541
4. Thakur A, Thakur P, Khurana S M P, (2022) Synthesis and Applications of Nanoparticles, Springer Singapore:1–544
5. Ilaria C, Iole V, Francesco T, Carlo P (2023) Environmental safety of nanotechnologies: the eco-design of manufactured nanomaterials for environmental remediation. Sci Total Environ 864:161181
6. Bhandari G, Dhasmana A, Chaudhary P, Gupta S, Gangola S, Gupta A, Rustagi S, Shende SS, Rajput VD, Minkina T, Malik S, Slama P (2023) A perspective review on green

nanotechnology in agro-ecosystems: opportunities for sustainable agricultural practices & environmental remediation. Agriculture 13(3):668
7. Feijoo S, Kamali M, Dewil R (2022) A review of wastewater treatment technologies for the degradation of pharmaceutically active compounds: carbamazepine as a case study. Chem Eng J 455:140589
8. Firoozi M, Hashemi M, Narooie MR, Daraei H (2023b) Evaluation of phenol degradation rate using advanced oxidation/reduction process (AO/RP) in the presence of sulfite and zinc oxide under UV. Optik 279:170787
9. Barroso-Martínez JS, Romo AIB, Pudar S, Putnam ST, Bustos E, Rodríguez-López J (2022) Real-time detection of hydroxyl radical generated at operating electrodes via redox-active adduct formation using scanning electrochemical microscopy. J Am Chem Soc 144(41): 18896–18907
10. Armstrong DG, Huie RE, Koppenol WH, Lymar SV, Merenyi G, Neta P, Ruscic B, Stanbury DM, Steenken S, Wardman P (2015) Standard electrode potentials involving radicals in aqueous solution: inorganic radicals (IUPAC Technical Report). Pure Appl Chem 87(11–12):1139–1150
11. Barakat M, Kumar R (2021) Nanomaterials for environmental applications. CRC Press, Boca Raton
12. Huang L, Huang X, Yan J, Liu Y, Jiang H, Zhang H, Steenari B, Liu Q (2022) Research progresses on the application of perovskite in adsorption and photocatalytic removal of water pollutants. J Hazard Mater 442:130024
13. Martemucci G, Costagliola C, Mariano MS, D'Andrea LD, Napolitano P, D'Alessandro A (2022) Free radical properties, source and targets, antioxidant consumption and health. Oxygen 2(2):48–78
14. Zeng G, Shi M, Dai M, Zhou Q, Luo H, Lin L, Zang K, Meng Z, Pan X (2023) Hydroxyl radicals in natural waters: light/dark mechanisms, changes and scavenging effects. Sci Total Environ 868:161533
15. Kerkez Đ, Bečelić-Tomin M, Gvoić V, Mandić AK, Maćerak AL, Pilipović DT, Pešić V (2023) Pyrite cinder as an effective fenton-like catalyst for the degradation of reactive azo dye: effects of process parameters and complete effluent characterization. Catalysts 13(2):424
16. Poonia K, Patial S, Singh P, Ahamad T, Khan AAP, Van Le Q, Nguyen VT, Hussain CM, Singh P (2023) Recent advances in Metal Organic Framework (MOF)-based hierarchical composites for water treatment by adsorptional photocatalysis: a review. Environ Res 222: 115349
17. Shan A, Lu Y, Cheng L, Hou Z, Xili D, Li Y, Liu J, Ma H, Yang J (2023) Enhanced direct degradation of photogenerated holes induced by dissolved O2 chemisorbed on oxygen-deficient nano-TiO2. Mater Today Sustain 22:100391
18. Lin L, Liu C, Dang VD, Fu H (2022) Atomically dispersed Ti-O clusters anchored on NH2-UiO-66(Zr) as efficient and deactivation-resistant photocatalyst for abatement of gaseous toluene under visible light. J Colloid Interface Sci 635:323–335
19. Sinha R, Ghosal PS (2023) A comprehensive appraisal on status and management of remediation of DBPs by TiO2 based-photocatalysts: insights of technology, performance and energy efficiency. J Environ Manage 328:117011
20. Akhter P, Nawaz S, Shafiq I, Nazir A, Shafique S, Jamil F, Park Y, Hussain M (2023) Efficient visible light assisted photocatalysis using ZnO/TiO2 nanocomposites. Mol Catal 535:112896
21. Suhan MBK, Shuchi SB, Al-Mamun MR, Roy H, Islam MS (2023) Enhanced UV light-driven photocatalytic degradation of methyl orange using MoO3/WO3-fluorinated TiO2 nanocomposites. Environ Nanotechnol Monit Manag 19:100768
22. Wang Q, Zhao Y, Zhang Z, Liao S, Deng Y, Wang X, Ye Q, Wang K (2022) Hydrothermal preparation of Sn3O4/TiO2 nanotube arrays as effective photocatalysts for boosting photocatalytic dye degradation and hydrogen production. Ceram Int 49(4):5977–5985
23. Cuadra JG, Molina-Prados S, Mínguez-Vega G, Estrada AC, Trindade T, Oliveira C, Seabra MP, Labrincha J, Porcar S, Cadena R, Fraga D, Carda JB (2023) Multifunctional silver-coated

transparent TiO2 thin films for photocatalytic and antimicrobial applications. Appl Surf Sci 617:156519
24. Sepúlveda M, Saldan I, Mahnaz A, Cicmancova V, Michalicka J, Hromadko L, Bulánek R, Sopha H, Macak JM (2022) Magnetically guidable single TiO2 nanotube photocatalyst: Structure and photocatalytic properties. Ceram Int 49(4):6764–6771
25. Wang C, Liu Y, Li Y, Sun X, Xu L, Huang W (2023) Facile defect construction of TiO2 nanotube for excellent photocatalytic degradation of tetracycline under visible light. J Photochem Photobiol A Chem 437:114475
26. Rizwan K, Bilal M, Slimani Y, Show PL, Rtimi S, Roy A, Iqbal HM (2022) Hydrogen-based sono-hybrid catalytic degradation and mitigation of industrially-originated dye-based pollutants. Int J Hydrogen Energy 48(17):6597–6612
27. Dey C, Nandi M, Goswami MM (2022) pH dependent enhanced synchronous photocatalytic removal of cationic and anionic dyes by CoFe2O4 magnetic nanoparticles. J Mol Struct 134859
28. Franz S, Bestetti M (2023) Kinetic models in photoelectrocatalysis. Elsevier, Amsterdam, pp 217–263
29. Ahmed MA, Ahmed MA, Mohamed AA (2023) Synthesis, characterization and application of chitosan/graphene oxide/copper ferrite nanocomposite for the adsorptive removal of anionic and cationic dyes from wastewater. RSC Adv 13(8):5337–5352
30. Hu Q, Lan R, He L, Liu H, Pei X (2023) A critical review of adsorption isotherm models for aqueous contaminants: Curve characteristics, site energy distribution and common controversies. J Environ Manage 329:117104
31. Sharma A, Kumar N, Mudhoo A, Garg VK (2023) Phytobiomass-based nanoadsorbents for sequestration of aquatic emerging contaminants: an overview. J Environ Chem Eng 11(2): 109506
32. Candido JDC, Weschenfelder SE, Ferraz HC (2023) A review on the synthesis and application of magnetic nanoadsorbents to the treatment of oilfield produced water. Braz J Chem Eng. https://doi.org/10.1007/s43153-023-00331-0
33. Slimani Y, Hannachi E (2021) Green chemistry and sustainable nanotechnological developments: principles, designs, applications, and efficiency. Apple Academic Press, Palm Bay, pp 1–18
34. Shi Y, Ni R, Zhao Y (2023) Review on multidimensional adsorbents for CO_2 capture from ambient air: recent advances and future perspectives. Energy Fuel 37:6365–6381
35. Naghdi S, Shahrestani MM, Zendehbad M, Djahaniani H, Kazemian H, Eder D (2022) Recent advances in application of metal-organic frameworks (MOFs) as adsorbent and catalyst in removal of persistent organic pollutants (POPs). J Hazard Mater 442:130127
36. Ahmed Y, Zhong J, Yuan Z, Guo J (2022) Roles of reactive oxygen species in antibiotic resistant bacteria inactivation and micropollutant degradation in Fenton and photo-Fenton processes. J Hazard Mater 430:128408
37. Kumar V, Sharma N, Umesh M, Chakraborty P, Kaur K, Duhan L, Sarojini S, Thazeem B, Pasrija R, Vangnai AS, Maitra SS (2023) Micropollutants characteristics, fate, and sustainable removal technologies for landfill leachate: a technical perspective. J Water Process Eng 53: 103649
38. Zhu L, Sun Z, Wen J, Wang X, Liu Y, Zhao X, Liu Y (2022) Quantitative structure-activity relationship study on the degradation of polyhalogenated carbazoles by sulfidated zero-valent iron/peroxymonosulfate system. J Environ Chem Eng 10(2):107244
39. Barka E, Noutsopoulos C, Galani A, Panagou I, Kalli M, Koumaki E, Malamis S, Mamais D (2023) Removal of contaminants of emerging concern from wastewater using an integrated column system containing zero valent iron nanoparticles. Water 15(3):598
40. Vu KA, Mulligan CN (2023) Remediation of organic contaminated soil by Fe-based nanoparticles and surfactants: a review. Environ Technol Rev 12:60–82
41. Annapureddy V, Na S, Hwang G, Kang MH, Sriramdas R, Palneedi H, Yoon W, Hahn B, Kim JM, Ahn C, Park D, Choi J, Jeong D, Flatau AB, Peddigari M, Priya S, Kim KS, Ryu J (2018)

Exceeding milli-watt powering magneto-mechano-electric generator for standalone-powered electronics. Energy Environ Sci 11(4):818–829
42. Sun H, Lee S, Park S (2022) Bimetallic CuPd alloy nanoparticles decorated ZnO nanosheets with enhanced photocatalytic degradation of methyl orange dye. J Colloid Interface Sci 629: 87–96
43. Ayodhya D, Sumalatha V, Gurrapu R, Sharath Babu M (2023) Catalytic degradation of HIV drugs in water and antimicrobial activity of Chrysin-conjugated Ag-Au, Ag-Cu, and Au-Cu bimetallic nanoparticles. Results Chem 5:100792
44. Elsaid K, Olabi A, Abdel-Wahab A, Elkamel A, Alami AH, Inayat A, Chae K, Abdelkareem MA (2023) Membrane processes for environmental remediation of nanomaterials: Potentials and challenges. Sci Total Environ 879:162569
45. Liao Y, Loh CH, Tian M, Wang R, Fane AG (2017) Progress in electrospun polymeric nanofibrous membranes for water treatment: fabrication, modification and applications. Prog Polym Sci 77:69–94
46. Gao Y, Yang J, Song X, Li N, Guo C, Shi H, Chadha U, Selvaraj SK, Thanu SV, Cholapadath V, Abraham AM, Zaiyan MM, Manoharan M, Paramsivam V (2022) A review of the function of using carbon nanomaterials in membrane filtration for contaminant removal from wastewater. Mater Res Express 9:012003
47. Abushawish A, Bouaziz I, Almanassra IW, Al-Rajabi MM, Jaber L, Khalil AKA, Takriff MS, Laoui T, Shanableh A, Atieh MA, Chatla A (2023) Desalination pretreatment technologies: current status and future developments. Water 15:1572
48. Ma B, Ulbricht M, Hu C, Fan H, Wang X, Pan Y, Hosseini SS, Panglisch S, Van Der Bruggen B, Wang Z (2023) Membrane life cycle management: an exciting opportunity for advancing the sustainability features of membrane separations. Environ Sci Technol 57(8): 3013–3020
49. Slimani Y, Hannachi E, Yasin G (2022) Air pollution management by nanomaterials. Apple Academic Press, Palm Bay, pp 409–441
50. Lavate S, Kumar S, Seena S, Srivastava R (2023) Recent advances in environmental nanotechnology. Elsevier, Amsterdam, pp 293–318
51. Ku C, Chung C (2023) Advances in humidity nanosensors and their application: review. Sensors 23(4):2328
52. Slimani Y, Hannachi E (2020) Magnetic nanosensors and their potential applications. In: Nanosensors for smart cities. Elsevier, pp 134–155
53. Tasic L, Stanisic D, Barros CHN, Covesi LK, Bandala ER (2022) Inactivation of Escherichia coli using biogenic silver nanoparticles and ultraviolet (UV) radiation in water disinfection processes. Catalysts 12(4):430
54. Basumatary S, Kumar KJ, Daimari J, Mondal A, Kalita S, Dey KS, Deka AK (2023) Biosynthesis of silver nanoparticles using Antidesma acidum leaf extract: its application in textile organic dye degradation. Environ Nanotechnol Monit Manag 19:100769
55. Chen HS, Qiu C, Jiang Y, Liao X, Wu D, Shen M, Ding T (2022) Silver nanoparticles on UiO-66 (Zr) metal-organic frameworks for water disinfection application. Food Sci Human Wellness 11(2):269–276
56. Hu G, Jin W, Chen Q, Cai Y, Zhu Q, Zhang W (2016) Antibacterial activity of silver nanoparticles with different morphologies as well as their possible antibacterial mechanism. Appl Phys A 122:874
57. Moond M, Singh S, Sangwan S, Devi P, Beniwal A, Rani J, Kumari A, Rani S (2023) Biosynthesis of silver nanoparticles utilizing leaf extract of Trigonella foenum-graecum L. for catalytic dyes degradation and colorimetric sensing of Fe3+/Hg2. Molecules 28(3): 951. PMID: 36770623; PMCID: PMC9919385.
58. Elbadawy HA, Elhusseiny AF, Hussein SM, Sadik WA (2023) Sustainable and energy-efficient photocatalytic degradation of textile dye assisted by ecofriendly synthesized silver nanoparticles. Sci Rep 13(1):2302

59. Sartori P, Delamare APL, Machado G, Devine DM, Da Silva Crespo J, Giovanela M (2023) Synthesis and characterization of silver nanoparticles for the preparation of chitosan pellets and their application in industrial wastewater disinfection. Water 15(1):190
60. Sultan MA, Nagieb ZA, El-Masry HM, Taha GR (2022) Physically-crosslinked hydroxyethyl cellulose-g-poly (acrylic acid-co-acrylamide)-Fe3+/silver nanoparticles for water disinfection and enhanced adsorption of basic methylene blue dye. Int J Biol Macromol 196:180–193
61. Li L, Yu S, Zheng R, Li P, Li Q, Liu J (2022) Removal of iodide anions in water by silver nanoparticles supported on polystyrene anion exchanger. J Environ Sci China 128:45–54
62. Yu D, Liu L, Ding B, Yu J, Si Y (2023) Spider-Web-Inspired SiO2/Ag nanofibrous aerogels with superelastic and conductive networks for electroporation water disinfection. Chem Eng J 461:141908
63. Al-Hamoud K, Shaik MR, Khan M, Khan M, Adil SF, Kuniyil M, Assal ME, Al-Warthan A, Tahir M, Khan ST, Mousa AA (2022) Pulicaria undulata extract-mediated eco-friendly preparation of TiO2 nanoparticles for photocatalytic degradation of methylene blue and methyl orange. ACS Omega 7(6):4812–4820
64. Kong CPY, Suhaimi NSM, Shahri NNM, Lim JW, Nur M, Hobley J, Usman A (2022) Auramine O UV photocatalytic degradation on TiO2 nanoparticles in a heterogeneous aqueous solution. Catalysts 12(9):975
65. Balaraman P, Balasubramanian B, Liu W, Kaliannan D, Durai M, Kamyab H, Alwetaishi M, Maluventhen V, Ashokkumar V, Chelliapan S, Maruthupandian A (2022) Sargassum myriocystum-mediated TiO2-nanoparticles and their antimicrobial, larvicidal activities and enhanced photocatalytic degradation of various dyes. Environ Res 204:112278
66. Mousa SM, Shalan AE, Hassan H, Ebnawaled A, Khairy SA (2021) Enhanced the photocatalytic degradation of titanium dioxide nanoparticles synthesized by different plant extracts for wastewater treatment. J Mol Struct 1250:131912
67. Da Silva MV, Abrantes N, Nogueira V, Gonçalves F, Pereira R (2016b) TiO 2 nanoparticles for the remediation of eutrophic shallow freshwater systems: Efficiency and impacts on aquatic biota under a microcosm experiment. Aquat Toxicol 178:58–71
68. Sukhadeve GK, Bandewar H, Janbandhu SY, Jayaramaiah J, Gedam RS (2023) Photocatalytic hydrogen production, dye degradation, and antimicrobial activity by Ag-Fe co-doped TiO2 nanoparticles. J Mol Liq 369:120948
69. Slimani Y, Almessiere MA, Sadiq MMJ, Hannachi E, Caliskan S, Akhtar S, Baykal A, Gondal MA (2023) Synthesis of Ce and Sm Co-doped TiO2 nanoparticles with enhanced photocatalytic activity for rhodamine B dye degradation. Catalysts 13(4):668
70. Niknam H, Sadeghzadeh-Attar A (2023) Mg-doped TiO2 nanorods-SrTiO3 heterojunction composites for efficient visible-light photocatalytic degradation of basic yellow 28. Opt Mater 136:113395
71. Zhao Q, Ren Y, Huang L, Chen YS, Bian Z (2022) In situ Fe(III)-doped TiO2 mesocrystals catalyzed visible light photo-Fenton system. Catal Today 410:309–316
72. Chen Y, Li A, Fu X, Peng Z (2022) Electrospinning-based (N,F)-co-doped TiO2-δ nanofibers: an excellent photocatalyst for degrading organic dyes and heavy metal ions under visible light. Mater Chem Phys 291:126672
73. Mathiarasu RR, Manikandan A, Panneerselvam K, George MA, Raja KK, Slimani Y, Baykal A, Asiri AM, Kamal T (2021) Photocatalytic degradation of reactive anionic dyes RB5, RR198 and RY145 via rare earth element (REE) lanthanum substituted CaTiO3 perovskite catalysts. J Mater Res Technol 15:5936–5947
74. Xiongbo C, Cen C, Zhixiong T, Zeng W, Chen D, Fang P, Chen Z (2013) The key role of pH value in the synthesis of titanate nanotubes-loaded manganese oxides as a superior catalyst for the selective catalytic reduction of NO with NH3. J Nanomater 2013:1–7
75. Djellabi R, Ordonez MF, Conte F, Falletta E, Bianchi CL, Rossetti I (2022) A review of advances in multifunctional XTiO3 perovskite-type oxides as piezo-photocatalysts for environmental remediation and energy production. J Hazard Mater 421:126792

76. Rasalingam S, Peng R, Koodali RT (2014) Removal of hazardous pollutants from wastewaters: applications of TiO2-SiO2 mixed oxide materials. J Nanomater 2014:1–42
77. Lissarrague MHS, Alshehri S, Alsalhi A, Lassalle VL, López-Corral I (2023) Heavy metal removal from aqueous effluents by TiO2 and ZnO nanomaterials. Adsorp Sci Technol 2023:1–15
78. Sharma S, Kumar N, Makgwane PR, Chauhan NS, Kumari K, Rani M, Maken S (2022) TiO2/SnO2 nano-composite: new insights in synthetic, structural, optical and photocatalytic aspects. Inorg Chim Acta 529:120640
79. Navarro-Gázquez P, Muñoz-Portero M, Blasco-Tamarit E, Sánchez-Tovar R, García-Antón J (2022) Synthesis and applications of TiO_2/ZnO hybrid nanostructures by ZnO deposition on TiO_2 nanotubes using electrochemical processes. Rev Chem Eng. https://doi.org/10.1515/revce-2021-0105
80. Litynska M, Dontsova T, Gusak A (2023) Nanostructured iron-based sorption materials for drinking water and wastewater treatment. In: Springer proceedings in physics. Springer Nature, Singapore, pp 243–255
81. Gindaba GT, Demsash HD, Jayakumar M (2022) Green synthesis, characterization, and application of metal oxide nanoparticles for mercury removal from aqueous solution. Environ Monit Assess 195(1):9
82. Wu S, Xiong W, Li H (2022) Insights into the Fe oxidation state of sphere-like Fe2O3 nanoparticles for simultaneous Pb2+ and Cu2+ detection. J Alloys Compd 934:167863
83. Ashraf I, Singh N, Agarwal A (2022) Green synthesis of iron oxide nanoparticles using Amla seed for methylene blue dye removal from water. Mater Today Proc 72:311–316
84. Isik Z, Bouchareb R, Arslan H, Özdemir S, Gonca S, Dizge N, Deepanraj B, Prasad S (2022) Green synthesis of iron oxide nanoparticles derived from water and methanol extract of Centaurea solstitialis leaves and tested for antimicrobial activity and dye decolorization capability. Environ Res 219:115072
85. Manikandan A, Thanrasu K, Dinesh A, Raja KK, Durka M, Almessiere MA, Slimani Y, Baykal A (2022) Photocatalytic applications of magnetic hybrid nanoalloys and their nanocomposites. Springer, Berlin, pp 1–33
86. Wang S, Wang J (2023) Bimetallic and nitrogen co-doped biochar for peroxymonosulfate (PMS) activation to degrade emerging contaminants. Sep Purif Technol 307:122807
87. Xing X, Ren X, Alharbi NS, Chen C (2022) Efficient adsorption and reduction of Cr (VI) from aqueous solution by Santa Barbara Amorphous-15 (SBA-15) supported Fe/Ni bimetallic nanoparticles. J Colloid Interface Sci 629:744–754
88. Kai B, Zhang G, Zhu J, Wang X, Wang M, Lou F, Liu Y, Song C, Guo X (2022) Promoting propane dehydrogenation with CO2 over the PtFe bimetallic catalyst by eliminating the non-selective Fe (0) phase. ACS Catal 12(11):6559–6569
89. Malik MA, Alshehri AA, Patel R (2021) Facile one-pot green synthesis of Ag–Fe bimetallic nanoparticles and their catalytic capability for 4-nitrophenol reduction. J Mater Res Technol 12:455–470
90. Bahadoran A, Liu Q, Liu B, Gu J, Zhang D, Fakhri A, Gupta VK (2021) Preparation of Sn/Fe nanoparticles for Cr (III) detection in presence of leucine, photocatalytic and antibacterial activities. Spectrochim Acta A Mol Biomol Spectrosc 253:119592
91. Lien HL, Zhang W (2005) Hydrodechlorination of chlorinated ethanes by nanoscale Pd/Fe bimetallic particles. J Environ Eng 131(2005):4–10
92. Wang X, Chen C, Chang Y, Liu H (2009) Dechlorination of chlorinated methanes by Pd/Fe bimetallic nanoparticles. J Hazard Mater 161(2–3):815–823
93. Khan SB, Jain M, Pandey A, Pant KK, Ziora ZM, Blaskovich MAT, Shetti NP, Aminabhavi TM (2022) Leveraging the potential of silver nanoparticles-based materials towards sustainable water treatment. J Environ Manage 319:115675
94. Yu Y, Zhou Z, Huang G, Cheng H, Han L, Zhao S, Chen Y, Meng F (2022) Purifying water with silver nanoparticles (AgNPs)-incorporated membranes: Recent advancements and critical challenges. Water Res 222:118901

95. Deshmukh SP, Patil S, Mullani SB, Delekar SD (2019) Silver nanoparticles as an effective disinfectant: A review. Mater Sci Eng C 97:954–965
96. Gogoi SK, Gopinath P, Paul A, Ramesh A, Ghosh SS, Chattopadhyay A (2006) Green fluorescent protein-expressing Escherichia coli as a model system for investigating the antimicrobial activities of silver nanoparticles. Langmuir 22(22):9322–9328
97. Tripathi N, Goshisht MK (2022) Recent advances and mechanistic insights into antibacterial activity, antibiofilm activity, and cytotoxicity of silver nanoparticles. ACS Appl Biomater 5(4):1391–1463
98. Jameel MS, Aziz AA, Dheyab MA, Khaniabadi PM, Kareem AA, Alrosan M, Ali AS, Rabeea MA, Mehrdel B (2022) Mycosynthesis of ultrasonically-assisted uniform cubic silver nanoparticles by isolated phenols from Agaricus bisporus and its antibacterial activity. Surf Interfaces 29:101774
99. Hoai PTT, Huong NTM (2023) Latest avenues on titanium oxide-based nanomaterials to mitigate the pollutants and antibacterial: Recent insights, challenges, and future perspectives. Chemosphere 324:138372
100. Aravind M, Amalanathan M, Aslam S, Noor A, Jini D, Majeed S, Velusamy P, Alothman AA, Alshgari RA, Mushab MSS, Sillanpaa M (2023) Hydrothermally synthesized Ag-TiO2 nanofibers (NFs) for photocatalytic dye degradation and antibacterial activity. Chemosphere 321:138077
101. Park J, Lee I (2014) Photocatalytic degradation of 2-chlorophenol using Ag-doped TiO2 nanofibers and a near-UV light-emitting diode system. J Nanomater 2014:1–6
102. Hannachi E, Slimani Y, Nawaz M, Sivakumar R, Trabelsi Z, Vignesh RV, Akhtar S, Baykal A, Yasin G (2022) Preparation of cerium and yttrium doped ZnO nanoparticles and tracking their structural, optical, and photocatalytic performances. J Rare Earths 41(5): 682–688
103. Hannachi E, Slimani Y, Nawaz M, Trabelsi Z, Yasin G, Bilal M, Almessiere MA, Baykal A, Thakur A, Thakur P (2022) Synthesis, characterization, and evaluation of the photocatalytic properties of zinc oxide co-doped with lanthanides elements. J Phys Chem Solid 170:110910
104. Geetha GV, Sivakumar R, Slimani Y, Sanjeeviraja C, Kannapiran E (2022) Rare earth (RE: La and Ce) elements doped ZnWO4 nanoparticles for enhanced photocatalytic removal of methylene blue dye from aquatic environment. Phys B Condens Matter 639:414028
105. Maqbool H, Bibi I, Nazeer Z, Majid F, Ata S, Raza MQ, Iqbal M, Slimani Y, Khan MI, Fatima M (2022) Effect of dopant on ferroelectric, dielectric and photocatalytic properties of Co–La-doped dysprosium chromite prepared via microemulsion route. Ceram Int 48(21): 31763–31772
106. Manikandan A, Slimani Y, Dinesh A, Asiri AM, Thanrasu K, Baykal A, Jaganathan SK, Dzudzevic-Cancar H (2021) Perovskite's potential functionality in a composite structure. Elsevier, Amsterdam, pp 181–202
107. Ajeesha T, George AAM, Manikandan A, Mary JA, Slimani Y, Almessiere MA, Baykal A (2020) Nickel substituted MgFe2O4 nanoparticles via co-precipitation method for photocatalytic applications. Phys B Condens Matter 606:412660
108. Nawaz M, Slimani Y, Almofty SA, Gungunes CD, Baykal A (2019) Exploration of catalytic and cytotoxicity activities of CaxMgxNi1-2xFe2O4 nanoparticles. J Photochem Photobiol B Biol 196:111506
109. George MA, Ajeesha TL, Manikandan A, Anantharaman A, Jansi R, Kumar ER, Slimani Y, Baykal A (2021) Evaluation of Cu–MgFe2O4 spinel nanoparticles for photocatalytic and antimicrobial activates. J Phys Chem Solid 153:110010
110. Ata S, Shaheen I, Majid F, Bibi I, Ijaz-ul-Mohsin J, K., Slimani, Y. & Iqbal, M. (2021) Hydrothermal route for the synthesis of manganese ferrite nanoparticles and photocatalytic activity evaluation for the degradation of methylene blue dye. Z Phys Chem 235(11): 1433–1445
111. Al-Zahrani SA, Manikandan A, Thanrasu K, Dinesh A, Raja KK, Almessiere MA, Slimani Y, Baykal A, Bhuminathan S, Jayesh SR, Ahmed J, Alorfi HS, Hussein MA, Khan I, Asiri AM

(2022) Influence of Ce3+ on the structural, morphological, magnetic, photocatalytic and antibacterial properties of spinel MnFe2O4 nanocrystallites prepared by the combustion route. Crystals 12(2):268
112. Dinesh A, Raja KK, Manikandan A, Almessiere M, Slimani Y, Baykal A, Alorfi HS, Hussein MA, Khan A (2022) Sol–gel combustion synthesis and photocatalytic dye degradation studies of rare earth element Ce substituted Mn–Zn ferrite nanoparticles. J Mater Res Technol 18: 5280–5289
113. Aamir M, Bibi I, Ata S, Majid F, Alwadai N, Almuqrin AH, Albalawi H, Slimani Y, Bashir M, Iqbal M (2021) Micro-emulsion approach for the fabrication of La1−xGdxCr1−yFeyO3: Magnetic, dielectric and photocatalytic activity evaluation under visible light irradiation. Results Phys 23:104023
114. Ghafoor A, Bibi I, Ata S, Majid F, Kamal S, Rehman F, Iqbal S, Aamir M, Slimani Y, Iqbal M, Mailk A (2021) Synthesis and characterization of magnetically separable La1−x Bi x Cr1−y Fe y O3 and photocatalytic activity evaluation under visible light. Z Phys Chem 235(11): 1413–1431
115. Bibi I, Muneer MA, Iqbal M, Alwadai N, Almuqrin AH, Altowyan AS, Alshammari FH, Almuslem AS, Slimani Y (2021) Effect of doping on dielectric and optical properties of barium hexaferrite: Photocatalytic performance under solar light irradiation. Ceram Int 47(22): 31518–31526
116. Zhang X, Xu H, Xi M, Jiang Z (2023) Removal/adsorption mechanisms of Cr(VI) and natural organic matter by nanoscale zero-valent iron-loaded biochar in their coexisting system. J Environ Chem Eng 11(3):109860
117. Guerra FD, Attia MI, Whitehead DC, Alexis F (2018) Nanotechnology for environmental remediation: materials and applications. Molecules 23(7):1760
118. Hooshyar Z, Bardajee GR, Ghayeb Y (2013) Sonication enhanced removal of nickel and cobalt ions from polluted water using an iron based sorbent. J Chem 2013:1–5
119. Tang C, He Z, Liu Y, He X, Chen G, Xie C, Huang J (2022) AuPd nanoporous dendrites: High electrocatalytic activity and surface plasmon-enhanced stability for ethanol electrooxidation. Chem Eng J 453:139962
120. Naz S, Kalsoom R, Ali F, Amjed N, Younas U, Pervaiz M, Alsaiari MA, Faisal M, Harraz FA, Luque R (2022) Photocatalytic removal of dyes from aqueous medium by Fe, Mn and Fe-Mn nanoparticles synthesized using Cannabis sativa leaf extract. Water 14(21):3535
121. Poguberović SS, Krčmar D, Maletić S, Kónya Z, Pilipović DT, Kerkez Đ, Rončević S (2016) Removal of As(III) and Cr(VI) from aqueous solutions using "green" zero-valent iron nanoparticles produced by oak, mulberry and cherry leaf extracts. Ecol Eng 90:42–49
122. Tripta N, Rana PS (2022) Structural, optical, electrical, and photocatalytic application of NiFe2O4@NiO nanocomposites for methylene blue dye. Ceram Int 49(9):13520–13530
123. Rahman ZU, Shah U, Alam A, Shah Z, Shaheen K, Khan SB, Khan SA (2022) Photocatalytic degradation of cefixime using CuO-NiO nanocomposite photocatalyst. Inorg Chem Commun 148:110312
124. Rehman WU, Khattak MTN, Saeed A, Shaheen K, Shah Z, Hussain S, Bakhsh EM, Alraddadi HM, Fagieh TM, Akhtar K, Khan SB, Khan SB (2023) Co3O4/NiO nanocomposite as a thermocatalytic and photocatalytic material for the degradation of malachite green dye. J Mater Sci Mater Electron 34(1):15
125. Ayub A, Ahsan S, Meeroff DE, Lashaki MJ (2022) Amine-grafted mesoporous silica materials for single-stage biogas upgrading to biomethane. Chem Eng J 445:136497
126. Anyanwu J, Wang Y, Yang RT (2022) Tunable amine loading of amine grafted mesoporous silica grafted at room temperature: Applications for CO2 capture. Chem Eng Sci 254:117626
127. Lin Y, Xu J, Sudhakar BS, Gu J, Hong R (2019) Preparation of spherical aminopropyl-functionalized MCM-41 and its application in removal of Pb(II) ion from aqueous solution. Nanotechnol Rev 8(1):275–284
128. Anyanwu J, Wang Y, Yang RT (2020) Amine-grafted silica gels for CO2 capture including direct air capture. Ind Eng Chem Res 59:7072–7079

129. Bai F, Liu X, Liu Y, Li M, Sani S, Guo W, Sun C (2022) CO2 capture from dilute sources using triamine functionalized MCF silica at ambient temperature. Micropor Mesopor Mater 349:112370
130. Okonkwo CN, Lee JT, Thybaut J, Chiang Y, Jones CW (2020) Selective removal of hydrogen sulfide from simulated biogas streams using sterically hindered amine adsorbents. Chem Eng J 379:122349
131. Nomura A, Jones CW (2013) Amine-functionalized porous silicas as adsorbents for aldehyde abatement. ACS Appl Mater Interfaces 5(12):5569–5577
132. Drese JH, Talley AD, Jones CW (2011) Aminosilica materials as adsorbents for the selective removal of aldehydes and ketones from simulated bio-oil. ChemSusChem 4(3):379–385
133. Zhang Y, Qian C, Shi R, Li S, Shi F, Hu J, Khan S, Yang X (2022) In-situ synthesis of metal-organic framework embedded in ordered mesoporous silica functionalized with carboxyl groups for 4-nitrophenol to 4-aminophenol. Appl Surf Sci 597:153720
134. Zhang Y, Qian C, Duan J, Liang Y, Luo J, Han Y, Hu J, Shi F (2022) Synthesis of HKUST-1 embedded in SBA-15 functionalized with carboxyl groups as a catalyst for 4-nitrophenol to 4-aminophenol. Appl Surf Sci 573:151558
135. Wang S, Ning S, Zhang W, Zhang S, Zhou J, Wang X, Wei Y (2020) Synthesis of carboxyl group functionalized silica composite resin for strontium removal. Mater Des 185:108224
136. Ehgartner CR, Werner V, Selz S, Hüsing N, Feinle A (2021) Carboxylic acid-modified polysilsesquioxane aerogels for the selective and reversible complexation of heavy metals and organic molecules. Micropor Mesopor Mater 312:110759
137. Xu R, Tan H, Guo M, Zuo S, Sun X (2022) Ultra-fast synthesis of thiol-functionalized organosilica (OS-SH) for adsorption of Hg(II) from aqueous solution. Russ J Phys Chem A 96(7):1561–1566
138. Tang N, Liu X, Jia M, Shi XH, Fu J, Guan D, Ma QL (2022) Amine- and thiol-bifunctionalized mesoporous silica material for immobilization of Pb and Cd: Characterization, efficiency, and mechanism. Chemosphere 291:132771
139. Liang R, Zou H (2020) Removal of aqueous Hg(ii) by thiol-functionalized nonporous silica microspheres prepared by one-step sol–gel method. RSC Adv 10(31):18534–18542
140. Lian M, Wang L, Feng Q, Niu L, Zhao Z, Alam T, Song C, Li X, Zhang Z (2021) Thiol-functionalized nano-silica for in-situ remediation of Pb, Cd, Cu contaminated soils and improving soil environment. Environ Pollut 280:116879
141. Xu S, Ning S, Wang Y, Wang X, Dong H, Chen L, Yin X, Fujita T, Wei Y (2023) Precise separation and efficient enrichment of palladium from wastewater by amino-functionalized silica adsorbent. J Clean Prod 396:136479
142. Zhang X, Wen H, Huang Q, Tan Y, Sun Z, Hua Y, Wu X, Li M (2023) Selective removal and long-term immobilization of uranium by ultralow content Fe0 in the pores of amino functionalized silica gel. Chem Eng J 461:142015
143. Shafqat SS, Khan A, Zafar M, Alhaji MM, Sanaullah K, Shafqat SS, Murtaza S, Pang SC (2019) Development of amino-functionalized silica nanoparticles for efficient and rapid removal of COD from pre-treated palm oil effluent. J Mater Res Technol 8(1):385–395
144. Perrotti TC, De Freitas NC, Alzamora M, Sanchez DR, Carvalho NM (2019) Green iron nanoparticles supported on amino-functionalized silica for removal of the dye methyl orange. J Environ Chem Eng 7(4):103237
145. Han Y, Chen J, Gu X, Chen J (2021) Adsorption of multi-bivalent heavy metal ions in aqueous solution onto aminopropyl-functionalized MCM-48 preparation by co-condensation. Sep Sci Technol 56(11):1819–1829
146. Mohapatra L, Cheon D, Yoo S (2023) Carbon-based nanomaterials for catalytic wastewater treatment: a review. Molecules 28(4):1805
147. Baraneedharan P, Paul B, Anil AC, Mohammed SB, Rajendran S (2022) Advances in preparation, mechanism and applications of various carbon materials in environmental applications: A review. Chemosphere 300:134596

148. Devi MK, Yaashikaa PR, Kumar PS, Manikandan S, Oviyapriya M, Varshika V, Rangasamy G (2023) Recent advances in carbon-based nanomaterials for the treatment of toxic inorganic pollutants in wastewater. New J Chem 47:7655–7667
149. Li Y, Zhang P, Du Q, Peng X, Liu T, Wang Z, Xia Y, Zhang W, Wang K, Zhu H, Wu D (2011) Adsorption of fluoride from aqueous solution by graphene. J Colloid Interface Sci 363(1): 348–354
150. Pang H, Wu Y, Wang X, Hu B (2019) Recent advances in composites of graphene and layered double hydroxides for water remediation: a review. Chemistry 14(15):2542–2552
151. Janwery D, Memon FH, Memon AA, Iqbal M, Memon FN, Ali W, Choi K, Thebo KH (2023) Lamellar graphene oxide-based composite membranes for efficient separation of heavy metal ions and desalination of water. ACS Omega 8:7648–7656
152. Lin X, Gan L, Owens G, Chen Z (2022) Removal of cadmium from wastewater using biofunctional reduced graphene oxide synthesized by Lysinibacillus sphaericus. J Clean Prod 383:135369
153. Das TK, Ghosh SK, Das NC (2023) Green synthesis of a reduced graphene oxide/silver nanoparticles-based catalyst for degradation of a wide range of organic pollutants. Nano-Struct Nano-Objects 34:100960
154. Wang W, Zhang J, Xiao G, Liu X, Qu H, Zhou S (2023) Preparation of magnetic porous graphene oxide by intercalating rigid molecule and subsequent magnetization for enhancing pharmaceuticals removal from water. Mater Today Commun 34:105119
155. Chu Z, Xiao M, Dong Q, Li G, Hu T, Zhang Y, Zhang H (2022) Porous reduced graphene oxide for ultrasensitive detection of nitrogen dioxide. Chin Chem Lett 34(1):107197
156. Jayawardena R, Eldridge DS, Malherbe F (2022) Sonochemical synthesis of improved graphene oxide for enhanced adsorption of methylene blue. Colloids Surf A Physicochem Eng Asp 650:129587
157. Li J, Jin Z, Chao Y, Wang A, Wang D, Chen S, Qian Q (2023) Synthesis of graphene-oxide-decorated porous ZnO nanosheet composites and their gas sensing properties. Chemosensors 11(1):65
158. Verma N, Chundawat TS, Chandra H, Vaya D (2022) An efficient time reductive photocatalytic degradation of carcinogenic dyes by TiO2-GO nanocomposite. Mater Res Bull 158:112043
159. Rajkumari NP, Rouf A, Dutta P, Goswami P (2023) Synchronizing charge-carrier capacity and interfacial morphology of green rGO modified ZnO and TiO2 heterojunctions and study of their photocatalytic behaviour towards UV and visible light active drug and dye. Mater Sci Eng B 287:116094
160. Mashhoor H, Eftekhari M, Rezazadeh N, Nazarabad MK (2022) Graphene oxide–tungsten oxide (GO–WO3) adsorbent for the removal of copper ion. Nanotechnol Environ Eng 8(1): 75–86
161. Afza N, Shivakumar A, Waqas M, Kumar AN, Bhatt AS, Murthy HA, Ravikumar C, Mylarappa M, Selvanandan S (2022) Facile hydrothermal synthesis of cerium oxide/rGO nanocomposite for photocatalytic and supercapacitor applications. Appl Surf Sci Adv 11: 100307
162. Rajapaksha P, Orrell-Trigg R, Shah D, Cheeseman S, Vu K, Ngo ST, Murdoch B, Choudhury N, Yin H, Cozzolino D, Truong Y, Lee AF, Truong VK, Chapman J (2023) Broad spectrum antibacterial zinc oxide-reduced graphene oxide nanocomposite for water depollution. Mater Today Chem 27:101242
163. Lingamdinne LP, Reddy LV, Koduru JR, Chang Y (2022) Insights into manganese ferrite anchored graphene oxide to remove Cd(II) and U(VI) via batch and semi-batch columns and its potential antibacterial applications. Chemosphere 310:136888
164. Kadari M, Makhlouf M, Khaoua OO, Kesraoui M, Bouriche S, Benmaamar Z (2023) The removal efficiency of cadmium (Cd2+) and lead (Pb2+) from aqueous solution by graphene oxide (GO) and magnetic graphene oxide (α-Fe2O3/GO). Chem Afr 6:1515

165. Sidiqua MA, Priya VS, Begum N, Aman N (2023) Synthesis of magnetic ZnFe2O4- reduced graphene oxide nanocomposite photocatalyst for the visible light degradation of cationic textile dyes. Nanotechnol Environ Eng 579:336
166. Wang X, Guo D, Zhang J, Yao Y, Liu Y (2023) Insights into the electron transfer mechanisms of permanganate activation by carbon nanotube membrane for enhanced micropollutants degradation. Front Environ Sci Eng 17(9):106
167. Fu Q, Lou J, Shi D, Zhou S, Hu J, Wang Q, Huang W, Wang K, Yan W (2022) Adsorption and removal mechanism of Pb (II) by oxidized multi-walled carbon nanotubes. J Iran Chem Soc 19(7):2883–2890
168. Sajid M, Asif M, Baig N, Kabeer M, Ihsanullah I, Mohammad AW (2022) Carbon nanotubes-based adsorbents: Properties, functionalization, interaction mechanisms, and applications in water purification. J Water Process Eng 47:102815
169. Li Z, Liu F, Zhu S, Liu Z, Zhang F, Ni P, Chen C, Jiang Y, Lu Y (2022) Carbon nanotubes regulated by oxidizing functional groups as peroxidase mimics for total antioxidant capacity determination. Biosens Bioelectron 11:100190
170. Rodríguez C, Leiva E (2019) Enhanced heavy metal removal from acid mine drainage wastewater using double-oxidized multiwalled carbon nanotubes. Molecules 25(1):111
171. Tian S, He Y, Zhang L, Li S, Bai Y, Wang Y, Wu J, Yu J, Guo X (2022) CNTs/TiO2- loaded carbonized nanofibrous membrane with two-type self-cleaning performance for high efficiency oily wastewater remediation. Colloids Surf A Physicochem Eng Asp 656:130306
172. Yang Y, Liu K, Sun F, Liu Y, Chen J (2022) Enhanced performance of photocatalytic treatment of Congo red wastewater by CNTs-Ag-modified TiO2 under visible light. Environ Sci Pollut Res 29(11):15516–15525
173. Abo-Dief HM, Hussein OBH, Ihsan A, El-Bahy SM, Raslan AM, Shahid M, Warsi MF (2022) Ternary metal oxide WO3.NiO.ZnO nanoparticles and their composite with CNTs for organic dye photocatalytic degradation. Ceram Int 48(15):22228–22236
174. Zhuang W, Zheng Y, Xiang J, Zhang J, Wang P, Zhao C (2022) Enhanced Hydraulic-driven piezoelectric ozonation performance by CNTs/BaTiO3 nanocatalyst for Ibuprofen removal. Chem Eng J 454:139928
175. Alimohammadi E, Mahdikhah V, Sheibani S (2022) Type-II band alignment in CNT-modified $SrTiO_3$-Fe_2TiO_5 heterostructure nanocomposite for photocatalytic degradation of organic dyes. Appl Surf Sci 598:153816
176. Jia Y, Zhang Y, Zhang X, Cheng J, Xie Y, Zhang Y, Yin X, Song F, Cui H (2022) Novel CdS/PANI/MWCNTs photocatalysts for photocatalytic degradation of xanthate in wastewater. Sep Purif Technol 309:123022
177. Al-Musawi TJ, McKay G, Rajiv P, Mengelizadeh N, Balarak D (2022) Efficient sonophotocatalytic degradation of acid blue 113 dye using a hybrid nanocomposite of CoFe2O4 nanoparticles loaded on multi-walled carbon nanotubes. J Photochem Photobiol A Chem 424:113617
178. De La Flor MPP, Camarillo R, Martínez F, Jiménez C, Quiles RA, Rincón J (2022) Synthesis and characterization of bimetallic TiO2/CNT/Pd-Cu for efficient remediation of endocrine disruptors under solar light. J Environ Chem Eng 10(2):107245
179. Onyancha RB, Aigbe UO, Ukhurebor KE, Muchiri PW (2021) Facile synthesis and applications of carbon nanotubes in heavy-metal remediation and biomedical fields: A comprehensive review. J Mol Struct 1238:130462
180. Zhong R, Tao J, Yang X, Fu Z, Yang H, Zhoumeng P, Bao R, Liu Y, Yi J (2022) Preparation of carbon nanotubes with high filling rate of copper nanoparticles. Micropor Mesopor Mater 344:112231
181. Tungittiplakorn W, Lion LW, Cohen C, Kim J (2004) Engineered polymeric nanoparticles for soil remediation. Environ Sci Technol 38(5):1605–1610
182. Morsy M, Abdel-Salam AH, Gomaa IA, Moustafa H, Kalil H, Helal A (2022) Highly efficient photocatalysts for methylene blue degradation based on a platform of deposited GO-ZnO nanoparticles on polyurethane foam. Molecules 28(1):108

183. Wu J, Ma X, Gnanasekar P, Wang F, Zhu J, Yan N, Chen JM (2022) Superhydrophobic lignin-based multifunctional polyurethane foam with SiO2 nanoparticles for efficient oil adsorption and separation. Sci Total Environ 860:160276
184. Zuliani A, Chelazzi D, Mastrangelo R, Giorgi R, Baglioni P (2023) Adsorption kinetics of acetic acid into ZnO/castor oil-derived polyurethanes. J Colloid Interface Sci 632:74–86
185. Khezraqa H, Etemadi H, Salami-Kalajahi M (2023) Investigating the effect of polyamidoamine generation 2 (PAMAM-G2) polymeric nanostructures dendrimer on the performance of polycarbonate thin film nanocomposite membranes for water treatment. J Polym Environ 31:3604–3618
186. Afshar EG, Taher MA (2022) New fabrication of CuFe2O4/PAMAM nanocomposites by an efficient removal performance for organic dyes: Kinetic study. Environ Res 204:112048
187. Khosravi MJ, Hosseini SM, Vatanpour V (2022) Polyamidoamine dendrimers-Mil-125 (Ti) MOF embedded polyethersulfone membrane for enhanced removal of heavy metal, antibiotic and dye from water. J Environ Chem Eng 10(6):108644
188. Ranjbaran H, Ameri E, Dehghani B (2023) Preparation of dendrimer/TiO2 polysulfone nanofiltration membrane to improve antibacterial, antifouling and separation performance of contaminants (heavy metals, salts, dyes). Polym Bull. https://doi.org/10.1007/s00289-023-04785-7
189. Zaoui F, Daouadji BE, Beldjilali M, Hachemaoui M, Sebba FZ, Mokhtar A, Boukoussa B (2023) Classical and ultrasonic preparation of new dendrimer-CuNPs nanocomposites: application in the catalytic reduction, antioxidant and antimicrobial activities. Appl Organomet Chem 37(6):e7099
190. Sugashini S, Gomathi T, Devi RA, Sudha P, Rambabu K, Banat F (2022) Nanochitosan/carboxymethyl cellulose/TiO2 biocomposite for visible-light-induced photocatalytic degradation of crystal violet dye. Environ Res 204:112047
191. Fijoł N, Aguilar-Sanchez A, Ruiz-Caldas M, Redlinger-Pohn J, Mautner A, Mathew AP (2023) 3D printed polylactic acid (PLA) filters reinforced with polysaccharide nanofibers for metal ions capture and microplastics separation from water. Chem Eng J 457:141153
192. Amjad M, Mohyuddin A, Nadeem S, Ulfat W, Saeed S, Asghar N, Ahmed S (2023) Development of biodegradable vinyl acetate and acrylic acid grafted gelatin copolymer for dye adsorption. Polym Bull. https://doi.org/10.1007/s00289-023-04782-w
193. Krishnamoorthi R, Anbazhagan R, Tsai H, Wang C, Lai J (2022) Biodegradable, superwettable caffeic acid/chitosan polymer coated cotton fibers for the simultaneous removal of oils, dyes, and metal ions from water. Chem Eng J 427:131920
194. Sarkhel R, Ganguly P, Das P, Bhowal A, Sengupta S (2022) Synthesis of biodegradable PVA/cellulose polymer composites and their application in dye removal. Environ Qual Manag 32(3):313–323
195. Meena PL, Saini JK (2023) Synthesis of polymer-metal oxide (PANI/ZnO/MnO2) ternary nanocomposite for effective removal of water pollutants. Results Chem 5:100764
196. Sawunyama L, Oyewo OA, Seheri N, Onjefu SA, Onwudiwe DC (2023) Metal oxide functionalized ceramic membranes for the removal of pharmaceuticals in wastewater. Surf Interfaces 38:102787
197. Izadkhah V, Ghorbani-Vaghei R, Alavinia S, Asadabadi S, Emami N, Jamehbozorgi S (2023) Fabrication of zirconium metal-organic-framework/poly triazine-phosphanimine nanocomposite for dye adsorption from contaminated water: Isotherms and kinetics models. J Mol Struct 1275:134691
198. Gelaw TB, Sarojini BK, Kodoth AK (2022) Chitosan/hydroxyethyl cellulose gel immobilized polyaniline/CuO/ZnO adsorptive-photocatalytic hybrid nanocomposite for congo red removal. J Polym Environ 30(10):4086–4101
199. Marwani HM, Ahmad S, Rahman MM (2022) Catalytic reduction of environmental pollutants with biopolymer hydrogel cross-linked gelatin conjugated tin-doped gadolinium oxide nanocomposites. Gels 8(2):86

200. Ibeaho WF, Chen M, Shi J, Chen C, Duan Z, Wang C, Xie Y, Chen Z (2022) Multifunctional magnetic hydrogels fabricated by iron oxide nanoparticles mediated radical polymerization. ACS Appl Polym Mater 4(6):4373–4381
201. Wan T, Tang Q, Wang T, Wang J, He S, Jia Y, Yu M (2022) Adsorption behaviours of copper (II), lead(II), and cadmium(II) ions from aqueous solution by polyethylenimine -modified magnetic hydrogel nanocomposites. J Polym Res 29(12):520
202. Sethi S, Medha N, Thakur S (2022) Synthesis and characterization of nanocomposite chitosan-gelatin hydrogel loaded with ZnO and its application in photocatalytic dye degradation. Mater Today Proc 78:815–824
203. Campbell ML, Guerra FD, Dhulekar J, Alexis F, Whitehead DC (2015) Target-specific capture of environmentally relevant gaseous aldehydes and carboxylic acids with functional nanoparticles. Chemistry 21(42):14834–14842
204. Zhang M, Ling X, Zhang X, Han G (2022) A novel alginate/PVA hydrogel -supported Fe3O4 particles for efficient heterogeneous Fenton degradation of organic dyes. Colloids Surf A Physicochem Eng Asp 652:129830

Nanotechnological Applications in Food and Agriculture

18

A. V. Trukhanov, Preeti Thakur, and Atul Thakur

Abstract

Agricultural production, sometimes known as "farming," is the process of cultivating certain plants and rearing livestock to provide food, feed, fiber, and a variety of other desirable items. Agriculture is the core of most emerging countries, providing both direct and indirect nourishment for humans. Agri-food production is critical since it is one of the key economic drivers. It can also provide access to value-added crops. Food nanotechnology is a booming topic of study that opens up a whole new world of prospects for the food sector. In addition, nanotechnology is rapidly introducing new opportunities for innovation in the food business, but uncertainties and health concerns are also arising with it. Agricultural practices are frequently in the spotlight because global warming, energy, resource restrictions, and the world's fast rising population are putting unprecedented strain on food and water supplies. Many challenges are faced with food production capacity, including a declining ratio of arable land for cultivation. Agriculture is becoming increasingly crucial as a source of food in a world of dwindling resources and an ever-increasing world population. It is critical to employ current technologies in agricultural and food sciences such as nanotechnology and nanobiotechnology. The present focus of nano-agriculture is on target farming, which involves the use of nanosized particles with particular features to increase crop and animal output. Agri-food nanotechnology is an

A. V. Trukhanov (✉)
Laboratory of Magnetic Film Physics, Scientific-Practical Materials Research Centre of NAS of Belarus, Minsk, Belarus

P. Thakur
Amity School of Applied Sciences, Amity University, Gurugram, Haryana, India

A. Thakur
Amity Centre for Nanotechnology, Amity University Haryana, Gurgaon, Haryana, India

D. Suhag et al. (eds.), *Integrated Nanomaterials and their Applications*,
https://doi.org/10.1007/978-981-99-6105-4_18

interdisciplinary field, when compared to its usage in medicine delivery and pharmaceuticals, the use of nanotechnology in agriculture and food is pretty recent. For "sustainable intensification," nanoscience has the capability to safeguard and monitor plant development, identify plant and animal illnesses, boost global food supply, improve food quality, and minimize waste. This chapter provides an overview about the importance of agriculture and food production in nanotechnology.

Keywords

Targeted farming · Agri-food · Value-added crops · Sustainable intensification

18.1 Introduction

Nanotechnology provides novel agrochemical agents and delivery systems for increasing crop yield while decreasing pesticide consumption. Furthermore, in the foodservice industry, nanoscience technologies may be used to detect pathogens in packaging, improve flavor and color quality, and increase safety by strengthening barrier qualities. Nanotechnology holds great promise for providing benefits not only within food products but also in the environment around food products. The fundamental application and functional categories of nanotechnology that are currently being created for food packaging are as follows: The development of plastic product barriers, the addition of active ingredients that can provide features that are essential and go beyond those of conventional active packaging, and the detection and signaling of pertinent information are all examples of improvements. Nano food packaging materials have the potential to increase food shelf-life, enhance food safety, warn customers when food is contaminated or rotten, mend packaging tears, and even release additives to prolong the life of the food in the container.

Nanobiotechnology has the potential to increase crop yields through the following applications: nano-formulations of agro-chemicals for utilizing insecticides and fertilizers for crop improvement; the use of nano-sensors/nano-biosensors in crop improvement for the detection of diseases and agrochemical residues; nanodevices for plant gene modification; crop diseases diagnostics; animal care, livestock breeding, poultry production; and post-harvest management [1]. Nanotechnology applications include gene or DNA transfer via nanoparticles in plants for the development of insect-resistant cultivars, food processing and storage, nano-feed additives, and extended product shelf-life. Nanotechnology has the potential to speed the production of biomass-to-fuels technologies. When compared to traditional encapsulating agents, nanomaterials allow for better encapsulation and release efficiency of active food ingredients, and the development of nano-emulsions, liposomes, micelles, biopolymer complexes, and cubosomes has resulted in improved properties for bioactive compound safety, controlled delivery systems, food matrix integration, and concealing undesired flavors. Also, the growing public concern about food quality and health benefits is driving academics to develop a technique to improve food quality while affecting the nutritional content of the product as little as possible. The food sector has expanded its need for nanoparticle-based products since many of them include critical nutrients and have been shown to be non-toxic [2, 3].

But concerns about the safety of nanoparticles have emerged based on only a few toxicological studies, and academics and industry will need to demonstrate that these nanotechnologies do not have a greater harmful influence on the environment and on human beings.

18.2 Applications of Nanotechnology in Agriculture

Green methods for synthesizing nanoparticles with plant extracts are advantageous because they are simple, convenient, environmentally friendly, and require less reaction time. Nanomaterials created using environmentally friendly and green methods have the potential to improve agriculture by improving fertilization, plant growth regulators, and pesticides. Furthermore, they reduce the amount of harmful chemicals that pollute the environment. Nanobiotechnology in farming has gained a significant amount of attention in the last decade, thanks to a lot of public financing, the level of progress is good, and many innovative methods have been incorporated in agriculture [4]. This could be attributed to the unique essence of farmlands, which operates as an open platform wherein matter and energy are readily exchanged. In contrast to commercial nanoproducts, the scale of demand for input materials is always large due to the lack of control over the input of nanomaterials. Nanotech offers new agrochemical agents and delivery mechanisms to boost crop productivity while reducing application of pesticides. Nanotechnology can boost agricultural production by using: (1) nano-formulations of agrochemicals for applying pesticides and fertilizers for crop improvement; (2) nano-sensors in crop protection for identifying diseases and agrochemical residues; (3) nanodevices for plant genetic engineering; (4) plant disease diagnostics; (5) animal health, animal breeding, poultry production; and (6) postharvest management. Precision farming techniques may be used to increase crop yields while minimizing soil and water damage. It can also reduce nitrogen loss due to leaching and emissions, as well as soil-microorganisms.

According to experts, the potential benefits of nanotechnology for agriculture, food, fisheries, and aquaculture should be balanced against concerns for the soil, water, and environment, as well as worker occupational health. Insecticide resistance is a prime example of evolution on an ecological time scale. Insecticide resistance research is necessary because it leads to a better understanding of mechanisms that operate in real time, as well as because it is economically significant. Insects have become a growing problem for agriculture and public health.

18.3 Agriculture Intensification Systems for Long-Term Sustainability

Sustainable intensification refers to a production system that aims to increase yield while minimizing environmental impact while cultivating the same agricultural area [5]. This paradigm provides the best combination of agricultural production approaches in light of the current biophysical, social, cultural, and economic situation. To increase productivity, novel nanomaterials based on the use of inorganic, polymeric, and lipid nanoparticles synthesized using various techniques (e.g., emulsification, ionic gelation, polymerization, oxydo-reduction, etc.) have been developed [6]. They can be used to develop intelligent nano-systems for the immobilization and release of nutrients in soil, for example, such systems have the advantage of minimizing leaching while improving plant nutrient uptake and mitigating eutrophication by reducing nitrogen transfer to groundwater (Fig. 18.1). Furthermore, nanomaterials could be used to improve the structure and function of pesticides by increasing solubility, increasing resistance to hydrolysis and photodecomposition, and/or providing a more specific and controlled release toward target organisms (Fig. 18.2) [7].

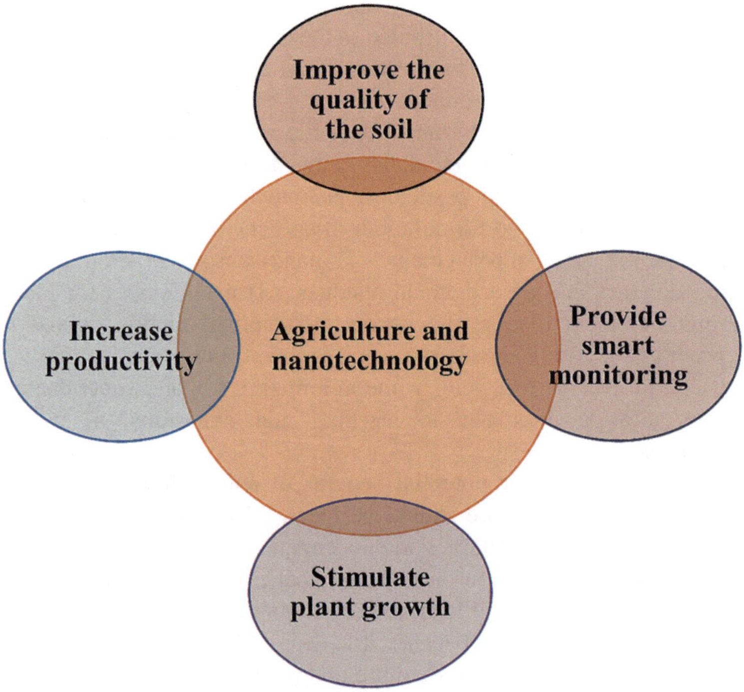

Fig. 18.1 Possible applications of nanotechnology in agriculture: Using nanomaterials (SiO_2, TiO_2, and carbon nanotubes), we can (a) boost productivity by using nano-pesticides and nano-fertilizers, (b) enhance soil quality by using nano-zeolites, (c) stimulate plant growth by using nanomaterials, and (d) offer smart monitoring by using nano-sensors by wireless communication devices

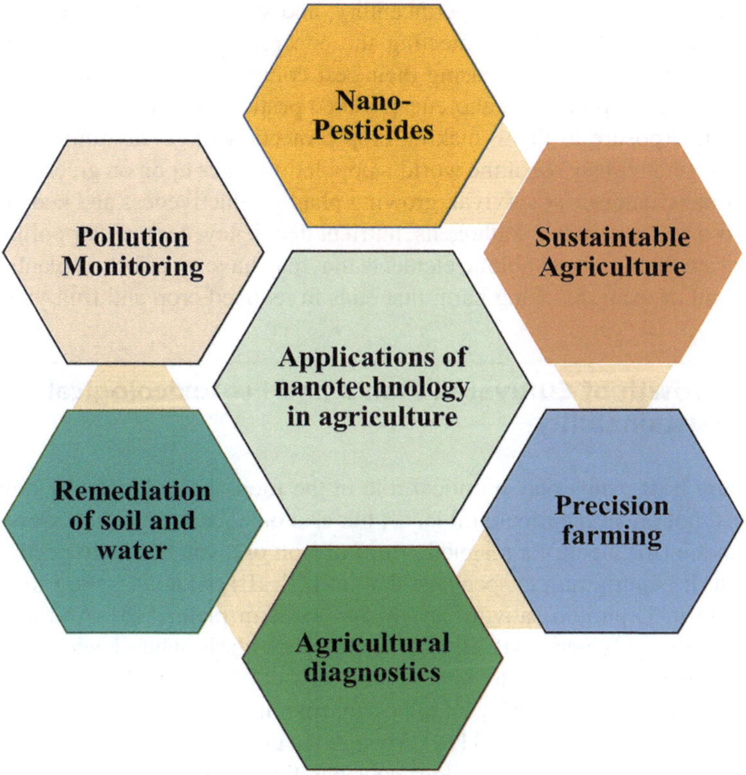

Fig. 18.2 Applications of nanotechnology in agriculture

18.4 Systems to Improve the Quality of the Soil

Hydrogels, nano-clays, and nano-zeolites have been shown to improve soil's water-holding capacity, acting as a slow-release source of water and reducing hydric shortage periods during crop season. The use of such systems is advantageous for agricultural purposes as well as reforestation of degraded areas. Organic nanomaterials, such as polymers and carbon nanotubes, and inorganic nanomaterials, such as nano metals and metal oxides, have also been used to absorb pollutants, increasing soil remediation capacity and reducing treatment time and costs [8, 9].

18.5 Pesticides with Nanoparticles

NPs may be crucial in the management of insect pests and host infections because they are common in agricultural fields and products. A nano-encapsulated pesticide formulation that was recently developed has slow releasing properties as well as improved solubility, specificity, permeability, and stability [10]. These benefits are primarily obtained by either protecting the encapsulated active ingredients from premature degradation or increasing their pest control efficacy for a longer period of time. The development of nano-encapsulated pesticides reduced pesticide dosage and human exposure to them, making crop protection more environment friendly [11]. Also, greater than 1/2 of the world's population relies upon on green vegetation and their merchandise for survival, growing plant productiveness and keeping them healthy is a prime concern. Pathogens, nutrient deficiency, and soil/air pollution, for example, are all biotic and abiotic elements that may have an effect on plants growth with the aid of using inducing harm that ends in reduced crop and fruit yield [12].

18.6 Growth of Cultivated Plants and Ecotoxicological Sustainability

Agriculture host plants play a critical role in the food chain. Plants are now being developed not only on agricultural lands, but also on aqueous medium. Several NPs of iron oxide (magnetite), a magnetic form of iron ore, can naturally deposit in the plant host. It is intriguing to speculate that iron (II, III) oxide NPs (Fe_3O_4-NPs) can accumulate in Lepidium sativum and Pisum sativum plants [13]. And apart from that, the use of polymeric NPs in agriculture, particularly when loaded with plant-derived insecticides, appears to be novel and forward-thinking. There is no reason to doubt that microorganisms play an important role in the maintenance of soil health, ecosystem health, and crop yield [14]. Though the understanding of ecotoxicological aspects of the agricultural field under consideration is critical. If agricultural plant nanomaterials are free of toxic nanocomposite, there is a unique opportunity for increased crop production. Therefore, the introduction of engineered (either chemical or green) NPs in the agricultural field should always be followed by a routine check-up in order to maintain an eco-friendly agricultural field [15].

18.7 Nano-Fertilizers: A Sustainable Source of Crop Nutrition

In order to increase crop output and soil fertility, nutrient supplementation, also known as element fertilization, is often necessary. However, proper fertilizer management is seen to be one of the most important requirements for long-term agricultural success. Contrarily, access to food is a fundamental human right. Everywhere in the world, there is a threat to global food security. Natural resource shortage is a factor that threatens food security. Approximately nine billion people will live on Earth by the year 2050, up from the present seven billion [16]. The globe will need 60–100% more food to feed its expanding population. Intensive farming is being used to fulfill the growing demand for food, but this technique eventually creates a vicious cycle of declining agricultural yields and dwindling soil fertility. According to estimates, 40% of all agricultural land worldwide has suffered serious degradation, which has led to a major decline in soil fertility as a result of such intensive farming methods. As a result, enormous amounts of fertilizer are required to increase agricultural output and soil fertility. Additionally, it has been clearly shown that only one-third of crop yield is influenced by fertilizers, with the remaining half coming from how well other agricultural inputs are employed [17].

However, only 30–40% of the nutrients in standard fertilizers are really used. For instance, the nitrogen (N) 30–35%, phosphorus (P) 18–20%, and potassium (K) 35–40% nutrient utilization efficiency of traditional fertilizers has hardly changed over a number of decades [18]. Furthermore, the final fertilizer concentration at the target places has a significant impact on how fertilizers, sprayed on leaves or applied directly to the soil, utilize nutrients. In practical scenarios, significantly less than the required minimum concentration often reaches the desired destination due to various factors such as chemical leaching, drift, runoff, hydrolysis, evaporation, photolysis, or even microbial degradation. As a result, the regular application of too many fertilizers has a detrimental effect on the soil's natural nutritional equilibrium. In addition to this, drinking water has been seriously affected by harmful compounds that have leached into rivers and reservoirs. Through increased nutrient usage effectiveness and addressing the enduring issue of eutrophication, nano-fertilizers may be the greatest choice for resolving macro- and micronutrient deficiencies. The potential for nano-fertilizers, which are made to control nutrient delivery depending on crop demands while reducing differential losses, is immense [19].

The use of excellent seeds, irrigation, and appropriate fertilizer management can boost agricultural output by 35–40%. The productivity of crops may rise dramatically with the application of fertilizers with nano-formulated compositions. Carbon nanoparticles, for instance, can boost the grain yields of rice (10.29%), spring maize (10.93%), soybean (16.74%), winter wheat (28.81%), and vegetables (12.34–19.76%) when mixed with fertilizer [20]. As a result of the great porosity of plant root and leaf surfaces at the nanoscale, nanoparticles stimulate several crucial facets of plant life. Therefore, applying nano-fertilizers can enhance plant nutrient absorption through existing holes, or the procedure can make new pores or use endocytosis or ion channels to assist complexation with molecular transporters or root exudates. In sustainable agriculture, a new phrase known as "control loss fertilizer" is utilized. When these fertilizers come into touch with water, they self-assemble in the soil to form a nano network, which reduces nonpoint pollution of agricultural inputs. In contrast, the soil network is accessed by imprisoned fertilizer nutrients through hydrogen bonds, surface tension, molecular force, or viscous force. Their spatial scale enlarges as a result, making them readily blocked by soil filtration and remaining fixed in soil close to crop roots. This makes it easier for plants to absorb nutrients to satisfy their needs throughout the growth cycle. For instance, an innovative strategy has been employed to successfully lower the rate of nitrogen transfer into the environment. The use of control loss fertilizer improves a 9.8% rise in soil residual mineral nitrogen and a 5.5% increase in wheat output over the use of conventional fertilizers. It also decreases nitrogen runoff and leaching loss by 21.6% and 24.5%, respectively (Fig. 18.3).

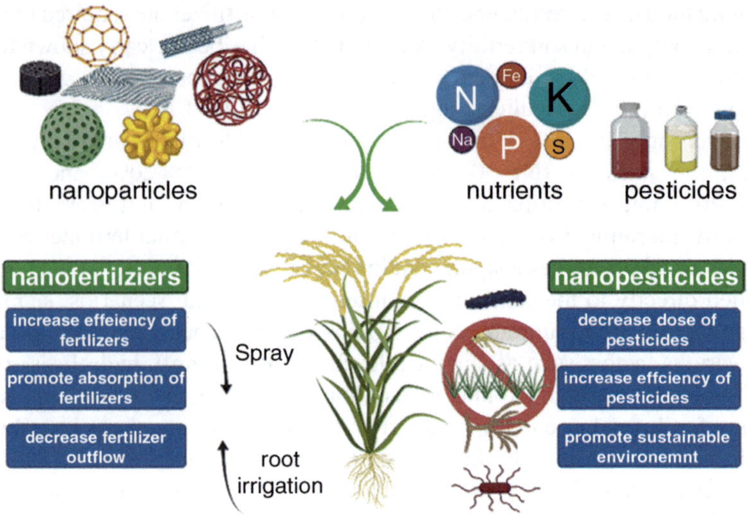

Fig. 18.3 Nanofertilizers and Nano-pesticides for crop production [21]

18.8 Crop Growth, Seed Germination, and Quality Enhancement, Utilizing Nanomaterials

The creation of low-cost nanotech applications for better seed germination, plant growth, development, and environmental adaptation is the main goal of the emerging scientific innovation platform known as nanoscience. The development, survival, and population dynamics of a plant are all facilitated by the germination of the seed [21]. On the other side, seed germination is affected by environmental conditions, genetic features, moisture availability, and soil fertility. The use of nanomaterials enhances plant growth and development as well as germination. Multiwalled carbon nanotubes (MWCNTs), for instance, enhance seed germination in a number of crop species, including tomato, corn, soybean, barley, wheat, maize, peanut, and garlic. Similar to this, agricultural plants benefit from the application of nano SiO_2, TiO_2, and Zeolite to encourage seed germination [22]. Additionally, it has been found that Fe/SiO_2 nanomaterials have a great deal of promise for enhancing the germination of maize and barley seeds. The underlying processes by which nanomaterials can increase germination are still unclear, despite a sizable amount of research on their beneficial effects on germination. Nanomaterials including ZnO, TiO_2, MWCNTs, FeO, ZnFeCu-oxide, and hydroxy fullerenes, in addition to promoting germination, have to promote crop growth and development while enhancing crop quality in a number of crop species, including peanut, soybean, mungbean, wheat, onion, spinach, tomato, potato, and mustard [23].

18.9 Plant Adaptation to Climate Change Factors Using Nanomaterials

As the world's population growing and resources getting scarcer, food security is becoming a difficult problem. Climate baseline changes over time, such as those in temperature, water shortage, cold, salinity, alkalinity, and toxic metal pollution, are referred to as progressive climate change. In order to help plants, adjust to environmental challenges more quickly, it is important to avoid damaging already fragile ecosystems [24]. This task requires a multi-pronged approach that involves stimulating the plant's enzymatic system, hormonal control, stress gene expression, toxic metal absorption regulation, and reducing the plant's life cycle to prevent water shortage stress or flash floods. Researchers have worked hard to create technology and methods that would enable sustainable agricultural systems while reducing their detrimental effects on the environment [25].

Recent developments in nanomaterial engineering imply that nano-fertilizers can increase crop productivity in currently unfavorable conditions. In around 23% of the world's farmed lands, salinity stress significantly restricts crop yield. The application of nano-SiO_2 to tomato and squash plants under NaCl stress, on the other hand, has been shown to enhance seed germination, increase plant fresh weight, dry weight, and chlorophyll content and stimulate proline accumulation. Similar to this, Torabian et al. [26] shown that foliar spraying iron sulfate ($FeSO_4$) nanoparticles increase the tolerance of sunflower cultivars to salt stress. Recently, it was shown that silicon nanoparticles (SiNPs) might successfully reduce the stress caused by UV-B on wheat. The nutrient availability and plant germination and growth can both be enhanced over time by nano zeolite. Excellent work was done by Abdel-Aziz et al. [27] to research the uses of nanomaterials (Fig. 18.4).

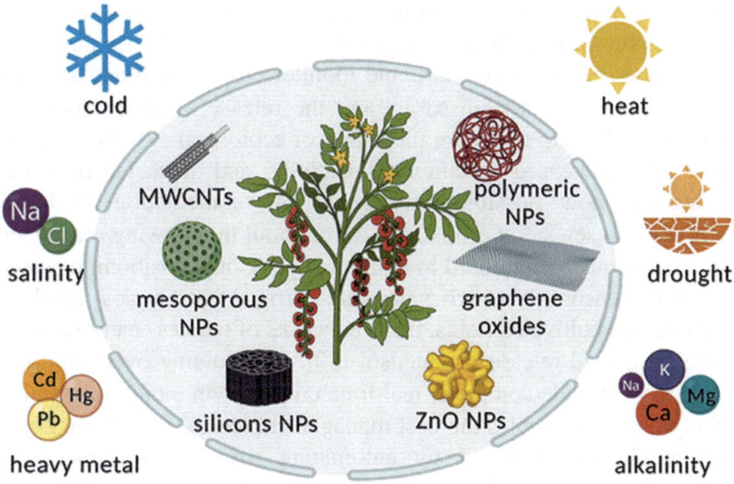

Fig. 18.4 Nanoparticles for stress tolerance which helps in sustainable agriculture [21]

They found, for the purpose of producing yield from the date of planting, the life cycle of wheat plants treated with nano-fertilizers was 23.5% shorter than that of plants treated with conventional fertilizer (130 vs. 170 days).

In places prone to drought or even abrupt flash floods, where crop maturity is a crucial component of sustainable crop production, the use of nano-fertilizers to speed up plant development and productivity reveals their potential as useful instruments in agricultural operations. Additionally, the nanoparticles are useful in the detoxification or remediation of dangerous contaminants, such as heavy metals. Biotic variables like pests and diseases have a big impact on crop output. Pesticides have been frequently used by farmers to prevent crop losses, which have a severe impact on environmental sustainability and human health. Pest and disease risks may be successfully decreased by using nanomaterials, minimizing the severity of crop losses and environmental dangers. Numerous plant diseases brought on by Fusarium solani, Fusarium oxysporum fsp Radicis Lycopersici, Verticillium Dahliae, Phytophthora infestans, and Ralstonia solanacearum can be effectively controlled by metal oxide nanomaterials such as CuO and ZnO.

Increased enzyme activity can be used to explain how nanoparticles affect crop development in challenging conditions. For instance, the use of nanomaterials like nano-SiO_2 or nano-ZnO improves plant tolerance to extreme climate events by increasing the accumulation of free proline and amino acids, the uptake of nutrients and water, and the activity of antioxidant enzymes like superoxide dismutase, catalase, peroxidase, nitrate reductase, and glutathione reductase [28].

18.10 Nanomaterials as Nano-Sensors: Perturbation Measurement and Monitoring

Nanomaterial engineering is the most cutting-edge research area for sustainable agriculture. Nanomaterials are used in precision farming to lower prices and labor requirements, boost productivity, and promote environmentally friendly growth. The creation of nano-sensors to measure and monitor crop growth and soil conditions, nutrient deficiency, toxicity, diseases, and the release of agrochemicals into the environment would help to ensure the safety of ecological systems and sustainable agriculture [29]. Increased specificity, sensitivity, and quick reactions to identify deficiencies are now possible because to the combination of biology and nanomaterials into sensors. For instance, throughout the growing season, cultivated fields are continuously monitored in real time by a global positioning system (GPS) based on nano-sensors. Through nanoscale carriers that employ wireless signals dispersed over the cultivated fields, these networks of wireless nano-sensors keep an eye on the controlled release mechanism [30]. By avoiding overusing agricultural inputs, this may provide complete, real-time crop growth monitoring and effective, high-quality data that enables the best management techniques. Water consumption efficiency might be increased by automating irrigation systems with sensor technologies. Nano-sensors that measure soil water tension in real time in conjunction with autonomous irrigation control are used when there is a water shortage. The timely use of insecticides or fertilizers to prevent crops from infestation would also benefit from early identification of insects or diseases. In order to track insect assaults, Afsharinejad et al. [31] developed a wireless nano-sensor. This sensor can distinguish between various insect species and the volatile chemical compounds that numerous host plant species produce. A nano-gold immunosensor has been shown to be effective at detecting the Karnal bunt disease in wheat plants by Sing et al. [32]. Additionally, the creation of bionic plants using nanoparticles inserted into living plants' cells and chloroplasts for sensing or imaging objects in their environment and communicating as infrared devices, or even self-powering plants as light sources, has significant potential in precision farming.

By keeping track on crop and soil health, the clever application of nano-sensors in agriculture is a new instrument that promotes sustainable growth. Although there are surprisingly few applications of nano-sensors, particularly in field investigations, suggesting a new direction for future study [33].

18.11 Food Science's New Frontier: Nanotechnology

The potential of nanotechnology in food science has not yet been fully exploited due to a lack of knowledge. Most current applications of nanotechnology are in the fields of electronics, automation, super-materials, or life sciences like medicine [34]. However, humankind's experience accumulated through decades of scientific discoveries plainly indicates that such a new science will not be restricted inside a limited number of applications, but would have a significant influence on human existence in its whole. Nanoscience has the ability to transform the world's food industry. Nanotechnology will most likely dramatically improve four important areas of the food industry: the creation of novel functional materials, micro- and nanoscale processing, product development, and the design of procedures and instruments for food safety and biosecurity. With the assistance of nanotechnology nonpolluting, less expensive, and more efficient procedures will be created; lighter and more accurate food production equipment will be produced; and nutritious, nutraceutical-loaded meals will be packaged in lighter, more functional, and stronger packaging materials. Food safety might be considerably enhanced by developing "bacteria-repellent" surfaces or packaging materials that change color in the presence of hazardous microbes or toxins. As a result, food science and technology should harness the strong powers of nanotechnology for the benefit of humanity.

The uses of nanotechnology in the foodservice industry may be divided into two categories: food nanostructured ingredients and food nanosensing [35]. Food nanostructured substances include a wide range of applications, from food processing to food packaging. These nanomaterials can be employed in food processing as food additives, carriers for smart nutrition delivery, anti-caking agents, antibacterial agents, fillers to improve mechanical strength and durability of packaging materials, and so on. Food nanosensing, however, can be used to improve food quality and safety evaluation.

18.12 Two Methods of Nanofabrication

Nowadays, nanostructures are readily accessible, which has prompted the creation of various antibacterial drugs. To create them naturally, polymer nanoparticles (nanocapsules and nanospheres), nanotubes, nanofibers, and liposomes can be mixed with antimicrobial, plant extracts, peptides, and essential oils. Due to the suitability of the nano-structural system, the characteristics of the nanomaterial and antibacterial agent are highlighted. In order to create nanomaterials and nanostructures, top-down and bottom-up methods are employed. The top-down approach involves nanoscale grinding of bulk materials and stabilization of the nanoparticles using colloidal stabilizers. The top-down approach's main drawback is that it cannot be employed for mass manufacturing due to its slow output pace. Surface flaws, contamination, and stress introduction can potentially result in errors. Additionally, this method produces waste and is constrained by the equipment available. The process of making materials consisting of individual atoms that spontaneously self-assemble and self-regulate is known as nanotechnology. Bottom-up approaches result in less flaws, more homogeneity, and better short- and long-range order (Fig. 18.5).

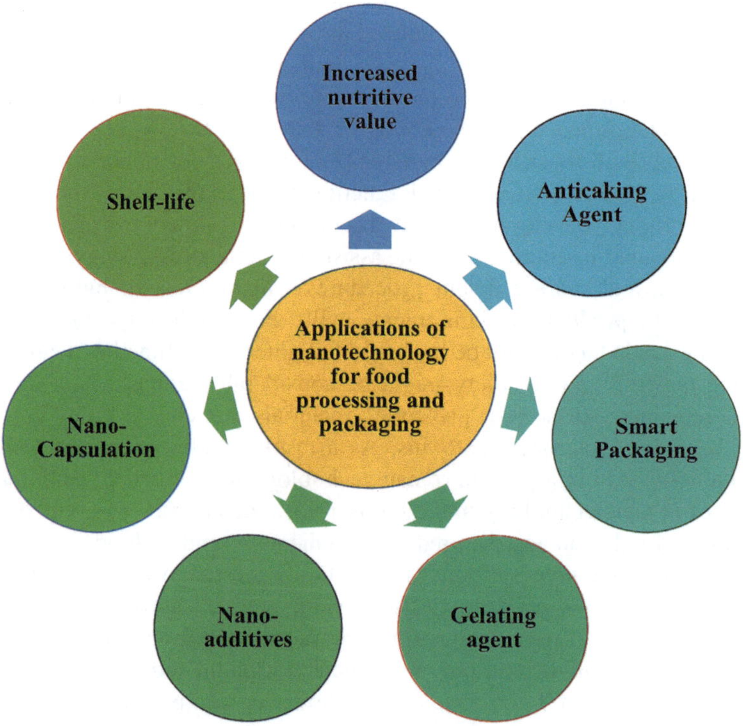

Fig. 18.5 Applications of nanotechnology for food processing and packaging

18.13 Four Nanostructures in Food Systems

Some food items contain substances that are nanoscale in size yet are not lab-made. While some dietary proteins have a more complex structure, others have globular forms with sizes between 10 nm and 100 nm. Most lipids and polysaccharides are linear polymers with a thickness of 1 nm, or one-dimensional nanostructures. Natural nanostructures may be found in milk and goods produced from it, including casein. Liposomes, nano-emulsions, microemulsions, and polymeric nanoparticles are a few examples of synthetic nanostructured food systems. These polymers enable greater control over the release and preservation of bioactive components while also improving solubility and bioavailability. The bulk of bioactive substances have a very acidic property, including vitamins, proteins, lipids, and carbohydrates. This is considerably worse in the duodenum and stomach where there is a high concentration of digestive enzymes. Encapsulation protects these bioactive molecules from damage while also facilitating their quicker absorption in food, which is a significant disadvantage for bioactive chemicals that are not encapsulated. The development of newer, smarter nanoparticle-based tiny consumable capsules aims to enhance the daily food consumption of vital micronutrients, vitamins, and medications.

18.14 Nano-Emulsions

Nano-emulsions, which are colloidal dispersions with diameters ranging from 50 nm to 1000 nm, are employed in the manufacture of beverage, sweetener, salad dressing, and other processed goods [36]. Nano-emulsions can be used to clean equipment and increase clarity without changing the product's appearance or flavor. Functional nano-emulsified and encapsulated nano-compounds are utilized to deliver lutein, lycopene, b-carotene, vitamins A, D, and E3, coenzyme Q10, and omega-3 fatty acids to certain places [37]. Capsaicin-loaded nano-emulsions were stabilized using natural polymers including chitosan and alginate. Regularly adding elements of functional meals into droplets can change the interfacial layer's characteristics, which can slow down chemical breakdown processes. Additionally, bottled milk and water that have been supplemented with vitamins, antioxidants, and minerals are made using nano-emulsions [38].

18.14.1 Nanostructured Proteins-Biopolymeric Particles

Micelles, which are protein-biopolymeric spheres with diameters of 5–100 nm, can include lipids, antioxidants, and vitamins. To make insoluble components soluble in water, microemulsions are created. Glycerides for food items have also been made with the use of microemulsions. Together, lipophilic and hydrophilic antioxidants increase the efficiency of microemulsion antioxidation. Food-grade biopolymers, such as polysaccharides or proteins, are used to make nanoparticles. Numerous biodegradable biopolymeric nanoparticles include polylactic acid. Many manufacturers actively promote this product. Proteins, vaccinations, and medications are frequently packaged and transported in polylactic acid. There are some negative consequences of polylactic acid exist, such as quick bloodstream clearance and liver and kidney buildup. It is useless as a nanoparticle for dispersing active ingredients elsewhere in the body unless it is linked to an associative molecule, such as polyethylene glycol.

18.14.2 Nanocomposites

In nanocomposite food packaging, polymers are a practical substitute for conventional packaging materials (glass, metals, and paper). Nanocomposites are polymer matrices reinforced with nanofillers, such as carbon nanotubes, cellulose microfibrils, nano-clays, and nano-oxides, with a minimum feature size of less than 100 nm. Although both synthetic and natural polymers have been widely employed in food packaging, the latter is more common. However, the demand for biodegradable packaging composed of polyvinyl alcohol, polyglycolic acid, and polylactide has surged due to environmental concerns.

18.14.3 Liposomes

Phospholipid bilayer vesicles having a spherical form are known as liposomes. In the food sector, liposomes have been utilized to encapsulate useful substances. More recently, their capacity to absorb food preservatives has been examined, which may help with food preservation. Food solubility, stability, and bioavailability may all be enhanced by fat-based nanoencapsulation while harmful interactions with other food ingredients are avoided. It looks like a family of lipid-based antioxidant carriers has a lot of potential. Nanoliposomes aid in the targeted delivery of vitamins, minerals, enzymes, additives, and nutraceuticals. A higher interfacial area and smaller size of liposomes increase the bioavailability of encapsulated compounds.

18.15 Food Processing

Nanostructured food products are being developed with the objective of improving taste, appearance, and consistency (Cientifica Report, 2006) [39]. Nanotechnology is increasing the shelf-life of various types of food materials while also reducing food waste caused by microbe infestation [40]. Nanocarriers are now being used as delivery systems for food additives in food products without affecting their basic morphology. Furthermore, the significance of nanotechnology in food processing can be assessed by looking at its role in improving food products in terms of (1) food texture, (2) food appearance, (3) food taste, (4) nutritive value of the food, and (5) food shelf-life. Surprisingly, nanotechnology not only affects all of the aforementioned aspects, but it has also led to significant changes in food products, providing them with novel qualities.

18.15.1 Food Texture, Taste, and Appearance

Numerous possibilities are available, thanks to nanotechnology for boosting food quality and flavor. The release and preservation of flavors, as well as the creation of a balanced flavor profile, have all been improved with the application of nanoencapsulation methods [41]. Anthocyanins, a highly reactive and unstable plant pigment with a range of biological functions, were enclosed in nanoparticles by Zhang et al. [42]. By encasing cyanidin-3-*O*-glucoside (C3G) molecules into the inner cavity, apo recombinant soybean seed H-2 subunit ferritin (rH-2)'s heat stability and photostability were enhanced. The creation of multifunctional nanocarriers for the transport and preservation of bioactive compounds is described here. Rutin is a typical dietary flavonoid with significant pharmacological properties, but because of its low absorption, its use in food is restricted. The absorbance, thermal stability, and UV radiation stability of ferritin trapped rutin over free rutin were all increased by the inclusion of ferritin nanocages [43]. Nano-emulsions are being utilized increasingly often to deliver lipid-soluble bioactive compounds because they can be manufactured using natural components and simple production procedures, and because they can be customized to maximize water-dispersion and bioavailability.

18.15.2 Nutritive Value

The majority of bioactive compounds, such as lipids, proteins, carbohydrates, and vitamins, are sensitive to the stomach and duodenum's high acidic environment and enzyme activity. Encapsulation of these bioactive compounds not only allows them to withstand such harsh conditions but also allows them to easily assimilate in food products, which is difficult to achieve in non-capsulated form due to their low water solubility [44]. To provide significant health benefits, nanoparticle-based tiny edible capsules with the goal of improving delivery of medicines, vitamins, or fragile micronutrients in daily foods are being developed. The different techniques used to encapsulate substances in miniature forms to more effectively deliver nutrients like protein and antioxidants for precisely targeted nutritional and health benefits are nanocomposite, nano-emulsification, and nano-structuration. Polymeric nanoparticles have been discovered to be suitable for encapsulating bioactive compounds (e.g., flavonoids and vitamins) in order to protect and transport bioactive compounds to target functions [45].

18.15.3 Shelf-Life or Preservation

By postponing or preventing deterioration until the product reaches the target site, nanoencapsulation of these bioactive components extends the shelf-life of food products, which are frequently degraded and ultimately rendered inactive by harsh environments in food supplements. A protection against moisture and gas exchange, as well as the delivery of tastes, colors, antioxidants, enzymes, and anti-browning agents, might all be provided by edible nano-coatings on a variety of food components. These coatings could also help to extend the shelf-life of processed meals even after the container has been opened [46, 47].

18.16 Food Packaging with Nanotechnology

Methods of food packaging are used to guarantee that the food's quality is maintained while still being packaged in a way that makes it safe to consume. By scavenging oxygen and other gases that cause food to deteriorate, packaging primarily helps to protect food from external shocks and vibration, microbial infection, and temperature [48]. Packaging materials should be composed of biodegradable materials to decrease environmental damage. This idea has increased the use of nanotechnology to the food packaging sector. A good packing material must be both robust and biodegradable, as well as gas and moisture permeable. The advantages of nano-based "smart" and "active" food packagings over conventional packaging techniques include better packaging materials with improved mechanical strength, barrier properties, and antimicrobial films, as well as nanosensing for pathogen detection and alerting consumers to the safety status of food [49]. In order to provide a barrier against high temperatures and mechanical damage and hence increase food shelf-life, nanocomposite and nanolaminates are actively used in food packaging. The use of nanoparticles in packaging materials allows for the provision of high-quality food with a longer shelf-life. In order to provide better mechanical and thermostable packing materials, polymer composites are being created. To make superior materials, a lot of organic or inorganic fillers are employed. A number of products, including antimicrobial food packaging, nanocomposite, and nanolaminates that shield food from heat and mechanical damage, are made with nanoparticles and used in food packaging. Although these nanoparticles are usually regarded as harmless, there is a lot of worry about how easily they may be absorbed via food. Knowing more about these nanoparticles' immunogenicity, potential for damage, and distribution inside the human body is therefore crucial. Making stronger and more resilient packing materials is the aim of polymer composite materials. Inorganic and organic fillers are used to strengthen polymer composites. Cost-effective polymeric packaging materials have been created for developments in the integration of polymeric nanoparticles. Nanocomposites based on graphene nanoplates can be utilized to make heat- and barrier-resistant food packaging. Carbon nanotubes and nanofibers are not commercially feasible options for food packaging due to their high cost and difficulties in processing dispersions. Fresh produce should not be totally sealed to prevent gas migration and penetration caused by cellular respiration. Fizzy beverage packaging should be designed to minimize oxidation and decarbonation. When food matrices or packaging materials are utilized, the flow of oxygen, carbon dioxide, and water vapor is altered. Consequently, a variety of nanocomposite materials, including polymers, may be employed for this. Adding nano-bio-composites to food packaging enhances its ability to act as a gas barrier [50].

18.17 Nano-Sensors

Nano-sensors help in the detection of gasses released by food deterioration as well as any changes in the color of the food. The sensors are often sensitive to gasses including ammonia, hydrogen, hydrogen sulfide, nitrogen oxides, and sulfur dioxide. They consist of an electronic data processing component and a sensor component that can recognize changes in light, heat, humidity, gas, and chemicals and transform those changes into electrical signals [51]. Due to their great sensitivity and selectivity, nano-sensors are more effective than conventional sensors. Palladium, platinum, and gold are some of the metals used to make these gas sensors. Toxins like aflatoxin B1 are also found in milk, and gold-based nanoparticles are employed to find them. The sensitivity of the sensors is increased by the use of DNA and single-walled carbon nanotubes in their construction. Agriculture uses nano-sensors to monitor the soil conditions essential for agricultural growth. They also help with the pesticide surface detection on fruits and vegetables. Nano-sensors for locating carcinogens in food products have also been created, in addition to pesticides. In gas sensors, conducting polymers are also utilized. These electroactive conjugated polymer-based sensors have conductive particles inserted within an insulating polymer matrix (Fig. 18.6) [52].

In fact, a form of detector called an electronic nose uses a number of chemical sensors and is connected to a system for data processing. The sensor is known as an electronic nose because it functions similarly to a human nose. There have been reports of electronic tongue detecting devices that use the same principles as the electronic nose in addition to the electronic nose. It changes color when it comes into

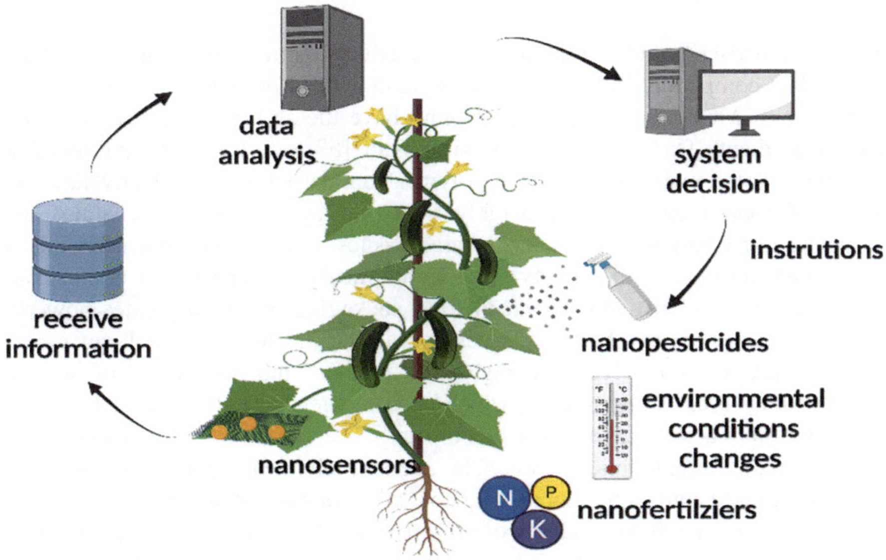

Fig. 18.6 Nano-sensors for monitoring different aspects of plants [21]

touch with any traces of food decomposition, letting you know that the food is no longer safe to eat. These days, nano-biosensors are proven to be highly effective and beneficial for the detection of bacteria and viruses. Biosensors and biomimetic sensors, both are efficient in finding mycotoxins and other harmful substances. Biomimetic membranes and proteins are employed to make sensors. The sensors serve as fictitious cell surfaces to help identify and get rid of infections. Because nano-sensors have been designed to have better levels of selectivity and sensitivity, their usage has actually helped to reduce detection limits.

18.18 Nano-Coatings for Smart Surfaces

Combining biopolymers with nanostructured materials (NSMs) can enhance the characteristics of polymers. NSMs made from biopolymers can be utilized to enhance the practical qualities of intelligent and active packaging. These three forms of packaging are referred to as intelligent, active, and enhanced packaging since a single piece of packing material may serve numerous functions. There is a lot of talk about "thick and homogeneous layers"in nano-coatings. Effective ways for determining the safety and quality of meals after they have been created are non-invasive techniques for measuring air content and gas amount beyond the phases of production. The presence of oxygen inside packaging has the potential to shorten the shelf-life of food because it encourages the growth of potentially harmful microbes.

18.18.1 Antibacterial Nanoparticles in Active Packaging

The active barrier features of the packaging facilitate the release of antibacterial and antioxidant compounds, therefore eliminating unwanted pollutants like oxygen and water vapor. Molecules in active packaging have the ability to remove or release food ingredients from or into the environment [53]. At the time, antimicrobial packaging is the main use for active polymer nanoparticles. The effectiveness and utility of food items are considerably increased by nanoparticle coatings and capsules. For example, the sensory characteristics and esthetic appeal should be under control to assure the quality of packaged meals. Antibacterial active packaging, a specific and distinctive kind of food packaging that incorporates powerful antibacterial agents, can be used to achieve this. Exudates from fresh-cut melon, steak, chicken, and silver asparagus all prevent the formation of aerobic psychotropics (yeasts and molds), which need oxygen, in order to function. They also have antibacterial effects on E. coli and S. aureus. Without compromising sensory quality, orange juice, zinc oxide, and egg albumen in liquid form successfully lower mold counts, Salmonella, Lactobacillus plantarum, and yeast [54, 55]. In order to preserve the bioavailability of bioactive substances during food preparation, transit, and storage in the gastrointestinal tract, functional antibacterial components and nutraceuticals must be encapsulated. Additionally, adding nanoparticles to food enhances the absorption of polyphenolic compounds including resveratrol, curcumin, and epigallocatechin-3-gallate. These polyphenols will not be destroyed by the stomach's acidity. Research has proven that edible nano-coatings are a feasible alternative to microbial degradation and preservatives in terms of food preservation and shelf-life extension. Gelatin coatings containing cellulose or chitosan nanocrystals have been shown to be particularly successful at maintaining the freshness of food over prolonged storage using nanomaterials like nano-silica. Additionally, techniques for preserving fruit have been developed that mix polyethylene with nanopowders such as rutile TiO_2, anatase TiO_2, kaolin, and silver (Ag). The antibacterial activity of chitosan-coated nanoparticles may be mediated through the interaction between positively charged chitosan with negatively charged cell membranes. The membrane becomes permeable as a result of these interactions, allowing chemicals inside the cell to escape and rupturing the membrane. Antimicrobial food packaging sheets made of nanocellulose, polyvinyl alcohol (PVA), and silver nanocomposite were created and put on the market.

18.19 Nanotechnology-Enabled Food Production, Safety, and Quality Solutions

Dietary quality deteriorates as a result of several chemical interactions between environmental factors and dietary ingredients. These undesired processes can be stopped using a variety of nanomaterials. Although some metal and metal oxide nanoparticles can generate ROS and result in oxidative stress, these nanomaterials also include nano-toxins that also exploit this mechanism to create oxidative stress, leading to an imbalanced redox state [56]. As a result, less reactive nanoparticles are employed as carriers for antioxidants. Encapsulated flavonoids and vitamins can release their contents into a somewhat acidic environment, such as the stomach [57]. Nanotechnology can remove the negative effects of dietary components. The carrier is infused with specific nutraceuticals to help the body reduce cholesterol accumulation. The general population is aware that there are not many nanostructures in food that can start nano-effects on their own [58]. For instance, green tea includes nano-selenium, which enhances the body's absorption of selenium. Nanoencapsulation is a crucial part of nanotechnology since it wraps the components and assures product function, including the controlled release of the core. Large numbers of active molecules are frequently combined in encapsulated compounds, which have a pH-triggered controlled release, enhanced stability, and a longer shelf-life [59]. The physical characteristics of materials used in packaging and food have been proven to be greatly improved by nanoparticles. Silicate nanocomposites showed a variety of desired characteristics, such as flame and UV light resistance. To enhance the look and flavor of food, many creative natural items have been created, such as colors and esthetic appeal [60].

18.20 Safeguarding against Chemical Ingredient/Promote better Physical Properties

18.20.1 Flavors

Flavors are an important aspect of the food system because they enable sensory awareness of minor tastes, which improves the eating experience overall. Nanoencapsulation methods are commonly employed to increase taste retention and release while also providing culinary harmony [61]. Silica dioxide (SiO_2) nanoparticles have been shown to transport scents or flavors in food and non-food goods. Because nano-silica particle exposure through food is reliant on the nano-silica content of processed foods, estimating the amount of nano-silica that will be eaten is crucial. The nano-silica content of freshly brewed coffee was measured in order to compute the nano-silica concentration in a cup of coffee. Fumed silica is used to keep coffee creamer from clumping [62].

18.20.2 Anti-Caking Substances

SiO_2 is used in food and non-food goods to thicken pastes, as anti-caking agents in powdered products, and to convey scents or tastes. This is classed as a food additive (E551) in the European Union. It has been revealed that a percentage of the SiO_2 in E551-containing dietary components is nanosized. Eating E551-containing foods exposes the gastrointestinal epithelium to nanosized SiO_2. Despite the fact that silica nanoparticles can be deadly when exposed to human lung cells, they are routinely used [63, 64].

18.20.3 Colorants

Food color additives must be authorized by the FDA's Office of Cosmetics and Colors, according to FDA rules. Many different nanoscale color additives are now being investigated and created. Several nanomaterials have been authorized for use as food coloring ants, which is important for product marketing. TiO_2 was permitted as a food coloring additive by the US Food and Drug Administration, but its usage in concentrations larger than 1% w/w in food was forbidden. Food color additive mixes containing TiO_2 may contain up to 2% SiO_2 or aluminum oxide as dispersion aids [65]. Carbon black, however, has been phased out as a food color. Although beta-carotene is a provitamin A and may be used to color foods, its hydrophobicity and chemical stability prevent it from being employed.

18.20.4 Other Advantages

Nanomaterials are being created on a regular basis to increase the mechanical and physical qualities of packaging, such as tensile strength, gas permeability, stiffness, flame, and water resistance. Polymer nanocomposites are the most recent materials with significant promise for application in active food packaging. In the 1990s, multilayer silicate-polymer nanocomposites were introduced. Because of their improved strength, flame resistance, and possible utility in UV-shielding applications, these new polymer nanoparticles are generally predicted to radically revolutionize the food packaging business.

18.21 Concerns about Safety

Opportunities and exposure to these chemicals increase as food nanotechnology research improves. Human exposure to nanoparticles is rising, and various research have been conducted to investigate the toxicity of nanomaterials in meals, namely food additives/ingredients and food packaging. After exposure, the bioavailability, biodistribution, routes, and toxicity of nanomaterials are all unclear. There has been a rising emphasis on the use of nanoparticles as food additives due to public and government interest. More than 93% of the TiO_2 in chewing gum is at the nanoscale level. TiO_2 is a chemical that may be emitted into the environment, absorbed by someone chewing gum, and stored in the body. Foods containing E551 may expose the gastrointestinal epithelium to SiO_2 nanoparticles [66]. The destiny and toxicity of nanomaterials in food packaging and foodstuffs are determined by their physico-chemical characteristics and dose. Nanotechnology in the food sector demands rigorous analysis and assessment in silico, in vitro, and in vivo. It is feasible to quantify and analyze absorption, distribution, metabolism, excretion, and ultimate toxicity by combining all of these physical, osmotic, pH, chemical, and biological processes. Food nanotechnology requires a uniform worldwide regulatory framework.

18.22 Conclusion

Nanotechnology has been used in agriculture and the food industry to increase crop production with food quality enrichment by improving farming systems. The emergence of engineered nanomaterials and their actions within the context of sustainable agriculture have dramatically changed the world agriculture canvass in terms of novelty, rapid growth, and enormity in order to meet projected global food demand. Agriculture in the twenty-first century faces a series of challenges, including the need to increase production of food and fiber to feed the world with a shrinking rural labor force, change in climate, and urbanization. These issues will be exacerbated when we have to serve over 9 billion people by 2050. As a result, there will be an increase in supply for agricultural products. Raw materials will soon be viewed as the foundation of commercial activity and manufacturing, particularly in developing countries with larger populations. To deal with this scenario, agriculture-dependent countries must adopt more efficient methodologies, labor-saving techniques, and environmentally friendly production methods. Despite the availability of a wealth of information on individual nanomaterials, the toxicity level of many NPs remains unknown, limiting their use due to a lack of understanding of risk assessments and human health effects. For this technology to be exploited, a comprehensive database and alarm system, as well as international cooperation for regulation and legislation, is required.

The benefits of using nanotechnology in the food industry are numerous and expected to grow. This new, rapidly evolving technology has an impact on every aspect of the food system, including production, processing, packaging, transportation, shelf-life, and bioavailability. Because of their unique and novel properties, commercial applications of nanomaterials in the food industry will grow. The human population's exposure to nanomaterials will continue to rise. As a result, the public's primary concern is the health impact of nanomaterials in food. The ability to quantify nanomaterials throughout the food life cycle is critical for manufacturing consistency, product safety, and potential benefits. The safety of food and food-related products containing nanomaterials will determine public acceptance. A unified international regulatory framework for food nanotechnology is required.

References

1. Dwivedi S, Saquib Q, Al-Khedhairy AA, Musarrat J (2016) Understanding the role of nanomaterials in agriculture. Springer, Berlin, pp 271–288
2. Thakur A, Thakur P (2021) Ferrite nanoparticles for agriculture-related activity. In: Singh JP, Chae KH, Srivastava RC, Caltun OF (eds) Applications of nanostructured ferrites. Woodhead Publishing, Cambridge, pp 315–330. https://doi.org/10.1016/B978-0-443-18874-9.00005-9
3. Kaur S, Sharma K, Singh R, Kumar N (2022) Advancement in crops and agriculture by nanomaterials. In: Thakur A, Thakur P, Khurana SP (eds) Synthesis and applications of nanoparticles. Springer, Singapore, pp 319–335. https://doi.org/10.1007/978-981-16-6819-7_14
4. Dineshkumar R, Paul A, Gangopadhyay M, Singh N, Sen R (2017) Smart and reusable biopolymer nanocomposite for simultaneous microalgal biomass harvesting and disruption: integrated downstream processing for a sustainable biorefinery. ACS Sustain Chem Eng 5(1): 852–861
5. Sadik OA, Du N, Kariuki V, Okello V, Bushlyar V (2014) Current and emerging technologies for the characterization of nanomaterials. ACS Sustain Chem Eng 2:1707–1716
6. Grillo R, Abhilash PC, Fraceto LF (2016) Nanotechnology applied to bio-encapsulation of pesticides. J Nanosci Nanotechnol 16:1231–1234
7. Singh R, Thakur P, Thakur A, Kumar H, Chawla P, Rohit JV, Kaushik R, Kumar N (2021) Colorimetric sensing approaches of surface-modified gold and silver nanoparticles for detection of residual pesticides: a review. Int J Environ Anal Chem 101(15):3006–3022
8. Sekhon BS (2014) Nanotechnology in agri-food production: an overview. Nanotechnol Sci Appl 7:31–53
9. Khin MM, Nair AS, Babu VJ, Murugan R, Ramakrishna S (2012) A review on nanomaterials for environmental remediation. Energy Environ Sci 5:8075–8109
10. Khota LR, Sankarana S, Majaa JM, Ehsania R, Schuster EW (2012) Applications of nanomaterials in agricultural production and crop protection: a review. Crop Prot 35:64–70

11. Bhattacharyya A, Duraisamy P, Govindarajan M, Buhroo AA, Prasad R (2016) Nano-biofungicides: emerging trend in insect pest control. In: Prasad R (ed) Advances and applications through fungal nanobiotechnology. Springer, Cham, pp 307–319
12. Nuruzzaman M, Rahman MM, Liu Y, Naidu R (2016) Nanoencapsulation, nano-guard for pesticides: a new window for safe application. J Agric Food Chem 64:1447–1483
13. Abbas SS, Haneef M, Lohani M, Tabassum H, Khan AF (2016) Nanomaterials used as a plants growth enhancer: an update. Int J Pharm Sci Rev Res 5:17–23
14. Perlatti B, Bergo PLS, Silva MFG, Fernandes JB, Forim MR (2013) Polymeric nanoparticle-based insecticides: a controlled release purpose for agrochemicals. In: Trdan S (ed) Insecticides-development of safer and more effective technologies. InTech, Rijeka, pp 523–550
15. Thakur A, Thakur P, Khurana SP (eds) (2022) Synthesis and applications of nanoparticles. Springer, Singapore pp 1–554
16. Prasad R, Kumar V, Prasad KS (2014) Nanotechnology in sustainable agriculture: present concerns and future aspects. Afr J Biotechnol 13:705–713
17. Marzbani P, Afrouzi YM, Omidvar A (2015) The effect of nano-zinc oxide on particleboard decay resistance. Maderas Cienc Tecnol 17:63–68
18. Subramanian KS, Manikandan A, Thirunavukkarasu M, Rahale CS (2015) Nano-fertilizers for balanced crop nutrition. Springer, Berlin, pp 69–80
19. Solanki P, Bhargava A, Chhipa H, Jain N, Panwar J (2015) Nano-fertilizers and their smart delivery system. Springer, Berlin, pp 81–101
20. Shukla, P. J., Chaurasia, P., Younis, K., Qadri, O. S., Faridi, S. A., & Srivastava, G. (2019). Nanotechnology in sustainable agriculture: studies from seed priming to post-harvest management. Nanotechnology for Environmental Engineering, 4(1). https://doi.org/10.1007/s41204-019-0058-2
21. Liu C, Zhou H, Zhou J (2021) The applications of nanotechnology in crop production. Molecules 26(23):7070
22. Lu C, Zhang C, Wen J, Wu G, Tao MX (2002) Research of the effect of nanometer materials on germination and growth enhancement of Glycine max and its mechanism. Soybean Sci 2002(21):168–171
23. Banerjee J, Kole C (2016) (2016). Plant nanotechnology: an overview on concepts, strategies, and tools. In: Kole C, Kumar D, Khodakovskaya M (eds) Plant nanotechnology. Springer, Cham, pp 1–14
24. Vermeulen SJ, Aggarwal PK, Ainslie A, Angelone C, Campbell BC, Challinor AJ, Hansen JD, Ingram J, Jarvis A, Kristjanson PM, Lau CB, Nelson GC, Thornton PK, Wollenberg EK (2012) Options for support to agriculture and food security under climate change. Sci Direct 15(1): 136–144
25. Haghighi, M., Afifipour, Z., & Mozafarian, M. (2012). The Effect of N-Si on Tomato Seed Germination under Salinity Levels. Journal of Biological and Environmental Sciences, 6.
26. Torabian S, Zahedi M, Khoshgoftar AH (2017) Effects of foliar spray of nano-particles of FeSO4 on the growth and ion content of sunflower under saline condition. J Plant Nutr 40(5): 615–623
27. Abdel-Aziz H, Hasaneen MNA, Omer AM (2016) Nano chitosan-NPK fertilizer enhances the growth and productivity of wheat plants grown in sandy soil. Span J Agric Res 14(1):e0902
28. Wang S, Wang F, Gao S (2015) Foliar application with nano-silicon alleviates cd toxicity in rice seedlings. Environ Sci Pollut Res 22(4):2837–2845
29. Sharma P, Sharma A, Sharma M, Bhalla N, Estrela P, Jain A, Thakur P, Thakur A (2021) Nanomaterial fungicides: in vitro and in vivo antimycotic activity of cobalt and nickel nanoferrites on phytopathogenic fungi. Glob Chall 1:1700041
30. Tudi M, Daniel RH, Wang L, Lyu J, Sadler R, Connell D, Chu C, Phung DT (2021) Agriculture development, pesticide application and its impact on the environment. Int J Environ Res Public Health 18(3):1112. PMID: 33513796; PMCID: PMC7908628
31. Afsharinejad A, Davy A, Jennings B, Brennan C (2016) Performance analysis of plant monitoring nanosensor networks at THz frequencies. IEEE Internet Things J 2016(3):59–69

32. Singh R, Singh R, Singh D, Mani JK, Karwasra SS, Beniwal MS (2010) Effect of weather parameters on Karnal bunt disease in wheat in Karnal region of Haryana. J Agrometeorol 2010(12):99–101
33. Pal K, Khurana SMP, Debnath N, Das S (2022) Micronutrient nanoparticles: synthesis, properties and application in agriculture. In: Thakur A, Thakur P, Khurana SP (eds) Synthesis and applications of nanoparticles. Springer, Singapore
34. Thakur P, Chahar D, Taneja S, Bhalla N, Thakur A (2020) A review on MnZn ferrites: Synthesis, characterization and applications. Ceramics International 46:15740-15763. https://doi.org/10.1016/j.ceramint.2020.03.287
35. Ezhilarasi PN, Karthik P, Chhanwal N, Anandharamakrishnan C (2013) Nanoencapsulation techniques for food bioactive components: a review. Food Bioprocess Technol 6:628–647
36. Anton N, Vandamme TF (2011) Nano-emulsions and micro-emulsions: clarifications of the critical differences. Pharm Res 28:978–985
37. Gupta A, Eral HB, Hatton TA, Doyle PS (2016) Nanoemulsions: formation, properties and applications. Soft Matter 12:2826–2841
38. Tadros TF, Izquierdo P, Esquena J, Solans C (2004) Formation and stability of nanoemulsions. Adv Colloid Interf Sci 108-109:303–318
39. Singh T, Shukla S, Kumar P, Wahla V, Bajpai VK (2017) Application of nanotechnology in food science: perception and overview. Front Microbiol 8:1501. Erratum in: Front Microbiol. 2017 Dec 12;8:2517. PMID: 28824605; PMCID: PMC5545585
40. Pradhan N, Singh S, Ojha N, Srivastava A, Barla A, Rai V et al (2015) Facets of nanotechnology as seen in food processing, packaging, and preservation industry. Biomed Res Int 2015: 365672
41. Nakagawa K (2014) Nano- and micro-encapsulation of flavor in food systems. In: Kwak H-S (ed) Nano- and microencapsulation for foods, chap 10. Wiley, Oxford, pp 249–272
42. Zhang T, Lv C, Chen L, Bai G, Zhao G, Xu C (2014) Encapsulation of anthocyanin molecules within a ferritin nanocage increases their stability and cell uptake efficiency. Food Res Int 62: 183–192
43. Yang R, Zhou Z, Sun G, Gao Y, Xu J, Strappe P et al (2015) Synthesis of homogeneous protein-stabilized rutin nanodispersions by reversible assembly of soybean (Glycine max) seed ferritin. RSC Adv 5:31533–31540
44. Koo OM, Rubinstein I, Onyuksel H (2005) Role of nanotechnology in targeted drug delivery and imaging: a concise review. Nanomed Nanotechnol Biol Med 1:193–212
45. Langer R, Peppas NA (2003) Advances in biomaterials, drug delivery, and bionanotechnology. AICHE J 49:2990–3006
46. Bradley EL, Castle L, Chaudhry Q (2011) Applications of nanomaterials in food packaging with a consideration of opportunities for developing countries. Trends Food Sci Technol 22: 603–610
47. Sari P, Mann B, Kumar R, Singh RRB, Sharma R, Bhardwaj M et al (2015) Preparation and characterization of nanoemulsion encapsulating curcumin. Food Hydrocol 43:540–546
48. Thakur A, Sharma N, Bhatti M, Sharma M, Trukhanov AV, Trukhanov SV, Panina LV, Astapovich KV, Thakur P (2021) Synthesis of barium ferrite nanoparticles using rhizome extract of Acorus Calamus: characterization and its efficacy against different plant phytopathogenic fungi. Nano Struct Nano Objects 24:100599
49. Mihindukulasuriya SDF, Lim LT (2014) Nanotechnology development in food packaging: a review. Trends Food Sci Technol 40:149–167
50. Othman SH (2014) Bio-nanocomposite materials for food packaging applications: types of biopolymer and nano-sized filler. Agric Agric Sci Proc 2:296–303
51. Pathania et al (2019) Development of tungsten doped Ni-Zn nano-ferrites with fast response and recovery time for hydrogen gas sensing application. Results in Physics 15: 102531. https://doi.org/10.1016/j.rinp.2019.102531
52. Jianrong C, Yuqing M, Nongyue H, Xiaohua W, Sijiao L (2004) Nanotechnology and biosensors. Biotechnol Adv 22:505–518

53. Bradley EL et al (2011) Applications of nanomaterial in food packaging with a consideration of opportunities for developing countries. Trends Food Sci Technol 22:604–610
54. Llorens A et al (2012) Metallic-based micro- and nanocomposites in food contact materials and active food packaging. Trends Food Sci Technol 24:19–20
55. Pyrgiotakis G et al (2015) Inactivation of foodborne microorganisms using engineered water nanostructures (EWNS). Environ Sci Technol 49(6):3737–3745
56. Kumari S et al (2023) Nano Ca–Mg–Zn ferrites as tuneable photocatalyst for UV light-induced degradation of rhodamine B dye and antimicrobial behavior for water purification. Cer Int 49: 12469-12480. https://doi.org/10.1016/j.ceramint.2022.12.107
57. Wang Y, Nowack B (2018) Environmental risk assessment of engineered nano-SiO2, nano iron oxides, nano-CeO2, nano-Al2O3 and quantum dots. Environ Toxicol Chem 37:1387–1395
58. Ndlovu N, Mayaya T, Muitire C, Munyengwa N (2020) Nanotechnology applications in crop production and food systems. Int J Plant Breed Crop Sci 7(1):603–613
59. Khanra A, Sangam S, Shakeel A, Suhag D, Mistry S, Rai MP, Chakrabarti SK, Mukherjee M (2017) Sustainable growth and lipid production from Chlorella pyrenoidosa using N-doped carbon Nanosheets: unravelling the role of graphitic nitrogen, vol 6. ACS Sustain Chem Eng, p 774
60. Singh T, Shukla S, Kumar P, Wahla V, Bajpai VK, Rather IA (2017) Application of nanotechnology in food science: perception and overview. Front Microbiol 8:1501
61. Nakagawa K (2014) Nano- and microencapsulation of flavor in food systems. In: Kwak H-S (ed) Nano- and microencapsulation for foods. Wiley, Oxford, pp 249–272
62. Wang Y, Kalinina A, Sun T, Nowack B (2016) Probabilistic modeling of the flows and environmental risks of nano-silica. Sci Total Environ 545–546:67–76
63. Peters R, Kramer E, Oomen AG, Herrera Rivera ZE, Oegema G, Tromp PC, Fokkink R, Rietveld A, Marvin HJ, Weigel S, Peijnenburg AA, Bouwmeester H (2012) Presence of nano-sized silica during in vitro digestion of foods containing silica as a food additive. ACS Nano 6:2441–2451
64. Athinarayanan J, Periasamy VS, Alsaif MA, Al-Warthan AA, Alshatwi AA (2014) Presence of nanosilica (E551) in commercial food products: TNF-mediated oxidative stress and altered cell cycle progression in human lung fibroblast cells. Cell Biol Toxicol 30:89–100
65. U.S. FDA (2020) CFR e Code of Federal Regulations Title 21. Food and Drug. Chapter I. Food and Drug Administration, Department of Health and Human Services. Part 73. Listing of color additives exempt from certification. Subpart A. Foods. Sec. 73.575. Titanium dioxide
66. Giubilato E, Cazzagon V, Amorim MJB, Blosi M, Bouillard J, Bouwmeester H, Costa AL, Fadeel B, Fernandes TF, Fito C, Hauser M, Marcomini A, Nowack B, Pizzol L, Powell L, Prina-Mello A, Sarimveis H, Scott-Fordsmand JJ, Semenzin E, Stahlmecke B, Stone V, Vignes A, Wilkins T, Zabeo A, Tran L, Hristozov D (2020) Risk management framework for nano-biomaterials used in medical devices and advanced therapy medicinal products. Materials 13:4532

Nano/Mesoporous Materials as State-of-the-Art Military Applications

19

Larrisa V. Panina, Atul Thakur, and Preeti Thakur

Abstract

Nanotechnology is rapidly gaining prominence across various fields, ranging from medicine to military applications. Considered as the forefront of innovation, nanotechnology holds immense potential for military use, with a multitude of applications on the horizon. The deployment of miniaturized military systems would offer a substantial strategic advantage against adversaries. Examples include miniature drones or swarms of artificial bees, which enhance battlespace awareness and situational visibility. Moreover, employing miniaturized AI-equipped bots at the frontline provides a comprehensive understanding of evolving battlefields. As nanotechnology progresses, it will unlock the development of a new class of lethal weapons, reshaping the geopolitical landscape. It is imperative for the world to acknowledge and embrace the advanced capabilities of nanotechnology, integrating them into contemporary warfare strategies.

Keywords

Military applications · Nanotechnology · Miniaturized drones · Battlespace

L. V. Panina
National University of Science and Technology, MISiS, Moscow, Russia

Immanuel Kant Baltic Federal University, Kaliningrad, Russia

A. Thakur (✉)
Amity Centre for Nanotechnology, Amity University Haryana, Gurgaon, Haryana, India
e-mail: athakur1@ggn.amity.edu

P. Thakur
Amity School of Applied Sciences, Amity University, Gurugram, Haryana, India

D. Suhag et al. (eds.), *Integrated Nanomaterials and their Applications*,
https://doi.org/10.1007/978-981-99-6105-4_19

19.1 Introduction

Nanotechnology research and development has a substantial influence on military materials and systems (including national defense and homeland security) [1–4]. The United States Department of Defense (DoD) is a key research funder, contributing a considerable share of federal funding to nanotechnology research, which includes materials research, electronic device research, and bio-chemical protection research [1, 2]. The overall Department of Defense expenditure in nanotechnology has varied but stayed substantial (hundreds of millions of dollars) since the formation of the National Nanotechnology Institute in 2000. The Department of Defense recognized nanotechnology as one of the six "Strategic Research Areas" in the mid-1990s. The Department of Defense nanotechnology program is split into seven program component areas that correlate to the National Nanotechnology Initiative (NNI) of the USA (Fig. 19.1).

Medicine, biological and chemical sensors, explosives, electronics for computers, power production and storage, and structural materials for vehicles, coatings, filters, and textiles are all military uses [3, 4]. Many of these uses are also useful for non-military purposes. Although the vast majority of military nanotechnology research should enhance our lives, there may be certain hazards linked with these

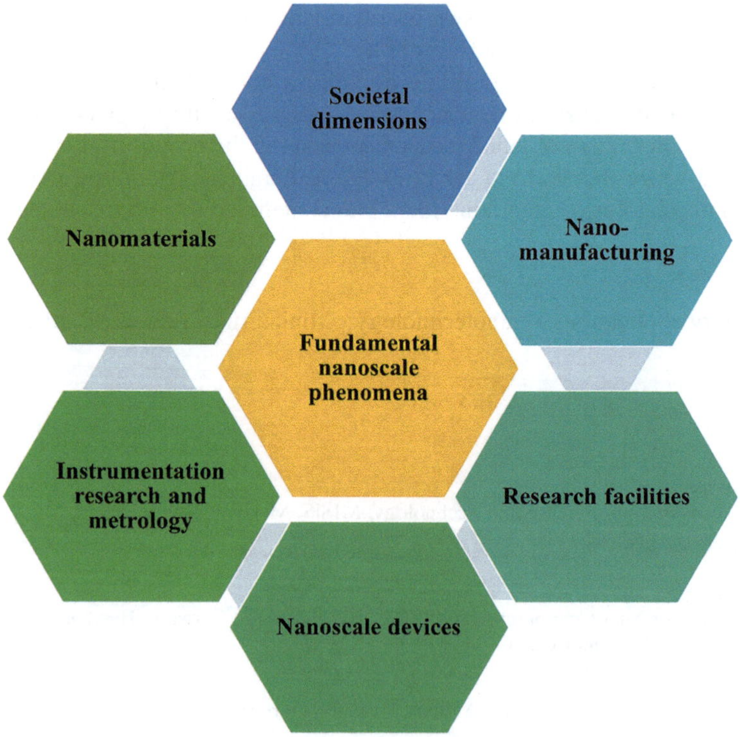

Fig. 19.1 The DoD Nanotechnology Program's seven distinct areas for nanotechnology

created nanoparticles, which can potentially harm the environment. Military activity, for example, can cause explosions, and high-tech weaponry blasts can emit poisonous nanoparticles. Furthermore, the widespread usage of nanotech sensors may have an influence on the environment when the sensors decay and designed nanoparticles seep into the soil. Nanotechnology has the potential to benefit the military in several ways. Advances in material science have led to the development of exceptionally strong armor materials like tungsten and carbon nanotubes, enhancing ballistic protection in applications such as personal body armor, bulletproof vests, and protective enclosures. The pursuit of invisibility technmology remains an active area of research, aiming to manipulate electomagnetic waves to render objects effectively invisible through innovative techniques like metamaterials and cloaking devices [5, 6]. Scientists are experimenting with light manipulation to make soldiers appear to vanish. When "electrochromic camouflage" is utilized, fabric colors may change rapidly to fit in with their environment.

19.2 Origin

Nanotechnology has historically been used rapidly and extensively in warfare and defense. Several countries, including China, the UK, Russia, and, most notably, the USA, have funded military applications of this technology over the last two decades. The US government has long been regarded as a national leader in research and development in this field, but it is now being challenged by international competition as the importance of nanotechnology grows. As this sphere expands, it will have a dominant platform at the forefront of military interests in the use, or misuse, of its power [7].

19.2.1 United States National Nanotechnology Initiative

In order to steer funds toward the advancement of nanoscience and nanotechnology, with a focus on realizing the promise of nano-weapons, the US government launched the National Nanotechnology Initiative in 2000. In addition to coordinating the employment of nanotechnology across all military branches, including the Air Force, Army, and Navy, this initial US concept has now been broadened. The US government invested over $19.4 billion in nanoscience between 2001 and 2014, as well as in the creation and production of nano-weapons for military defense. According to the twenty-first Century Nanotechnology Research and Development Act of 2003, national cooperation, productivity, and competitiveness will help the USA continue its leadership in the field of nanotechnology.

Around 60 nations worldwide developed national nanotechnology programs between 2001 and 2004. Research and development in this field "has now become a socio-economic target in an area of strong international collaboration and competitiveness," claims R.D. Shelton, a global technology assessor. Data as of 2017 showed that the USA alone had published 4725 patents in the USPTO, preserving their position as the industry leader in nanotechnology for more than 20 years. The most recent military nanotechnological weaponry research is the development of defensive military equipment with the aim of improving existing designs of light, flexible, and resilient materials. Through the use of sensors and the modification of electromechanical properties, these cutting-edge designs are endowed with capabilities that also improve offensive tactics.

19.3 Nanotechnology and the Military

One-dimensional nanoscale formations, such as thin films and sheets, share similarities with two-dimensional structures like carbon nanotubes [8]. Additionally, three-dimensional objects at the nanoscale are also possible, such as nano-porous structures where the bulk material remains macro-sized while the cavities are nano-sized. Examples of such structures include nano sponges and nano-porous membranes. It is crucial to recognize that nanotechnology encompasses more than just manipulating the behavior of matter at the micro- and macro-levels. At the nanoscale, the fundamental building blocks of both matter and life exist. It is within this realm that cellular functions, genetic reproduction, molecular chemistry, and cutting-edge electronics all converge [9]. Significant new advancements in a wide range of sectors may be possible if it is possible to comprehend how these processes function and to alter occurrences at this level to achieve certain results. This includes biotechnology, electronics, and material science. The properties of matter in a larger form can be drastically changed by nanostructures due to their small size. Because the surface area of nanomaterials is increased, surface characteristics may become dominant and increase their reactivity.

At the nanoscale, quantum physics plays a crucial role, and as a result, material properties—including those of individual atoms and bulk matter—can vary greatly and exhibit traits that are not visible at higher sizes. To give an example, gold nanoparticles can be orange, purple, red, or green in color [10]. Similar to this, carbon atoms in the form of carbon nanotubes can exhibit tensile strengths 100 times greater than those of steel and, depending on their arrangement, can either be metallic or semi-conducting. For instance, titanium dioxide and zinc oxide are both frequently found in sunscreen. When reduced to the nanoscale, both particles, which are composed of macro-particles, appear translucent. Nanotechnology can envision new goods, tools, and solutions to address pressing global issues. When it comes to military platforms, equipment, weapons, and missiles—where weight reduction is crucial and material quality is crucial—nano characteristics have a significant impact. It has several military uses and can remove many limitations by lessening its weight and bulk [11].

That might improve our ability to battle, especially in high-altitude and mountainous places, by allowing us to move large amounts of equipment in response to emergencies. Undoubtedly, nanotechnology helps the globe to overcome many obstacles. There are numerous products that may be developed to simplify living. For instance, solar cells might be printed on paper and smaller-sized medications may be constructed to improve distribution. Electricity-free water filters could also be developed. The use of nanotechnology to reduce weight and volume would have numerous civil and military uses [12].

19.4 Mesoporous Nanomaterials for Military Applications

A number of fields, including optics, catalysis, drug delivery systems, coatings, cosmetics, bio-separation, diagnostics, gas separation, and the military, heavily rely on materials having particularly specified porous characteristics on the nanoscale [13]. Nano-porous materials have void regions that can take on cylindrical or cage-like shapes within an amorphous or crystalline structure. These materials can be classified into three main types based on their pore size: microporous, mesoporous, and macro-porous [14]. Microporous materials, such as MOFs (metal-organic frameworks), zeolites, carbons, and amorphous glasses, have incredibly small pore size distributions ranging from 0.5 to 2 nm [15]. These chemicals have good thermal stability and catalytic activity that are helpful in the cracking processes, as well as the capacity to serve as ion exchange media, drying agents, and gas separation materials. Metal organic frameworks (MOFs) are one of the subtypes of microporous solids with the quickest development [16].

Zeolites and similar crystalline molecular sieves have an inherent limitation on the size and accessibility of their pores because of the pore templates available for their production. Conversely, macro-porous substances, such as porous polymer beads, which contain pores with a size range of 50–1000 nm, provide for easy access to the interior pores at the price of selectivity. These restrictions led to the development of mesoporous materials, which have holes that range in size from 2 to 50 nm [17].

19.4.1 Mesoporous Materials Have the Following Major Benefits

1. Large surface areas (>500 m^2/g) and narrow pore size distributions.
2. Framework/wall replacements with different metal oxides (MO_2), such as titania, alumina, and silica.
3. Simple organic functionalization techniques.
4. Low toxicity and biocompatibility.

19.4.2 Mesoporous Material's Properties and Characterization

It is possible to classify ordered mesoporous materials based on their size and pore shape, including whether they are cylindrical, cage-like (2D or 3D), or cylindrical. Examples of cylindrical structures with uniform pore diameter that show promise for usage in catalysis, adsorption, and as drug administration systems are MCM-48, AMS-6 (Iad), MCM-41, SBA-15, and NFM-1 (p6mm). In contrast, cage-type meso-caged solids like FDU-1(Imm), SBA-1(Pmn), and AMS-8(Fdm) are constructed from spherical or elliptical cages that are joined in three dimensions by microscopic cage-connecting windows. The mass transfer of active compounds may be controlled using these materials. One of the most versatile classes of mesoporous materials is porous silicas, which combine well-defined pore sizes with silica's well-known biocompatibility in a number of applications. An essential quality of a porous silica material is its ability to integrate functional organics (R) on the silica wall. Porous silica that has undergone calcination has a high density of surface silanol (Si-OH) groups. When these silanols react with other silanes, different functional groups (Si-R) are formed on the silica framework, and these groups can be employed to conjugate different molecules of interest [18].

19.5 Military Applications of Nanotechnology

As previously said, several countries are aware that applications relating to military are significantly impacted by nanotechnology. Nanotechnology can influence combat technology in a variety of ways. In order to enable mobility over all sorts of terrain, including mountains, all types of war-like equipment, weapon platforms, and missiles may be constructed from lighter, harder, heat-resistant nanomaterial. The military's use of nanotechnology will give troops on the ground stronger protection, more lethality, more endurance, and superior repair capabilities in a combat situation.

19.5.1 Areas Impacted

The automotive industry will be at the forefront of direct nanotechnology impact across all vehicle classes. The primary objective is to achieve overall weight reduction, enhance fuel efficiency, and improve ergonomic features. This encompasses various vehicles, including tanks, infantry combat vehicles, gun towers, engineer vehicles, and cargo transports. Additionally, nanotechnology contributes to the reinforcement of automobiles, providing increased strength and safety in the event of collisions. MXene is found to be exceptional for EMI shielding [19]. With nanotechnology, the military may concentrate on five different fields. These are listed below:

1. Focus on the combat soldier,
2. Ability to introduce nanotechnology in information dominance,
3. Weapons,
4. As discussed above, vehicles and platforms,
5. Logistics.

The combat soldier worries about the weight of his gear. Additionally, his sensors might accomplish the same thing with less electrical power. Future Indian Army soldiers will wear the barest minimum of protective gear and will be equipped with a Battlefield Management System on their wrists that will provide them access to all the information they want, including the common operating picture. In addition, his weapon would be lighter and he would be undetectable by infra-red (IR) technology. Thanks to nanotechnology combined with micro- or macro-fiber, future troops will be outfitted with an all-impact suit that provides protection against bullets, grenade fragments, biological agents, and chemical agents. Both his weapon and ammo would be lightweight and precise. He will have a lightweight smart helmet as well as a solar system for preserving food and water.

In terms of sensors and other communication devices, nanoelectronics will fundamentally lead to higher technology and quality. Microchips used in the electronics sector will consume less electricity as a consequence. Furthermore, it would lead to quicker processing and less noise. Future nano-electronic technology will have the capacity for cognitive functions, allowing computers to carry out higher-level human jobs. These will increasingly increase, allowing robots to carry out command duties more effectively. These changes would have a big impact on the Indian Army, which has a good monitoring system.

Nanomaterials will provide smoother energy release control and shorter diffusion paths for high explosive ammunition bursts. As a result, the total effect at the target end will be improved. Additionally, ammunition containing nanoparticles can improve the penetration of kinetic rounds into objects, particularly armor. To make the material stealthier, the architecture can be made more flexible to scatter light that hits these particles. The effects of nanotechnology would result in lighter-weight frameworks for guns, rifles, and automated shooting systems over the next 10 years. Other armed forces utilize nanotechnology to alter the bullet's trajectory while firing to account for body vibration. Whether they are tanks or other ground vehicles, future platforms should be lighter so that more of them may be loaded onto an IL-76, C-130, or Globe Master transport aircraft. The tank should move more quickly and be deadlier. The long-term objective is to fabricate thin, light nano-composite panels to cover exposed sections of tanks, infantry combat vehicles, and other vehicles. Nanoparticles' capacity to scatter light and infrared would enable stealth, and electronic systems and other components would also be lightweight. In a similar vein, autonomous vehicles built on nanomaterials might be more effective in a combat environment.

Future warfare will be heavily dependent on logistics. For the Indian Army, supply chain management is crucial to logistics. This would make it simpler to get gasoline, food, water, and ammunition supplies to the required locations at the proper times during field operations. Such procedures necessitate the use of modular containers. The military may transfer these in a variety of sizes to the places where they are needed. Each of these containers would be equipped with RFID (radiofrequency identification) tags to facilitate distribution, transportation, and collection. Thanks to nanotechnology, these containers would be lightweight and simple for transport planes and utility helicopters to move. All of these elements would make supply distribution on the battlefield simpler when combined with artificial intelligence. Nanoferrites have been found quite useful as RADAR absorbing materials [20].

All of the main armed forces in the world are actively working to create materials and systems that include nanotechnology. At the moment, nanotechnology research is focused on improving medical facilities and creating powerful, light, and multi-functional materials that may be utilized as armor to give protection and enhance communication in a network-centric military environment. Nanotechnology undoubtedly provides the military with a wide range of fresh options.

19.6 The Following Are a Few Domains Where Nanotechnology Can Be Used Extensively

19.6.1 Electronics and Computers

Electronics and computers will get much smaller, quicker, and less power-hungry with the use of NT [21–23]. These systems, which will be improved by higher levels of artificial intelligence, will be used by the military as a whole and even built into extremely small components (such as guns, eyewear, uniforms, mini- and microrobots, and explosives). Conversely, massive battle-management and strategy-planning systems will have many levels and a high degree of autonomous decision-making. They would establish a pervasive network that could be used, for example, to logistics and be utilized in conjunction with sensors, wireless communication tools, and mall light displays (Fig. 19.2).

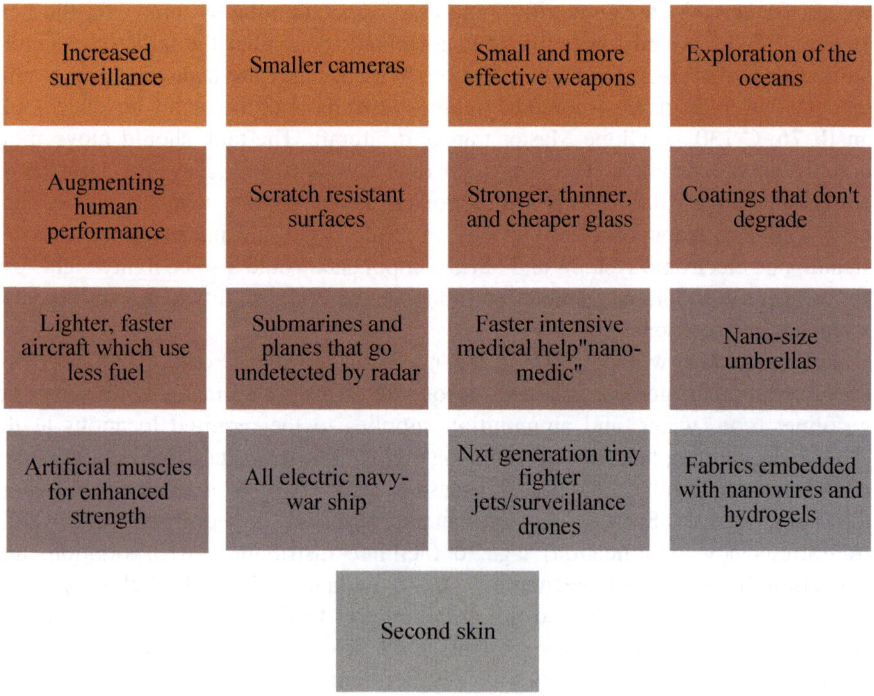

Fig. 19.2 Potential applications of nanotechnology in Military

19.6.2 Smaller Sensors

Nanotechnology allows for smaller sensors, often down to micrometer sizes rather than centimeter levels, however in many applications, signal strength relies on the size of the receiving component, possibly limiting the capacity to shrink the entire system. There might be a reduced size constraint due to the power supply or the communication antenna. However, low-cost mass production may enable the placement of tens of thousands of sensors in a single area [24].

19.6.3 Lighter, Stronger, and Heat-Resistant Materials

Lighter, stronger, and heat-resistant materials will allow conventional land, sea, and air vehicles to move more quickly and nimbly. New designs for smaller cars with more potent engines, though, may also be made viable. With lighter energy-storage and conversion technologies, such as fuel cells with membranes constructed of nanoparticles or hydrogen tanks based on nanotubes or fullerenes, all-electric military vehicles, including electromagnetic guns, may become practical. Small systems may exhibit muscle-like action due to materials deforming or contracting molecules. Applying moveable pigment particles to surfaces with locally variable color can be used to camouflage an object. Surfaces having the ability to specifically absorb electromagnetic radiation can reduce the amount of radar and infrared fingerprints. Nanostructured materials can make stronger light armor, yet it is uncertain if they would be superior to heavy armor. Wearers of nanofiber-based clothing can more successfully protect themselves from gunfire.

19.6.4 Advance Materials for Weapons and Explosives

Modern explosives and materials allowed conventional weapons to fire with long ranges and less mass. Precision in targeting could experience substantial enhancements through the integration of nanotube-based guidance systems. this holds true not only for conventional weaponry but also extents to missile technology, where nano-tube based systems have the potential to greatly improve accuracy. But it is uncertain whether NT-designed materials and structures will outperform the characteristics of tungsten or (depleted) uranium in penetrating projectiles. Also, it is unclear whether NT-based armor will defeat NT penetrators or vice versa.

19.6.5 Smart Systems Worn by Soldiers

They could detect the wearer's health and either release medications or compress wounds using smart materials. The human body movements can produce the energy needed for communication.

19.6.6 Autonomous Mini- and Micro-Robots

With NT, autonomous mini- and microrobot size might be drastically decreased to less than 0.1 mm. These could use either conventional propulsion principles (wheel, track, propeller, wing, jet, rotor) or biomimetic ones (using legs, wriggling like a snake, hopping, using flapping wings, etc.) that are familiar from more sophisticated technical systems to move in all media, including on land, in water, and in the air. A broad variety of functions may be accomplished by mini/microrobots, including as reconnaissance, communication with bigger weapons, locating targets for small-scale actuation, and producing weapons. Even while mobility might be made easier, say by the wind, or even if they were initially carried by grenades, rockets, etc., their speed and range would be limited. Sensing, communication, and actuation skills would all suffer as size shrank. Close quarters or broad use, however, could compensate for such limitations. One scenario may include the robot approaching a central node while inside a target object or person, where it might listen in, change, or destroy. Acting in big groups, blocking windows and air intakes, abrading mechanisms, etc. are further options.

Biological–technological hybrids such as insects or tiny animals that are controlled by nerve/brain electrodes could potentially achieve comparable objectives within the realm of miniaturized and microscale robotics. Consequently they can be classified as part of the mini/microrobot category.

19.6.7 Smart Systems Implanted into the Human Body

Nanotechnology offers greater potential for implanted systems in human bodies, surpassing the capabilities of microsystems technology. One application involves monitoring the health and stress levels of soldiers and administering drugs or hormones as needed. Another application entails the use of electrodes connected to specific areas of the cerebral cortex, sensory organs, sensory nerves, motor nerves, or muscles. This technology could be employed to enhance the reflexes of pilots, for instance. However, achieving complex sensory perceptions or cognitive abilities requires significant advancements in brain research and a reduced ethical barrier for human testing.

19.6.8 For Military Uses of Outer Space

Smaller launch vehicles and notably smaller satellites are only two of the many alternatives that NT (with MST) will make possible for military space missions. Small satellites might be launched in swarms for radar, communication, or intelligence purposes. The system's massive effective antenna size would result in great target resolution, but the entire antenna and solar-cell area would dictate signal intensity. Small spacecraft may manipulate other satellites after rendezvous and docking or strike them directly with high relative velocities in order to damage or kill them. NT-enabled electromagnetic acceleration might also be used by kinetic-energy space weapons.

19.6.9 For Nuclear Weapons

Nuclear weapons will not cause the anticipated qualitative shift. NT will facilitate smaller guiding or safety/arming/fusing systems. However, this would not change a weapon's essential properties, such as the requirement for a critical mass of uranium or plutonium to trigger a fission explosion and produce a total mass of at least 70 kg. The exception would be hypothetical pure-fusion bombs without a fission primary, in which case the fusion fuel would be ignited at a micro-spot using MST/NT. It would be challenging to distinguish them from conventional weapons because of the yield, size, and mass potential arbitrarily low values. One idea is to release a little amount of antimatter onto lithium deuteride after carefully trapping it in a microtrap. These concepts should be thoroughly studied even if they are currently speculative.

19.6.10 Chemical and Biological Weapon

In contrast, NT will provide qualitatively fresh alternatives to chemical and biological weapons. Use of capsules that more securely house the active chemicals and only release them when necessary is one option. NT may enable the development of chemical or biological weapons that exclusively respond to particular protein or gene patterns, as well as genetic developments that conflate the two. Such weapons may be used to harm certain plant or animal breeds or to target specific ethnic groups or even lone individuals. NT can also be used to more readily infiltrate the body or its cells. It is possible to design mechanisms that limit or prevent damage to one's own forces, such as self-destruction after a certain period of time or reliable vaccination. All of these would lead to biological warfare being far more effective and manageable. Conversely, NT could make it possible for more sensitive sensors to be used with chemical or biological weapons. High active surface area NT materials can be used to bind and neutralize chemicals, decontaminate polluted regions, and make protective clothing.

Most applications mature over a period of 3–5 years. It is not necessary for all of them to demonstrate sufficient resistance to countermeasures, cost-effectiveness, technical feasibility, and commercial viability. When a mass market is envisaged, many will be developed concurrently or primarily in the civilian sector. Military uses in this case may be influenced by civilian technologies. In many perilous situations, however, the military will continue to take the lead, occasionally opening the door for later civilian uses. Numerous combat countermeasures that would heavily rely on NT itself are to be predicted given the wide spectrum of effects if the necessary technology is currently available. Armed forces might introduce more autonomous systems, actively safeguard them (using reactive surfaces to deter items from clinging to them, "guard" micro/nanorobots), or passively shield them (using molecular-scale sieves). Which is more effective, weapons or countermeasures, is not yet clear. However, there are no indications of defense dominance, making counterattack and preventative assault likely to be vital in armed conflict.

19.7 Potential Threats to Consumer, Worker, and Environmental Safety

It is essential that consumers and employees are provided the proper protection and safety when nanotechnology is used in the military (or any other context, for that matter), and it should not have any detrimental impacts on them. Although there are many environmental advantages to nanotechnology, there are also some disadvantages that must be addressed. One of these is the growing toxicological contamination of the environment as a result of some nanotechnology products' or nanomaterials' ambiguous size, shape, and chemical composition [25]. Nanoparticle production also requires energy-intensive techniques. Finally, the infrastructure for recovering and recycling harmful nanomaterials is still missing, and their release might harm the environment [26]. However, nanoparticles have large potential to be utilized for the remediation of environmental pollution especially water pollution [27, 28].

19.8 Possible Human Rights Issues

Nanotechnology and human rights issues include uneven access to technology's advantages, unequal distribution of its financial rewards, the technology's negative impacts on society and culture, privacy, and informed consent. All of these topics are intimately related to one another. As was already said, society may have concerns about nanotechnology, including potential risks to the environment, human health, and safety. All of these have a detrimental indirect effect on the economy, especially in less developed countries. There are also some short-term effects, such as the displacement of traditional sectors when nanotechnology-based goods take over. The development of better and more sophisticated surveillance systems due to nanotechnology poses privacy concerns. As with any new technology, it is critical to assess the benefits and drawbacks of nanotechnology research. Nanotechnology is unquestionably a boon for the military, and research conducted for military purposes may also be quite useful in civilian situations [29].

19.9 Conclusion

Modern, cutting-edge technology today permeates many facets of life and is not only the preserve of a small group of people [30]. It is important to be aware of the likelihood that should a country miss the next technological advancement, the enemy would still have it. This would have difficult-to-imagine consequences for the future of military warfare. It is essential to acquire this knowledge in order to guarantee the security of any state or country in the future. Furthermore, the versatility of nanotechnology has increased interest in this field of study.

With enormous potential benefits and equally significant potential risks, nanotechnology (NT) is predicted to bring about revolutionary advancements in a variety of sectors. There may be specific hazards as the military develops, and it will take specialist study and research to prevent them. In NT, the development of military science and technology is increasing. Such uses in warfare may become commonplace in the future. Risks to stability and arms control may specifically arise from new biological weapons and microrobots. Non-medical body implants present many problems for individuals as well as society as a whole, which might make them more acceptable through the military. More study is needed to discover the best defense against these dangers. For the short- and medium-term, a number of constraints and restrictions are suggested. As a starting step, greater transparency and international cooperation are required.

References

1. Ramsden JJ (2012) Nanotechnology for military applications. Nanotechnol Percept 8(2): 99–131
2. Chahar D, Thakur P, Sun AC, Thakur A (2023) Investigation of structural, electrical and magnetic properties of nickel substituted Co-Zn nanoferrites. J Mater Sci Mater Electron 34: 901. https://doi.org/10.1007/s10854-023-10273-5
3. Dhanda N, Thakur P, Kumar R, Fatima T, Hameed S, Slimani Y, Sun AD, Thakur A (2023) Green synthesis of Ni-Co nanoferrites using aloe vera extract: structural, optical, magnetic, antimicrobial studies. Appl Organomet Chem 2023:1–13. https://doi.org/10.1002/aoc.7110
4. Dhanda N, Thakur P, Sun AC, Thakur A (2023) Structural, optical, and magnetic properties along with antifungal activity of Ag-doped Ni-Co nanoferrites synthesized by eco-friendly route. J Magn Magn Mater 572:170598
5. Shipbaugh C (2006) Offense-defense aspects of nanotechnologies: a forecast of potential military applications. J Law Med Ethics 34(4):741–747
6. Wang S, Jiang W, Jiang W, Ye F, Mao Y, Xuan S, Gong X (2014) Multifunctional polymer composite with excellent shear stiffening performance and magnetorheological effect. J Mater Chem C 2(34):7133–7140
7. Sikder AK, Sikder N (2004) A review of advanced high performance, insensitive and thermally stable energetic materials emerging for military and space applications. J Hazard Mater 112(1–2):1–15
8. Kala D, Gupta S, Kaushal A (2022) Nanotechnology in healthcare. In: Thakur A, Thakur P, Khurana SP (eds) Synthesis and applications of nanoparticles. Springer, Singapore
9. Cameron DS, N. M. (2006) Nanotechnology and the human future: policy, ethics, and risk. Ann N Y Acad Sci 1093(1):280–300
10. Edwards E, Brantley C, Ruffin PB (2017) Overview of nanotechnology in military and aerospace applications. Wiley, Hoboken, pp 133–176
11. Abed MS, Jawad ZA (2022) Nanotechnology for defence applications. In: Nanotechnology for electronic applications. Springer, Singapore, pp 187–205
12. Altmann J (2007) Military nanotechnology: potential applications and preventive arms control. Routledge, New York

13. Ivanets A, Prozorovich V, Kouznetsova T (2022) Advancement in water purification by nanomaterials. In: Thakur A, Thakur P, Khurana SP (eds) Synthesis and applications of nanoparticles. Springer, Singapore
14. Wan Y, Zhao D (2007) On the controllable soft-templating approach to mesoporous silicates. Chem Rev 107(7):2821–2860
15. Thakur, A., Thakur, P., Khurana, S.P. (eds) Synthesis and Applications of Nanoparticles. Springer, Singapore. https://doi.org/10.1007/978-981-16-6819-7
16. Yaghi OM, Li G, Li H (1995) Selective binding and removal of guests in a microporous metal–organic framework. Nature 378(6558):703–706
17. Beck JT, Vartuli JC, Roth WJ, Leonowicz ME, Kresge CT, Schmitt KD, Chu CT, Olson DL, Sheppard E, McCullen S, Higgins J, Schlenker J (1992) A new family of mesoporous molecular sieves prepared with liquid crystal templates. J Am Chem Soc 114(27):10834–10843
18. Doadrio J, De Sousa EMB, Izquierdo-Barba I, Doadrio AL, Pérez-Pariente J, Vallet-Regí M (2006) Functionalization of mesoporous materials with long alkyl chains as a strategy for controlling drug delivery patterns. J Mater Chem 16(5):462–466
19. Verma R, Thakur P., Chauhan A, Jasrotia R., ThakurA (2023), A review on MXene and its' composites for electromagnetic interference (EMI) shielding applicationsCarbon208:170–190. https://doi.org/10.1016/j.carbon.2023.03.050
20. Rana K, Thakur P, Tomar M, Gupta P, Thakur A (2018) Investigation of cobalt substituted M-type barium ferrite synthesized via co-precipitation method for radar absorbing material in Ku-band (12–18 GHz), Ceramics International, 44(6): 6370–6375. https://doi.org/10.1016/j.ceramint.2018.01.028
21. Kumari S, Dhanda N, Thakur A, Gupta V, Singh S, Kumar R, Hameed S, Thakur A (2023) Nano Ca-Mg-Zn ferrite as tuneable photocatalysts for UV light induced degradation of rhodamine B dye and antimicrobial behaviour for water purification. Ceram Int 49(8):12469–12480
22. Verma R, Thakur P, Sun AC, Thakur A (2023) Investigation fo structural, microstructural and electrical characteristics of hydrothermally synthesized Li0.5-0.5xCoxFe2.5-0.5xO4 ($0.0 \leq x \leq 0.4$) ferrite nanoparticles. Phys B Condens Matter 661:414926. https://doi.org/10.1016/j.physb.2023.414926
23. Bharti MK, Gupta S, Chalia S, Garg I, Thakur P, Thakur A (2020) Potential of magnetic nanoferrites in removal of heavy metals from contaminated water: mini review. J Supercond Nov Magn 33:3657–3665
24. Kurahatti RV, Surendranathan AO, Kori SA, Singh N, Kumar AR, Srivastava S (2010) Defense applications of polymer nanocomposites. Def Sci J 60(5):561–563
25. Thakur P, Verma Y, Thakur A (2022) Toxicity of nanomaterials: an overview. In: Thakur A, Thakur P, Khurana SP (eds) Synthesis and applications of nanoparticles. Springer, Singapore. https://doi.org/10.1007/978-981-16-6819-7_25
26. Kosal M, Kosal M (2009) The need to Foster revolutionary science. In: Nanotechnology for chemical and biological defense. Springer, New York, pp 121–133
27. Verma R, Chauhan A, Kumari S, Jasrotia R, Ali A, Gopalakrishnan C, Kumar R, Ghotekar S (2023) Green synthesis of ZnO NPs using Timbur (*Zanthoxylum armatum* DC.) plant extract for antimicrobial and dye degradation applications. Chemical paper 1–11
28. Thakur A, Thakur P (2023) Ferrite nanoparticles for agriculture related activity. In: Applications of nanostructured ferrites. Woodhead Publishing Series in Electronic and Optical Materials pp 315–330. https://doi.org/10.1016/B978-0-443-18874-9.00005-9
29. Roco MC (2011) The long view of nanotechnology development: the National Nanotechnology Initiative at 10 years. J Nanopart Res 13:427–445
30. Chen CH, Woo EM (2018) Polymer science and technology. Chem Phys 15(7):2495–2506

Toxicology and Toxicity Studies of Nano-Biomaterials

20

Raksha Rathore, Deepa Suhag, Fayu Wan, Atul Thakur, and Preeti Thakur

Abstract

The subject of nanotechnology has significant changes in the last 10 years as new applications of nanoparticles (NPs) and other nano-biomaterials have been discovered. However, the toxicity of nanoparticles and their interactions with biological systems are still poorly understood. The toxicity of a nanoparticle is affected by its size, shape, surface area, and surface chemistry. Nanoparticles can be ingested, breathed, or absorbed via the skin. With greater manufacturing and deliberate (sunscreens, medicine delivery) or inadvertent (environmental, occupational) exposure to these particles, the chance of unfavorable health impacts will most likely grow. To ensure the risk-free and long-term implementation of nanotechnology, the possible health concerns of new nanomaterials must be

R. Rathore · D. Suhag
Amity Centre for Nanotechnology, Amity University Haryana, Gurgaon, Haryana, India
e-mail: dsuhag@ggn.amity.edu

F. Wan
School of Electronics and Information Engineering, Nanjing University of Information Science & Technology, Nanjing, China

Centre for Nanotechnology, Amity University Haryana, Gurugram, India

A. Thakur (✉)
Amity Centre for Nanotechnology, Amity University Haryana, Gurgaon, Haryana, India

School of Electronics and Information Engineering, Nanjing University of Information Science & Technology, Nanjing, China
e-mail: athakur1@ggn.amity.edu; 100108@nuist.edu.cn

P. Thakur
Amity Centre for Nanotechnology, Amity University Haryana, Gurgaon, Haryana, India

Amity School of Applied Sciences, Amity University, Gurugram, Haryana, India
e-mail: pthakur@ggn.amity.edu

D. Suhag et al. (eds.), *Integrated Nanomaterials and their Applications*,
https://doi.org/10.1007/978-981-99-6105-4_20

physiologically defined. As a result, there is increased scientific curiosity about the impacts of nanomaterials found in the surroundings of humans. Humans have been exposed to many nanoscale materials since infancy, and the recently developed area of nanotechnology has turned into yet another threat to human existence. NPs may penetrate the human body, get past numerous biological barriers, and potentially even reach the most sensitive organs because of their small size. Scientists have proposed suggestions that NPs smaller than 10 nm act like gases, are easily absorbed into human tissues, and might disrupt the normal biochemical environment of cells. NPs are more dangerous to human health than their large-sized chemical particle equivalents, it is commonly accepted that NP toxicities are inversely related to NP size. Scientists were concerned about the unpredictable health effects of NPs because of their unique physicochemical characteristics in many biological systems. As a result, it is believed that new methods of thinking that seek to contribute to the safe use of NPs are critical to narrow the knowledge gap and focus entirely on NP toxicity problems.

Keywords

Nanoparticles toxicity · Health risks · Nanoscale materials · Biochemical environment

20.1 Introduction

The exact methods via which NPs and biological systems interact are still unknown. The intricacy of the particles is increased by the fact that they can interact and connect with biological matter, as well as modify their surface features in response to their surroundings. A growing scientific understanding of the mechanisms governing NPs' interactions with cells suggests that cells can readily uptake NPs through either active or passive methods [1, 2].

The area of nanoscience known as nanotoxicology is concerned with the investigation of the harmful impact of nanomaterials or nanoparticles on living organisms. The safety of manufactured nanoparticles in humans is a serious problem as their usage in biological applications increases. Nanoparticles (NPs) are frequently utilized as drug carriers and in nanomedicine because of their tiny size and distinctive characteristics [3, 4]. Their size, morphology, surface functional groups, and dose-dependent characteristics may also contribute to their toxicity to normal, healthy human cells, tissues, and organs [5]. Numerous studies have demonstrated that biosynthesized nanoparticles with biocompatible surface functional groups are less hazardous to human cells than chemically produced NPs. This is because synthetic compounds are used as surface functional and capping agents in chemically synthesized materials [6, 7]. Contrarily, certain biosynthesized nanoparticles can become harmful when they interact with cells, break down into simpler shapes, or amass. The four types of nanomaterials that are present in the human environment are carbon-based nanomaterials, metal-based nanoparticles, dendrimers, and composites [8, 9]. Carbon-based nanomaterials like fullerenes and nanotubes are employed in films, coatings, and electronics. Metal-based nanomaterials are employed in food, cosmetics, and medicinal applications. Examples include nano-silver, nanogold, and metal oxides like titanium dioxide (TiO_2). There is currently little information available concerning the detrimental consequences of nanomaterial exposure on human health. They could be able to stay in the human environment and body for a longer amount of time than their bigger counterparts because of their tiny size and, in certain situations, superior stability. Some nanomaterials could be easier to breathe in, consume, or absorb via the skin. Additionally, like viruses, these NPs seem to be able to exploit an endocytotic mechanism to hijack an already-existing transport system across the body. As a result, any toxicological hazards must be handled carefully because it appears that broad translation of NPs throughout the body is possible if the body is exposed [10, 11].

20.2 Nanotoxicity: Are We Heading in the Right Direction?

The more important query is: How dangerous are nanomaterials at the conceivable amounts at which they may be used? Almost everything can be poisonous in high enough quantities. The underlying material, size, shape, and coatings will all have an impact on any potentially harmful consequences of nanoparticles. Nanotechnology is anticipated to form the foundation for many of the significant technological advancements of the twenty-first century [12–15]. The potential release of nanomaterials into the environment and consequent consequences on the health of ecosystems are becoming a rising worry that must be addressed given the rate at which they are being produced. Understanding the destiny and behavior of produced nanomaterials in the environment is a prerequisite for achieving this. A necessary step toward achieving this is understanding the fate and behavior of created nanomaterials in the environment. b) Do they affect the environment and human health differently than bigger particles of the same substance? Through both deliberate and accidental releases, such as air emissions and solid or liquid waste streams from production plants, nanomaterials are introduced into the environment. Additionally, the amount of nanomaterials that are used in paints, textiles, and personal care items like sunscreen and cosmetics that are released into the environment is inversely proportional to their utilization. Designed nanoparticles are so tiny, it is feasible to breathe in airborne particles made of nanomaterials with diameters ranging from a few nanometers to several micrometers. Nanomaterials can aggregate into larger particles or longer fiber chains, changing their characteristics and impacting how they behave in both indoor and outdoor settings as well as how they could be exposed to and enter the human body. Solid nanoscale particles can accidentally or purposely come into contact with skin. The stratum corneum, a thick layer of dead keratinized skin cells on the surface of the skin, prevents particles, ionic chemicals, and water-soluble compounds from passing through. Intentional cutaneous exposure to nanoscale compounds may occur when lotions, creams, wound dressings, detergents, and socks are used. Nanoscale ZnO and TiO_2 materials can be exposed as sunscreen ingredients or as coatings on fibrous materials to make them water- and stain-resistant. Dermal contact with human materials created during the creation or burning of nanoparticles may result in unintentional exposure. The ability of certain nanomaterials to permeate the skin and cause toxicological consequences such as cytotoxicity in the skin, or other organs, accumulation in the skin or metabolism to even smaller particles, or photoactivated nanoparticles is unknown. Numerous chemical classes, including metals, metal oxides, carbon, and semiconductor nanoparticles, can be used to categorize the wide range of nanotechnology goods [16, 16, 18].

20.3 Potential Routes of Human Exposure to Nanomaterials

Humans are exposed to nanoparticles primarily through ingestion, either directly through food or indirectly through food container nanoparticle disintegration or subsequent absorption of breathed particles. Once within the body, they may be carried by blood circulation throughout. Their size and surface characteristics, including polarity, hydrophilicity, lipophilicity, and catalytic activity, may have an impact on how they are distributed throughout the body. As particle size reduces, surface area per unit mass rises, hence it is believed that nanoparticles would have more chemical and biological activity within the body [19]. It is therefore hypothesized that smaller nanoparticles may be more hazardous than their bigger equivalents. It is generally accepted that cells absorb smaller nanoparticles more quickly than bigger ones. Inhalation of nanoparticles in the air might also expose people to them. Skin absorption is a significant method of human exposure to nanoparticles [20].

20.4 Nanomaterial Characterization

Determining the toxicity of nanoparticles requires accurate characterization of those materials. The quality of the dispersion of nanoparticles is influenced by the suspension medium, which may have an impact on the biological activity of the particles. The preparation and solubilization of the nanoparticles affect these characteristics. Nanomaterials must be dispersed stably with precise control over particle size and distribution before being tested for toxicity in an aqueous media. Their biological impact may be influenced by how they disperse in the medium. The degree of clustering of particles depends on the dispersion condition of the nanomaterial. The form, size, surface area, and closeness of the nanoparticles to one another all affect how they aggregate [21–23]. Their dispersion status may be impacted by the physiological and biological media frequently utilized in toxicity research. Therefore, it is necessary to measure the particle size distribution of both “as-received” and “as-dosed” (Table 20.1).

As cellular absorption plays a major role in determining nanomaterial toxicity, the protein corona may have a considerable effect on their toxicological profile.

Table 20.1 Possible risks of nanomaterials

Nanomaterials	Possible Risks
1. Silica nanoparticles, carbon nanomaterials	Pulmonary inflammation, granulomas, and fibrosis
2. Carbon, silver, and gold nanomaterials	Distribution into other organs including the central nervous system
3. Quantum dots, carbon and TiO_2 nanoparticles	Skin penetration
4. MnO_2, TiO_2, and carbon nanoparticles	May enter the brain through nasal epithelium olfactory neurons
5. TiO_2, Al_2O_3, carbon black, co, and Ni nanoparticles	May be more toxic than micron sized particles

For toxicity and characterization of nanoparticles in aqueous environments, there is no well-defined battery of experiments. Studies are carried out utilizing various nanoparticle sources and kinds, cell lines, mediums, tissues, animal models, and/or delivery techniques, according to a study of the current literature.

20.5 Nanomaterials' In Vitro Toxicity Assessment

In vitro toxicity tests are essential to examining the mechanistic impacts of nanomaterials on biological entities. The use of in vitro testing to complement and/or augment the more expensive and time-consuming in vivo animal studies has many benefits, including inexpensive costs, quick turnaround times for findings, and no ethical concern. To evaluate the cytotoxicity of different compounds and the effects of nanomaterials on cellular physiology, typical in vitro models include monolayer cell cultures, cocultures, multilayer cocultures, xenograft models, and tissue slices. Systems using human cells have the potential to do away with the requirement for interspecies extrapolation, enhancing testing effectiveness and minimizing the usage of animals. In vitro systems, however, have drawbacks. Due to restricted metabolic capacity, in vitro biotransformation of a substance may be significantly less than in vivo systems. The development of high-throughput, predictive, and mechanism-based tests is crucial and desirable for determining the potential toxicity of nanomaterials that humans are exposed to but for which there is less evidence of their toxicity.

20.6 Nanomaterials Made of Carbon

As its distinctive one-dimensional hollow nanostructure and peculiar capabilities, carbon nanotubes (CNTs) are emerging as an essential new class of multifunctional building blocks for the development of nanotechnology. The urgent necessity for large-scale manufacture of carbon nanotubes for application in commercial goods has been reaffirmed by the nanotechnology industry's recent fast achievements. In the upcoming years, as industrial scale facilities for the relatively inexpensive manufacturing of multi-walled carbon nanotubes (MWCNTs) proliferate, so will the exposure of professionals and the general public to MWCNTs. Several study teams have looked at the absorption of CNTs and their possible risks to people and other biological systems, especially MWCNTs. For instance, it has been demonstrated that CNTs cause human T-cells to exhibit inflammatory and apoptotic reactions. Through gene expression analysis, it was found that MWCNTs activated genes related to cellular transport, metabolism, cell cycle control, and stress response in human skin fibroblasts. Researchers found indications of cytotoxicity for carbon-based nanomaterials, although MWCNTs were the least hazardous of the carbon nanotubes, carbon nanofibers, and carbon nanoparticles studied [24, 25]. Water-soluble CNTs functionalized with polyethylene glycol chains did not have any harmful effects when tested in a variety of immune system cells.

According to several researches ultrafine carbon particles may traverse the blood–brain barrier to disrupt the central nervous system and can enter the lungs more easily than bigger particles. They found that carbon nanoparticles move from the lungs to the circulation rather than releasing clotting factors and that harmful effect appears shortly after exposure. Given that asbestosis, lung cancer, and malignant mesothelioma of the pleura are known to be caused by asbestos fiber inhalation, there seems to be a strong likelihood that CNTs, due to their structural similarities to asbestos, will likewise have major harmful effects on human health. According to several studies, carbon nanotubes show high cytotoxicity in vitro, including the ability to cause oxidative stress, halt cellular growth, and trigger apoptosis and necrosis.

20.6.1 CNTs and Phenomena at Bio-Interfaces

The carbon allotrope known as carbon nanotubes (CNTs) has special qualities that make it possible to employ it in nanotechnology-related fields including optics, electronics, and medicine. They stand out for their durability, electrical conductivity, and heat conductivity. The singularity of the atoms is due to their extraordinarily strong bonding structure, which also displays high extreme aspect ratios. Carbon nanotubes (CNTs) are divided into single-walled and multi-walled varieties depending on the number and configuration of their sidewalls. Exposure to these nanomaterials in the workplace and the general population has resulted from the usage of carbon nanotubes to improve the performance of various items, notably those used in healthcare. It has always been difficult to make carbon nanotubes biocompatible with biological systems. After being injected into the body, CNTs interact with organs and bodily fluids [26]. Numerous experimental data suggest that CNTs are harmful to different organ systems. The adsorption and transportation of CNTs in organisms via various ways of administration are used to assess the efficacy of CNTs as medication carriers. Following administration, CNTs are carried to many organs through the lymphatic or systemic circulation. Several methods, including oral, intravenous, inhalation, transdermal, subcutaneous, and intraperitoneal, are used to introduce CNTs into the body. The most often mentioned methods of administration in vivo are intratracheal and inhalation because of the more distinct and certain findings in the tests that have been reported. Compared to intravenous injection, oral, or topical delivery, breathing CNTs, according to some accounts, induces significant irritation [27]. CNTs interact with proteins or cells as they enter the body, which may cause metabolism or cause the CNTs to retain their original shape. They can also be sent to other bodily organs, where they may remain localized, move, or be expelled. Depending on their content, components, structure, size, and functionalization, CNT biodistribution in the body may produce varied levels of toxicity.

20.6.2 Quantum Dots

Nanocrystals called quantum dots (QDs) have 1000–100,000 atoms and show peculiar "quantum effects" including protracted fluorescence. Due to their distinctive optical and electrical characteristics, QDs are now utilized in the biomedical imaging and electronics sectors. One of QDs most advantageous characteristics is their wide range of fluorescence, which makes them the perfect fluorophores for biomedical imaging. To mark neoplastic cells, DNA, and cell membrane receptors, for instance, fluorescent QDs can be coupled with bioactive moieties to target particular biologic processes and cellular structures. Each form of QD has unique physicochemical characteristics that influence how harmful it could be [28]. The following factors have all been shown to be determining factors for QD toxicity: size, charge, concentration, outer coating bioactivity, oxidative, photolytic, and mechanical stability. This is significant and may be misleading, as QD toxicity is determined by multiple factors derived from both individual QD physicochemical properties and environmental conditions.

20.7 Metabolism of Nano-Biomaterials

The metabolism of nanoparticles is not well understood. The total metabolism of nanomaterials depends on hepatocytes. Understanding and tracking what, when, and how much nanostructures degrade are crucial because the breakdown of nanomaterials may cause unanticipated chemical reactions. The majority of nanomaterials have been observed to collect in the liver after oral administration in vivo mouse models, where the nanoparticles have been maintained and have been demonstrated to cause tissue damage. Nanomaterials can travel to different sections of the body due to their tiny size, either directly upon application or indirectly through inhalation, ingestion, and cutaneous exposure after entering the circulation. The metabolism of nanoparticles is not well understood. The total metabolism of nanomaterials depends on hepatocytes. Understanding and tracking what, when, and how much nanostructures degrade are crucial because the breakdown of nanomaterials may cause unanticipated chemical reactions. The majority of nanomaterials have been observed to collect in the liver after oral administration in

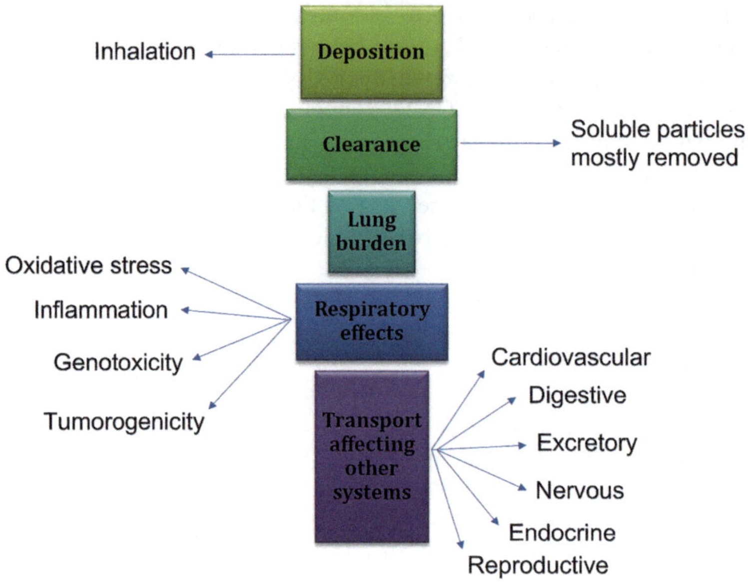

Fig. 20.1 The route of nanoparticles from inhalation to the disease site

in vivo mouse models, where the nanoparticles have been maintained and have been demonstrated to cause tissue damage. Nanomaterials can travel to different sections of the body due to their tiny size, either directly upon application or indirectly through inhalation, ingestion, and cutaneous exposure after entering the circulation (Fig. 20.1).

20.8 Nanoparticle Properties that Influence Toxicities

Size, surface area, shape, aspect ratio, surface coating, crystallinity, dissolution, and agglomeration are NPs properties that influence toxicity.

20.8.1 Dimensions and Surface Area

The surface area to volume ratio exponentially rises with decreasing nanoparticle size, enhancing biological and chemical reactivities [29]. Nanomaterial cytotoxicity results from the surface-to-cellular component contact. Therefore, the cytotoxicity of nanoparticles can vary greatly depending on surface area and particle size even if they have the same chemical make-up. In other words, compared to bigger particles with identical contents, nanoparticles are more hazardous. NPs are more numerous per unit mass and have a bigger surface area than larger particles. The body responds to a comparable mass made up of billions of NPs differently than it does to numerous microparticles. The huge surface area and high surface reactivity of the designed nanoparticles might cause more reactive oxygen species to be produced, which could lead to cytotoxicity and genotoxicity.

20.8.2 Shape

Nanoparticles' biological reactivity and toxicity are significantly influenced by their form. Common shapes for nanoparticles include spheres, cylinders, cubes, sheets, and rods. The cellular uptake of the nanoparticle is influenced by its form. Spherical nanoparticles are ingested by cells more frequently than the other types. Compared to gold nanospheres, gold nanorods do not accumulate as many autophagosomes. The cytotoxic effects of gold nanoparticle in stars, rods, and spheres were examined on human fetal osteoblast, osteosarcoma, and pancreatic duct cell lines using the MTT test [30, 31]. Star-shaped gold nanoparticles are the ones that are most hazardous to human cells. Gold nanoparticles' cytotoxicity and antitumor potential are influenced by their form. Mesoporous silica nanoparticles have shown promise as oral medication delivery systems for drugs. Needle-shaped nanoparticles are more hazardous than spherical nanoparticles due to their enhanced numerous endocytic processes, internalization rates, and more effective adhesiveness to the surface of the target cell.

20.8.3 The Aspect Ratio

A nanoparticle's aspect ratio is defined as its width to height ratio. A nanotube has an aspect ratio that is very near to zero, whereas a spherical particle has an aspect ratio of one. The more poisonous the NPs are, the greater their aspect ratio. Aspect-ratio-dependent toxicity is typically seen in the lung. Lung cancer, mesothelioma, and asbestosis are all brought on by nanofibers that have lengths of 2, 5, and 10 meters and thicknesses of around 150 nm. The pulmonary toxicity induced by high aspect ratio carbon nanotubes was investigated in Sprague-Dawley rats when administered directly into the trachea. On the peritoneal side of the diaphragm, carbon nanotubes generated considerable protein exudation and granulomas. Mesoporous silica NPs with aspect ratios of 5, 1.75, and 1 were studied to determine their in vivo toxicity, excretion, and biodistribution following oral administration. Systematic absorption through organs like the small intestine rises and urine excretion reduces when the aspect ratio is lowered. According to reports, kidney toxicity of silica nanoparticles depends on their form.

20.8.4 Crystallinity

The degree of poisonousness of nanomaterials could depend on their crystalline structure. Polymorphs, or several crystal forms with the same chemical composition, displayed a variety of chemical and physical traits. In both in vitro and in vivo experiments, it was discovered that 10-hydroxycamptothecin (HCPT) nanoparticle dispersions had cytotoxicity that is dependent on the polymorph. Pancake, prismatic, and needle morphologies of three 10-hydroxycamptothecin polymorphic nanoparticle dispersions were produced and described. The cytotoxicity data showed that the different HCPT nanoparticles' cellular toxicity varied depending on their size and shape. Despite similar cellular uptakes, needle-shaped HCPT nanoparticles are more potent in inducing an apoptotic response in cancer cells.

20.8.5 Surface Coating or Surface Functionalization

Their characteristics are changed via surface coatings of nanoparticles [32]. A layer or layers (the "shell") surround a particle's surface (the "core"). The objective of the surface coating may be to customize its functioning, wettability, stability, or dissolution. Surface coating can change poisonous particles into harmless ones due to bioavailability, whereas less harmful particles can become more toxic [32].

20.8.5.1 Dissolution

Nanoparticles' capacity to dissolve is a crucial characteristic that affects their safety, absorption, and related harmful mechanisms. Depending on the surface modification, two identical NPs of comparable composition and size may disintegrate in entirely different ways. The ideal cellular entrance route for nanoparticles that dissolve in media before being taken up by organisms is through clear ion channels and ion transporters.

20.8.5.2 Agglomeration

Nanomaterials have a high free surface energy, which makes them prone to aggregation in solution. To avoid aggregation, protective agents are used to shield nanomaterials. The degree of aggregation has an impact on the toxicity of nanoparticles as well. The aggregation of nanoparticles may result in inflammatory lung diseases in people. Carbon nanotubes and oxide nanoparticles are the nanomaterials where the toxicity is most frequently seen to be agglomeration-dependent. It has been noted that agglomerated carbon nanotubes are more hazardous than well-dispersed carbon nanotubes. Studies have demonstrated the significance of effective agglomeration management by revealing that larger silver nanoparticles exhibited significantly lower hemolytic toxicity compared to their smaller agglomerates.

20.9 Nanoparticle Toxicology

The precise mechanisms through which NPs and biological systems interact are unknown. The ability of the particles to interact and link with biological matter, as well as change their surface properties in reaction to their environment, adds to their complexity. Recent advances in science have increased our understanding of how NPs interact with cells, which shows that either active or passive methods are frequently employed by cells to absorb NPs. It is more difficult to comprehend intracellular systems and processes. Even identical particles of the same substance may behave quite differently due to minute differences in surface coating, charge, or size [33]. Due to this, it is challenging to classify how NPs behave when they interact with biological systems, making it challenging to pinpoint nanoparticle dangers.

20.9.1 Cytotoxicity

Depending on the method of administration and the location of the deposition, toxicity might vary in intensity. A system-based approach to toxicity information presentation that emphasizes lung, cutaneous, liver, and nervous system targets preserves clinical relevance. Each eukaryotic cell, including lung cells, contains functionally unique compartments that are encompassed by membranes. The two primary categories are the nucleus and organelles, which are composed of the mitochondria, endoplasmic reticulum, Golgi apparatus, peroxisomes, lysosomes, and endosomes. The nucleus and organelles are encircled by a lipid bilayer made up of several proteins. NPs can traverse organelle membranes since they have been discovered in lysosomes, mitochondria, and the nucleus; however, the internalization processes are not understood.

20.9.1.1 The Lung Cells

The epidermis, gastrointestinal tract, and respiratory tract are biological compartments that are "designed" to function as barriers against the entry of nanosized materials when nanoparticles come into contact with them. This overview will concentrate on the lung as a possible barrier for inhaled NPs since it serves as the body's single most significant entrance site for NPs. However, there is proof that NPs can deposit on the olfactory epithelium and go straight to the brain, according to published research. The main focus of current research is on how NPs affect bronchial epithelial cells, fibroblasts, and alveolar macrophages. Studies have already demonstrated that NPs drastically lower macrophage phagocytic capability, which lowers lung clearance capacity. Over the past three decades, numerous synthetic polymers have been developed for use as drug carrier devices, including the biocompatible and biodegradable poly (lactic-co-glycolic acid, or PLGA). Even though these drug-loaded nanostructures provide potential alternatives to chemotherapy for cancer, cytotoxicity needs to be considered. Chitosan and albumin, two biological capping substances, minimize cytotoxicity by simulating the physiological environment and "hiding" from the immune system [34]. However, more investigation is needed into the potential for enzymatic breakdown brought on by biophysical similarity. Numerous industries use silica (SiO_2) nanoparticles as additions in foods, medicines, cosmetics, printer toners, varnishes, and mechanical and chemical polishing (Fig. 20.2)

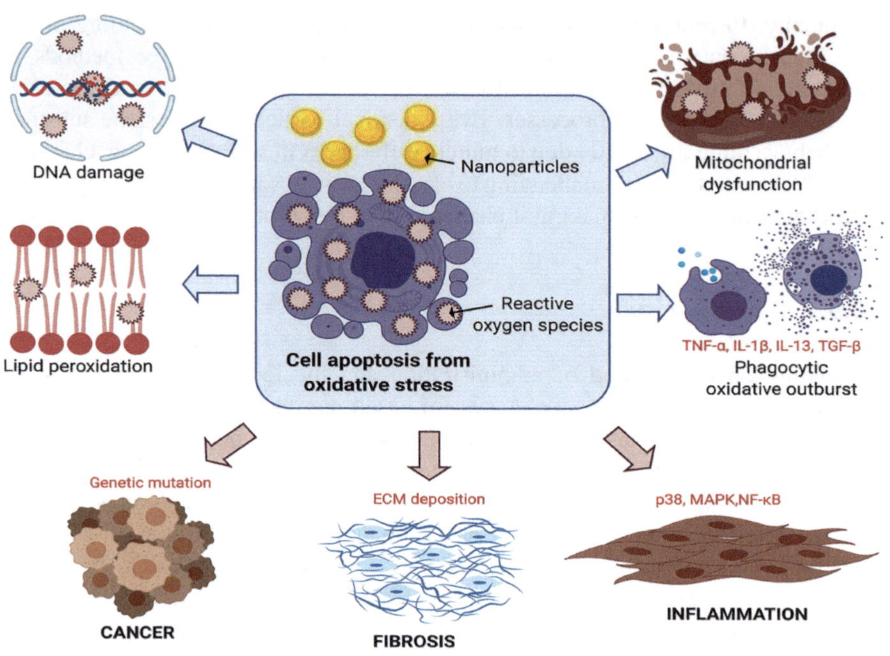

Fig. 20.2 Nanoparticles induced toxicity [35]

Recent uses of silica NPs include cancer therapy, DNA delivery, medication administration, enzyme immobilization, biosensors for simultaneous assessment of glucose, lactate, l-glutamate, and hypoxanthine levels in rat striatum, and biomarkers for leukemia cell detection using optical microscope imaging [36]. Silica nanoparticles have been demonstrated to have minimal toxicity when used in modest amounts. Sadly, silica NPs may also aggregate and, at a dosage of 25 g/mL, have been reported to induce protein aggregation in vitro. It has been suggested that oxidative stress may be to blame for the cytotoxicity of silica NPs both in vitro and in vivo. According to recent studies, utilizing silica NPs at moderate doses is theoretically possible and within acceptable safety limits; nonetheless, high-dose toxicity profiles call for more research.

20.9.1.2 The Skin Cells

The biggest organ in the body, the skin acts as the first barrier of defense between the inside organs of the human body and the outside world. As a result, it is exposed to some general environmental assaults in the air as well as specialized and possibly harmful chemicals found in creams, sprays, or garments. NPs used topically have the capacity to reach the systemic circulation after penetrating the skin, where they may cause extensive injury. The cytotoxic properties of gold and TiO_2 nanoparticles were discussed in this section. TiO_2 nanoparticles are an appealing element in commercial sunscreens and cosmetics due to a number of their characteristics. They enhance cream clarity and appearance and provide UV light inhibiting qualities. The effects of TiO_2 toxicity on biological processes such as cell proliferation, differentiation, motility, and death are shown to be cell type specific. Such detrimental effects, however, were not reproducible in vivo. Although it might be suggested that the number of follicular holes in the skin that permit particle absorption, as well as the surface coatings and functionalization of TiO_2 NPs, are associated with varying levels of penetration and toxicity [37].

Due to their simplicity in production and potential for biofunctionalization, gold nanoparticles (AuNP) are being investigated for therapeutic uses such as cutaneous medication administration. Au compounds are often utilized in therapeutic settings and are widely recognized as safe, for instance, in the treatment of rheumatoid arthritis. However, as particles are shrunk to the nanoscale scale, significant metabolic changes occur, demanding further investigation into their cytotoxic character. Contradictory results continue to be the principal obstacle to clinical use despite the relative amount of toxicity research concentrating on AuNP [38].

20.9.1.3 The Liver Cells

The body's primary organ for detoxification is the liver. It possesses the ability to deoxidize, store glycogen, and synthesize secretory proteins. It serves as a biological barrier by separating and removing diverse external substances by phagocytosis. Several MNPs (metal nanoparticles), including nano-metal monomers and nano-metal oxides, have been shown to accumulate in the liver and have negative effects in earlier in vivo experiments. Once MNPs are ingested, inflammatory cytokines shift. NiO NP decreased the levels of the anti-inflammatory cytokines IL-4 and IL-10 while increasing the levels of the pro-inflammatory cytokines IL-1 and IL-6. The liver undergoes structural modifications as a result of MNP-induced liver dysfunction. Inflammation brought on by MNPs may have had an impact on liver coefficients. The presence of liver damage was suggested by a considerable drop in total bilirubin and a rise in alkaline phosphatase (ALP) and aspartate aminotransferase (AST) in blood serum. The liver underwent structural alterations as the primary sign of the injury, which led to metabolic dysfunction. AgNP increased relative spleen weight and affected the liver by causing extensive and severe hepatocyte necrosis and hemorrhage, as well as multifocal peribiliary microhemorrhages and sporadic portal vein endothelial injury [39]. TiO_2 NP produced structural alterations in the liver, including hepatic inflammatory cell infiltration, a rise in the collagen content of the liver tissue, the start of fibrosis, and a thickening of the Glisson capsule. It was shown that AuNP activated liver macrophages, which accelerated the development of experimental immunological hepatitis and liver damage. MNPs promote liver cell toxicity by causing oxidative stress, inflammation, and perhaps even multiple forms of cell death. The liver is particularly susceptible to NP toxicity because it is the location of first-pass metabolism, and it has been demonstrated that given drugs can persist even after exposure has terminated. Even so, it is crucial to continue researching NP-mediated hepatocellular damage. Lipid peroxidation tests, comet assays, and oxidative DNA damage are frequently used to examine how NPs affect liver cells.

20.9.2 Nanoparticles' Impact on the Cardiovascular System

NPs can enter our bodies through inhalation (intranasal or intratracheal) or through oral exposure to the gastrointestinal tract, where they can then enter the circulatory system. These can have a significant impact on the system's normal operation. Some of the first effects of NPs on the cardiovascular system include an increase in blood pressure, a decrease in heart rate, and altered vascular tone and dysfunction [40]. NPs can have angiogenic/antiangiogenic, vasodilation/vasoconstriction, pro-oxidant/ antioxidant, cytotoxic, apoptotic, and phagocytic effects on our body depending on their physical properties, concentration, and retention time. People who have pre-existing cardiovascular disease are more vulnerable to these changes. Because of the assimilation of NPs in the bloodstream, such people are more prone to sudden cardiac arrests, heart attacks, and blood clotting. Epidemiological studies have found evidence of NPs' acute and chronic effects. The main cause of these maladies due to increased oxidative stress has been identified as increased ROS generation caused by nanoparticles. Silver nanoparticles, which are currently used in pacemakers, drugs, and coupled antibodies, have a dual and inverse effect on blood composition, angiogenesis (vessel formation), and membrane permeability.

Long-term TiO_2 NP exposure can result in their accumulation in the heart, resulting in sparse cardiac muscle fibers, inflammation, cellular necrosis, and cardiac biochemical dysfunction.

20.9.3 Neurotoxicity

Nanomaterials can reach the brain accidentally through blood circulation from the treatment of peripheral illnesses or through inhalation of air in addition to being used purposely for the treatment of brain disorders. Consequently, despite significant advancement in the industry, the employment of nanomaterials has led to a brand-new cause for worry: nanomaterial neurotoxicity. The brain and spinal cord make up the central nervous system. Both are sensitive human organs that require defense against xenobiotic damage. Following intravenous injection, it has been demonstrated that several NPs, including pegylated PLA immuno-nanoparticles and polysorbate 80-coated PBCA NPs, may pass the blood–brain barrier (BBB) and aggregate in the brain. But due to their distinct physicochemical characteristics, such as their huge surface area, NPs could result in neurotoxicity. Due to the lack of defined mechanisms and routes via which NPs may exert their toxic effects, it is necessary to analyze the possible neurotoxic consequences of these NPs on CNS function [41]. A few publications, albeit not many, have shown that NPs are neurotoxic both in vitro and in vivo. Common neurotoxicity mechanisms include oxidative stress, increased cell apoptosis and autophagy, immunological response, and inflammation, which all activate certain signaling pathways that impact blood–brain barrier activities. Additionally, a neurotoxic impact may affect neuronal structure or function directly or may set off a chain reaction through glial activation and glial-neuronal connections. It could show up right away or after some time.

Different regions of the central nervous system or the entire system may be affected by neurotoxicity, which can have temporary or permanent effects. Brain-specific cell types such as blood–brain barrier and blood–liquid barrier cells, glial cells or neuroglia, and neurons with and without myelin sheaths, a membrane covering responsible for accelerating signal conduction speed, should be utilized to test the neurotoxicity of nanomaterials. Larger neurons have thicker myelin sheaths than smaller ones; therefore, tiny neurons are unmyelinated and have very sluggish signal conduction velocity. Unmyelinated neurons are also helpful for in vitro evaluations of neurotoxicity since they are prevalent in illnesses linked to demyelination pathways, such as multiple sclerosis.

Nanoparticles have a much greater surface area per unit volume compared with the larger particles. It leads to nanoparticles more chemically reactive [42]. Concluding possible risks and causes of nanomaterial toxicity using conventional methodologies and procedures is risky since they are not sufficiently precise and dependable. As a result, since it is crucial for understanding how nanomaterials interact with the brain, the unique field that integrates nanotoxicology and neurology has to be improved by allocating original and specialized experimental models and methods. There may be safer nanocarrier systems with fewer side effects as a result of a better understanding and unraveling of neurotoxic mechanisms.

20.9.4 Immunotoxicity of Nanoparticles

Phagocytic immune cells including macrophages, neutrophils, and dendritic cells aid in the absorption of ENM (engineered nanomaterials). Size and surface modification of ENMs, for example, can affect how they interact with immune system cells such as phagocytes, lymphocytes, and mast cells. According to the surface functional group, there have been reports of both enhanced and decreased absorption of single-walled carbon nanotubes, suggesting that surface alterations may play a role in facilitating immunological recognition and/or uptake of ENMs. Furthermore, the activation of the complement system has been connected to ENM size. Although the relationship between NPs and the immune system has been shown, there is not much information at this time. The ability of an NP to bind or coat surface species, its solubility, lipophilicity and hydrophilicity, as well as its composition and electronic structure, are crucial characteristics in terms of how well it interacts with biological tissues, especially the immune system. Some NPs accumulate in local lymph nodes, where they might be taken up and processed by dendritic cells, interact with self-proteins to change their antigenicity, and then trigger different immunological reactions, including autoimmunity. Other NPs, like Pd allergic contact dermatitis, may induce allergic sensitization. In vitro studies have demonstrated that NPs can direct the production of cytokines toward Th1 (Pl, Pd, Ni, Co) or Th2 (Ti, mw, and sw Carbon) patterns. However, it is unlikely that NPs can act as a hapten, inducing specific IgE production; instead, they are more likely to act as an adjuvant, inducing a particular pattern of cytokines, antibodies, and cells that favor allergic sensitization to environmental allergens. Some NPs have been linked to allergic sensitization.

Additionally, in the lung of test animals, NPs raised the production of IL-1beta, MIP-1alpha, MCP-1, MIP-2, keratinocyte chemoattractant, TARC, GM-CSF, MIP-1alpha, and activation of the stress-activated MAPKs p38 and JNKs [43].

In addition, the exposure to ENMs might compromise healthy immune function; ENM-immune cell interactions could cause unintended activation of one or more immune cell types. Immune cell activation might have serious effects, including toxicological repercussions for other vital systems. There is growing evidence that suggests a specific immune cell, the mast cell, may contribute to the onset of both pulmonary and cardiovascular problems following ENM exposure. Mast cells have long been recognized to contribute to allergic inflammation in conditions including anaphylaxis, allergic rhinitis, allergic dermatitis, asthma, and atopic dermatitis. In fact, the activation of mast cells by ENM could be a major impediment to the creation of safe nanotechnologies that prevent not only the emergence of these allergic diseases but also any unintended anaphylactic-type responses linked to nanotherapeutics.

20.9.5 Genotoxicity of Nanomaterials

Modifications to the genetic code can have a gravely detrimental effect on human health since somatic cell mutations can cause cancer and other chronic disorders. Genotoxic events can affect the complete genome (during mitosis), chromosomes, or DNA. It is crucial to identify compounds that could be genotoxic and to research the processes of genotoxicity. A major concern is the possibility that exposure to NM might be genotoxic. The word "genotoxicity" refers to an agent's capacity to alter the genetic code by chemical or physical means. Genotoxic events can alter the quantity or make up of a cell's genetic material in a way that is either short-term (repairable damage) or long-term (mutations).

The two main processes that might result in NM genotoxicity are primary (direct or indirect) and secondary genotoxicity. Some NMs may be affected by one of these processes, while for some NMs, both mechanisms may operate concurrently following NM exposure. The primary direct genotoxicity of NMs is caused by their direct interaction with the genome after coming into physical contact with the DNA of the nucleus. Chromosome damage, significant DNA deformities, DNA breaks, and other DNA lesions might be caused by this interaction. The major mechanisms of indirect action include intracellular ROS formation through the Fenton-type reaction and reactive oxygen species (ROS) generated by NMs or hazardous ions released during NM breakdown. Oxidative stress is a significant mechanism for primary indirect genotoxicity. Free radicals may interact negatively with biological macromolecules like DNA and others. DNA strand breakage and a purine or pyrimidine oxidation lesions can result from ROS damage. Despite being repairable, the damage may potentially result in gene mutations and even more severe chromosomal damage. NMs can also induce DNA damage by using extra compounds that either interact with DNA or hinder DNA replication and cell division. For instance, protein kinases, which regulate cell cycle events including DNA replication and cell division, can interact with NMs. Determining whether primary genotoxicity is direct or indirect requires research into the absorption process and if NMs may enter the nucleus. The smallest NMs (with a size of barely a few nanometers) were able to penetrate the nucleus through nuclear pores. In certain investigations, larger NMs were discovered in the nucleus, indicating that there may be additional pathways for nuclear absorption, such as intracellular processes resembling endocytosis. NMs may interact with DNA directly once they are inside the nucleus or with DNA that is organized into chromosomal structures (direct genotoxicity) depending on the stage of the cell cycle. NMs may interact or bind to DNA molecules during interphase, preventing DNA replication and the conversion of DNA to RNA. NMs can also disrupt the process of mitosis by interfering with the splicing machinery.

Secondary genotoxicity is assumed to be the main mechanism of NM genotoxicity. ROS from inflammatory cells is responsible for it. Phagocytes can be activated by NMs, which causes an oxidative burst. The initial line of defense against the entry of bacteria or other materials involved in a clearing is inflammation. Ineffective clearance of chemicals, such as inhaled NMs, may trigger a persistent immune cell response. It has only recently been examined in vivo following persistent inflammation caused by the activation of immune cells like neutrophils or macrophages since this sort of secondary genotoxicity cannot be studied using normal in vitro techniques. The capacity of NMs to cause genotoxicity via inflammation cannot be assessed using normal monoculture methods, making it impossible to research secondary genotoxicity in vitro. As a result, advances in novel models focus on co-culturing immune system cells with target cells. Numerous studies also show that the size, surface characteristics, chemical composition, shape, dosage, and length of exposure all affect the NMs' ability to cause genotoxicity.

20.10 The Effects of Nanoparticles on Plants and Their Toxicity Mechanisms

Germination percentage, root elongation, biomass, and leaf count are the main physiological markers of NP toxicity in plants. NPs can have serious adverse consequences, including inhibiting plant growth and germination and even killing plants. Inhibition of plant species like soybean, maize, wheat (Triticum aestivum), ryegrass, and barley was found to be caused by exposure to NPs (MWCNTs, single-wall carbon nanotubes, ZnO NPs, Ag NPs, and Fe NPs). Several aspects of plant growth were also impacted, including seed germination, shoot length, biomass, and gene expression. Around ten years ago, Moore, Hund-Rinke, and Simon proposed that NP may be able to affect biota by the production of reactive oxygen species (ROS) that may cause biological processes to be disrupted. Moore also raised the prospect that NP may serve as a vehicle for other contaminants. No one toxicity mechanism can be said to apply to all NP. However, oxidative stress is a phenomenon that is frequently documented. Modifying hormones or hatching enzymes, to mention a few additional pertinent pathways, have been documented to have physiological repercussions, such as reproductive failure. These outcomes raise questions about population growth and the potential for generational consequences. Additionally, photosystem II was impacted by changes in the photosynthetic pigment content of algae and aquatic plants. Similar to this, some recent studies have focused on NP accumulation in terrestrial plants, which can lead to biochemical and physiological alterations. For instance, the exposure to CeO_2 NP has an impact on carbon fixation as well as water usage efficiency during photosynthesis.

20.10.1 The Impact of Nanoparticles on Plant Physiological Indices

Germination percentage, root elongation, biomass, and leaf count are the main physiological markers of NP toxicity in plants. NPs can have serious adverse consequences, including inhibiting plant growth and germination and even killing plants. Inhibition of plant species like soybean, maize, wheat (Triticum aestivum), ryegrass, and barley was found to be caused by exposure to NPs (MWCNTs, single-wall carbon nanotubes, ZnO NPs, Ag NPs, and Fe NPs). Several aspects of plant growth were also impacted, including seed germination, shoot length, biomass, and gene expression [44]. Bacillus thuringiensis (BT)-transgenic cotton's growth was slowed when it was exposed to SiO_2 NPs. When cultivated in a sand substrate, CuO NPs hinder wheat plant development and alter the root structure (Fig. 20.3).

Copper oxide NPs dramatically lower the fresh weights and root lengths of Arabidopsis seedlings, as well as the germination rate and biomass of rice seeds. The growth of radish, tomato, lettuce, wheat, cabbage, cucumber, and maize was hindered when large quantities of rare earth oxide nanoparticles (CeO_2, La_2O_3, Gd_2O_3, and Yb_2O_3) were administered to the roots of the plants. Catalase (CAT)

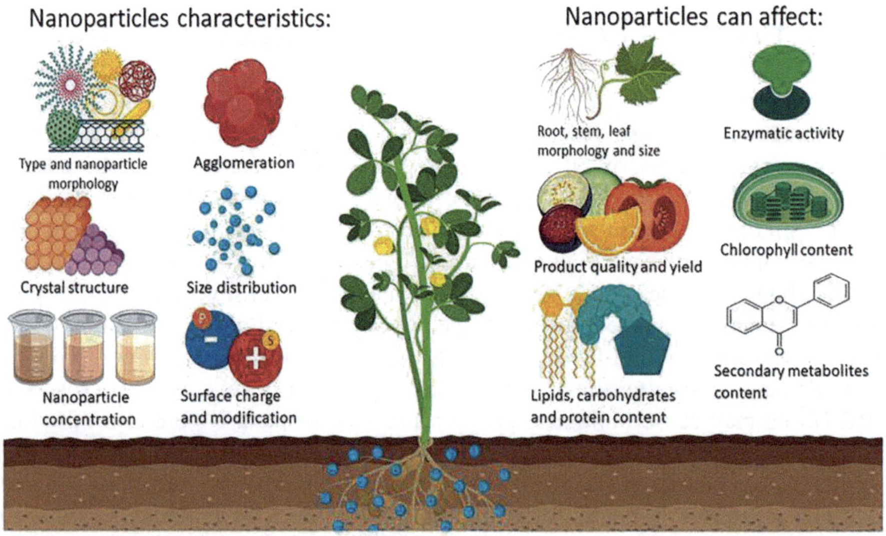

Fig. 20.3 The primary properties of nanomaterials and their toxicological effects on crops [44]

and total chlorophyll content are both increased by TiO_2 nanoparticles, but ascorbate peroxidase (APX) level is decreased.

20.11 Toxicology of ENMs in Aquatic Ecosystems

ENMs are used and disposed of in such large quantities daily; the ecosystem, and particularly the aquatic ecology, has become a significant victim of environmental contamination. In recent years, nanotechnology has become a significant inventive strategy for advancing both science and the economy. The aquatic ecology has suffered as a result of its extensive usage and introduction of high doses. Additionally, the uncontrolled discharge of these created ENMs from sewage sludge and industrial waste is a diverse and mostly unrecognized phenomenon, raising questions about their ecological toxicity. With varied particle morphology and elemental content, ENMs are synthesized in a wide range of shapes, sizes, and surface functions and enter the environment either knowingly or unknowingly. The extensive usage of ENMs has toxicological consequences on the aquatic ecosystem, leading to significant toxicity in aquatic plants, aquatic microorganisms, and aquatic animals, according to prior research. Carbon-based substances including carbon nanotubes and graphene oxide have also been discovered to be harmful to aquatic ecology, in addition to metallic ENMs. These compounds will very probably be released into the environment and ecosystem as a result of increased uses and unchecked manufacturing, causing cellular harm. Graphene oxide is also very unstable in natural surface water and rapidly interacts with other environmental surfaces, both of which are crucial in the transformation of these new materials, according to long-term stability studies. Despite intensive study of graphene-based applications, an investigation into how they affect the environment is still in its early phases.

The scientific community has created reliable, stable, and environmentally friendly procedures for secure synthesis and disposal in response to the discrepancy between ENM synthesis and toxicological data. The transfer of ENMs into the environment is significantly influenced by their in-bound characteristics. Because of their high aggregation stability, low photobleaching, and delayed photodegradation, doped ENMs, for instance, have a toxicological impact. Additionally, recent research has demonstrated that the transport and toxicological effects of ENMs are completely attributable to the dissolved ion concentration as opposed to the nanomaterial itself or its aggregated form.

20.12 Quick Recap and Outlook

The effects of several kinds of nanomaterials on the environment and human health as well as their destiny, behavior, and toxicity have been investigated. Several study teams have found that the nanomaterials can be harmful, although the reasons why this is the case is mainly unclear. Still a lot has to be learnt about how nanoparticles interact with the environment and the human body work. Much more study is necessary to analyze the stability of these matrices in a variety of test techniques to fully assess the risk of human exposure to nanoscale components of current and potential goods. Numerous studies on the toxicity of nanoparticles using various cell lines and incubation times are being published, but it is difficult to determine whether the toxicity observed is physiologically relevant due to the wide range of nanoparticle concentrations, variety of cell lines, culturing conditions, and lack of mechanism knowledge. It is still unsure of the best way to gauge nanomaterial density, surface area, mass concentration, or any combination of these. Have these toxicological investigations revealed any novel biological toxicity pathways to us? The answer, with a few exceptions, is no, it does not. It was predicted that a new understanding of how cells and organs function, both internally and externally with others, will be attained if the biomedical community adopts nanoparticles as new instruments for in vivo imaging on longer time scales. Engineered nanomaterials come in a huge diversity of sizes, shapes, compositions, and coatings that are comparable to, if not greater than, that of conventional chemicals. Smart sensors should be created as a consequence of developments in information technology and sensor design and will be able to detect nanoparticle quantities and assess their potential toxicity, possibly offering early warning indications of danger. It is necessary to come to a worldwide agreement on a series of in vitro screening tests for toxicity to people and the environment. To evaluate the safety of complicated multi-component and multifunctional nanomaterials, scientists will need to create verified models that can foresee the release, transport, transformation, accumulation, and absorption of designed nanomaterials in the environment. By connecting the physical and chemical characteristics of nanoparticles to their behavior, these models should be able to forecast possible effects of designed nanomaterials and nanoproducts as well as estimate consequences among sensitive populations. Structure-activity connections must be created to forecast biological consequences, ecological impacts, and degradation at the end of life. To generate nanoparticles with the appropriate human health and environmental performance in addition to their physical features, each of these models is necessary. Outside of the scientific community, it is challenging to explain studies on the advantages and disadvantages of nanotechnology, but it is essential for discussions that are grounded in science. To do this, communication activities must be developed that enable technical knowledge to be summarized, analyzed, and finally synthesized for a range of stakeholders, including consumers and decision-makers.

20.13 Conclusion

A thermodynamically unstable system is a system of nanoparticles with high specific surface energy and a large surface area. The NPs will agglomerate, disintegrate, and interact with other changes brought on by the body's environmental effect after entering the organism through a variety of pathways (such as high protein concentration, high ionic strength, and high acidity). The physical and chemical characteristics of NPs may alter with aggregation, which may have an impact on the biological consequences. It is necessary to have a solid grasp of the biocompatibility and toxicity of nanoparticles in order to employ them safely in biomedicine. As it was observed, information on NP toxicity is growing steadily, but more work is still needed to make real progress in this area. These investigations may now be carried out from the following angles: taking into account different particle types, taking into account the dose–response relationship, conducting in vivo and in vitro tests, and creating a toxic nanoscale database to better explain the division of harmful nanosize range. Nanoparticles have many uses in biomedicine because of their distinctive physicochemical and behavioral properties; nonetheless, worries concerning their hazardous effects on the biological system are now attracting the attention of the international health community. This highlights the significance of researching and understanding the cellular and molecular processes behind the impacts in question. Among the harmful mechanisms discovered are those that induce ROS, cause cell cytotoxicity, and have genotoxic and neurotoxic consequences. Nanoparticles' size, surface area, shape, aspect ratio, surface coating, crystallinity, dissolution, and agglomeration all affect how poisonous they are. For instance, it has been demonstrated in animal studies that smaller nanoparticles have higher acute toxicity. It was also shown that a nanoparticle's crystallinity or form might affect how poisonous it is. The harmful consequences of nanoparticles must now be taken into account while producing nanoparticles. To find the ones that operate the best without having negative side effects, their size, shape, and other essential characteristics should be altered. The creation of safer and more efficient nanoparticles will be made possible by taking into account the hazardous effects of nanoparticles. Studies in toxicology lay the groundwork for safeguarding both people and the environment. Because of this, it could be challenging to classify some of the more useful NPs as being more harmful to biological systems based on the experimental models that are now available and vice versa. Given the wide range of possible uses for NPs and the need to address information gaps, the pertinent harmful effects of NPs should be evaluated using globally accepted, impartial in vivo toxicological models that focus on critical systems.

References

1. Khan HA, Shanker R (2015) Toxicity of nanomaterials. Biomed Res Int 2015:1–2
2. Bahadar H, Maqbool F, Niaz K, Abdollahi M (2016) Toxicity of nanoparticles and an overview of current experimental models. Iran Biomed J 20(1):1–11
3. Thakur P, Verma Y, Thakur A (2022) Toxicity of nanomaterials: An overview. In: Thakur A, Thakur P, Khurana SP (eds) Synthesis and applications of nanoparticles. Springer, Singapore
4. Mohanta D, Patnaik S, Sood S, Das N (2019) Carbon nanotubes: evaluation of toxicity at biointerfaces. J Pharm Anal 9(5):293–300
5. Dhanda N, Thakur P, Thakur A (2023) Green synthesis of cobalt ferrite: a study of structural and optical properties. Mater Today: Proc 73:237–240ISSN 2214-7853. https://doi.org/10.1016/j.matpr.2022.07.202
6. Sahu SC, Hayes AG (2017) Toxicity of nanomaterials found in human environment. Toxicol Res Appl 1:239784731772635

7. Dhanda N, Kumari S, Kumar R, Kumar D, An-Cheng S, Thakur P, Thakur A (2023) Influence of Ni over magnetically benign co ferrite system and study of its structural, optical, and magnetic behaviour. Inorg Chem Commun 151:110569., ISSN 1387-7003,. https://doi.org/10.1016/j.inoche.2023.110569
8. Sharma P, Sharma A, Sharma M, Bhalla N, Estrela P, Jain A, Thakur P, Thakur A (2017) Nanomaterial fungicides: in vitro and in vivo antimycotic activity of cobalt and nickel nanoferrites on phytopathogenic fungi. Glob Chall 1:1700041. https://doi.org/10.1002/gch2.201700041
9. Dhanda N, Thakur P, An-Cheng S, Thakur A (2023) Structural, optical and magnetic properties along with antifungal activity of ag-doped Ni-co nanoferrites synthesized by eco-friendly route. J Magn Magn Mater 572:170598.,ISSN 0304-8853,. https://doi.org/10.1016/j.jmmm.2023.170598
10. Egbuna C, Parmar VK, Jeevanandam J, Ezzat SM, Patrick-Iwuanyanwu KC, Adetunji CO, Khan J, Onyeike EN, Uche CZ, Akram M, Ibrahim MM, Mahdy NME, Awuchi CG, Saravanan K, Tijjani H, Odoh U, Messaoudi M, Ifemeje JC, Olisah MC, Ibeabuchi C (2021) Toxicity of nanoparticles in biomedical application: Nanotoxicology. J Toxicol 2021:1–21
11. Dhanda N, Thakur P, Kumar R, Fatima T, Hameed S, Slimani Y, Sun A-CA, Thakur A (2023) Appl Organomet Chem:e7110. https://doi.org/10.1002/aoc.7110
12. Li X, Wang L, Fan Y, Feng Q, Cui F (2012) Biocompatibility and toxicity of nanoparticles and nanotubes. J Nanomater 2012:1–19
13. Mathur P, Thakur A, Singh M (2008) Effect of nanoparticles on the magnetic properties of Mn-Zn soft ferrite. J Magn Magn Mater 320:1364–1369. https://doi.org/10.1016/j.jmmm.2007.11.008
14. Kharbanda S, Dhanda N, An-Sun TA, Thakur P (2023) Multiferroic perovskite bismuth ferrite nanostructures: a review on synthesis and applications, 170569ISSN 0304-8853. J Magn Magn Mater 572. https://doi.org/10.1016/j.jmmm.2023.170569
15. Thakur P, Gahlawat N, Punia P et al (2022) Cobalt Nanoferrites: a review on synthesis, characterization, and applications. J Supercond Nov Magn 35:2639–2669. https://doi.org/10.1007/s10948-022-06334-1
16. Thakur A, Thakur P, Hsu JH (2011) Smart magnetodielectric nano-materials for the very high frequency applications. J Alloys Compd 509:5315–5319. https://doi.org/10.1016/j.jallcom.2011.02.021
17. Sonwani S, Madaan S, Arora J, Suryanarayan S, Rangra D, Mongia N, Vats T, Saxena P (2021) Inhalation exposure to atmospheric nanoparticles and its associated impacts on human health: a review. Front Sustain Cities 3:690444
18. Thakur A, Mathur P, Singh M (2007) Controlling the properties of manganese-zinc ferrites by substituting In3+ and Al3+ ions Z. Phys Chem 221(6):837–845. https://doi.org/10.1524/zpch.2007.221.6.837
19. Yadav A, Rai M (2022) Phytosynthesis of silver nanoparticles and their role as antimicrobials. In: Thakur A, Thakur P, Khurana SP (eds) Synthesis and applications of nanoparticles. Springer, Singapore
20. Ray PC, Yu H, Fu PP (2009) Toxicity and environmental risks of nanomaterials: challenges and future needs. J Environ Sci Health A 27(1):1–35
21. Thakur A, Thakur P, Hsu J (2014) Structural, magnetic and electromagnetic characterization of In3+ substituted Mn-Zn nanoferrites. Z Phys Chem 228:663–672. https://doi.org/10.1515/zpch-2014-0477
22. Yao Y, Zang Y, Qu J, Tang M, Zhang T (2019) The toxicity of metallic nanoparticles on liver: the subcellular damages, Mechanisms, And Outcomes. Int J Nanomed 14:8787–8804
23. Thakur A, Mathur P, Singh M (2007) Controlling the properties of manganese-zinc ferrites by substituting In3+ and Al3+ ions. Z Phys Chem 221:837–845. https://doi.org/10.1524/zpch.2007.221.6.837
24. Teleanu DM, Chircov C, Grumezescu AM, Teleanu RI (2019) Neurotoxicity of nanomaterials: an up-to-date overview. Nano 9(1):96

25. Mathur P, Thakur A, Singh M, Harris G (2008) Preparation and characterization of Mn04NixZn06−xFe2O4 soft spinel ferrites for low and high frequency applications by citrate precursor method. Z Phys Chem 222:621–633
26. Di Gioacchino M, Petrarca C, Lazzarin F, Di Giampaolo L, Sabbioni E, Boscolo P, Mariani-Costantini R, Bernardini G (2011) Immunotoxicity of nanoparticles. Int J Immunopathol Pharmacol 24(1 Suppl):65S–71S. PMID: 21329568.
27. Xiang C, Zhang Y, Guo W, Liang X-J (2020) Biomimetic carbon nanotubes for neurological disease therapeutics as inherent medication. Acta Pharmaceutica Sinica B 10:39–248
28. Kohl Y, Rundén-Pran E, Mariussen E, Hesler M, Yamani NE, Longhin E, Dusinska M (2020) Genotoxicity of nanomaterials: advanced in vitro models and high throughput methods for human Hazard assessment—a review. Nano 10(10):1911
29. Mathur P, Thakur A, Singh M (2008) Low temperature synthesis of Mn0.4Zn0.6In0.5Fe1.5O4 nanoferrite for high-frequency applications. J Phys Chem Solids 69:187–192. https://doi.org/10.1016/j.jpcs.2007.08.014
30. Shahdeo D, Kesarwani V, Suhag D, Ahmed J, Alshehri SM, Gandhi S (2021) Self-assembled chitosan polymer intercalating peptide functionalized gold nanoparticles as nanoprobe for efficient imaging of urokinase plasminogen activator receptor in cancer diagnostics. Carbohydr Polym 266:118138. https://doi.org/10.1016/j.carbpol.2021.118138
31. Thakur P, Chahar D, Taneja S, Bhalla N, Thakur A (2020) A review on MnZn ferrites: synthesis, characterization and applications. Ceram Int 46:15740–15763. https://doi.org/10.1016/j.ceramint.2020.03.287
32. Pathania A, Bhardwaj S, Thakur SS, Mattei JL, Quefelec P, Panina LV, Thakur P, Thakur A (2018) Investigation of structural, optical, magnetic and electrical properties of tungsten doped Ni–Zn nano-ferrites. Phys B Condens Matter 531:45–50. https://doi.org/10.1016/j.physb.2017.12.008
33. Yang J, Cao W, Rui Y (2017) Interactions between nanoparticles and plants: phytotoxicity and defense mechanisms. J Plant Interact 12(1):158–169
34. Rana K, Thakur P, Tomar M, Gupta V, Thakur A (2018) Investigation of cobalt substituted M-type barium ferrite synthesized via co-precipitation method for radar absorbing material in Ku-band (12–18 GHz). Ceram Int 44:6370–6375. https://doi.org/10.1016/j.ceramint.2018.01.028
35. Damasco JA, Ravi S, Perez JVD, Hagaman D, Melancon MP (2020) Understanding nanoparticle toxicity to direct a safe-by-design approach in cancer nanomedicine. Nano 10(11):2186
36. Smith MR, Brown JM, Zamboni WC, Walker NJ (2014) From Immunotoxicity to Nanotherapy: the effects of nanomaterials on the immune system. Toxicol Sci 138(2):249–255
37. Thakur A, Thakur P, Hsu JH (2012) Magnetic behaviour of Ni0.4Zn0.6Co0.1Fe1.9O4 spinel nano-ferrite. J Appl Phys 111:07A305. https://doi.org/10.1063/1.3670606
38. Pal K, Khurana SMP, Debnath N, Das S (2022) Micronutrient nanoparticles: synthesis, properties and application in agriculture. In: Thakur A, Thakur P, Khurana SP (eds) Synthesis and applications of nanoparticles. Springer, Singapore
39. Kumar A, Sharma V (2010) Toxicity assessment of nanomaterials: methods and challenges. Anal Bioanal Chem 398(2):589–605
40. Kumari S, Dhanda N, Thakur A, Gupta V, Singh S, Kumar R, Hameed S, Thakur P (2023) Nano ca–mg–Zn ferrites as tuneable photocatalyst for UV light-induced degradation of rhodamine B dye and antimicrobial behavior for water purification. Ceram Int 49(8):12469–12480, ISSN 0272-8842,. https://doi.org/10.1016/j.ceramint.2022.12.107
41. Bundschuh M, Filser J, Lüderwald S, McKee MS, Metreveli G, Schaumann GE, Schulz R, Wagner S (2018) Nanoparticles in the environment: where do we come from, where do we go to? Environ Sci Eur 30(1):6

42. Kumari S, Dhanda N, Thakur A, Singh S, Thakur P (2023) Investigation of calcium substitution on magnetic and dielectric properties of Mg–Zn nano ferrites. Mater Chem Phys 297:127394.. ISSN0254-0584. https://doi.org/10.1016/j.matchemphys.2023.127394
43. Jahan S, Yusoff I, Alias Y, Bakar AFA (2017) Reviews of the toxicity behavior of five potential engineered nanomaterials (ENMs) into the aquatic ecosystem. Toxicol Rep 4:211–220
44. Paramo LA, Feregrino-Perez AA, Guevara RG, Mendoza S, Esquivel K (2020) Nanoparticles in agroindustry: applications, toxicity, challenges, and trends. Nano 10(9):1654

Nano-Remediation Perspectives

21

Fayu Wan, Atul Thakur, and Preeti Thakur

Abstract

A variety of organic and inorganic pollutants that are harmful to the environment and human health have been driven by many global events, such as industrial expansion and population rise. It is essential to find new technologies and apply them to the problem of environmental pollution. Nanotechnology is a relatively new topic of study with several applications. A particularly effective tactic is nano-remediation, which is described as using manufactured materials to clean the environment. Dealing with persistent pollutants, such as insecticides, chlorinated solvents, halogenated chemicals, or heavy metals, may be done quickly, effectively, and efficiently this way. A long-term option to eliminate new contaminants like medicines or personal care products is nano-remediation. Due to their diversity and versatility, nanomaterials can be employed in media such as soil, water, or the air. This chapter provides an overview of the usage of nanomaterials for media cleansing. It explores the state of the art for different

F. Wan (✉)
School of Electronics and Information Engineering, Nanjing University of Information Science & Technology, Nanjing, China
e-mail: fayu.wan@nuist.edu.cn

A. Thakur
School of Electronics and Information Engineering, Nanjing University of Information Science & Technology, Nanjing, China

Amity Centre for Nanotechnology, Amity University Haryana, Gurgaon, Haryana, India
e-mail: athakur1@ggn.amity.edu

P. Thakur
Amity Centre for Nanotechnology, Amity University Haryana, Gurgaon, Haryana, India

Amity School of Applied Sciences, Amity University, Gurugram, Haryana, India
e-mail: pthakur@ggn.amity.edu

D. Suhag et al. (eds.), *Integrated Nanomaterials and their Applications*,
https://doi.org/10.1007/978-981-99-6105-4_21

nanomaterials, such as metal, carbon, polymer, and silica, used to purify water, soil, and air.

Keywords

Nano-remediation · Engineered materials · Environment · Human health · Nanomaterials

21.1 Introduction

Environmental remediation effectively removes pollutants or toxins from a variety of environments, including soil, surface, groundwater, and sediment, for the protection of human health and the environment. For the purposes of drinking fresh water, inhaling clean air, and using high-quality water for business and agriculture, clean and pollution-free environments are a must. Unfortunately, a variety of anthropogenic and natural toxins are constantly a threat to many important natural resources, including air, soil, groundwater, and others. The term "nano-remediation" refers to the employment of nanoparticles in environmental cleanup. The primary idea of pollution control at the nanoscale is to separate certain components and molecules from a mixture of contaminants via nanofabrication, thermal partitioning, or nano structure. By improving sensing, treating, and remediating environmental toxins with nanoscale devices, nanotechnology is leading to significant advancements in environmental conservation technologies [1]. Contrarily, the unique properties of nanotechnology may also lead to unanticipated environmental consequences, thus we should exercise extreme caution. The chapter includes a brief discussion of all current advances in nano-remediation methods to address environmental contamination, such as cleaning up oil spills and air and water pollution [2, 3]. By treating persistent contaminants with few remediation choices, removing intermediaries related to degradation, and accelerating the pace of breakdown or stabilization, nanotechnologies have the potential to greatly increase remediation capabilities. Target-specific capture, cost-effectiveness, ease of synthesis, green chemistry, non-toxicity, biodegradability, recyclability, and the potential for recovery after usage (regeneration) are a few crucial considerations when developing new nanomaterials for environmental remediation (Fig. 21.1).

The innovative component of nanotechnology has been the clustering or assembly of molecules to tailor the properties of the resulting nanomaterials to the intended use. The investigation of semiconductor and metal properties that rely on size and shape has received considerable attention in this field.

21.2 Nanomaterials

Particles of a size between 1 and 100 nm are known as nanoparticles (NPs) [5, 6]. Because of their unique properties, NPs are very reactive with environmental contaminants that have been integrated into their macroscale structure. NPs have a

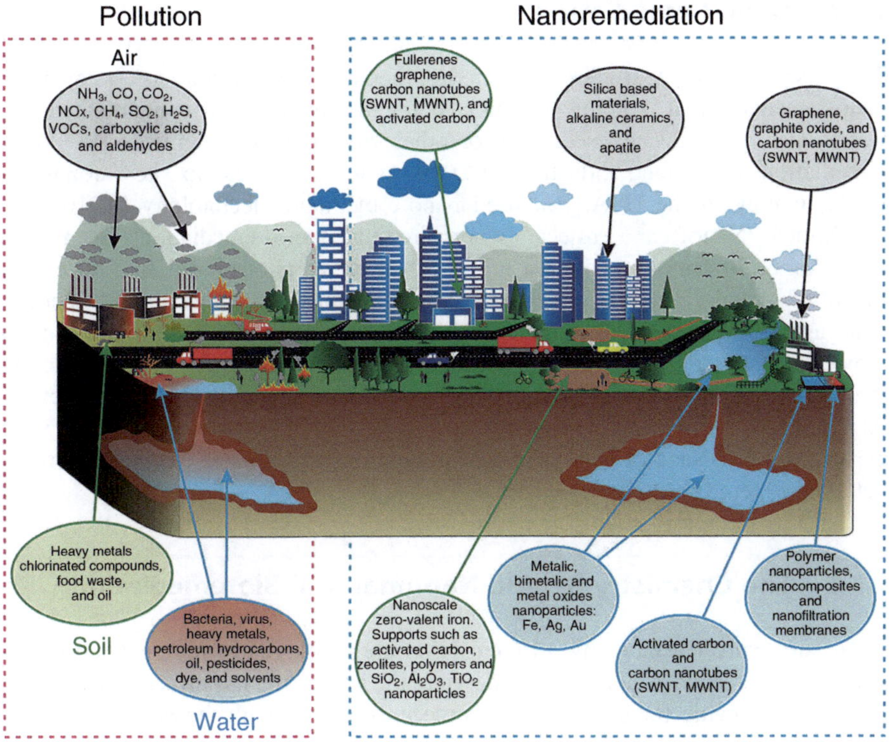

Fig. 21.1 Different types of nanomaterials employed for nano-remediation [4]

lot of reaction/adsorption sites on their surfaces. There are two types of nanomaterials: manufactured and natural. Naturally occurring NPs in soil include clay, iron oxide, and organic matter. Manufactured NPs are either created or synthesized with a unique attribute to enhance their industrial or technical uses. Nanomaterials can be produced in two broad methods. The first way entails dividing a significant area into smaller portions, either from top to bottom or from outside to inside. The second strategy involves building up tiny compounds into larger ones from bottom to top. Among the nanomaterials developed for pollution remediation are nanoscale zeolites, carbon nanotubes, metal oxides, bimetallic nanoparticles (BNPs), enzymes, and titanium dioxide (TiO_2).

The special characteristics of nanomaterials, such as large surface area, quantum size effect, ease of separation and recycling, etc., enable their usage as adsorbents. For instance, the ferromagnetism of iron-doped nanoparticles promotes reuse and recycling. Nanoparticles' potential for recycling makes them desirable from an economic standpoint. The primary factor in fullerene's capacity for absorption and ease of recycling is its hydrophobicity [7].

21.3 Nano-Remediation

The use of nanoparticles in environmental remediation is known as nano-remediation. It is being investigated for the treatment of contaminated soil, wastewater, groundwater, sediment, or other environmental elements [8]. Nano-remediation is a young industry; by 2009, at least 44 cleanup sites around the world, mainly in the USA, had used nano-remediation technology [9, 10]. The EC-funded NanoRem Project is researching nano-remediation in Europe [11]. About 70 nano-remediation initiatives at either pilot or full size have been identified globally, according to a report by the NanoRem collaboration. A nanoparticle agent must come into touch with the target contaminant during nano-remediation under circumstances that permit a detoxifying or immobilizing reaction. In situ application or a pump-and-treat procedure are frequently used in this method. At large-scale cleanup locations, some nano-remediation techniques have been used, particularly the usage of nano zero-valent iron for groundwater cleanup. There are yet undiscovered methods.

21.4 The Chemistry Behind Nanomaterial Bioremediation

There are many reasons why different NMs are employed in bioremediation. For instance, when a material is reduced down to the nanoscale, its surface area per mass increases, allowing for more of it to come into contact with the environment and impact the reactivity. Because NMs exhibit the quantum effect, less activation energy is needed to enable chemical reactions. Another phenomenon displayed by NPs that can be utilized to detect harmful substances is surface plasmon resonance [12]. Different metallic and nonmetallic NMs of varying sizes can be utilized for environmental cleanup, depending on their form and composition. Because NPs may diffuse or penetrate into a contaminated zone where microparticles cannot, as well as because they have higher reactivity to redox-amenable pollutants, it is possible, for instance, to use a variety of single metal NPs, bimetallic NPs, carbon base NMs, and other nanoparticles. It is noted that pollutants like carbon tetrachloride can cause oxide-coated FeO to form weak, outer-sphere complexes (CT). In batch trials and field evaluations, benzoquinone, by-trichloroethene, and other chlorinated aliphatic hydrocarbons can be converted into compounds with lesser toxicities while CT can be converted through electron transfer into methane, carbon monoxide, or format. TiO_2 nanotubes can be applied in the lab to break down pentachlorophenol (PCP) via a photo-electrocatalytic process in addition to field uses [13]. Additionally, biocatalysts for reductive dechlorination can be made from single metal NPs. In a bio-reductive test involving Pd, palladium, Pd (0) NPs can be deposited on the cell wall and inside the cytoplasm of Shewanella oneidensis and can be charged with radicals by adding various substrates as electron donors, such as hydrogen, acetate, and formate (II). The radical on the Pd (0) can catalytically combine with PCP when these charged Pd (0)-deposited S. oneidensis cells come into contact with chlorinated compounds, removing the chlorine molecule from the

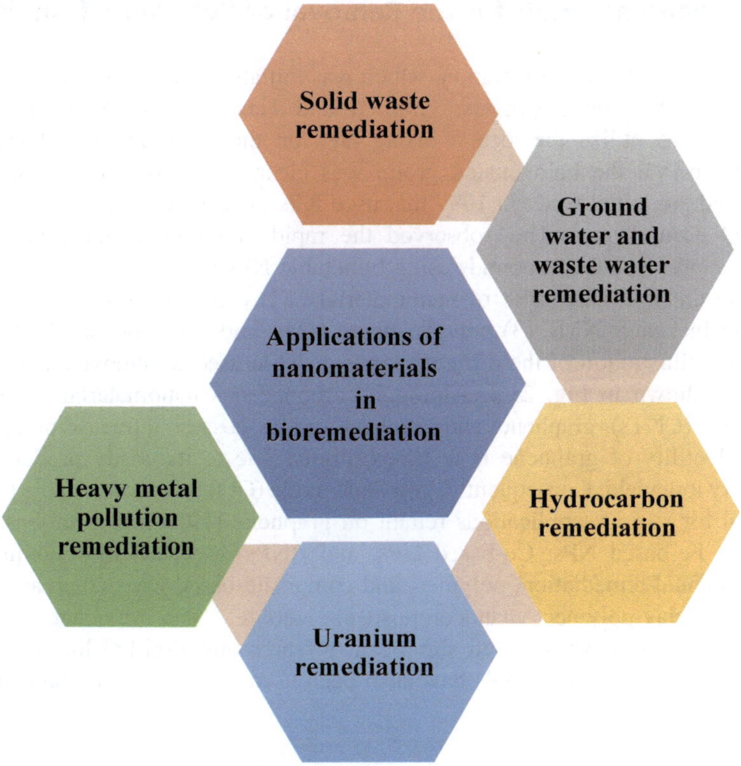

Fig. 21.2 Applications of NMs in bioremediation

chlorinated compounds [14]. Additionally, microbial cells with the ability to biodegrade or recover particular compounds can be immobilized using NPs. Instead of traditional cell immobilization on micron-sized medium or a fixed surface, Pseudomonas delafieldii has been coated with magnetic NPs (that is, Fe3O4) that have been functionalized using ammonium oleate. These magnetic NP-coated microbial cells were concentrated at a particular spot on the reactor wall, isolated from the bulk solution, and reused for the treatment of the same substrate by the application of an external magnetic field to the microbial cells. It has been shown that these microbial cells, which were fed to a bioreactor at a high biomass concentration, desulfurize organic sulfur from fossil fuel (specifically, dibenzothiophene) just as well as cells that are not NP-coated [15]. The upcoming discussions will specifically delve into nanomaterials (NMs) with a primary focus on their applications in addressing a wide range of waste remediation challenges (Fig. 21.2). In the coming decades, the ability of NMs to reduce pollution generation is expected to potentially spark the most revolutionary advances in the environmental field.

21.5 Nanomaterials for the Removal of Pollutants from Water

The use of NPs for water filtration, which was initially done in the 1990s, is one of the most recent developments. Gillham and Hannesin (1964) were the first researchers to utilize the idea of using NPs on the cleaning of polluted water. Utilizing nZVI, the halogenated group was cleaned up [16]. Wang and Zhang conducted the initial study in 1997 that used NPs to remove organo-chlorines from polluted groundwater. They observed the rapid and total eradication of many aromatic chlorinated compounds using bimetallic NPs [17].

There are four categories of nanomaterials: (1) carbonaceous NMs, (2) metal-based or inorganic NMs, (3) polymer-based NMs, and (4) composite NMs [18]. A schematic illustration of the different kinds of NMs used to remove environmental toxins is shown in Fig. 21.3. Although carbonaceous nanomaterials like carbon nanotubes (CNTs), graphene, and graphene oxides possess immense potential, the practical utility of graphene is at times limited due to its costly production and relatively low yield. Consequently, graphene oxide (GO) serves as a viable precursor material for various applications reliant on graphene [19]. And metal-based NMs (such as Fe-based NPs, Cu-based NPs, and BNPs) are used most frequently in environmental remediation, polymer- and composite-based NMs (such as chitosan, alginate, clay-polymer nanocomposites, zeolite, and biochar supported nanocomposites) have received significant research attention but limited practical application. [20] Fe-based NPs have also been used successfully in the area, along

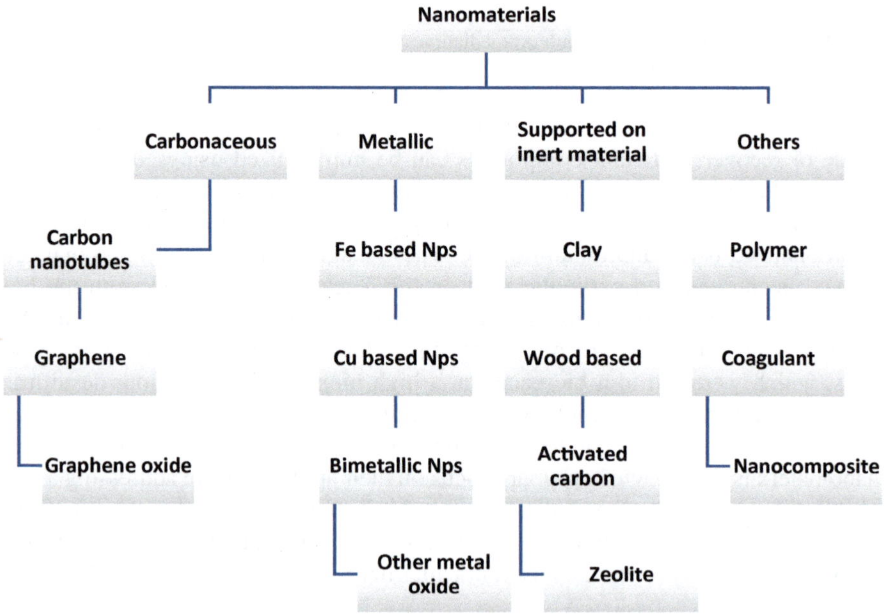

Fig. 21.3 Types of nanomaterials employed to remove water pollutants

with a few cases of Cu-based NPs, BNPs, various metal oxide NPs (such TiO_2, ZnO, and CNTs), and so on. The main benefits of using NMs for cleaning water and soil are their high selectivity, high adsorption capacity (because of a large specific surface area and numerous adsorption sites), and ease of regeneration after use [21].

21.5.1 Metal and Metal-Based Nanomaterials

Enhanced photocatalyst microparticles that incorporate gold nanoparticles can be used to purify water. Gold nanoparticles (AuNPs) are attractive probes for the attachment of peptides because of their unique physicochemical and optical properties, which make them well suited for conjugation with small biomolecules [22, 23]. Additionally, a variety of metal oxide nanoparticles, such as iron oxide (Fe_2O_3/Fe_3O_4), zinc oxide (ZnO), and titanium dioxide (TiO_2), are used for water purification due to their high reactivity, photolytic properties, as well as adsorbent properties derived from their vast surface area and affinity to different chemical groups [24, 25]. For instance, iron nanoparticles have been utilized to remove colors from wastewater from the textile, paint, and paper industries because of their stability in suspension medium and high adsorption capability [26]. Methyl orange and methylene blue, two of the most frequently used industrial dyes and pesticides with the greatest consequences on the environment and human health, have recently been shown to be extremely successful in being absorbed by these NPs [27–31]. The removal of methyl orange and phenol using magnetic iron oxide nanoparticles and carbon demonstrated that the nanocomposites had stronger interactions with the dye, with the amount of carbon being a crucial determinant in the NPs' adsorbent activity [32]. In addition to colors, several metals like chromium (VI) represent a significant class of water pollutants. According to recent studies, the use of iron oxide, zero-valent iron NPs, and organic acids (such citric acid) might lessen the harm that chromium does to the environment [33]. Titanium dioxide nanoparticles (TNPs) are an effective alternative for recently developing contaminants like pharmaceuticals and a common photocatalyst for the removal of micropollutants in water [34].

21.5.2 Nanomaterials Based on Carbon

Examples of nano-porous carbon-based materials that exhibit physicochemical properties that make them suitable for water treatment processes to remove contaminants like heavy metals, fluorides, textile dyes, or pharmaceutical products include activated carbons, carbon nanotubes (CNTs), including multi-walled and single-walled nanotubes (MWCNTs), graphene, and its oxide. Additionally, numerous organic and metal ion water pollutants have been shown to be attracted to the adsorption characteristics, high surface to volume ratio, conductivity, electrochemical stability, and catalytic activity of graphene oxide (GO) and reduced graphene oxide (rGO) nanosheets [35]. MWCNTs are used in the absorption of chromium in polluted groundwater [36]. Effect of absorption is impacted by factors like pH and adsorbent concentration. If the pH values are higher than 7, the adsorption reduced. MWCNTs have also been employed in water gasoline removal projects [37]. Due to the serious environmental damage that fluoride causes, many carbon-based substitutes have been utilized to defluoridate wastewater. The removal of fluoride by chemically and biologically reduced graphene oxide reveals that the former had an 87% reduction while the latter had a 94% capacity in this regard [38]. Similar to how much study has been done on the utilization of activated carbon in removing pharmaceutical chemicals due to its low cost, large pore size, and high porosity. For instance, a 2019 study that compared the adsorption of carbamazepine and sildenafil citrate onto granular and powdered activated carbon [39] indicates that 90% of the compounds were removed from the combination in just 10 h, as opposed to just 40% for granular activated carbon after 70 h, due to the higher surface area of powdered activated carbon. The evaluation of caffeine, ibuprofen, and triclosan adsorption using powdered activated carbon was also reported, and it indicated a substantial influence of pH [40].

21.5.3 Nanomaterials Based on Polymers

The use of nanoparticles, nanocomposites, or NF membranes might be among the polymer nanotechnology-based water treatment methods [41]. By detourning particles in the membrane pores and by the pollutants' chemical interaction with the membranes, which results in the pollutant being immobilized, polymeric nanomembranes are employed in particular to remove unwanted nanoparticles from the aqueous phase. In this sense, solvent casting is a common production technique that uses chitosan as a polymer to produce NF membranes. These membranes are a method for cleaning textile effluent, according to Long et al. (2020) [42], because they reject electroneutral and negatively charged colors less frequently than positively charged dyes. However, NF efficiency is also significantly influenced by the physical size of the dyes [43]. By using the membranes as a matrix or support for other types of materials, one might increase the durability and effectiveness of these nanofiltration membranes. Recently, a number of additives have been employed as membrane matrices to change synthetic and natural polymers including polyamide, cellulose, and chitosan. These additions include triethanolamine, metal oxide nanoparticles, and carbon nanotubes [44]. For instance, employing carboxylated MWCNTs in polyamide membranes has been shown to improve the rate at which salt is rejected, which is very useful for eliminating industrial salts from textile effluents [45]. Additionally, polyethersulfone membranes with MWCNTs, graphene, or other functionalized polymers demonstrated effective heavy metal and dye rejection in aqueous conditions [46].

21.6 Soil Nano-Remediation

The environment was irreversibly changed when humans settled during the transition from hunter-gatherers to farmers. The dominance of wheat, first as a source of food and then as a medium of trade, had an effect on the extinction of animal and plant species, the altering of river courses, soil erosion, and pollution. Due to the existence and expansion of industrialization and excessive urbanization, the deterioration and pollution of soil have worsened [47]. Nanomaterials have lately grown in favor for soil remediation because of their high reactivity, high surface-to-volume ratio, surface functionalization, and capacity to change physical characteristics including size, shape, porosity, and chemical composition. These qualities work together to trap impurities efficiently and selectively. The intercalation of nanoparticles in the soil allows for the cleaning of huge areas while spending less money and time, thanks to the in situ application. In soil pollution nano-remediation, zero-valent iron, carbon nanotubes, and magnetic nanoparticles have prevailed [48].

21.6.1 Nanomaterials Based on Metal

At the nanoscale, zero-valent iron (nZVI) is an electron donor with a negative reduction potential. nZVI use is one of the most popular in pilot tests because it enables the removal of polychlorinated biphenyls, organochlorine pesticides, and chlorinated organic solvents by oxidation–reduction transformation techniques sequestration [49]. Additionally, nZVI has shown effective cleaning percentages for the removal of DDT, lead, cadmium, nitrate, trichloroethene, and hexavalent chromium [50]. nZVI may be made in a variety of methods, including electrochemically, carbothermally, with ultrasonic assistance, and by green synthesis. nZVI has weak mobility, poor agglomeration dispersion stability, and is challenging to separate from treated soil even though it is reactive as a reducing agent [51]. The most widely used techniques for surface modification and material function maintenance involve alloying Pd, Pt, Ag, Cu, and Ni with other noble metals. Other techniques (ethylene glycol) include covering surfaces with synthetic polymers like polyethylene or biopolymers like starch, carboxymethyl cellulose, or guar gum. While the incorporation of nZVI on the surface of supports like silica, activated carbon, zeolites, or polymer membranes makes it easier to separate the nanomaterial from the cleansed soil. Additionally, nZVI can be "trapped" in particle emulsions or dispersions in biopolymers including calcium alginate, chitosan, and gum arabic using a "trapping" method. Other metal-based nanoparticles include those made of silica, aluminum, titanium, iron phosphate, goethite, and magnetic materials [50].

21.6.2 Nanomaterials Based on Carbon

Carbonaceous nanoparticles are unique in that they have a large surface area, a high degree of microporosity, outstanding sorption properties, and an ecologically friendly composition. Some architectural designs incorporate materials like activated carbon nanoparticles, fullerene C60, fullerene C540, SWCNTs, MWCNTs, graphene, and others [52]. Activation or functionalization of carbon-based nanomaterials also has additional advantages, much like in other environmental remediation applications. CNTs have lately become more well-liked because they have a greater capacity for adsorption than graphene, graphene oxides, biochar, and granular activated carbon. Adsorption is influenced by the exposure area and surface functional groups like -COOH and -OH. The adsorption capacity can be increased by joining functional groups like $-NH_2$, -SH, oxidation processes, coating with nonmagnetic metal oxides, and grafting magnetic iron oxides. Materials that are more prone to flocculation and less appropriate for nano-remediation have increased surface area, a high surface-to-volume ratio, and therefore high reactivity. The surfactant poloxamer 407 has successfully stabilized multi-wall carbon nanotubes [53]. The removal of heavy metals by CNTs is possible for Pb^{2+}, Cu^{2+}, Ni^{2+}, and ZN^{2+}, but it depends on the pH, the quantity of organic matter present, and the presence of silt and clay particles. Additionally, by eliminating pollutants including hexachlorocyclohexane, crude oil, Cr (VI), Cd, and DDT and promoting microbial activity and plant development, CNTs can enhance soil quality [54]. CNTs can be used in aqueous dispersion, separation columns, and membrane filtration processes.

21.7 Nano-Remediation in the Gas Phase

One of the biggest issues the world is presently dealing with is air pollution, which has an impact on both public health and climate change. Particulate matter (PM10 and PM2.5), nitrogen oxides (NOx), sulfur dioxide (SO_2), carbon monoxide (CO), lead, and ground-level ozone, which is produced by chemical interactions between NOx and volatile organic compounds (VOCs), are the six most prevalent and dangerous outdoor air pollutants [55]. The following are recognized as secondary particulate matter precursors: ammonia (NH_3), NOx, SOx, and VOCs. Despite being the most major greenhouse gas created by human activity, carbon dioxide (CO_2) is not a pollutant. Potential remedies for this problem include the use of graphene oxides (GOs), graphite oxides, CNTs with highly reactive surface sites, and mesoporous silica materials with ordered and adjustable porous structure, high surface area, large pore volume, and heat stability [56].

21.7.1 Materials Based on Carbon

Nanotechnology has benefits for reducing air pollution, including remediation and treatment, pollution prevention, monitoring, and sensing. By changing the reaction temperature with the addition of water, it is possible to control the amount of oxygen-containing functional groups that are present on the surface of graphite oxide. This material has been used in ammonia gas sensors that operate at different temperatures [57, 58]. Furthermore, nanoparticles based on carbon offer the chance to mix different nanomaterials to make nanocomposites, which combine numerous properties into a single new material [59, 60]. GO and zirconium hydroxide/graphene composites have been exploited as an environmental remediation method by the adsorption of SO_2. GO was also employed as a photocatalyst to break down VOCs and underwent photoreduction while being subjected to UV light [61]. A GO membrane with a sizable specific surface area and continuous pore structure was also used to collect PM2.5 [62]. Numerous studies have been conducted on CNTs in an effort to enhance their adsorption capacities. Typically, additional reactive materials with the appropriate functional groups or charges must be used to modify or coat CNTs. Modified MWCNTs or SWCNTs have been employed in investigations by Dong et al. (2013), [63] Park et al. 2019, [64] and Amade et al. (2014) [65] to detect H_2S and SO_2, CO and NH_3, NO_2 and NH_3, VOCs, and NOx.

21.7.2 Materials Based on Silica

Silica-based nanomaterials exhibit extraordinary flexibility due to their many advantageous properties, including their enormous surface area, variable pore size, and readily changeable surfaces. Additionally, in recent years, interest in using these nanomaterials to clean up dirty air and remove contaminants from the gas phase has increased because of their capacity for catalysis and adsorption. The physicochemical properties of silica nanoparticles may be improved through surface modification. For instance, hydroxyl groups on the surface of silica nanoparticles may facilitate the production of a number of surface phenomena, including gas adsorption and wetting. The process is effective for developing novel catalysts and adsorbents. The presence of amine groups on the surface of modified mesoporous silica was shown to assist the efficient capture of H_2S and CO_2 from natural gas in one of the earliest experiments examining the adsorbent capacity of the material [66]. According to the experts, the material swiftly eliminated up to 80% of the entire H_2S (in 35 min) and CO_2 in the sample, proving its outstanding efficacy in doing so (in 30 min). Similar findings from another research imply that amino-silicates may aid in the prevention of climate change by removing CO_2 from the environment [67]. Along with CO_2, these amine-modified silicates effectively remove organic contaminants including aldehydes and ketones [68]. In an industrial context, they may consequently be helpful for removing pollutants. Conversely, lead (Pb) atmospheric pollution is a problem that affects both the environment and human health worldwide, and it is not yet known how to totally eradicate Pb from the atmosphere. To solve this environmental issue, Yang et al. [69] created silica nanoparticles. The outcomes showed that their silica nanoparticles were capable of removing ambient Pb from contaminated air. As a result, silica-based nanoparticles may be desirable environmental agents to combat industrial pollution caused by lead and other heavy metals (Table 21.1).

Table 21.1 Advantages and possible risks of nanomaterials for remediation

Types of nanomaterials	Remediation mechanism	Advantages	Risks
Metal based	Adsorption, oxidation, reduction, photodegradation, photocatalysis [70]	High specific surface area, removal of diverse pollutants (chlorinated, organic solvents, polychlorinated biphenyls, organochlorine pesticides)	NPs can have adverse effects on pure cultures of bacteria, research of the risks in human and environmental health is missing
Carbon based	Adsorption	Compatibility with other treatments, high surface area	Different cell toxicity effects (reactive oxygen species production, lysosomal and DNA damage), rapid saturation, and high cost
Polymer based	Nanofiltration	Microporosity, sorption properties, eco-friendly nature, compatibility with other treatments, employment of polymer derived from waste materials	Denaturation by extreme temperature
Silica-based	Catalysis, adsorption	Versatility on surface modification, adaptable pore size, compatibility with other treatments [71]	Performance depends on pH Scattered size distribution

21.8 Risk to the Environment and Ecotoxicology

Despite the fact that they have been used effectively for soil and groundwater cleanup, exposure to nanoparticles may have harmful effects on both humans and the environment. For toxicological risk assessments, information on nanomaterial exposure, absorption, and immediate impacts is needed, as well as information on NPs' immediate effects after entering the human system [72]. To make judgments or suggestions, however, there is not enough information available in this area. A complex process and a number of factors affect environmental toxicity. Thus, the effects of synthetic NPs on living beings may depend on a number of factors, such as the dissolving potential, the properties of the particle surface, the potential for aggregation, the exposure environment characteristics, and the physiological, biological, and organism behavior when exposed to NPs [73]. Nanoparticles have an impact on both humans and the environment. For instance, iron oxide NPs can have a mutagenic impact by impairing an organism's ability to grow or reproduce. It can be demonstrated that exposure to subinhibitory levels of amoxicillin-bound iron oxide NPs encouraged bacterial growth in pseudomonas aeruginosa and Staphylococcus aureus in the presence of humic acid [74]. The combined impacts of NPs and other contaminants on terrestrial plants are currently the subject of scant investigation. To provide a firm foundation for risk assessment, more research is required to examine the cumulative implications under actual conditions [75]. The size and shape of NPs ultimately influence the amount of toxicity. However, it is not only unnecessary to monitor NPs in soil-plant systems, but additional information is also needed about how they are distributed in size and other ways [76]. There are various nano-risk techniques which are designed to be utilized as preliminary risk screening and/or research prioritization tools rather than to help regulatory decision-making. The conventional risk assessment framework is a helpful tool, but in the near future it might not be able to evaluate the risks that engineered nanomaterials pose to human health and the environment because of significant methodological limitations and epistemic uncertainty [77]. This section gives an overview of current studies on the environmental dangers of employing nanomaterials to remediate soil and groundwater. Gómez-Sagasti et al. [78] conducted a three-month experiment to investigate the impact of nZVI concentrations (ranging from 1 to 20 mg g DW soil^{-1}) on the microbiological traits of sandy-loam and clay-loam soils. The results provided evidence that the kind of soil may affect any negative impacts that nZVI may have on the microbial communities in the soil. The results showed that nZVI had a more strong inhibitory impact on soil microorganisms in sandy-loam soil compared to clay-loam soil. The high organic content of clay-loam soil, which acts as a protective agent when nZVI is applied to the soil by making it inactive and limiting contact with soil microorganism cells, explains this. The nZVI remediation of sandy-loam soil negatively impacted the bacterial biomass as well as the activity, diversity, and richness of arylsulfatases. They failed to discover any blatant concentration-response effects of nZVI therapy on the soil. In order to fully comprehend the kind and impact of soil attributes on the effects of nZVI on soil bacterial populations, additional research using a range of soil types and soil characteristics is reportedly required. The relationships between cytotoxicity and particle stability were studied by Dong et al. [79] in a separate inquiry, and they also looked at the impact of carboxymethyl cellulose (CMC) surface coating on the cytotoxicity and colloidal stability of nZVI against Escherichia coli (E. coli). The effects of particle cytotoxicity on CMC concentration, ionic strength (Ca^{2+}), and anti-aging treatment were also investigated. The results show that nZVI is time- and concentration-dependent toxic to E. coli. Conversely, the cytotoxicity of the nZVI was reduced when the particles were coated with CMC. This is because CMC-modified nZVI's cell membrane is intact, in contrast to naked nZVI, which showed cell membrane collapse when it came into contact with E. coli. The aged nZVI and CMC-nZVI showed no impact on E. coli because of the FeO conversion to less dangerous iron oxide. The toxicity of nZVI and CMC-nZVI in response to its presence was concentration-dependent because Ca^{2+} concentration might increase or decrease. The presence of Ca^{2+} may cause nZVI to settle and aggregate, which may lessen its toxicity. By encouraging NP adhesion to bacterial surfaces and producing a more toxic impact, Ca^{2+} might, however, conceivably make nZVI more hazardous. In a separate study, Chaithawiwat et al. [80] looked into how nZVI affected the different phases of bacterial development in four distinct bacterial strains. The results showed that during the lag and stationary phases, all bacterial strains were resistant to nZVI, but that resistance decreased during the exponential and decline phases. The results also show that as nZVI concentration was increased, bacterial inactivation increased. The results suggest that it is important to consider both the bacterial development phase and nZVI concentration when investigating the effects of nZVI on bacteria.

21.9 Conclusion

The latest advancements in soil and groundwater pollution nano-remediation are attempted to be described in this chapter. The main advantages of using nanomaterials in soil remediation are a decrease in overall costs and cleaning time, a near total eradication of pollutants at the site, and the removal of the need to dispose of contaminated soil. nZVI nanoparticles are frequently utilized in environmental remediation because of their high reactivity and higher capacity to immobilize heavy metals as Cd, Ni, and Pb. By altering and/or covering nZVI, it is feasible to reduce the cytotoxic effects on soil microorganisms. Carbon nanotubes (CNTs) are superior nanomaterials for both organic and inorganic cleaning because of their enormous surface area and great adsorption capability. More study is needed to better understand how CNTs affect the environment. Metal and MNP-based soil and groundwater remediation is now a possible strategy, thanks to the unique separation technique. Before using full-scale nano-remediation, further study is necessary, especially in respect to its efficacy and potential adverse impacts on the environment. Combining nano-remediation with other remediation technologies seems to hold the key to the future of soil remediation, as doing so brings soil remediation methods closer to "green" environmental standards.

More effort must be placed toward developing more intelligent nanomaterials in order to enhance soil cleanup. NPs having a high capacity for numerous jobs, such as interacting with hydrophilic and hydrophobic materials or catalyzing several pollutant processes on a single particle, could be produced as a consequence of more advanced research, for instance. To develop NPs that can remediate a range of pollutants and to enhance injecting systems, more research is necessary. The majority of currently conducted research on nano-remediation is limited to laboratory tests and modeling. It is difficult to translate these findings to in situ settings. In order to develop standardized application methods and dosages for nanomaterials at the field level, further study is required. In order to understand the fate and transport behavior of nanoparticles in soil, water, and sediments as well as how they affect environmental factors, efforts should also focus on the field application of nano-remediation. Nano-remediation has been created and studied over the past 20 years. However, there are concerns about how it may impact both the environment and humans. Given the quick development of nano-remediation technology, adequate evaluation must be done in order to eliminate or mitigate any potential environmental or ecological concerns. In order to comprehend the techniques for eliminating pollutants and the behavior of nanomaterials in nature, studies without contamination have also been carried out. Numerous research studies have examined how bacteria in soil and groundwater are affected by nano-remediation, but it is still unknown how compounds like nZVI interact with microbial activity.

References

1. Kamat PV, Meisel D (2003) Nanoscience opportunities in environmental remediation. C R Chim 6:999–1007
2. Punia P, Aggarwal RK, Kumar R, Dhar R, Thakur P, Thakur A (2003) Adsorption of cd and Cr ions from industrial wastewater using ca doped Ni–Zn nanoferrites: synthesis, characterization and isotherm analysis. Ceram Int 48(13):18048–18056
3. Chalia S, Bharti MK, Thakur P, Thakur A, Sridhara SN (2003) An overview of ceramic materials and their composites in porous media burner applications. Ceram Int 47(8): 10426–10441
4. Del Prado-Audelo M, Kerdan IG, Escutia-Guadarrama L, Reyna-González JM, Magaña JJ, Leyva-Gómez G (2021) Nanoremediation: nanomaterials and nanotechnologies for environmental cleanup. Front Environ Science 9:793765
5. Thakur A, Verma R, Wan F, Ravelo B, Edelman I, Ovchinnikov S, Thakur P (2023) Investigation of structural, elastic and magnetic properties of Cu2+ ions substituted cobalt nano ferrite. J Magn Magn Mater 581: 170980
6. Verma R, Thakur P, Sun AC, Thakur A (2023) Investigation of structural, microstructural and electrical characteristics of hydrothermally synthesized Li0.5-0.5xCocFe2.5-0.5xO4 ($0.0 \leq x \leq 0.4$) ferrite nanoparticles. Phys B: Cond Mater 661: 414926
7. Baby R, Saifullah B, Hussein MZ (2019) Carbon nanomaterials for the treatment of heavy metal-contaminated water and environmental remediation. Nanoscale Res Lett 2019(14):341
8. Singh A, Wan F, Yadav K, Salvi A, Thakur P, Thakur A (2023) Synergistic effect of ZnO nanoparticles with Cu2+ doping on antibacterial and photocatalytic activity. Inorg Chem Comm 157: 11425
9. Karn B, Kuiken T, Otto M (2009) Nanotechnology and in situ remediation: a review of the benefits and potential risks. Environ Health Perspect 117(12):1823–1831
10. Mueller NC, Braun J, Bruns J, Černík M, Rissing P, Rickerby D, Nowack B (2012) Application of nanoscale zero valent iron (NZVI) for groundwater remediation in Europe (PDF). Environ Sci Pollut Res 19(2):550–558
11. Bardos P, Bone B, Daly P, Elliott D, Jones S, Lowry G, Merly C (2014) "A risk/benefit appraisal for the application of nano-scale zero valent iron (nZVI) for the remediation of contaminated sites" (PDF). www.nanorem.eu. Accessed 3 Dec 2014
12. Himanshi, Jasrotia R, Prakash J, Verma R, Thakur P, Kandwal A, Wan F, Thakur A (2023) Synthesis, characterization and applications of doped barium hexaferrites: A Review. Phys B: Cond Mater 667: 415202
13. Quan X, Yang S, Ruan X, Zhao H (2005) Preparation of Titania nanotubes and their environmental applications as electrode. Environ Sci Technol 39(10):3770–3775
14. De Windt W, Aelterman P, Verstraete W (2005) Bioreductive deposition of palladium (0) nanoparticles on Shewanella oneidensis with catalytic activity towards reductive dechlorination of polychlorinated biphenyls. Environ Microbiol 7(3):314–325
15. Shan G, Xing J, Zhang H, Liu H (2005) Biodesulfurization of Dibenzothiophene by microbial cells coated with magnetite nanoparticles. Appl Environ Microbiol 71(8):4497–4502
16. Gillham RW, O'Hannesin SF (1994) Enhanced degradation of halogenated Aliphatics by zero-valent iron. Ground Water 32(6):958–967
17. Wang C-B, Zhang W-X (1997) Synthesizing nanoscale iron particles for rapid and complete Dechlorination of TCE and PCBs. Environ Sci Technol 1997(31):2154–2156
18. Guerra FD, Attia MF, Whitehead DC, Alexis F (2018) Nanotechnology for environmental remediation: materials and applications. Molecules 23(7):1760

19. Suhag D, Sharma A, Patni P, Garg S, Rajput SK, Chakrabarti SK, Mukherjee M (2016) Hydrothermally functionalized biocompatible nitrogen doped graphene nanosheet based biomimetic platforms for nitric oxide detection. J Mater Chem B 4(27):4780–4789
20. Kumari S, Dhnada N, Thakur A, Gupta V, Singh S, Kumar R, Hameed S, Thakur P (2023) Nano Ca–Mg–Zn ferrites as tuneable photocatalyst for UV light-induced degradation of rhodamine B dye and antimicrobial behavior for water purification. Ceram Int 49: 12469–12480
21. Pal K, Khurana SMP, Debnath N, Das S (2022) Micronutrient nanoparticles: synthesis, properties and application in agriculture. In: Thakur A, Thakur P, Khurana SP (eds) Synthesis and applications of nanoparticles. Springer, Singapore pp 337–356.
22. Shahdeo D, Kesarwani V, Suhag D, Ahmed J, Alshehri SM, Gandhi S (2021) Self-assembled chitosan polymer intercalating peptide functionalized gold nanoparticles as nanoprobe for efficient imaging of urokinase plasminogen activator receptor in cancer diagnostics. Carbohydr Polym 266:118138
23. Shahdeo D, Kesarwani V, Suhag D, Ahmed J, Alshehri SM, Gandhi S (2022) Corrigendum to" self-assembled chitosan polymer intercalating peptide functionalized gold nanoparticles as nanoprobe for efficient imaging of urokinase plasminogen activator receptor in cancer diagnostics"[carbohydrate polymers, 266, (2021) 118138]. Carbohydr Polym 284:119239
24. Aragaw TA, Bogale FM, Aragaw BA (2021) Iron-based nanoparticles in wastewater treatment: a review on synthesis methods, applications, and removal mechanisms. J Saudi Chem Soc 25: 101280
25. Krishnia L, Thakur P, Thakur A (2022) Synthesis of nanoparticles by physical route. In: Synthesis and applications of nanoparticles. Singapore, Springer Nature Singapore, pp 45–59
26. Singh R, Mehra R, Walia A, Gupta S, Chawla P, Kumar H, Kumar N (2021) Colorimetric sensing approaches based on silver nanoparticles aggregation for determination of toxic metal ions in water sample: a review. Int J Environ Anal Chem 103:1–16
27. Chahar D, Kumar D, Thakur P, Thakur A (2023) Visible light induced photocatalytic degradation of methylene blue dye by using Mg doped Co-Zn nanoferrites. Mater Res Bull 62: 112205.
28. Hannachi E, Slimani Y, Nawaz M, Trabelsi Z, Ghulam Y, Bilal M, Almessiere M, Baykal A, Thakur A, Thakur P (2022) Synthesis, characterization, and evaluation of the photocatalytic properties of zinc oxide co-doped with lanthanides elements. J Phys Chem Sol 70: 110910
29. Punia P, Bharti MK, Chalia S, Dhar R, Ravelo B, Thakur P, Thakur A (2021) Recent advances in synthesis, characterization, and applications of nanoparticles for contaminated water treatment-a review. Ceram Int 47(2):1526–1550
30. Singh R, Thakur P, Thakur A, Kumar H, Chawla P, Rohit VJ, Kumar N (2021) Colorimetric sensing approaches of surface-modified gold and silver nanoparticles for detection of residual pesticides: a review. Int J Environ Anal Chem 101(15):3006–3022
31. Bharti MK, Gupta S, Chalia S, Garg I, Thakur P, Thakur A (2020) Potential of magnetic nanoferrites in removal of heavy metals from contaminated water: mini review. J Supercond Nov Magn 33:3651–3665
32. Istratie R, Stoia M, Păcurariu C, Locovei C (2019) Single and simultaneous adsorption of methyl orange and phenol onto magnetic iron oxide/carbon nanocomposites. Arab J Chem 12: 3704–3722
33. Yang J, Zhong L, Liu L (2017) Chromium (VI) reduction in the nano- or micron-sized iron oxide - citric acid systems: kinetics and mechanisms. J Environ Chem Eng 5:2564–2569
34. Mahmoud WMM, Rastogi T, Kümmerer K (2017) Application of titanium dioxide nanoparticles as a Photocatalyst for the removal of micropollutants such as pharmaceuticals from water. Curr Opin Green Sustain Chem 6:1–10
35. Chakraborty A, Patni P, Suhag D, Saini G, Singh A, Chakrabarti SK, Mukherjee M (2015) N-doped carbon nanosheets with antibacterial activity: mechanistic insight. RSC Adv 5: 23591–23598
36. Mpouras T, Polydera A, Dermatas D, Verdone N, Vilardi G (2021) Multi wall carbon nanotubes application for treatment of Cr (VI)-contaminated groundwater; modeling of batch & column experiments. Chemosphere 269:128749

37. Lico D, Vuono D, Siciliano CB, Nagy J, De Luca P (2019) Removal of unleaded gasoline from water by multi-walled carbon nanotubes. J Environ Manag 237:636–643
38. Roy S, Manna S, Sengupta S, Ganguli A, Goswami S, Das P (2017) Comparative assessment on defluoridation of waste water using chemical and bio-reduced graphene oxide: batch, thermodynamic, kinetics and optimization using response surface methodology and artificial neural network. Process Saf Environ Prot 111:221–231
39. Delgado N, Capparelli A, Navarro A, Marino D (2019) Pharmaceutical emerging pollutants removal from water using powdered activated carbon: study of kinetics and adsorption equilibrium. J Environ Manag 236:301–308
40. Kaur H, Bansiwal A, Hippargi G, Pophali GR (2018) Effect of hydrophobicity of pharmaceuticals and personal care products for adsorption on activated carbon: adsorption isotherms, kinetics and mechanism. Environ Sci Pollut Res 25:20473–20485
41. Abdelbasir SM, Shalan AE (2019) An overview of nanomaterials for industrial wastewater treatment. Korean J Chem Eng 36:1209–1225
42. Long Q, Zhang Z, Qi G, Wang Z, Chen Y, Liu Z-Q (2020) Fabrication of chitosan Nanofiltration membranes by the film casting strategy for effective removal of dyes/salts in textile wastewater. ACS Sustain Chem Eng 8:2512–2522
43. Weng R, Chen L, Lin S, Zhang H, Wu H, Liu K et al (2017) Preparation and characterization of antibacterial cellulose/chitosan nanofiltration membranes. Polymers (Basel) 9:1–13
44. Lakhotia SR, Mukhopadhyay M, Kumari P (2018) Cerium oxide nanoparticles embedded thin-film nanocomposite Nanofiltration membrane for water treatment. Sci Rep 8:4976–5010
45. Al-Hobaib AS, Al-Sheetan KM, Shaik MR, Al-Suhybani MS (2017) Modification of thin-film polyamide membrane with multi-walled carbon nanotubes by interfacial polymerization. Appl Water Sci 7:4341–4350
46. Peydayesh M, Mohammadi T, Nikouzad SK (2020) A positively charged composite loose Nanofiltration membrane for water purification from heavy metals. J Membr Sci 611:118205
47. Kumar L, Ragunathan V, Chugh M, Bharadvaja N (2021) Nanomaterials for remediation of contaminants: a review. Environ Chem Lett 19:3139–3163
48. Mukhopadhyay R, Sarkar B, Khan E, Alessi DS, Biswas JK, Manjaiah KM et al (2021) Nanomaterials for sustainable remediation of chemical contaminants in water and soil. Crit Rev Environ Sci Techn 52:1–50
49. Cheng P, Zhang S, Wang Q, Feng X, Zhang S, Sun Y et al (2021) Contribution of Nano-zero-valent iron and arbuscular mycorrhizal fungi to phytoremediation of heavy metal-contaminated soil. Nano 11:1264
50. Stefaniuk M, Oleszczuk P, Ok YS (2016) Review on Nano Zerovalent iron (nZVI): from synthesis to environmental applications. Chem Eng J 287:618–632
51. Guerra FD, Attia MF, Whitehead DC, Alexis F (2018) Nanotechnology for environmental remediation: materials and applications. Molecules 23:1–23
52. Marcon L, Oliveras J, Puntes VF (2021) In situ nanoremediation of soils and groundwaters from the nanoparticle's standpoint: a review. Sci Total Environ 791:148324
53. Matos M, Correia AAS, Rasteiro MG (2017) Application of carbon nanotubes to immobilize heavy metals in contaminated soils. J Nanopart Res 19:126
54. Shan J, Ji R, Yu Y, Xie Z, Yan X (2015) Biochar, activated carbon, and carbon nanotubes have different effects on fate of (14)C-catechol and microbial Community in Soil. Sci Rep 5:16000–16011
55. Manisalidis I, Stavropoulou E, Stavropoulos A, Bezirtzoglou E (2020) Environmental and health impacts of air pollution: a review. Front Public Health 8:14–13
56. Srivastava N, Mishra G, Saxena SK (2022) Environmental Nanomedicine. In: Thakur A, Thakur P, Khurana SP (eds) Synthesis and applications of nanoparticles. Springer, Singapore pp 487–501
57. Luo L, Peng T, Yuan M, Sun H, Dai S, Wang L (2018) Preparation of graphite oxide containing different oxygen-containing functional groups and the study of ammonia gas sensitivity. Sensors 18:3745

58. Bannov AG, Prášek J, Jašek O, Zajíčková L (2017) Investigation of pristine graphite oxide as room-temperature Chemiresistive ammonia gas sensing material. Sensors (Switzerland) 17:1–10
59. Scida K, Stege PW, Haby G, Messina GA, García CD (2011) Recent applications of carbon-based nanomaterials in analytical chemistry: critical review. Anal Chim Acta 691:6–17
60. Seredych M, Bandosz TJ (2010) Effects of surface features on adsorption of SO2 on graphite oxide/Zr(OH)4 composites. J Phys Chem C 114:14552–14560
61. Tai XH, Chook SW, Lai CW, Lee KM, Yang TCK, Chong S et al (2019) Effective Photoreduction of graphene oxide for Photodegradation of volatile organic compounds. RSC Adv 9: 18076–18086
62. Jung W, Lee JS, Han S, Ko SH, Kim T, Kim YH (2018) An efficient reduced graphene-oxide filter for PM2.5 removal. J Mater Chem A 6:16975–16982
63. Dong KY, Choi J, Lee YD, Kang BH, Yu YY, Choi HH et al (2013) Detection of a CO and NH3 gas mixture using carboxylic acid-functionalized single-walled carbon nanotubes. Nanoscale Res Lett 8:12–13
64. Park S, Byoun Y, Kang H, Song Y-J, Choi S-W (2019) ZnO nanocluster-functionalized single-walled carbon nanotubes synthesized by microwave irradiation for highly sensitive NO2 detection at room temperature. ACS Omega 4:10677–10686
65. Amade R, Hussain S, Ocaña IR, Bertran E (2014) Growth and functionalization of carbon nanotubes on quartz filter for environmental applications. J Environ Eng Ecol Sci 3:1–7
66. Huang HY, Yang RT, Chinn D, Munson CL (2003) Amine-grafted MCM-48 and silica Xerogel as superior sorbents for acidic gas removal from natural gas. Ind Eng Chem Res 42:2427–2433
67. Choi S, Drese JH, Eisenberger PM, Jones CW (2011) Application of amine-tethered solid sorbents for direct CO2 capture from the ambient air. Environ Sci Technol 45:2420–2427
68. Nomura A, Jones CW (2013) Amine-functionalized porous Silicas as adsorbents for aldehyde abatement. ACS Appl Mater Inter 5:5569–5577
69. Yang X, Shen Z, Zhang B, Yang J, Hong W-X, Zhuang Z et al (2013) Silica nanoparticles capture atmospheric lead: implications in the treatment of environmental heavy metal pollution. Chemosphere 90:653–656
70. Punia P, Bharti MK, Dhar R, Thakur P, Thakur A (2022) Recent advances in detection and removal of heavy metals from contaminated water. ChemBioEng Rev 9(4):351–369
71. Ivanets A, Prozorovich V, Kouznetsova T (2022) Advancement in water purification by nanomaterials. In: Thakur A, Thakur P, Khurana SP (eds) Synthesis and applications of nanoparticles. Springer, Singapore pp 387–404
72. Thakur P, Verma Y, Thakur A (2022) Toxicity of nanomaterials: an overview. In: Thakur A, Thakur P, Khurana SP (eds) Synthesis and applications of nanoparticles. Springer, Singapore pp 535–544
73. Lee SP, Ali GAM, Algarni H, Chong KF (2019) Flake size-dependent adsorption of graphene oxide aerogel. J Mol Liq 2019(277):175–180
74. Dissanayake NM, Current KM, Obare SO (2015) Mutagenic effects of iron oxide nanoparticles on biological cells. Int J Mol Sci 2015(16):23482–23516
75. Current KM, Dissanayake NM, Obare SO (2017) Effect of iron oxide nanoparticles and amoxicillin on bacterial growth in the presence of dissolved organic carbon. Biomedicine 2017(5):55
76. Du W, Xu Y, Yin Y, Ji R, Guo H (2018) Risk assessment of engineered nanoparticles and other contaminants in terrestrial plants. Curr Opin Environ Sci Health 2018(6):21–28
77. Shrivastava M, Srivastav A, Gandhi S, Rao S, Roychoudhury A, Kumar A, Singhal RK, Jha SK, Singh SD (2019) Monitoring of engineered nanoparticles in soil-plant system: a review. Environ Nanotechnol Monit Manag 2019(11):100218
78. Gómez-Sagasti MT, Epelde L, Anza M, Urra J, Alkorta I, Garbisu C (2019) The impact of nanoscale zero-valent iron particles on soil microbial communities is soil dependent. J Hazard Mater 2019(364):591–599

79. Dong H, Xie Y, Zeng G, Tang L, Liang J, He Q, Zhao F, Zeng Y, Wu Y (2016) The dual effects of carboxymethyl cellulose on the colloidal stability and toxicity of nanoscale zero-valent iron. Chemosphere 2016(144):1682–1689
80. Chaithawiwat K, Vangnai A, McEvoy JM, Pruess B, Krajangpan S, Khan E (2016) Impact of nanoscale zero valent iron on bacteria is growth phase dependent. Chemosphere 2016(144): 352–359

Impact of Nano-Biomaterials on the World 22

Irina Edelman, Sergey Ovchinnikov, Atul Thakur, and Preeti Thakur

Abstract

People have been exposed to airborne nanosized materials (100 nm) for thousands of years, but since the industrial revolution, particularly with regard to combustion operations, the level of exposure has significantly increased. Since the advent of nanotechnology, two decades ago, the danger of exposure to nanomaterials by ingestion, absorption through the skin, absorption through the dermis, and medication administration utilizing designed nanoparticles, has grown. Nanomaterials acquire new mechanical, electrical, optical, catalytic, and, last but not least, biological capabilities when they shrink from bulk to nanoscale. The creation, usage, and disposal of the product, as well as the toxicological impacts of nanomaterials, must all be studied. Basic multidisciplinary research including materials scientists, toxicologists, medical professionals, and environmental engineers is necessary to better understand the health risks and safety concerns of nanoparticles.

I. Edelman · S. Ovchinnikov (✉)
Kirensky Institute of Physics, Federal Research Center KSC, Siberian Branch, Russian Academy of Sciences, Krasnoyarsk, Russia
e-mail: sgo@iph.krasn.ru

A. Thakur
Amity Centre for Nanotechnology, Amity University Haryana, Gurgaon, Haryana, India
e-mail: athakur1@ggn.amity.edu

P. Thakur
Amity Centre for Nanotechnology, Amity University Haryana, Gurgaon, Haryana, India

Amity School of Applied Sciences, Amity University, Gurugram, Haryana, India
e-mail: pthakur@ggn.amity.edu

D. Suhag et al. (eds.), *Integrated Nanomaterials and their Applications*,
https://doi.org/10.1007/978-981-99-6105-4_22

Keywords

Nanomaterials · Health hazards · Cytotoxicity · Human health · Environment

22.1 Introduction

Nanotechnology which combines engineering with biology, chemistry, medicine, and physics is seen by many experts as the next logical step in science [1]. A substance's physical and chemical characteristics can alter significantly from those of the same material in bulk form when its dimensions are reduced to extremely tiny levels. The development of microscopic or even molecular devices using current nanotechnology has the potential to advance energy, space exploration, medicine, and environmental protection [2]. It will not be long until the entire history can be compressed inside our pockets or the system extended by specially created molecules that resemble life systems, thanks to our rapidly expanding knowledge of nanoscience and our capacity to engineer new goods and services. Engineered nanomaterials, such as those used in cosmetics, food packaging, medication delivery systems, therapies, biosensors, and other products, are quickly encroaching our daily lives [3, 4]. Nanomaterials are widely employed for a variety of commercial products such as wound treatment, detergents, or antimicrobial coatings because their size scale is close to that of biological macromolecules and they have antibacterial and odor-fighting characteristics [5, 6].

As the use of nanomaterials is growing, so does the population that is exposed to them. Despite the clear advantages of using nanoscale materials, there are unanswered questions regarding the potential effects of nanoparticles on the environment. Before mass production of nanomaterials, one of the most important problems that must be resolved in the near future is how harmful they are to people and how they affect the environment [7]. There is much discussion about the possibility that the unique features of nanoparticles could have hazardous biological impacts. What biological reactions might be expected when nanoparticles undergo biodegradation in the cellular environment? For instance, biodegraded nanoparticles may gather inside cells and cause intracellular changes including organelle integrity disruption or gene alterations. One of the most pressing questions is whether nanomaterials are more dangerous than their non-nano counterparts. In the environment, can nanoparticles degrade into more dangerous forms? It is critical for nanotoxicology research to uncover and comprehend how nanomaterials affect the environment before they can be used in daily activities in order to avoid their negative effects [8]. This chapter states recent developments in nanomaterial toxicity and environmental impact to answer concerns about potential environmental implications of upcoming nanotechnologies.

While it is still a young field of study, nanotechnology has the potential to have a big impact on how we approach and solve problems. It is believed that nanotechnology would significantly affect research, food systems, agriculture, medicine, and the environment.

22.2 Knowledge Gaps in Fate, Exposure and Toxicity of Nanomaterials

The National Institute for Workplace Safety and Health (NIOSH) is the principal government organization conducting research on the potential effects of nanomaterials on occupational safety and health, whereas in India National Institute of Occupational Health (NIOH) is established to prevent and control occupation related health problems by creating a safer work environment through intensive research (basic/epidemiological/translational), development of appropriate technology for risk minimization. In this role, NIOSH is aware that the research community is at the forefront of developing new nanomaterials, evaluating their use in various contexts, and figuring out their toxicological and environmental effects. Despite the lack of widespread knowledge of the threats that nanoparticles pose to human health and the environment, it is advisable to be aware of the most recent information and handling guidelines for nanomaterials from NIOSH and other reliable sources. Although the causes of nanotoxicity are not fully understood, it is known that nanoparticles have hazardous effects on the lungs, reproductive system, heart, gastrointestinal tract, skin, and immune system [9]. Nanomaterials can also result in persistent contamination of the air, water, and—most importantly—soil [10, 11]. The uptake of nanomaterials by plants, their transportation in the environment, and indicators of ecological exposure can all be significantly impacted by changes to the fundamental features of nanomaterials (such as size, shape, oxidation state, and partitioning). Because these pollutants are too small to be easily detected, they have long-term impacts that are unknown. Before large-scale nanotechnology is fully established, larger and multicenter investigations are required to ascertain human reaction and the destiny of the nanoparticles in the environment [12].

We do not know enough about how prolonged, low-dose exposure to nanomaterials can damage human health. The lungs, gut, and skin are the three main organs via which humans are exposed. In addition to being used in food goods and packaging, nanomaterials may be ingested or inhaled by factory workers. Nanomaterials become lodged in the liver after entering the body, but the long-term risks they bring are unknown. The issues that can arise from new breakthroughs have already been seen in the world. It is crucial that advancements in nanotechnology do not lead to similar health crises given the world's experiences with asbestos (which, despite being used for thousands of years, was only discovered as a source of disease in the 1900s), the contentious development of genetically modified foods, and the highly relevant microplastics crisis. We can protect users, workers, and the surroundings from potential health and safety dangers by strengthening non-animal tests for nanotechnology. Our lives have already been improved by nanotechnology, and with a better understanding of their safety, we may more confidently make use of the advantages this new technology brings.

22.3 Impact of Nanomaterials on Human Health

The possible effects of employing nanotechnological components and tools on human health are known as the health repercussions of nanotechnology. There is a lot of discussion over whether or not nanotechnology will be good or bad for human health because it is a relatively young science. The potential medical uses of nanotechnological advancements to treat disease and the possible health risks provided by nanomaterial exposure make up the two areas of the health effects of nanotechnology. Researchers, engineers, and medical experts are using a highly developed set of nanoscience and nanotechnology methodologies to study the present worldwide epidemic and how it may possibly help the scientific, technological, and medical sectors combat the pandemic [13].

The majority of naturally occurring substances and chemicals that come into touch with the human body have created tolerance. It is less likely to be safe with novel drugs since it lacks natural tolerance to them. Skin, lungs, and the digestive system are three ways that nanoparticles might enter the body. The development of "free radicals," which can harm cells, may be facilitated by this. Concerns have also been raised about nanoparticles' potential to enter the circulation and pass across the blood–brain barrier. According to Kirchner et al. [14], there are three main reasons why nanoparticles become harmful when they come into touch with live cells:

1. "Chemical toxicity of the materials used to make them. Cd^{2+}, for example, is released from cadmium selenide nanoparticles.
2. Nanoparticles' small size allows them to adhere to cellular membranes and enter cells. Attachment of nanoparticles to membranes and storage of nanoparticles within cells can impair cellular functions, even if the nanoparticles are chemically inert and do not decompose or react with other matrix components.
3. Form. Carbon nanotubes, for example, can easily pierce cell membranes."

The environmental effects of nanotechnology are becoming a more active area of study. Until recently, the potential negative effects of nanomaterials on human health and the environment were speculative and unsubstantiated [15, 16]. Wiesner et al. [17] examined the risks of manufactured nanomaterials as well as potential human and animal health risks. They found that fullerene-based nanomaterials are cytotoxic to human cells, may damage DNA, accumulate in rat livers, and create reactive oxygen species in addition to having antibacterial capabilities. Although the Federal Drug Administration (FDA) in the USA has approved some non-toxic nanoparticles for use in paints and sunscreen lotions, there are toxic nanoparticles and chemicals that are known to accumulate in the food chain, such as asbestos, diesel particulate matter, ultra-fine particles, and lead. It is well acknowledged that it is extremely difficult to apply knowledge of bulk materials to nanoparticles because of how different their chemical characteristics are from those of bulk materials.

Anti-bacterial silver nanoparticles dissolve in acids that would not dissolve bulk silver, which is one sign of their higher reactivity [18]. A list of some exposure episodes involving people and the environment is shown in Table 22.1. The table

Table 22.1 Impact and future prospects of nanomaterials

	Positive impacts	Negative impacts	Future prospects
1	Development of medicine and Economy	Long term regulatory measures for safety or lack of unified set of global regulations	Tumor targeting delivery system
2	Better stability of unstable active ingredients	Potential to be genotoxic if unregulated	Drug discovery/ vaccine
3	Medical robotics (better diagnosis and treatment).	Cultural, economic, and military (defense) changes	Implantable delivery system
4	Good biocompatibility, biodegradability, and bioavailability	Products/byproducts for manufacturing nanomaterials released in surroundings or penetrating the human body	Nanoparticles for gene delivery
5	Targeted drug delivery	Costlier to be used by majority	Molecular diagnosis
6	Detection of nutrients and pathogens by biosensors, quantum dots Nano-clay for water retention	Nanotoxicity (uncertain cytotoxicity of the nanostructured materials incorporated or migrated in food products)	Biosensors, nanobots, xenobots, bio-labels

unequivocally demonstrates that nanoparticles may be released from both point sources and non-point sources, such as businesses, landfills, moist deposition from the atmosphere, storm-water runoff, and attrition from products containing nanomaterials. In the atmosphere, nanomaterials are subject to photochemical reactions. Within living organisms, they can accumulate, alter, or decay. Nanomaterials must be removed from the air using air filters and respirators due to diverse atmospheric transmission processes. Inhaling nanomaterials that are released into the air and eating drinking water or food (like fish) that has collected nanoparticles are alternative exposure situations. Manufacturing is the most frequent moment when humans come into contact with nanomaterials.

Dermal exposure via sunscreen and cosmetics is also possible. To make the particles mobile in the subsurface and efficiently transfer nano iron to cells, special surface coatings will be required. These changes may increase their risk of inadvertent exposure to people and other creatures. Many other types of nanomaterials (such as metal oxide nanoparticles and quantum dots used as contrast agents in magnetic resonance imaging) will also require surface coatings in order to perform their intended function. Despite the fact that there are many different types of nanomaterials, free nanoparticles have mostly been the subject of concerns [19]. The likely pathways through which nanoparticles enter and move throughout the human body are depicted in Fig. 22.1 [20].

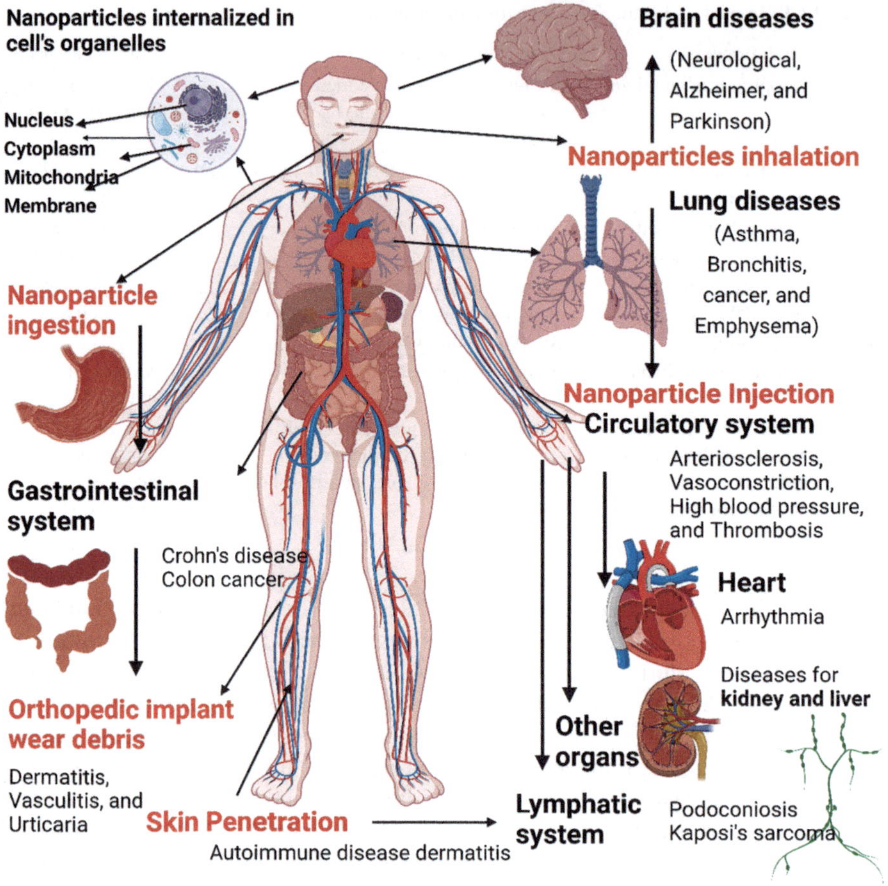

Fig. 22.1 Diagram showing the nanoparticle exposure routes in the human body [20]

22.4 Medical Applications

Nanomedicine is a medical application of nanotechnology [21]. Nanomedicine approaches range from medical applications of nanomaterials to nano-electronic biosensors and even potential future applications of molecular nanotechnology. In the near future, nanomedicine hopes to provide a valuable set of research tools and clinically useful devices [22, 23]. According to the National Nanotechnology Initiative, the pharmaceutical industry will see new commercial applications such as advanced drug delivery systems, new therapies, and in vivo imaging [24]. Another active area of research is neuro-electronic interfaces and other nanoelectronics-based sensors. Cell repair machines, according to the speculative field of molecular nanotechnology, could revolutionize medicine and the medical field in the future.

DNA oligonucleotide ligands possess the ability to identify other molecules and, in some cases, can even display catalytic properties. Aptamers are used in a large variety of biosensors [25, 26] and therapeutics [27]. Aptamer binding depends on its 3D structure. The problem now is that there are no aptamer single crystals, so the classical X-ray diffraction cannot help. To understand the molecular basis of the complex formation, knowledge about a main aptamer spatial conformation is very desirable but is often unavailable because of limitations of analytical methods for structure analysis. The most known method of small angle X-ray scattering (SAXS) with synchrotron source provides only the molecular shape, but not its structure. To solve this problem, recently a four-step approach to determine a true 3D structure of aptamers in solution using SAXS and molecular structure restoration (MSR) has been developed. The approach consists of (1) acquiring SAXS experimental data of an aptamer in solution, (2) building a spatial distribution of the molecule's electron density using SAXS results, (3) constructing a 3D model of the aptamer from its nucleotide primary sequence and secondary structure, and (4) comparing and refining the modeled 3D structures with the experimental SAXS model. This approach was used to analyze the 3D structure of RE31 aptamer to thrombin in a native free state at different temperatures. The resulting 3D structure of RE31 has the most energetically favorable conformation and the same elements such as a B-form duplex, non-complementary region, and two G-quartets. More broadly, this study demonstrates the complementary approach for constructing and adjusting the 3D structures of aptamers, DNA enzymes, and ribozymes in solution and could supply new opportunities for developing functional nucleic acids. The use of magnetic nanoparticles to destroy disease-causing neoplasms in the body is one of their newest applications. This direction is being actively developed, and the selective attachment of nanoparticles to cells that need to be destroyed is facilitated by their functionalization with an aptamer [28]. Aptamers are also widely used in cancer diagnostics and therapy [29]. Ehrlich carcinoma cell cultures are the most accessible for experiments, and recently they have become model objects for studying various effects on their destruction, including the action of a magnetic field [30]. It is observed that the magneto-dynamic therapy using the gold-coated magnetic NPs functionalized with DNA aptamers to selectively kill tumor cells in vivo [31]. The cell-specific DNA aptamer AS-14, binding to the fibronectin protein in Ehrlich carcinoma, delivered gold-coated magnetic nanoparticles to the mouse tumor. Applying an alternating magnetic field with a frequency of 50 Hz to the tumor site causes the nanoparticles to vibrate and attract fibronectin proteins and integrin's to the surface of the cell membrane. This leads to apoptosis followed by necrosis of tumor cells without heating the tumor, adjacent healthy cells and tissues. The core-shell Fe3O4@C NPs functionalized with aptamers were used successively in the experiments on the magneto-mechanical disruption of Ehrlich ascites carcinoma cell cultures [32].

22.5 Nanotoxicology

The study of possible health concerns caused by nanomaterials is known as nanotoxicology. Nanomaterials are easier for the human body to absorb than bigger particles since they are so tiny. How these nanoparticles act inside the organism is one of the key questions that has to be solved. The behavior of nanoparticles depends on their size, shape, and surface reactivity with the surrounding tissue. For instance, they may overstress the phagocytes that take in and get rid of foreign matter. Inflammation and stress responses would result from this, weakening the body's defenses against additional diseases. In addition to the possibility that non-degradable or slowly degradable nanoparticles build up in organs, another issue is how they might interact with biological processes occurring inside the body. Because nanoparticles have a large surface area, they will quickly adsorb some of the macromolecules they come into contact with when exposed to tissue and fluids. This could affect, for instance, how proteins and enzymes regulate one another. It is challenging to generalize about the health concerns associated with exposure to nanomaterials due to the multitude of factors determining toxicity; each new nanomaterial must be evaluated separately, and all material features must be taken into account. Health and environmental concerns coexist in the workplaces of companies that manufacture or use nanomaterials, as well as in nanoscience and nanotechnology laboratories (Fig. 22.2).

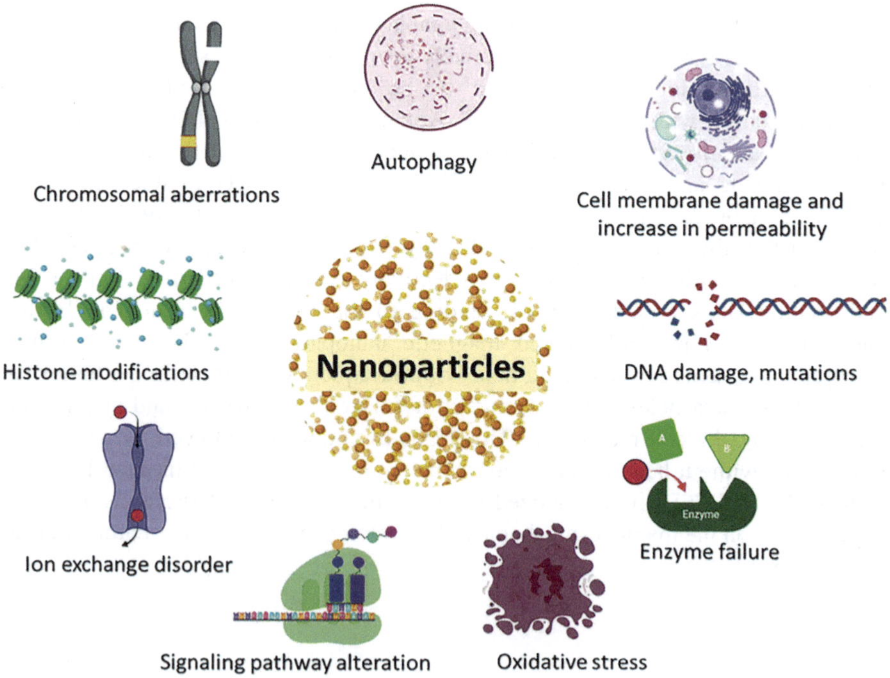

Fig. 22.2 Toxicity associated with nanoparticles [24]

22.6 Environmental Impact

The two primary sources of nanomaterials in the environment are those produced naturally and those produced by humans. Examples of natural sources include volcanoes, viruses, ocean spray, dust storms, bacteria, and bushfires. The two categories of anthropogenic sources include inadvertently formed nanomaterials, such as those released by combustion aerosols, including those from motor vehicle exhaust emissions, coal fly ash, and wielding operations, and purposely manufactured nanomaterials with specific features. These materials could encompass primarily metallic substances and metal oxides, such as fullerenes, quantum dots, nanowires, and nanotubes [33]. Furthermore, in each of these groups, nanoparticles can develop accidentally [34]. This evaluation concentrates on four important features of the deliberately created ENM group. These comprise (1) the description and use of ENM, (2) the estimation of ENM exposure, (3) the evaluation of hazards, and (4) environmental legislation.

The breakdown of nanomaterials (such as surface-bound nanoparticles or nanoscale coatings) can release free nanoparticles into the environment either directly or indirectly. The environment may be accessed in a variety of ways. Medicines are commonly coated, and studies have shown that these coatings can degrade either as a result of UV light-induced environmental alteration or as a result of human body metabolism [35]. About their transit and fate, aquatic nanomaterials have received comparably little research. This only serves to emphasize the need of investigating the many potential mechanisms that can alter the characteristics of nanoparticles once they are discharged into the environment. Nanomaterials or product spills during production, shipping, or disposal constitute additional routes for environmental exposure. Unless it is established that the benefits outweigh the risks, the Royal Society and The Royal Academy of Engineering warn against employing free nanoparticles in environmental remediation or other uses [36]. How much "nano-litter" exposure may impact living things and the precise toxicity pathways are currently unknown. Due to the numerous experimental difficulties and problems encountered when determining the toxicity of nanomaterials, Kumar and Sharma recently reviewed the toxicological studies on biological systems with metal and metal oxide nanoparticles, fullerenes, and carbon nanotubes [37] and came to the conclusion that additional research is still required to definitively establish their safety and toxicity. Chemical toxicology was a major consideration in the development and standardization of the majority of toxicity assessment techniques. Yet, due to their distinct physicochemical characteristics, nanoparticles might impede or complicate traditional toxicity assessments. In conclusion, unless the ambiguities surrounding their destiny, transport, and toxicity are resolved, applications of nanoparticles that involve their direct introduction to the environment promise to be problematic.

22.6.1 Environmental Risk Assessment for Nanoparticles

Nanoparticles may be emitted as aerosols into the atmosphere, surface water, and soil depending on the kind. Aggregates, functionalized nanoparticles, bare nanoparticles, and nanoparticles embedded in a matrix are all examples of nanoparticle releases into the environment [38]. When nanoparticles are released into the environment, whether on purpose or accidentally, they disperse and wind up in the air, water, and soil, where they can persist for a very long time or be consumed by living creatures. They can provide an ecotoxicological danger as they biodegrade or bioaccumulate in the food chain. There have been instances of diesel exhaust producing nanoparticles [39]. Automobiles continue to emit lead compounds and carbon nanoparticles due to incomplete combustion. The majority of contemporary cars are equipped with pollution control systems that utilize catalysts like platinum and other noble metals [40, 41].

It was discovered a few years ago that during the course of their lives, automobile catalysts release platinum nanoparticles with diameters ranging from 0.8 to 10 nm [42]. Although metal nanoparticles like platinum and palladium are included in the most recent catalysts created by the car industry, their sizes are different from those of conventional catalysts. It has been proven that nano-dispersed platinum group elements may penetrate mammalian tissues [43]. In biological systems, very little soluble or biodegradable nanoparticles may accumulate. Metal and metal oxide nanoparticle cytotoxicity can be reduced by utilizing organic coatings. Emerging nanotechnologies may raise environmental issues, which are related to the possible bioaccumulation of nanoparticles in natural systems. For instance, nanoparticles found in sunscreens have the potential to pollute water and soil and bioaccumulate in the food chain [44–46]. Blaser et al. found that the silver discharged is combined with sewage sludge and may move further over agricultural fields during their evaluation of the emission [47]. Engineered nanoparticles that disintegrate are more likely to do so because they are lipophilic and soluble in water (and so have the potential to penetrate fatty cell membranes) [48].

When analyzing environmental exposure, it is important to identify and quantify the sources, analyze the environmental release pattern, evaluate ambient concentrations, and assess the possibility of bioaccumulation. It is feasible to forecast how artificial nanoparticles will react in the environment using information about natural nanoparticles [49].

22.6.2 The Positive Impact of Nanomaterials on the Environment

Raising the standards for the air, soil, and water are among the global environmental problems. The business sector is actively focused on identifying pollutants (from chemical spills, fertilizer runoff, and pesticide runoff), modernizing industrial and mining sites, treating pollutants, and preventing additional pollution.

Nanomaterials could offer a practical solution to these problems. Nanoparticles may be used to improve environmental cleaning processes and even provide efficient energy solutions, such as solar cells built of nanoparticles. Additionally, nanoparticles improve the quality and functioning of many consumer items [50–52]. As a result, more people are exposed to manufactured nanoparticles every day. Nanotechnology, however, has both positive and negative consequences on the environment. Nanotechnology can be used to improve the quality of water. Examples of nanomaterials that can be used for water remediation include carbon nanotubes (CNTs), zeolites, nanoparticles of zero valent iron (ZVI), silver nanoparticles, and other nanomaterials [53]. Photocatalysts include other nanomaterials including tungsten oxide, titanium dioxide, and zinc oxide. These photocatalysts can convert organic pollutants into inert molecules by oxidizing them. TiO_2 is the preferred material due to its excellent photoconductivity, good photostability, accessibility, cost, and non-toxicity. Silver at the nanoscale has an antimicrobial impact. Numerous polymeric nanoparticles are furthermore used to remediate wastewater. The use of magnetic nanoparticles for water purification also deserves attention. An important advantage of using magnetic NPs in this field is the possibility to extract them easily from the medium by applying a magnetic field. Among others, Fe_3O_4@C NPs are the most attractive candidates for this purpose, since they combine good sorption properties of carbon with high magnetic properties of magnetite (Fe_3O_4). Many authors have devoted their efforts to application of Fe_3O_4@C NPs as adsorbents of various substances. These NPs were used as sorbents of heavy metals (Cu, Ni, Co, and Cd) [54]. It was shown that the carbon layer is responsible for the effective adsorption of the polycyclic aromatic hydrocarbons by Fe_3O_4@C NPs [55]. Fe_3O_4@C NPs obtained by the light hydrothermal reaction of glucose with iron were used as the magnetic solid-phase extraction sorbent of brominated flame-retardants and pentachlorophenol [56]. A significant volume of research publications has been dedicated to the investigation of dye adsorption by Fe_3O_4@C and Fe_3O_4@SiO_2 nanoparticles [57–59]. The high adsorption capacity of magnetic NPs together with easy magnetic separation and the possibility of their reusing noted in these works stimulate the search for new technological solutions for further improving the characteristics of adsorbents based on such kind of NPs.

Another cutting-edge technique that may be utilized to purify water in residential, commercial, and industrial settings is nanofiltration. Molybdenum disulfide (MoS_2) is used to create an energy-efficient membrane that filters five times as much water as conventional ones. To wipe up oil spills in bodies of water, a paper towel constructed of nano-fabric, which is weaved from tiny wires of potassium manganese oxide, has been developed. Nanotechnology thus offers a way to treat the polluted water and stop further pollution. It is possible to filter airborne dangerous gases using nanotechnology. But first, we must utilize accurate sensors to pinpoint the contaminants' molecular locations. Heavy metal ions and radioactive elements can both be detected by a sensor known as a nanocontact sensor. These sensors are portable, affordable, and simple to use on-site. Currently, NO_2 and NH_3 gas detection is done using single-walled nanotubes (SWNTs).

In addition, SWNTs sensors have the potential to attain high sensing activity at ambient temperature, in contrast to traditional sensors, which function between

200 and 600 °C. For the detection of pesticides, heavy metals, and VOCs, cantilever sensors have been created. Toxic gases including NO_x, SO_2, and CO_2 may be absorbed using a mixture of CNTs and gold particles. Due to its enormous surface area, manganese oxide, a distinct porous nanomaterial, provides outstanding hazardous gas adsorption. By deploying specific sensors to identify toxins, we may so contribute to the sustainability of both human health and the environment. Therefore, nanotechnology presents a novel approach for lowering waste production, greenhouse gas emissions, and the discharge of hazardous chemicals into waterways [60].

22.7 The Positive Impact of Nanomaterials on Health

The use of nanotechnology is expanding quickly in a variety of fields, including industrial applications, medical imaging, illness detection, medication delivery, gene therapy, and cancer treatment. Due to its many potential benefits for human health and the possible threats to human health, nanotechnology is at the forefront of the rapid development of healthcare products [61, 62].

When shrunk to a nano-molecular size, many tiny particles that are currently thought to be innocuous are likely to develop special properties and might show harmful biological effects. It is well established that the physical and chemical characteristics of the particle will affect the toxicity of designed NP. The same material that is not on the nanoscale can have different properties in nanomaterials. The size of the NP, together with its surface area, aggregation state, crystal structure, surface charge, and porosity, are crucial factors in figuring out how poisonous it is. Nanoparticles have a variety of special health advantages; molecular imaging employs them to identify, measure, and visualize cellular and molecular alterations that take place both in vitro and in vivo [63]. Fluorescent biological probes consisting of organic dyes are frequently used in biology because they are inert and may interact without losing sensitivity in a variety of cellular functions. Nanoparticles can be used as probes in vivo by attaching them to protein, antibody, and nucleic acid molecules. These nanoparticles might then be used as tools to see and quantify molecular processes taking place inside the body. In a broad spectrum, they have good brightness, photostability, and absorption coefficients (Fig. 22.3).

Using nanoparticles for site-specific, targeted medication administration increases bioavailability, has fewer adverse effects, is less hazardous to other organs, and is less expensive [65, 66]. The investigation of tumor-specific thermal scalpels to heat and burn tumors represents an attractive prospective application of nanoparticles in cancer therapy. It is possible to stop the growth of a tumor by utilizing near infrared-absorbing polyethylene-coated gold nano-shells of 130 nm. Gold and silver nanoparticles, which are utilized in wrinkle creams, deodorants, and burn treatments, have potent antifungal, antibacterial, and anti-inflammatory qualities. They are non-cytotoxic, very stable, biocompatible, and neutral [64].

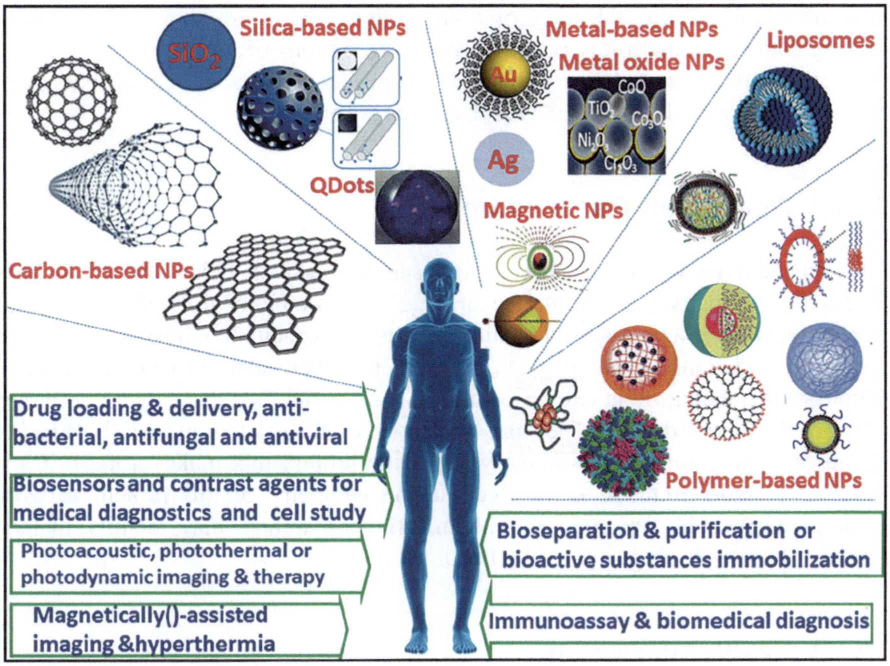

Fig. 22.3 Nanoparticles and their applications in biomedical fields [64]

22.7.1 Enhanced Therapeutics, Diagnostics, and Sensors Powered by Nanotechnology

A small chip may contain up to 64 nano-sensors, each of which is capable of detecting the tiniest levels of hazardous gases. It requires a very little amount of electricity and is made to plug into a cell phone. In order to monitor crop disease and moisture levels in agriculture, researchers are also creating sensors. These nanotechnology-enabled sensors will aid in securing firefighters, troops, and our food supply due to their small size and versatility [67]. The fact that a patient's body does not absorb the complete therapeutic amount is a significant challenge for modern medicine. Using nanotechnology, scientists can more accurately deliver pharmaceuticals to specific organs and tissues, and they can create medications such that the active component penetrates cell membranes more quickly, resulting in a reduction in the overall amount of medication needed.

Nanotechnology may also change medicine delivery by tackling concerns like how to maintain the release of pharmaceuticals in the body and improving bioavailability—the quantity of active substance per dosage. Some pharmaceuticals can now be transported via "nano-vehicles." Stavudine and zidovudine, two HIV drugs, have been encapsulated in liposomes, which may transport the drug payload by bonding with cell membranes, in carriers with a size range of 120–200 nanometers. Since the half-lives of both of these drugs are short, the liposome coating could help them stay active for extended periods of time.

Many targeted drug delivery methods utilize dendrimers, which are branching nano-molecules, and fullerene "buckyball" cages [68].

22.7.2 Fluorescent Quantum Dots

Glow-enabled nanoscale semiconductors can be used as biosensors to identify illness. In comparison to traditional organic dyes, quantum dots, also known as nanocrystals, have a number of benefits, such as the capacity to adjust their luminescence to a wide range of frequencies and a much slower rate of in vivo degradation. Fluorescent quantum dots can be combined with antibodies that target cancerous, TB, or HIV-infected cells. They might also be used to detect malaria by making fluorescent quantum dots particularly target the protein that makes a mesh in the inner membrane of blood cells. Because the geometry of this protein network changes when cells are infected with malaria, researchers may identify malaria infection by the shape generated by the dots.

22.7.3 Health Monitoring

Nanotubes and nanoparticles can be used as glucose, carbon dioxide, and cholesterol sensors for in situ monitoring of homeostasis, the body's system for maintaining metabolic equilibrium. Nanotechnology has also improved the treatment of tuberculosis. For the present TB therapy, a complex medication regimen that is spread out over several months is required. A lot of folks either do not take their meds as prescribed or stop taking the course altogether. Drug formulations based on nanotechnology break down more gradually and supply more of the active ingredient in fewer doses. To allow for continued delivery of the therapy, the drugs are encapsulated in biodegradable polymers such as microspheres and liposomes. Polylactide co-glycolide nanoparticles have been used effectively in drug carrier studies for TB because they degrade well and do not elicit an immunological response. Nanoparticles might potentially serve as the basis for an aerosol TB vaccination. There is no requirement for trained personnel to administer the vaccination because it is stable at room temperature without a needle. This is important in distant locations without a strong cold chain.

Nanotechnology may advance vaccination to new heights by creating substitutes for injectable vaccines for diseases that affect the underprivileged. Injectable vaccines could be helpful if the latent virus is combined with nanoparticles to strengthen the immune response. The development of a pandemic influenza vaccine is also being done using this method.

22.7.4 Drug Delivery Using Nano-Capsules

Nano-capsules, which have a millimeter-thin diameter, can have an antibody coating on their surface to help guide blood flow to artificial tumors. An instantaneous blast that occurs when the capsules arrive at the tumor causes them to instantly open and release their medicinal contents. There are small gold particles in the range of 6 nm, or 6 millionths of a millimeter, on the surface of the polymer. These particles cling across and are particular to the laser light, guiding the capsules to position their drug load capacity when it is needed. When the gold specks are exposed to near infrared light, they instantly melt without damaging the surrounding material, revealing the rupturing of the capsule [69].

22.7.5 Nano-Capsule Bandages to Combat Infection in the Future

The traditional dressings must be removed if the skin becomes damaged or the healing process is slowed. In contrast, when a wound becomes infected, nano-capsular dressings immediately release antibiotics. They do not need to be removed, which increases the likelihood that a wound will heal without leaving a scar. Additionally, ulcers can be treated with nano-capsular bandages, which are most frequently applied by soldiers on the battlefield. These medicinal dressings target the treatment before the infection worsens by releasing antibiotics from nano-capsules activated by the presence of disease-causing pathogenic or causative bacterial organism. The capsules containing the antibiotics, which serve as dressings, are ruptured by bacterial toxins.

The development of formulation methods, notably interfacial nano-deposition and interfacial polymerization, is aided by nano-capsules. The ability to release them as monodisperse particles with unique biological, electrical, optical, and magnetic characteristics is another advantage. Because they are created to generate contents in response to a specific bimolecular triggering action mechanism, they are restricted in drug delivery systems. Agrochemicals, wastewater treatment, genetic engineering, cosmetics, cleaning products, and adhesive components are just a few of the areas where nano-capsules are useful. Enzymes, adhesives, catalysts, polymers, oils, inorganic micro- and nanoparticles, latex particles, and even living cells are all encapsulated using them. They can be utilized to distribute active medicinal substances, to sum up (APIs). They offer cutting-edge, efficient drug delivery technologies for the foreseeable future [70]. The nano-capsules market has an interesting and promising future. This is a technology that will continue to offer answers for various major difficulties, as evidenced by the expected industry growth and the expanding number of economically relevant applications for the sector.

It will be far more effective, for example, to administer and target smart drugs that can be tailored to a patient's requirements. The stability and bioavailability of pharmaceuticals can be improved by developing transdermal drug delivery devices.

22.7.6 Nanobots in Medical Field

Nanobots are robots that employ nanotechnology. Emerging technology known as "nano robotics''produces machines or robots with parts that are at or very close to the scale of a nanometer. The nanoscale technical gaps in physics, chemistry, and biology will be bridged in part by nanobots. With the help of these nanobots, new techniques and items will be developed for use in technical and medical-pharmaceutical applications. Nanobots are also suitable candidates for complex therapies due to their small size. Drug delivery systems and contrast agents are both made of nanobots, they can be utilized for drug delivery in a very active way. Drugs typically go through the entire body before they reach the location where the sickness is present. The medicine can be targeted to a particular place using nanotechnology, which can lower the likelihood of any potential side effects [71]. More specifically, the term "nanorobotics" refers to the engineering field of nanotechnology that focuses on designing and creating nanobots with sizes starting at 0.110 micrometers and made of molecular or nanoscale components. The phrases nanobots, nanoid, nanite, nanomachine, or nano-mite are used to describe the nanodevices that are currently the subject of research and development. Although several crude molecular machines and nanomotors have been tried, nanomachines are largely used in the research and development stage.

These nanorobots frequently transport biochemical payloads throughout the system and block local tumor blood flow, which may result in tissue death and cause the tumor to shrink. The development of nanobots and their application to medicine is now under way. When injected intravenously, nanobots used in medicine target cancer cells and repair damaged tissue because they are built to perform specific biological tasks. When research projects and nanobots are joined, a new benchmark in the development of medical analysis is attained. Humanity has made significant progress in terms of procedure safety and dependability. Endoscopy is one of the methods that has been created to aid in diagnosis and provide a better understanding of the body's inner workings. But as we are all aware, technology must eventually be phased out. And much as historical practices have evolved to address the shortcomings of their forebears, nanorobotics will work to address the following flaws in current medical technologies:

1. Remove unhealthy tissue layers that require time to recover.
2. Although anesthesia can significantly reduce pain, it only has a temporary effect.
3. For delicate surgeries like eye surgery, there is still no 100% success rate.
4. The patient's life is entirely in the hands of the operator, surgeon, or physician in any invasive technique. It is dangerous because a single error could result in the patient's death.

The following are the key advantages of this technology:

1. No or little tissue trauma.
2. Much quicker healing time.

Fig. 22.4 Nanobots in medicine and drug delivery [71]

3. Less after-treatment care is necessary.
4. Rapid reaction to an unexpected development.
5. Monitoring and diagnosis from the interior.

Nanobots can carry payloads like medications or healthy cells to a precise spot in the body by being steered externally as instructed by a computer. These nanobots' ability to go through normal cellular channels is an extra benefit (Fig. 22.4).

The primary drawback of designing nanobots is that it is expensive and involves numerous complexities. The power supply is the biggest challenge. More research must be done in order for the bots to defeat the body's immune reaction. If terrorists employ nanobots improperly, they might potentially be turned into bioweapons and pose a threat to civilization. Nanobots are foreign to the body since they can have catastrophic consequences if nanobacteria are present. The body contains a lot of foreign objects, so biodegradability will be a big issue. Thus, it is necessary to take strong measures to address each of these shortcomings.

22.7.7 Development of a "Global Super-Brain"

A third exciting development in nanobot research is the use of nanotechnology to build a "global super-brain" where potentially, human cognition might be transmitted to an artificial interface, producing a "brain–cloud interface." Theoretically, in the future, nanobots might wirelessly transmit brain data to a network of cloud-hosted supercomputers, enabling data extraction in real-time and brain-state monitoring. But before it can be implemented, this use of nanotechnology must go through a rigorous period of study and design, as well as go beyond the technology's ethical and moral implications.

22.8 Xenobots

Recent research has paved the way for the creation of xenobots, small robots with a length of less than 1 mm and made of 500–1000 living cells. They may be found in many different straightforward forms, some of which include legs. Studies have shown that they can move tiny objects, travel effectively in a linear or circular pattern, work together with other xenobots to complete tasks, and live for around 10 days. Scientists think that these xenobots might be developed in various ways that could help human, animal, and environmental health, despite the fact that their position as "programmable living robots" made from live, organic tissue raises ethical questions. Xenobots future is somewhat uncertain due to ethical concerns, scientists working in this field are enthusiastic about their potential applications in removing microplastics from the ocean, removing toxins and radioactive materials from hazardous locations, improving the efficiency and effectiveness of targeted drug delivery, and repairing cells and tissues [71, 72].

Xenobots could be used to remove plaque from our arteries, remove radioactive waste from the environment, collect microplastics from the ocean, convey drugs inside of people, and clean up radioactive waste. The xenobots are suitable for internal drug administration since they may endure in aquatic settings for days or weeks without extra nutrition. In addition to these immediate useful activities, the xenobots may aid scientists in their understanding of cell biology, paving the way for future improvements in human health and lifespan.

22.9 Conclusion and Challenges

Nanotechnology is developing quickly, and this will undoubtedly have an impact on human health, the environment, and toxicology [73]. It is without question how important nanotechnologies are to our well-being, but it is even more important to research any potential negative effects. Nanomaterials' safety, toxicity, and health implications have lagged considerably behind their production.

This is believed to be due to the lack of specific guidelines and widespread consensus among researchers regarding appropriate experimental procedures or study designs. It also depends on the unique properties of nanoscale materials [74, 75], which provide difficulties for the toxicological assessment of novel nanomaterials. Despite a significant rise in publications in recent years, all of these issues lead to variable and erratic outcomes and obstruct the advancement of this field. This study attempts to critically compile and evaluate the many methods and challenges connected to the health and safety implications of nanomaterials for human health and the environment using the abundance of knowledge already accessible. The preparation procedures for typical nanomaterials and prospective applications have been briefly explored. The developing science of nanotoxicology will greatly contribute to the development of safe and sustainable nanotechnology. A deeper comprehension of the dangers posed by nanoparticles in the environment and on humans is crucial for the development and use of various nanomaterials in the future. A multidisciplinary team effort involving material scientists, molecular biologists, toxicologists, and physicists is necessary to create nanotoxicology. One of the numerous topics of nanotoxicology is the comprehension of biological responses to nanomaterial exposure and the processes underlying them. It is important to recognize and document in the literature any challenges that were experienced when conducting various in vitro and in vivo studies on nanomaterials. This would make it easier for beginners to figure out the problem and help them save time and effort. Evaluation of the safety aspects of the particles must be given top priority due to the anticipated large-scale production of the particles and the likelihood of human exposure, either directly through the use of cosmetics, drugs, or drug delivery devices or indirectly through distribution through the air, water, and soil [76, 77].

There is a need for better environmental risk and life cycle assessment, as well as defined sustainability targets, in the governance of technical innovation in the field of nanotechnology. In many aspects, nanotechnology is an example of a technologically based attempt to solve a problem that actually requires social, economic, and political solutions.

References

1. Shen JC, Zhongguo Y, Xue K, Xue Y, Xue B (2006) Nanobiomaterials. Chinese 28(4):472–474
2. Verma R, Thakur P, Sun AC, Thakur A (2023) Investigation for structural, microstructural and electrical characteristics of hydrothermally synthesized $Li_{0.5-0.5x}Co_xFe_{2.5-0.5x}O_4$ ($0.0 \leq x \leq 0.4$) ferrite nanoparticles. Phys B Condens Matter 661:414926
3. Thakur P, Thakur A (2022) Introduction to nanotechnology. In: Thakur A, Thakur P, Khurana SP (eds) Synthesis and applications of nanoparticles. Springer, Singapore
4. Thakur P, Thakur A (2022) Nanomaterials, their types and properties. In: Thakur A, Thakur P, Khurana SP (eds) Synthesis and applications of nanoparticles. Springer, Singapore
5. Kumari S, Dhanda N, Thakur A, Gupta V, Singh S, Kumar R, Hameed S, Thakur A (2023) Nano ca-mg-Zn ferrite as tuneable photocatalysts for UV light induced degradation of rhodamine B dye and antimicrobial behaviour for water purification. Ceram Int 49(8):12469–12480
6. Chahar D, Thakur P, Sun AC, Thakur A (2023) Investigation of structural, electrical and magnetic properties of nickel substituted co-Zn nanoferrites. J Mater Sci Mater Electron 34:901
7. Kala D, Gupta S, Kaushal A (2022) Nanotechnology in healthcare. In: Thakur A, Thakur P, Khurana SP (eds) Synthesis and applications of nanoparticles. Springer, Singapore
8. Hasirci V, Vrana E, Zorlutuna P, Ndreu A, Yilgor P, Basmanav FB, Aydin E (2006b) Nanobiomaterials: a review of the existing science and technology, and new approaches. J Biomater Sci Polym Ed 17(11):1241–1268
9. Ruiz-Hitzky E, Darder M, Wicklein B, Ruiz-García C, Martín-Sampedro R, Del Real G, Aranda P (2020) Nanotechnology responses to COVID-19. Adv Healthc Mater 9(19):2000979
10. Sharifi S, Behzadi S, Laurent S, Forrest ML, Stroeve P, Mahmoudi M (2012) Toxicity of nanomaterials. Chem Soc Rev 41(6):2323–2343
11. Thakur P, Verma Y, Thakur A (2022) Toxicity of nanomaterials: an overview. In: Thakur A, Thakur P, Khurana SP (eds) Synthesis and applications of nanoparticles. Springer, Singapore

12. Srivastava N, Mishra G, Saxena SK (2022) Environmental Nanomedicine. In: Thakur A, Thakur P, Khurana SP (eds) Synthesis and applications of nanoparticles. Springer, Singapore
13. Tetley TD (2007) Health effects of nanomaterials. Biochem Soc Trans 35(3):527–531
14. Kirchner C, Liedl T, Kudera S, Pellegrino T, Javier AM, Gaub HE, Stölzle S, Fertig N, Parak WJ (2005) Cytotoxicity of colloidal CdSe and CdSe/ZnS nanoparticles. Nano Lett 5(2): 331–338
15. Gogotsi Y (2003) How safe are nanotubes and other nanofilaments? Mater Res Innov 7(4): 192–194
16. Bharti MK, Gupta S, Chalia S, Garg I, Thakur P, Thakur A (2020) Potential of magnetic nanoferrites in removal of heavy metals from contaminated water: mini review. J Supercond Nov Magn 33:3657–3665
17. Wiesner MR, Lowry GV, Alvarez PJJ, Dionysiou D, Biswas P (2006) Assessing the risks of manufactured nanomaterials. Environ Sci Technol 40(14):4336–4345
18. Richter A, Sadowski JT (2011) Nanoscience and nanotechnology, vol 9. Open Physics, p 263
19. Marquis BJ, Love SA, Braun KN, Haynes CL (2009) Analytical methods to assess nanoparticle toxicity. Analyst 134(3):425
20. Barhoum A, García-Betancourt ML, Jeevanandam J, Hussien EA, Mekkawy SA, Mostafa M, Omran MM, et al. (2022) Review on Natural, Incidental, Bioinspired, and Engineered Nanomaterials: History, Definitions, Classifications, Synthesis, Properties, Market, Toxicities, Risks, and Regulations. Nanomaterials 12(2):177. MDPI AG. Retrieved from https://doi.org/10.3390/nano12020177
21. Gu N, Sheng J (2023) Introduction to nanomedicine. In: Micro/Nano technologies. Springer Nature, Singapore, pp 3–16
22. Wagner V, Dullaart A, Bock A, Zweck A (2006) The emerging nanomedicine landscapes. Nat Biotechnol 24(10):1211–1217
23. Freitas RA (2005) What is nanomedicine? Nanomedicine: nanotechnology. Biol Med 1(1):2–9
24. Gadag S, Sinha S, Nayak Y, Garg S, Nayak UY (2020) Combination Therapy and Nanoparticulate Systems: Smart Approaches for the Effective Treatment of Breast Cancer. Pharmaceutics 12(6):524. MDPI AG. Retrieved from https://doi.org/10.3390/pharmaceutics12060524
25. Labib M, Zamay AS, Kolovskaya OS, Reshetneva IT, Zamay GS, Kibbee RJ (2012) Aptamer-based impedimetric sensor for one very actively developing area in this field is related to aptamers that are synthetic RNA or bacterial typing. Anal Chem 84(19):8114–8117
26. Pang X, Cui C, Wan S, Jiang Y, Zhang L, Xia L (2018) Bioapplications of cell-SELEX-generated aptamers in cancer diagnostics, therapeutics, theranostics and biomarker discovery: a comprehensive review. Cancers (Basel) 10(2):47
27. Sun H, Zhu X, Lu PY, Rosato RR, Tan W, Zu Y (2014) Oligonucleotide aptamers: new tools for targeted cancer therapy. Mol Ther Nucleic Acids 3:e182
28. Tomilin FN, Moryachkov R, Shchugoreva I, Zabluda VN, Peters G, Platunov M, Spiridonova V, Melnichuk A, Atrokhova A, Zamay SS, Ovchinnikov SG, Zamay GS, Sokolov A, Zamay TN, Berezovski MV, Kichkailo AS (2019) Four steps for revealing and adjusting the3D structure of aptamers in solution by small-angle X-ray scattering and computer simulation. Anal Bioanal Chem 25:6723–6732
29. Zamay GS, Zamay TN, Lukyanenko KA, Kichkailo AS (2020) Aptamers increase biocompatibility and reduce the toxicity of magnetic nanoparticles used in biomedicine. Biomedicine 8:59–65
30. Hori S, Herrera A, Rossi JJ, Zhou J (2018) Current advances in aptamers for cancer diagnosis and therapy. Cancers 10:9–33
31. Belyanina IV, Zamay TN, Zamay et al (2017) In vivo cancer cells elimination guided by aptamer-functionalized gold-coated magnetic nanoparticles and controlled with low frequency alternating magnetic field. Theranostics 7:3326–3337

32. Efremova MV, Naumenko VA, Spasova M, Garanina AS, Abakumov MA, Blokhina AD, Melnikov PA, Pre-lovskaya AO, Heidelmann M, Li Z-A et al (2018) Magnet-ite-gold nanohybrids as ideal all-in-one platforms for theranostics. Sci Rep 8:1–19
33. Sadik OA (2013) Anthropogenic nanoparticles in the environment. Environ Sci Process Impacts 15(1):19–20
34. Kumar PS, Kumar A, Fernandes TF, Goonetilleke A (2014) Nanomaterials and the environment. J Nanomater 2014:1–4
35. Chow E, Pierstorff E, Cheng G, Ho D (2008) Copolymeric Nanofilm platform for controlled and localized therapeutic delivery. ACS Nano 2(1):33–40
36. Wijnhoven SWP, Willie JGM, Herberts C (2009) Nano-silver—a review of available data and knowledge gaps in human and environmental risk assessment. Nanotoxicology 3(2):109–138
37. Kumar A, Sharma V (2010b) Toxicity assessment of nanomaterials: methods and challenges. Anal Bioanal Chem 398(2):589–605
38. Nowack B, Bucheli TD (2007) Occurrence, behavior and effects of nanoparticles in the environment. Environ Pollut 150(1):5–22
39. Vaaraslahti K, Virtanen A, Ristimäki J, Keskinen J (2004) Nucleation mode formation in heavy-duty diesel exhaust with and without a particulate filter. Environ Sci Technol 38(18): 4884–4890
40. Astruc D (2020) Introduction: nanoparticles in catalysis. Chem Rev 120(2):461–463
41. Nischwitz V, Michalke B, Kettrup A (2003) Speciation of Pt(II) and Pt(IV) in spiked extracts from road dust using on-line liquid chromatography-inductively coupled plasma mass spectrometry. J Chromatogr A 1016:223–234
42. Artelt S, Creutzenberg O, Kock H, Levsen K, Nachtigall D, Heinrich U, Rühle T, Schlgl R (1999) Bioavailability of fine dispersed platinum as emitted from automotive catalytic converters: a model study. Sci Total Environ 228:219–242
43. Ek KH, Rauch S, Morrison GM, Lindberg P (2004) Platinum group elements in raptor eggs, feces, blood, liver and kidney. Sci Total Environ 334:149–159
44. Hannah W, Thompson PB (2008) Nanotechnology, risk and the environment: a review. J Environ Monitor 10:291–300
45. Nohynek GJ, Dufour EK, Roberts MS (2008) Nanotechnology, cosmetics and the skin: is there a health risk? Skin Pharmacol Physiol 21:136–149
46. Nohynek GJ, Lademan J, Ribau C, Robert MS (2007) Grey goo on the skin? Nanotechnology, cosmetic and sunscreen safety. Crit Rev Toxicol 37:251–277
47. Blaser SA, Scheringer M, MacLeod M, Hungerbühler K (2008) Estimation of cumulative aquatic exposure and risk due to silver: contribution of nano functionalized plastics and textiles. Sci Total Environ 390:396–409
48. Wu PG, Zhu JH, Xu ZH (2004) Template-assisted synthesis of mesoporous magnetic nanocomposite particles. Adv Funct Mater 14:345–351
49. Hasselv M, Readman JW, Rainville JF, Tiede K (2008) Nanoparticle analysis and characterization methodologies in environmental risk assessment of engineered nanoparticles. Ecotoxicology 17:344–362
50. Bharti MK, Gupta S, Chalia S, Garg I, Thakur P, Thakur A (2020) Potential of magnetic nanoferrites in removal of heavy metals from contaminated water: mini review. J Supercond Nov Magn 33:3651–3665
51. Singh R, Mehra R, Walia A, Gupta S, Chawla P, Kumar H, Thakur A, Kaushik R, Kumar N (2021) Colorimetric sensing approaches based on silver nanoparticles aggregation for determination of toxic metal ions in water samples: a review. Int J Environ Anal Chem 103:1–16
52. Punia P, Bharti MK, Dhar R, Thakur P, Thakur A (2022) Recent advances in detection and removal of heavy metals from contaminated water. ChemBioEng Rev 9(4):351–369
53. Xia T, Li N, Nel AE (2009) Potential health impact of nanoparticles. Annu Rev Public Health 30(1):137–150

54. Habila MA, Alothman ZA, El-Toni AM, Labis JP, Khan A, Al-Marghany A, Elafifi HE (2017) One-step carbon coating and polyacrylamide functionalization of Fe_3O_4 nanoparticles for enhancing magnetic adsorptive-remediation of heavy metals. Molecules 22:2074
55. Zhang S, Niu H, Hu Z, Cai Y, Shi Y (2010) Preparation of carbon coated Fe_3O_4 nanoparticles and their application for solid-phase extraction of polycyclic aromatic hydrocarbons from environmental water samples. J Chromatogr A 1217:4757–4764
56. Yang J, Li J-y, Qiao J-q, Cui S-h, Lian H-z, Chena H-y (2014) Magnetic solid phase extraction of brominated flame retardants and pentachlorophenol from environmental waters with carbon doped Fe3O4 nanoparticles. Appl Surf Sci 321:126–135
57. Wang J, Zhang Q, Liu J, Ji X, Ma J, Tian G (2020) Preparation and excellent adsorption of water pollution dyes over magnetic Fe3O4@C nanoparticles with hollow grape cluster morphology. J Nanoparticles Res 22:196
58. Xiang H, Ren G, Zhong Y, Xu D, Zhang Z, Wang X, Yang X (2021) Fe3O4@C nanoparticles synthesized by in situ solid-phase method for removal of methylene blue. Nano 11:330
59. Lin C-R, Ivanova OS, Edelman IS, Knyazev YV, Zharkov SM, Petrov DA, Sokolov AE, Svetlitsky ES, Velikanov DA, Solovyov LA, Chen YZ, Yaw-Teng Tseng YT (2022) Carbon double coated Fe3O4@C@C nanoparticles: morphology features, magnetic properties, dye adsorption. Nano 12:376
60. Gautam S, Mishra A, Koundal P (2022) Nanotechnology for functional/high-performance/smart textiles. In: Thakur A, Thakur P, Khurana SP (eds) Synthesis and applications of nanoparticles. Springer, Singapore
61. Sangam S, Garg P, Sanyal T, Pahari S, Khurana SMP, Mukherjee M (2022) Graphene quantum dots and their hybrid hydrogels: a multifaceted platform for theranostic applications. In: Thakur A, Thakur P, Khurana SP (eds) Synthesis and applications of nanoparticles. Springer, Singapore
62. Sabir F et al (2022) Functionalized nanoparticles in drug delivery: strategies to enhance direct nose-to-brain drug delivery via integrated nerve pathways. In: Thakur A, Thakur P, Khurana SP (eds) Synthesis and applications of nanoparticles. Springer, Singapore
63. Pate S, Sahoo S (2015) Nanotechnology in healthcare: applications and challenges. Med Chem 05(12):2161
64. Zhao F, Yao D, Guo R, Deng L, Dong A, Zhang J (2015) Composites of Polymer Hydrogels and Nanoparticulate Systems for Biomedical and Pharmaceutical Applications. Nanomaterials 5(4):2054–2130. MDPI AG. Retrieved from https://doi.org/10.3390/nano5042054
65. Dhanda N, Thakur P, Kumar R, Fatima T, Hameed S, Slimani Y, Sun AD, Thakur A (2023) Green synthesis of Ni-co nanoferrites using aloe vera extract: structural, optical, magnetic, antimicrobial studies. Appl Organomet Chem 2023:1–13
66. Dhanda N, Thakur P, Sun AC, Thakur A (2023) Structural, optical, and magnetic properties along with antifungal activity of ag-doped Ni-co nanoferrites synthesized by eco-friendly route. J Magn Magn Mater 572:17059
67. Chen S, Liu T, Zheng Z, Ishaq M, Fan P, Chen T, Tang J (2022) Recent progress and perspectives on Sb2Se3-based photocathodes for solar hydrogen production via photoelectrochemical water splitting. J Energy Chem 67:508–523
68. Ray PC, Yu H, Fu PP (2009b) Toxicity and environmental risks of nanomaterials: challenges and future needs. J Environ Sci Health A 27(1):1–35
69. Kothamasu P, Kanumur H, Ravur N, Maddu C, Parasuramrajam R, Thangavel S (2012) Nanocapsules: the weapons for novel drug delivery systems. Bioimpacts 2(2):71–81
70. Bakshi MS (2020) Impact of nanomaterials on ecosystems: mechanistic aspects in vivo. Environ Res 182:109099
71. Cavalcanti A, Shirinzadeh B, Zhang M, Kretly L (2008) Nanorobot Hardware Architecture for Medical Defense. Sensors 8(5):2932–2958. MDPI AG. Retrieved from https://doi.org/10.3390/s8052932
72. Wang X, Cui X, Zhao Y, Chen C (2020) Nano-bio interactions: the implication of size-dependent biological effects of nanomaterials. Sci China Life Sci 63(8):1168–1182

73. Suhag D, Bhatia R, Das S, Shakeel A, Ghosh AB, Singh A, Sinha OP, Chakrabarti SK, Mukherjee M (2015b) Physically cross-linked pH-responsive hydrogels with tunable formulations for controlled drug delivery. RSC Adv 5(66):53963–53972
74. Suhag D, Chauhan M, Shakeel A, Das S (2020) Emerging trends in Nanotheranostics. In: Saxena S, Khurana S (eds) NanoBioMedicine. Springer, Singapore
75. Chakraborty A, Patni P, Suhag D, Saini G, Singh A, Chakrabarti SK, Mukherjee M (2015) N-doped carbon nanosheets with antibacterial activity: mechanistic insight, vol 5. RSC Adv, p 23591
76. Ivanova OS, Edelman IS, Lin C-R, Svetlitsky ES, Sokolov AE, Lukyanenko KA, Sukhachev AL, Shestakov NP, Chen Y-Z, Spivakov AA (2023) Core–shell Fe3O4@C nanoparticles for the organic dye adsorption and targeted magneto-mechanical destruction of ehrlich ascites carcinoma cells. Materials 16:23
77. Blackiston DJ, Lederer EK, Kriegman S, Garnier S, Bongard JC, Levin M (2021) A cellular platform for the development of synthetic living machines. Sci Robot 6(52):eabf1571

Correction to: Nanomaterials for Environmental Applications

Yassine Slimani and Essia Hannachi

Correction to:
Chapter 17 in: D. Suhag et al. (eds.), *Integrated Nanomaterials and their Applications*, https://doi.org/10.1007/978-981-99-6105-4_17

The original version of this chapter was inadvertently published with an incorrect author name in a few references in the References list. The author name "Almessiere, MA" has been replaced with "Slimani, Y" in the following references: 26, 33, 49, 69, 73, 102, 103, 104, 105, 106, 108, 109, 111, 113, 114, and 115.

The updated version of this chapter can be found at
https://doi.org/10.1007/978-981-99-6105-4_17

D. Suhag et al. (eds.), *Integrated Nanomaterials and their Applications*,
https://doi.org/10.1007/978-981-99-6105-4_23

GPSR Compliance

The European Union's (EU) General Product Safety Regulation (GPSR) is a set of rules that requires consumer products to be safe and our obligations to ensure this.

If you have any concerns about our products, you can contact us on ProductSafety@springernature.com

In case Publisher is established outside the EU, the EU authorized representative is:

Springer Nature Customer Service Center GmbH
Europaplatz 3
69115 Heidelberg, Germany

Batch number: 10725823

Printed by Printforce, the Netherlands